To Toshiko

Bacteria in Oligotrophic Environments

Chapman & Hall Microbiology Series

Physiology/Ecology/Molecular Biology/Biotechnology

SERIES EDITORS

C.A. Reddy, Editor-in-Chief
Department of Microbiology
Michigan State University
East Lansing, MI 48824-1101

A.M. Chakrabarty
Department of Microbiology and Immunology
University of Illinois Medical Center
835 S. Wolcott Avenue
Chicago, IL 60612

Arnold L. Demain
Department of Biology, Rm 68-223
Massachusetts Institute of Technology
Cambridge, MA 02139

James M. Tiedje
Center for Microbial Ecology
Department of Crop and Soil Sciences
Michigan State University
East Lansing, MI 48824

Other Publications in the Chapman & Hall Microbiology Series

Methanogenesis; James G. Ferry, ed.
Acetogenesis; Harold L. Drake, ed.
Gastrointestinal Microbiology, Volume 1; Roderick I. Mackie and Bryan A. White, eds.
Gastrointestinal Microbiology, Volume 2; Roderick I. Mackie, Bryan A. White, and
Richard E. Isaacson, eds.

Forthcoming Titles in the Chapman & Hall Microbiology Series

Mathematical Modeling in Microbial Ecology; Arthur L. Koch,
Joseph A. Robinson, and George L. Milliken, eds.
Oxygen Regulation of Gene Expression in Bacteria; Rob Gunsalus, ed.
Metal Ions in Gene Regulation; Simon Silver and William Walden, eds.

Chapman & Hall Microbiology Series

Bacteria in Oligotrophic Environments

Starvation-Survival Lifestyle

Richard Y. Morita

Department of Microbiology
College of Science and
College of Oceanic and Atmospheric Sciences
Oregon State University
Corvallis, Oregon

CHAPMAN & HALL

INTERNATIONAL
THOMSON PUBLISHING

New York • Albany • Bonn
Boston • Cincinnati • Detroit
London • Madrid • Melbourne
Mexico City • Pacific Grove
Paris • San Francisco • Singapore
Tokyo • Toronto • Washington

Join Us on the Internet
WWW: http://www.thomson.com
EMAIL: findit@kiosk.thomson.com

thomson.com is the on-line portal for the products, services and resources available from International Thomson Publishing (ITP). This Internet kiosk gives users immediate access to more than 34 ITP publishers and over 20,000 products. Through *thomson.com* Internet users can search catalogs, examine subject-specific resource centers and subscribe to electronic discussion lists. You can purchase ITP products from your local bookseller, or directly through *thomson.com*.

Visit Chapman & Hall's Internet Resource Center for information on our new publications, links to useful sites on the World Wide Web and an opportunity to join our e-mail mailing list. Point your browser to:
http://www.chaphall.com or **http://www.thomson.com/chaphall/lifesce.html** for Life Sciences.

Cover design: Curtis Tow Graphics

Copyright © 1997 by Chapman & Hall

Printed in the United States of America

Chapman & Hall
115 Fifth Avenue
New York, NY 10003

Chapman & Hall
2-6 Boundary Row
London SE1 8HN
England

Thomas Nelson Australia
102 Dodds Street
South Melbourne, 3205
Victoria, Australia

Chapman & Hall GmbH
Postfach 100 263
D-69442 Weinheim
Germany

International Thomson Editores
Campos Eliseos 385, Piso 7
Col. Polanco
11560 Mexico D.F
Mexico

International Thomson Publishing–Japan
Hirakawacho-cho Kyowa Building, 3F
1-2-1 Hirakawacho-cho
Chiyoda-ku, 102 Tokyo
Japan

International Thomson Publishing Asia
221 Henderson Road #05-10
Henderson Building
Singapore 0315

1 2 3 4 5 6 7 8 9 10 XXX 01 00 99 98 97

Library of Congress Cataloging-in-Publication Data

Morita, R. Y. (Richard Y.)
    Bacteria in oligotrophic environments : starvation-survival life
  style / Richard Y. Morita ; with contributions from Paul A. Blum ...
  [et al.].
        p.   cm. -- (Chapman & Hall microbiology series)
    Includes bibliographical references and index.
    ISBN 0-412-10661-2 (alk. paper)
    1. Bacteria--Ecology. 2. Bacteria--Metabolism. 3. Energy
  metabolism. I. Title.   II. Series.
  QR100.M67   1997
  579.3'17--dc21   96-50399
                                                       CIP

British Library Cataloguing in Publication Data available

To order this or any other Chapman & Hall book, please contact **International Thomson Publishing, 7625 Empire Drive, Florence, KY 41042.** Phone: (606) 525-6600 or 1-800-842-3636. Fax: (606) 525-5778. e-mail: order@chaphall.com.

For a complete listing of Chapman & Hall titles, send your request to **Chapman & Hall, Dept. BC, 115 Fifth Avenue, New York, NY 10003.**

Contents

Preface

This book was written because of my concern that the subject of microbial ecology does not adequately address the most important environmental factor—the availability of energy (or lack thereof) for heterotrophic microbes. I first became interested in this subject as a graduate student when I could not explain the existence of bacteria in deep strata of the various cores obtained during the Mid-Pacific Expedition of 1950. At the marine microbiology meetings at Princeton University in 1967 (Morita, 1968), I first questioned whether some of the marine bacteria are transients or are not metabolizing in the marine environment. Survival of the vegetative microbial forms has always been neglected by microbiologists over the years (Gray and Postgate, 1976).

There can be no doubt that most of the biosphere is oligotrophic (i.e., lacks available energy to support microbial growth) and that this should be considered the normal state of most environments. Because the normal state of most environments is oligotrophic, it follows that the starvation-survival lifestyle is also the normal physiological state of microorganisms in nature. There are degrees of starvation-survival. Therefore, we must address the concept of the lack of energy in ecosystems and the starvation-survival processes not only of microbes but many other organisms in oligotrophic environments. Because I believe that the concept of starvation-survival is important to all phases of microbial ecology, I have covered both terrestrial and aquatic environments. Starvation-survival is also important to many areas of microbiology. I deal mainly in the marine area of microbiology, so the book is naturally slanted toward the marine environment more than the terrestrial and freshwater environments. The subject matter invites many questions and criticisms, and there is much research to be done before more complete answers can be formulated. This will result in a better understanding of microbial ecology.

In order to fully understand the oligotrophic nature of ecosystems, I have reviewed the literature in soil literature (microbiology, biochemistry, and chemistry), in limnology, and in oceanography (microbiology, chemistry, and geochemistry). I have purposely crossed disciplines in order to make the subject matter more complete, therefore, the book should be sufficiently detailed so that it can serve as a subject matter reference.

I wish to thank Dr. Ian Dundas for the many fruitful conversations dealing with starvation-survival and his critical review of several of these chapters. Special thanks go to my wife, Dr. Toshiko Nishihara Morita, who has given considerable effort to correcting and proofreading my manuscripts over the years. I am most appreciative of the excellent editing of this book by Dr. Ramon Seidler and his wife, Katherine.

R. Y. Morita
Corvallis, Oregon, 1996

About the Author

Richard Yukio Morita, Ph.D., is Professor Emeritus of Microbiology and Oceanography at Oregon State University in Corvallis, Oregon.

Professor ("Dick") Morita received his undergraduate training in Bacteriology, Zoology, and Chemistry at the University of Nebraska, receiving his B.S. degree in both Chemistry and Bacteriology in 1947. He obtained his M.S. degree in Bacteriology from the University of Southern California in 1949 and his Ph.D. degree in Microbiology and Oceanography from Scripps Institution of Oceanography in 1954. He has served Oregon State University since 1962 as Associate Professor, since 1964 as Professor, and since 1989 as Professor Emeritus in both the Department of Microbiology and the College of Oceanic and Atmospheric Sciences. He has served on the editorial boards for *Geomicrobiology Journal* (1977 to present), *Marine Ecology Progress Series* (1979 to 1992), *Applied and Environmental Microbiology* (1980 to 1982), and *The Journal of Marine Biotechnology* (1991 to present). In addition to serving on numerous professional committees, from 1968 to 1969 he was the Program Director for Biochemistry at the National Science Foundation and is presently (since 1989) the scientific advisor to the Japan Marine Science and Technology Center (JAMSTEC). He was elected an Honorary Member of the American Society for Microbiology, among several other distinguished awards throughout his career.

Professor Morita has lectured nationally and internationally and has attended several conferences, workshops, and symposia as an invited speaker both before and after his retirement. Previously, he has published 48 chapters in various books, 134 scientific papers, 4 review articles, 2 laboratory manuals, edited 3 books, and authored another. His major scientific contributions include: (1) the discovery of bacteria that require pressure, i.e., barophiles, with Dr. Claude Zobell, (2) the discovery of the existence of phage particles in the marine environment, with Dr. Francisco Torrella (3) the finding that nitrifiers (e.g., bacteria that oxidize ammonia to nitrite) could also use methane and carbon monoxide, with Dr. Ron Jones, (4) the discovery of a new thermal group of bacteria, i.e., psychrophiles, and (5) the concept of starvation-survival in microbiology. He was the first microbiologist to initiate open ocean (or blue water) microbiology during this century. His most outstanding research expeditions were as microbiologist for the Mid-Pacific Expedition of 1950, and the Royal Danish Navy "Galathea" Deep Sea Expedition (1952). He has continually focused on the study of environmental extremes (i.e., always testing the limits) in terms of microbial processes and potentials. As a result, he has studied the microbiology of deep-sea sediments and the deepest regions of the oceans, both in cold and polar environments. He has continued to be a leader in marine microbiology and has succeeded in accomplishing the modern development of this field of research for those who would follow in his footsteps.

Contributors

Chapter 8

Paul A. Blum, Department of Biological Sciences, University of Nebraska at Lincoln, Lincoln, NB 68588

Chapter 9

Nicholas B. Bruggeman, Tyler A. Kokjohn, Holly S. Schrader, John O. Schrader, Julie J. Shaffer, James M. Vanderloop, Jeremy J. Walker, School of Biological Sciences, University of Nebraska at Lincoln, Lincoln, NB 68588

Bacteria in Oligotrophic Environments

1

Oligotrophic Environments—Problems and Concepts

> Ecology has sometimes been defined as that branch of biology entirely abandoned to terminology. Ecology can also be defined as an art; specifically, as the art of talking about what everybody knows about in a language that nobody understands. Microbial ecology has sometimes appeared to be the art of talking about what nobody really knows about in a language that everyone pretends to understand. The challenge in microbial ecology is to seek out the factors determining the growth of micro-organisms in their natural habitats and to talk about these factors in a language that everybody can understand.
>
> —Epilogue (Clark, 1968)

1.1 Introduction

All ecosystems are energy driven and most natural environments lack available energy to support microbial growth. When one addresses questions related to energy in biological systems, the type, amount, quality of the energy, and the turnover rate and replenishment of the energy must be addressed. In oligotrophic ecosystems, available energy is at a premium. This book explores the problem of the metabolic status of bacteria in oligotrophic environments. Data in the literature do not truly address this question. Scientists have been mainly concerned with the response of bacteria (usually grown in rich media) to oligotrophic conditions, whereas *from an ecological viewpoint, we should be concerned with the response of starved bacteria to eutrophic conditions and their rate of activity in the natural environment.* It is well known that the chemical composition of a microorganism depends on the environmental conditions in which it is grown. This becomes important when we recognize that most of the biosphere is oligotrophic and most of the bacteria are in the starvation-survival mode (Morita, 1986). Hence, organisms in oligotrophic environments would probably have no reserve energy sources (storage polymers). Any ecosystem has biochemically and physiologically diverse communities, all competing for fluctuating concentrations of unevenly distributed substrates, many of which are insoluble. The fluctuation of substrates results in unbalanced growth of microbes as well as periods of nongrowth between reproduction events. This condition is coupled with the fact that many ecosystems have fluctuating chemical and physical composition. The situation is grossly different from laboratory rich media systems.

1

Odum (1971) defined an ecosystem as a biological community interacting with its abiotic surroundings in a manner that results in energy flow, trophic structure, and materials cycling. Energy flow may be taking place in a specific ecosystem, but not all microorganisms are participating due to the lack of specific energy sources or specific nutrients, and/or physical conditions needed to satisfy all the species present. Furthermore, the rate of energy flow may be exceedingly slow. Thus, many microbes are found in a spore or dormant stage.

> Many years ago, Pfeffer remarked that the entire world and all the friendly and antagonistic relationships among different organisms are primarily regulated by the necessity for obtaining food. Microbial ecologists have expressed essentially this idea for "supply of food materials in soil can be said to be perennially inadequate" and "there are many micro-organisms in the soil and they are always hungry." (Clark, 1968)

Clark (1968) further emphasizes that as a consequence of this ever-continuing struggle for food, the specializations developed among microorganisms for utilizing energy-yielding materials have become enormous. This specialization, coupled with other ecological factors, is the reason why one or another microorganism becomes more or less active or suddenly disappears. Thus, it should be emphasized that the ability of a specific bacterium to utilize an energy source may be influenced by other organisms and the physical and chemical environmental factors.

Because the vast majority of publications dealing with oligotrophic bacteria employ some type of peptone (not a defined medium), no delineation can be made between nutrients and the energy sources. I will use the term *nutrients* when it is used in the original citations, where, in fact, the term *energy* should be used.

1.2 Historical Background on Starvation-Survival

Postgate and Hunter's (1962) publication was probably the first modern research paper on the survival of starved bacteria. During discussions dealing with the elucidation of marine bacteria at the Princeton meetings on marine microbiology, Morita (1968) first questioned whether some marine bacteria are transients, or bacteria not actually metabolizing in the marine environment. Clark (1967) had previously suggested that many organisms in soil are in a resting or dormant state. Subsequently, Sieburth et al. (1974) proposed that bacteria in a water column are transients starving to death. However, it is hard to determine if transient or indigenous microbes are metabolizing or not metabolizing in oligotrophic environments. After Novitsky and Morita (1976) published data showing that ultramicrocells can be formed as part of a strategy to survive starvation conditions,

Stevenson (1978) wrote a review providing various citations supporting the case for bacterial dormancy in aquatic environments. Many researchers working with soil have also concluded that most bacteria are in a dormant state. Free-living bacteria, suggested to be dormant, were shown to change in size with both enrichment and dilution of nutrient (Wright, 1978; Wright and Coffin, 1983). Wright (1978) presented a method V_{max}/AODC (specific activity for [^{14}C] glucose total uptake [assimilated plus respired]/acridine orange direct count) to measure the average physiological state and metabolic role of bacteria in the aquatic environment. He concluded that marine bacteria are adapted to conditions of nutrient starvation by becoming relatively inactive or "dormant," existing for weeks or months in a reversible physiological state that reflects the nutrient availability. Starved cells are not dormant. Dormancy represents an inactive metabolic state. Starved cells have the ability to take up utilizable substrates immediately. The state of dormancy exist because there are no utilizable substrates present. For most microorganisms in natural environments, the term starved should be used in place of dormant.

Atlas and Barta (1981) and Andrews (1984) adapted Pianka's (1970) survival classification for plants and animals as r and K strategists to microbes. In this situation an r strategist microbe displays rapid growth and dominates situations where resources are temporarily abundant (Winogradsky's [1924] zymogenous or opportunists); whereas the K strategists would be Winogradsky's autochthonous microbes. Oligotrophs would then be an example of K strategists; whereas the eutrophs (growth on high concentrations organic matter) would be an example of r strategists. However, comparison of oligotrophs with autochthonous bacteria is attractive but unjustified (Semenov, 1991).

1.3 Oligotrophic Environment

The term *oligotrophic* was first used to define an environment that had low nutrients, mainly certain lakes. *Eutrophic, mesotrophic,* and *oligotrophic* (from the German forms *nährstoffreichere [eutrophe] dan mittlereiche [mesotrophe] und zuletzt nährstoffarme [oligotrophe]*) were introduced by Weber (1907) to describe the general nutrient conditions as determined by the chemical nature of soil solution in German bogs (Hutchinson, 1973). In these German bogs, the succession proceeded from eutrophic to oligotrophic since the raised bog is subjected to continual leaching. Plant association characterizing the eutrophic low bog was described as eutraphent, or well nourished, whereas that on the oligotrophic raised bog was oligotraphent (Hutchinson, 1973). Naumann (1919), adapting Weber's (1907) terms, introduced the general concepts of oligotrophy and eutrophy into limnology, and the difference between them was based on phytoplankton populations. Few planktonic algae occurred in oligotrophic lakes. The pros and cons concerning these terms as they are applied to lakes is discussed by Hutchinson (1973) and Rohde (1973). However, the word "eutrophy" is greatly misused in contemporary dialogue (Wetzel, 1983). Since the term originally was applied to plant nutrients other than the energy source (light), its adaptation to describe heterotrophic bacteria was probably not given too much thought. Limnologists do not use dissolved organic matter (DOM) and/or particulate organic matter (POM) to delineate between the trophic levels. Probably the best delineation between trophic levels in the literature is shown in Table 1.1, where phosphorus concentration, chlorophyll concentration, Secchi depth, and primary production are employed. Thus the trophic level is assigned based on all the data available for the lake. By limnological definition, oligotrophic waters contain low concentrations of inorganic nutrients—not DOC (dissolved organic carbon) and POC (particulate organic carbon). In aquatic systems, low nutrients (phosphate and nitrogen) are generally considered characteristic of oligotrophic waters and not the organic matter (POC and DOC) (see Table 1.1). For a discussion of this subject matter, Flynn (1988a,b, 1989) should be consulted. Nowhere is there a description, in terms of organic matter content, in natural systems to delineate among eutrophic, mesotrophic, and oligotrophic environments in nature for heterotrophic bacteria.

The flux of organic carbon in oligotrophic lakes does not exceed 0.1 mg C/l per day (Hood, 1970); similar fluxes are observed in the ocean. Soil is considered grossly oligotrophic, mainly because nutrients are not bioavailable (Williams, 1985). Within the category of low nutrients, energy must be included and, in many cases, the carbon source is the energy source.

1.4 The Concept of Oligotrophic or Oligocarbophilic Bacteria

Bacteria capable of growth on media containing only minerals but obtaining their carbon and energy requirements from trace amounts of organic matter found in the air are termed "oligocarbophilic" bacteria (Beijerinck and Van Delden, 1903; Hirsch, 1964). However, some studies with oligocarbophilic bacteria subject the organisms to an atmosphere of hydrocarbons. The oligocarbophilic bacteria are the first described oligotrophs, but why this term ceased to be employed is not clear. However, Ishida and Kadota (1979), in a footnote, concede that oligotrophs are in a sense oligocarbophilic.

The term *oligotrophs* was adapted to describe heterotrophic bacteria that had the ability to grow at minimal organic substrate concentrations (1 to 15 mg C/l), even though they can grow on richer media (Kuznetsov et al., 1979). No mention is made as to whether the organic substrate is an energy source and/or nutrient(s). Oligotrophs are defined by their ability to grow only at low nutrient concentrations (Akagi et al., 1977; Mallory et al., 1977; Yanagita et al., 1978; and Ishida and

Table 1.1 Trophic classification system

Trophic State	Trophic State Index	Phosphorus Concentration (μg/liter)	Chlorophyll Concentration (μg/liter)	Secchi Disk Depth (meters)	Primary Productivity (mgC/m²/day)
Ultraoligotrophic	<20	<3	<0.3	>16	<50
Oligotrophic	20–35	3–9	0.3–2	7–16	50–250
Mesotrophic	35–50	9–24	2–6	2–6	250–1000
Eutrophic	51–65	24–75	6–40	0.75–2	1000–2000
Hypereutrophic	>65	>75	>40	<0.75	>2000

Source: Carlson 1977, Wetzel 1983. Reprinted with permission from *Atlas of Oregon Lakes*, D. M. Johnson, R. R. Petersen, D. R. Lycan, J. W. Sweet, M. E. Neuhaus, and A. L. Schaedel, 1985; copyright 1985, Oregon State University Press, Corvallis, OR

Kadota, 1981a). The definition is arbitrary. There is little or no evidence that oligotrophs prefer low concentrations of nutrients (Williams, 1985). Low nutrient media may help minimize competition on isolation plates (Poindexter, 1979). Poindexter (1984) found freshwater oligotrophs live in a phosphate-limited habitat (Poindexter, 1984), hence she introduces the term *phosphate oligotrophs*. When dealing with phytoplankton on the ocean surface, energy (sunlight) may not be limiting but other nutrients (phosphate, nitrogen, silicate, etc.) may be in short supply. This author does not agree that the term oligotrophs should be used to describe any bacteria present in low nutrient environments, but the term is retained in this book—mainly because it appears in the literature.

In his discussion concerning the effect of the energy supply, Russell (1950) states that

> The energy in organic matter differs fundamentally from the nutrients in it. Nutrients can be used over and over again by unending succession of organisms. . . . But energy cannot be used in this way. It is as indestructible as matter, but once transformed to heat, it cannot be used by microorganisms or any other living things; whatever energy is dissipated by organisms becomes out of reach of the others.

Thus we are forced to think in terms of the supply of energy and the turnover of nutrients. Unfortunately, in the microbiological literature, there has not been any effort to distinguish an organic energy source from organic nutrients. Thus, in the review of the literature we are faced with the situation where we must accept organic matter as energy for the heterotrophic bacteria and ignore the fact that the organic matter may be recalcitrant or not bioavailable.

Researchers working with oligotrophic bacteria all state that under natural conditions, these organisms have a high growth rate at low concentrations of organic substances. How is a high growth rate defined? Does the growth rate compare favorably with that of Pseudomonads? If oligotrophic bacteria have a fast growth rate, then they will deplete the substrates they utilize readily, hence the number of cells per unit volume will be low compared to organisms like *Pseudomonas aeruginosa* growing in nutrient broth at room temperature. The foundation for the existence of oligotrophic bacteria appears to be rather weak, mainly because investigators have not looked at the myriads of substrates as an energy source. In addition, the possible use of catalase in the system (Burton and Morita, 1964), the proper pH or Eh, the right gas tension, or other growth-limiting substances must also be checked. This is a Herculean task but until all the various substrates (and their combinations and concentrations) and the various physical and chemical environmental factors are investigated, the scientific foundation for the term *oligotroph* will remain weak. It could easily be that the oligotrophs are only organisms that have adapted to a low nutrient environment and have the ability to adapt back to an environment with higher concentrations of substrates. Morita (1986) states that most organisms in nature are in the starved state, which is the normal mode because most environments are oligotrophic. Microbes are known to be the oldest organisms in the evolution of Earth, yet the ancient amount of organic matter for the heterotrophic bacteria was probably not as great as it is today—mainly because green plant photosynthesis evolved many years later. According to Koch (1971), microorganisms have not only been selected for ability to grow under chronic starvation, but also for ability to respond quickly to unannounced and irregular windfalls of food. Thus, it could be that oligotrophs are organisms in the starved state capable of utilizing low concentrations of organic matter but having the capability to adapt to higher concentrations of organic matter. For those organisms termed *obligate oligotrophs,* the conditions may simply be one of not finding the correct substrate or combination of substrates and the correct physical and chemical factors to support optimal growth of that specific organism.

The possible use of catalase was previously mentioned as a factor to be checked in determining if an organism is an oligotroph. The case to support the idea that oligotrophic bacteria do not reproduce comes from the data of Dubinina and Lapteva (in Kuznetsov et al., 1979). From water samples taken from Parovoye Lake, 2600 colonies per ml developed on beef extract medium; whereas on Pringsheim medium, 160,000 colonies per ml developed. All the organisms growing on beef extract agar possessed catalase activity. About 40% of the organisms that developed on Pringsheim medium in the last dilution had no catalase. The growth of the oligotrophic bacteria *Leptothrix pseudoochraceae, Siderocapsa eusphaera,* and *Metallogenium personatum* on rich media lyse due to the accumulation of hydrogen peroxide in the media. The addition of catalase in the media permits good growth in the rich media by these organisms as shown by Dubinina (Kuznetsov et al., 1979). Catalase is produced by most gram-positive organisms, hence the development of oligotrophic bacteria in a nonoligotrophic environment can occur due to the catalase produced by associated organisms. Catalase activity is known to decrease drastically when microorganisms are grown in the presence of significant concentrations of glucose or beef extract (Jones et al., 1964; Yoshpe-Purer and Henis, 1976). Catalase may be responsible for growth inhibition by high substrate concentrations (Kuznetsov et al., 1979) but all the oligocarbophilic bacteria studied by Witzel et al. (1982a) possessed catalase and were able to grow in nutrient-rich media after adaption. Thus far, the action of catalase has not been studied on oligotrophs with the exception of the above named species.

Kuznetsov et al. (1979) defined oligotrophs as bacteria that develop upon first cultivation on media with a minimal amount of organic matter (1–15 mg C/l). If they are not able to grow when subsequently cultured on a higher concentration of organic matter, they were then termed obligate oligotrophs; otherwise they were termed facultative oligotrophs. Poindexter (1981a) used 5.0 to 10 mg C/l for the cultivation of oligotrophic bacteria. Hood and MacDonell (1987) arbitrarily employed media containing 5.5 mg C/l to demonstrate the presence of oligotrophic bacteria; whereas those cells that had the ability to grow on media containing 5.5 mg C/l as well as 5.5 g C/l were then termed facultative oligotrophs. Subsequent cultivation of oligotrophs on this low concentration to richer media was possible either by adaptation or mutation. However, it is possible that growth, initially, on low nutrient concentrations is a resuscitation mechanism for starved cells. Therefore, cells that initially grow out at the low carbon medium may not be oligotrophs. Fry (1990a) states that "obligate oligotrophs are comparatively rare in nature and can only grow on low concentrations of carbon (1–6 g C/l)." (This value stated in Fry's conclusion may be a typographical error and presumably should have read 1–6 mg C/l). Japanese researchers defined oligotrophs as organisms with the capability to grow on a medium containing organic carbon at a concentration of 1 ppm (1mg/l) (Suwa and Hattori, 1984). This concentration of carbon requirement for oligotrophs was later modified by the Japanese researchers (Yanagita, Kadota, Rage, Simidu, Ishida, Akagi, Ohta, and Hattori) to include those organisms capable of growing on medium containing less than 1 mg of organic carbon per liter (cf. Whang and Hattori, 1988). No distinction is made whether the organic carbon is bioavailable or not. Oligotrophs are also defined as organisms that have a low substrate saturation (Km) and low maximum growth rate (μm) (Jannasch, 1979; Hirsch et al., 1979). The difference between counts on 1/100 dilution of nutrient broth (DNB) and nutrient broth (NB) are considered oligotrophs (Hattori, 1980), because some isolates that grew on DNB will not grow on NB. Most investigators appear to define oligotrophic bacteria in terms of their ability to utilize low concentrations of media, usually containing peptone. There are all types of peptones (eg., Bacto-peptone [and other brands], neopeptone, proteose peptone, polypeptone). Unfortunately, peptone will not satisfy all the nutrition and energy requirements of all the physiological types of bacteria in nature. There is some evidence that peptone may be inhibitory to certain oligotrophs (Yoshinaga and Ishida, personal communication). It has been known for quite some time that nutrients can inhibit the growth of bacteria (Buck, 1974). High concentrations of amino acids can also inhibit the growth of certain organisms (Martin and MacLeod, 1984). In addition to looking for the proper substrate for the "oligotrophic bacteria," one must consider the physical and chemical parameters of its growth. Recognizing that substrate is one of the most important factors in controlling the density of bacteria (Wright, 1984), Wright (1988a) designed a model for growth of a bacterial population as a function of the concentration of substrate in a saturated type equation, based on the kinetic model introduced by Monod (1949). The point is that substrate input is the main driving function in the model and represents new labile organic carbon that enters into the model from all sources (Wright, 1988a). According to Wright (1988b), it is reasonable to expect that substrate concentration would change under conditions of changing bacterial density, and if a system is strictly substrate limited (no grazing), substrate will be low and each cell will get enough to meet maintenance needs. On the other hand, if grazing occurs, substrate would rise in concentration and begin to drive bacterial growth, which at steady state would be equal to the grazing rate.

Schut et al. (1993) defined obligate oligotrophic bacteria as those not capable of growth upon initial isolation on ordinary media but requiring very low nutrient media. Using this definition, 15 isolates were classified as obligate oligotrophs. Facultative oligotrophs and eutrophs were less than 1% of the total microbial population in Resurrection Bay, Alaska, and the central North Sea off the Dutch coast. Yet these 15 isolates could be adapted to growth under laboratory conditions after several months of nutrient deprivation prior to isolation. It may be that this period was necessary for resuscitation of the isolates as they obtained their resuscitation energy from gaseous compounds in the air. Yet these investigators recognized that there may be no demarcation line between oligotrophic and eutrophic bacteria in view of the fact that eutrophs could be trained to become obligate oligotrophs (Hood and MacDonell, 1987) and obligate oligotrophs can be converted into facultative oligotrophs (Yanagita et al., 1978; MacDonell and Hood, 1982; Martin and MacLeod, 1984). Ishida et al. (1989) tentatively defined obligate oligotrophs as organisms that cannot grow in rich media and facultative oligotrophs as organisms that grow in both poor and rich media. Furthermore, these obligate oligotrophs cannot form colonies on agar media and the maximum cell yields are not over 10^7 cells/ml. Ishida et al. (1989) claim that the obligate oligotrophs are the dominant heterotrophs in pelagic freshwater and seawater. Until the existence of obligate oligotrophs has definitely been established, these findings are based mainly on definition of the experimental procedures employed.

Prosthecate bacteria are considered to be obligate oligotrophs (Poindexter, 1981a; Kuznetsov et al., 1979), especially when they are observed to be in oligotrophic environments coupled with the observation that the lengths of the prosthecae normally increase as nutrient concentrations decrease (Schmidt and Stanier, 1966). However, Dow and Morgan (1986) question whether these prosthecate bacteria are uniquely adapted to low-nutrient environments, citing that they can be found in eutrophic waters (Morgan and Dow, 1986), oil-polluted seawater (Murakami et al., 1976), groundwater (Hirsch and Rades Rohkohl, 1983), human pus (Pongratz, 1957), hydrothermal vents in deep ocean (Jannasch and Wirsen, 1979), pulp-mill waste oxidation lagoons (Stanley et al., 1979), and an activated sludge treatment plant (Meyers and Meyers, 1986). Thus Morgan and Dow (1986) state that the prosthecate bacteria are not uniquely adapted to low-nutrient environments in either their distribution or their response to nutrient limitation and that prosthecate and other bacteria are scavenging very low nutrient concentrations.

However, in the literature, various terms (oligotrophic, hypotrophic, oligocarbophilic, and low nutrient bacteria) have been used to describe bacteria capable of growing in low-nutrient environments. Organisms that grow on high concentrations of nutrients have been termed eutrophs (Fry, 1990a), heterotrophs (Akagi and Taga, 1980; Akagi et al., 1977), saprophytes (Kuznetsov et al., 1979), hypotrophs (Baxter and Sieburth, 1984), and copiotrophs (Poindexter, 1981a,b). All bacteria are osmotrophs in that they can only utilize organic matter in solution.

Although Fry (1990a) reviewed the topic of oligotrophs, I do not believe a strong enough case has been made for the existence of oligotrophic bacteria. West and Fry (in Fry, 1990a) surveyed 40 pure cultures of laboratory bacteria; 95% were found to be facultative oligotrophs and none were in the obligate oligotroph category. Of the genera West and Fry (in Fry, 1990a) surveyed, the facultative oligotrophs found were *Acinetobacter, Bacillus, Beneckia, Corynebacterium, Escherichia, Klebsiella, Micrococcus, Proteus, Providencia, Pseudomonas, Serratia, Staphylococcus,* and *Streptococcus,* while Kuznetsov et al. (1979) lists *Agrobacterium, Photobacterium, Vibrio, Micrococcus, Staphylococcus, Microcyclus, Leptothrix, Ochrobium, Metallogenium,* and *Pasteuria.* Most of the oligotrophs isolated by Akagi et al. (1980) were mainly *Pseudomonas, Vibrio,* and *Acinetobacter-Moraxella.*

The concept of a model oligotroph was put forth during the Dahlem Conference in 1979 (Shilo, 1979). Unfortunately, most of the presumed adaptations of model oligotrophs to their niches have been demonstrated to be equally applicable to *Escherichia coli* (Koch, 1979) and it is doubtful that anyone would consider *E. coli* to be an oligotroph. *E. coli* is the only organism tested that has been shown to use its energy source preferentially for uptake at low nutrient concentrations (Purdy and Koch, 1976). The ability to use energy preferentially for uptake at low nutrient concentration is considered by Hirsch et al. (1979) in a group report to be an oligotroph property. In addition, microorganisms seem to be able to use K strategies sometimes and r strategies at other times (Koch, 1979). Real organisms need to have sensing systems and switching mechanisms to selected alternative responses (Koch, 1979). Although the group report by Hirsch et al. (1979) summarizes the characteristics for a model oligotroph, it is doubtful that these characteristics apply only to oligotrophs. This group report is rebutted by Koch (1979) and this author agrees with Koch's (1979) rebuttal. Nikitin and Zlatkin (1989) agree with the concept of "model oligotrophs." One could also ask, if bacteria from low-nutrient environments are suddenly exposed to the high concentration of nutrients, is the toxicity just a manifestation of substrate-accelerated death (Strange and Dark, 1965)? Another possibility is that the toxicity is due to a rapid influx of nutrients into cells previously adapted to scavenge them (Koch, 1971, 1979). Adaptation to higher concentrations of nutrients has been shown to occur in many bacteria characterized as oligotrophs, yet Nikitin and Zlatkin (1989) claim that full-strength broth totally inhibits the growth of oligotrophic bacteria. Most oligotrophs can be easily converted to growth on rich media (Mo-

Table 1.2   Capacity of various compounds when added at two concentrations to the basal medium to support growth of the oligotrophic and two eutrophic marine bacteria.

Compound[a]	Concn (mg of C per liter)	Growth of Organism[b] Oligotroph 486	Eutroph RP-303	Eutroph RP-250
Alanine	1000	−	+ (8)	+ (8)
	10	+ (2)	−	+ (5)[c]
Phenylalanine	1000	−	+ (8)	+ (8)[c]
	10	+ (5)	−	+ (8)[c]
Proline	1000	+ (6)	+ (4)	+ (5)
	10	+ (3)	−	+ (5)
Acetate	1000	−	+ (4)	+ (5)
	10	+ (5)		+ (5)
Aspartate	1000	+ (2)	+ (2)	+ (3)
	10	+ (3)	+ (2)	+ (2)
Glutamate	1000	+ (3)	+ (3)	+ (3)
	10	+ (5)	+ (3)	+ (3)
Malate	1000	+ (5)	+ (4)	+ (6)
	10	+ (8)	+ (8)	+ (6)
Succinate	1000	+ (10)	+ (3)	+ (3)
	10	+ (4)	−	+ (3)

Source: Reprinted by permission from *Applied and Environmental Microbiology* 47:1017–1022, 1984, P. Martin and R. A. MacLeod, 1974; copyright 1974, American Society for Microbiology.
[a] Acetate, aspartate, glutamate, malate, and succinate were added as their Na^+ salts.
[b] +, Growth; −, no growth as detected spectrophotometrically with an uninoculated control as a blank. The numbers in parentheses are the days required to achieve maximum growth with the compound tested.
[c] Bacteria were clumping and formed flocs.

aledj, 1978; Kuznetsov et al., 1979). *Caulobacter* (Krasil'nikov and Belyaec, 1967; Poindexter, 1979, 1981a) and *Hyphomicrobiium* (Poindexter, 1979) are considered to be oligotrophs, but the use of low-nutrient media may help to minimize competition on the isolation plates (Poindexter, 1979).

Because many of the isolated oligotrophs are capable of growth on higher concentrations of organic matter, they were divided into two categories: oligotroph and facultative oligotroph (Yanagita et al., 1978) or facultative and obligate oligotroph (Ishida and Kadota, 1981b). The obligate oligotroph cannot grow on conventional media. However, Hirsch et al. (1979) emphasized the existence of facultative oligotrophy. Morgan and Dow (1986) stated that the current concept of model oligotrophs and their presumed adaptation to such a mode of existence is questionable. I agree with Dow and Morgan.

The physiological bases for oligotrophy, according to Semenov (1991), are (1) the greater substrate affinity of the oligotrophs' transport system, (2) "economical" metabolism, and (3) the existence of a "master reaction" or "rate-determining steps" controlling the rate of metabolism. Thus, they are adapted to exploit ecological niches characterized by low substrate concentrations that occur in the nano- and picomolar range and low energy flow by use of their efficient utilization systems characterized by a unique metabolic regulation. Yet Semenov (1991) states that they may develop in rich as well as poor environments, either naturally or experimentally. To this author it implies that if organisms are faced with low substrate environments, the physiological bases for oligotrophy previously mentioned come into play.

In order to isolate oligotrophs, Hirsch and Conti (1964a,b) kept water in an open vessel for up to 2–3 months, because all easily assimilated substances were thereby used, and the saprophytic bacteria and the oligotrophs remained viable due to the organics suspended in the air. They then cultivated the bacteria on a liquid medium, which was placed in an atmosphere of methanol vapors as well as other volatile organic substances. Final isolation of the oligotrophic bacteria was made on the same medium to which 1.5% agar was added. It is interesting to note that isolates of Geller (1983a) were able to grow on the contaminants in the air. Thus, laboratory air does have sufficient organics present to satisfy the growth of various bacteria. With modifications of this method Moaledj (1978) and Mallory et al. (1977) also isolated a number of oligotrophic and oligocarbophilic bacteria. Kuznetsov (1970) assumed that many of the bacteria that failed to grow in the laboratory were oligocarbophilic.

The best documentation for the existence of oligotrophic bacteria was presented by Button et al. (1993). In order to prevent organic matter contamination, all glassware was heated to 550°C and screw-capped test tubes were fitted with acid-washed teflon liners. The seawater, used as the culture medium, was collected from the site to be examined. This seawater was passed through a fired Gelman A/E glass fiber filter, autoclaved, and refiltered and then placed (50 ml) in the screw-capped test tubes. Untreated seawater was then used as the inoculum and the dilution technique was employed to isolate the bacteria. The inoculum size was statistically calculated to range from 0.1 to 10^3 organisms per tube with 10 to 30 tubes at each dilution. Incubation was at 10°C, which was within 3° of the sample temperature, for a period between 3 and 8 weeks. Growth was evident in a number of tubes. Unfortunately, the DOC of the water was not measured, but the data indicates that bacteria could grow on the DOC present in the seawater. The main item in this investigation, to this author, is the exclusion of gaseous compounds in the air. However, an organism designated RB2256 isolated by the same technique by the same group (Schut et al., 1993) was later found not to be an oligotroph (Eguchi et al., 1996).

According to Gottschal (1985) and suggested by Poindexter (1981b), it would be tempting to assume that oligotrophs would dominate in a chemostat at low dilution rates and that copiotrophs would dominate at high dilution rates. Yet, there is no direct evidence presently available to warrant such an assumption (Gottschal, 1985). Careful examination of the field data by Gottschal (1985) indicates that the distinction between low and high nutrient-adapted species is not at all sharp and easy to define. Species considered to be copiotrophs may, in continuous culture, outgrow putative oligotrophs in the presence of substrate concentrations considered only to support good growth of excellent oligotrophs. The data of Jannasch (1968), as well as Matin and Veldkamp (1978), tend to verify this. Matin and Veldkamp (1978) concluded the *Pseudomonas* sp. may be considered the copiotroph and the *Spirillum* sp. the oligotroph.

Carlucci et al. (1986) employed the term *low-nutrient bacteria* for marine bacteria that grow well in the dilute nutrient concentrations equivalent to the organic matter in natural seawater. These low-nutrient bacteria are heterotrophs capable of responding to added organic substrates, including glucose, proline, acetate, peptone, algal extract, and yeast extract. Carlucci and Shimp (1974) and Mallory et al. (1977) isolated low-nutrient bacteria on agar plates that had no nutrient added. However, as pointed out by Akagi et al. (1977), agar has sufficient organic matter for microorganisms to use. A low-nutrient bacterium, designated as 15U by Carlucci and Shimp (1974), grew in unsupplemented seawater faster at 20° than at 5°C, and more rapidly in surface than in deep (3000 m) water. However, this bacterium seldom grew in seawater to concentrations where turbidity could be seen.

Various oligotrophic bacteria isolated by Akagi et al. (1977) and examined by Martin and MacLeod (1984), were found subsequently to be capable of growth in elevated concentrations of organic matter. The ability of an oligotroph and two eutrophs to use different amino acids is shown in Table 1.2. But many of the so-called oligotrophs have the ability to adapt to nutrient-rich media, including oligotroph 486. MacDonell and Hood (1982) and Hood and McDonell (1987) show most of the isolates capable of passing through a 0.2-μm filter were not capable of growing in various full strength media but had the ability to grow on nutrient broth that was diluted 10-fold. Adaptation to higher strength media was possible. Fry (1990a) found only one culture lost its capacity to grow on low-nutrient media within his culture collection of oligotrophs and stated that "all the other obligate and facultative oligotrophs handled have shown very stable growth profiles on media of different carbon concentrations." This has been true whether they have been maintained on high- or low-nutrient media. The copiotrophs have equal stability. The ability of microbial cells to survive is not necessarily correlated with the ability to grow rapidly. Stationary phase cells of *Aerobacter aerogenes* remain viable for relatively long periods compared to logarithmic phase cells (Harrison and Lawrence, 1963). Suboptimal temperatures in nature may permit a greater viability.

As Martin and MacLeod (1984) pointed out, "one could not conclude that because an organism is able to grow at low levels of peptone that it possessed an intrinsic capacity to grow at a low concentration of organic matter" and "the fact that such a peptone at high concentration inhibited growth of the organism, however, would not mean that this organism was intrinsically sensitive to a high concentration of organic material and therefore classifiable as an obligate oligotroph." Thus the data of Martin and MacLeod (1984) refuted the concept of oligotrophic bacteria. According to Kuznetsov et al. (1979) many bacteria isolated on low-nutrient media have or acquire the ability to grow on nutrient-rich media. It should also be pointed out that the copiotrophic or oligotrophic nature of species depends strongly on the substrate (and its concentration) used. Note that organism 486 is an oligotroph isolated by Akagi et al. (1980). MacDonell and Hood (1982) isolated numerous ultramicrobacteria from seawater and found that initially they grew best in a medium that had a concentration of 0.4 to 0.5 g of trypticase/l (Fig. 1.1) but they also had the ability to adapt to higher concentration of nutrients (Fig. 1.2) after 40 days of conditioning to culture media. After adaption to higher concentrations of media, the cells became large. Hence initial isolation of bacteria from oligotrophic environments is best conducted with very dilute media. Much more research is needed to determine if the oligotrophic bacteria do exist.

Morgan and Dow (1986) questioned the concept, proposed by Hirsch et al. (1979), of model oligotrophs and this presumed adaptation to such a mode of existence. They would just as soon forget about the definitions of oligotrophs, facultative oligotrophs, and copiotrophs. These authors suggested that the terms oligotrophic and copiotrophic be indicators of the type of environment and the organisms within, and not as distinct and excluded microbial types. According to Williams (1985) the concept of oligotrophy sensu stricto should be used with caution. Euryheterotrophy, rather than oligotrophy, was the major factor that permitted bacteria to grow on low-nutrient concentration (Harowitz et al., 1983). E. coli, which is definitely not an oligotroph, would fall into the category of a euryheterotroph because it has the capacity to grow on low-nutrient media (Koch, 1979). On the other hand, there are data that support the concept of oligotrophic bacteria. A generally accepted and precise definition of oligotrophic bacteria does not appear to exist (Fry, 1990a).

Hirsch et al. (1979) emphasized the existence of facultative oligotrophs. The obligate oligotroph is defined as an organism not capable of growth in 5 g of trypticase/l and incapable of producing visable turbidity in medium containing 0.5 g trypticase/l (Ishida and Kadota, 1981a). In the latter situation, growth had to be determined microscopically. Later, Yoshinaga and Ishida (1992) employed peptone in the same concentration to delineate between facultative and obligate

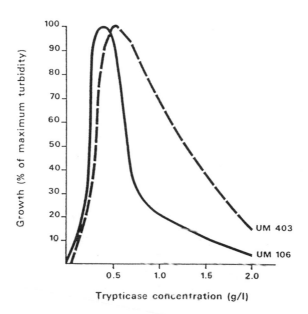

Figure 1.1 Growth response of isolates UM403 and UM106 in Trypticase broth. Reprinted by permission from *Applied and Environmental Microbiology* 43:566–571, M. T. MacDonell and M. A. Hood, 1982, copyright 1982, American Society for Microbiology.

oligotrophs. These obligate oligotrophs do not have the ability to become adapted to rich media. Oligotrophs growing under low-nutrient conditions have high amino acid uptake systems with high substrate affinity (see Chapter 7) and the energy produced is used preferentially for substrate uptake and protein synthesis rather than for DNA synthesis and reproduction (Yoshinaga and Ishida, 1992). The inability to grow on rich medium is a stable property of obligate oligotrophs, and they have the ability to utilize glutamate, glycine, serine, and glycolate, but not acetate, proline, or leucine, and the cell size is smaller than the facultative oligotrophs (Ishida and Kadota, 1981a,b). However, Ishida and Kadota (1981a,b) claimed that good growth (10^5 to slightly more than 10^6/ml, determined microscopically) of the obligate oligotrophs also occurred in artificial lake water containing only magnesium sulfate, calcium chloride, and potassium chloride. Is this "growth" due to fragmentation or due to the use of nutrients in the laboratory air? This latter situation was noted by Geller (1983b) on isolates from lake water. Obligate oligotrophs that are inhibited by high concentrations of organic nutrients have also been isolated (Hattori and Hattori, 1980). Yanagita et al. (1978) point out the division between oligotrophs and eutrophs into two groups is not sufficiently distinct. After reviewing the literature on oligotrophs, Kuznetsov et al. (1979) concluded that the division between oligotrophs and eutrophs is not exact.

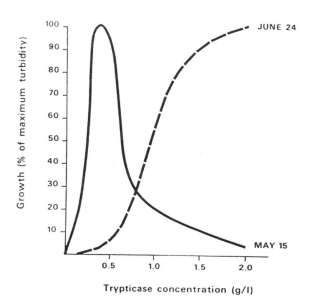

Figure 1.2   Growth response of UMl06 newly isolated (May 15) and after 40 days of adapting on culture medium (June 24). Reprinted by permission from *Applied and Environmental Microbiology* 43:566–571, M. T. MacDonell and M. A. Hood, 1982, copyright 1982, American Society for Microbiology.

As for obligate oligotrophs, Poindexter (1981a) indicated her uncertainty as to their existence. Morgan and Dow (1986) also stated that the existence of strict oligotrophs is rather doubtful. Hattori (1986) states that the concepts of oligotrophs and copiotrophs are not clearly defined.

Should obligate oligotrophic bacteria be defined as organisms capable of growth under the conditions of severe nutrient limitations only? How does one determine whether an organism is oligotrophic? Yanagita et al. (1978) defined obligate oligotrophic bacteria as organisms that grow in 10 mg polypeptone/l but not in 10 mg polypeptone and 5 g yeast extract/l. Recognizing that in nature there are organisms that have the ability to grow at low-nutrient concentrations or at relatively high concentrations, Yanagita et al. (1978) tentatively classified the former as oligotrophs and the latter as eutrophs. (Note the word "tentatively" employed by these investigators because the demarcation among the trophic groups is not necessarily clear.) They found that among the oligotrophs there were organisms that grew only in the presence of a minor amount of nutrients and others that grew in a wide concentration of nutrients. The former they assigned the term *oligotrophs* (others use the term *obligate oligotrophs*) and the latter, *facultative oligotrophs*. They found that oligotrophs isolated from natural water samples are converted to eutrophs (facultative oligotrophs) by adaptation, which further illustrates the complexity of the trophic characteristics of bacteria. It is well known that the various types of bacteria have different growth requirements. For instance, some bacteria have certain vitamin requirements, some have amino acid(s) requirements, some have trace element requirements, etc. If an obligate oligotroph is inoculated into a single medium, at a set temperature, Eh, and pH and good growth does not occur, why is it not possible to get better growth in the same medium having a greater concentration of all its ingredients? Some of the possible reasons may be (1) the concentration of substrates is too high; (2) a sudden shift from a nutrient-poor environment to a rich nutrient environment is deleterious to the organism and its metabolic processes are not geared for the sudden input of energy; (3) the organisms can grow to a limited extent but cannot take care of the peroxides formed during metabolism, and, as a result, expire; (4) the organism needs to be resuscitated in some manner before it can adapt to the richer medium. Does this mean that an obligate oligotrophic bacterium would survive but not grow in a eutrophic environment, whereas, a facultative oligotrophic bacterium would be able to grow in both eutrophic and oligotrophic environments?

Do oligotrophic bacteria exist? It seems illogical that evolution over time has not produced the obligate oligotroph. Any microorganism isolated and proposed as an obligate oligotroph must be tested for the previously mentioned item in the preceding paragraph tested under all different chemical (mixtures of various substrates, etc. and not only amino acids, peptone, and yeast extract) and physical conditions (pH, Eh [aerobe, anaerobe, microaerophile], various temperatures). In addition, organics from laboratory or atmospheric air must be excluded. If one is found, then the rule of priority should be followed and the organism should be called an "oligocarbophile" and not an obligate oligotroph.

1.5 Copiotrophic Bacteria

Poindexter (1981a) coined the term "copiotrophs" for organisms that have the ability to grow better at 100 times the concentration of organic matter than is found in the oligotrophic habitats. These organisms would include organisms referred to as "eutrophs" (Kuznetsov et al., 1979), "heterotrophs" (Akagi et al., 1977; Mallory et al., 1977), "saprophytes" (Kuznetsov et al., 1979), or "organisms that grow on rich media" (Poindexter, 1981a). However, Gottschal (1985), after reviewing the literature, considers it very doubtful whether on the basis of growth or no-growth that copiotrophs should be distinguished at all as a separate group of organisms. Would one consider *E. coli* a copiotroph? Yet in the intestinal tract its generation time is 20 h (Gibbons and Kapsimalis, 1967), but Koch (1971) and Neijssel (1980) state that *E. coli* divides in the human gut only twice a day. As

Koch (1971) points out, the intestinal tract is a harsh environment and there are irregularities in nutrient supply, competition from other microorganisms, presence of growth-inhibiting substances, and/or adverse physical and chemical conditions.

Copiotrophic bacteria have the ability to grow at 100 times the concentration of nutrients in their original habitat and this ability is a nutritional trait that would be hard to meet in ecological studies. (Unfortunately, in microbial studies dealing with environmental samples, the organic matter content is not determined.) Nearly all bacteria can grow better in rich organic media. Recognizing that facultative oligotrophs grow better on richer media, a question arises as to what is the difference between a facultative oligotroph and a copiotroph? Copiotrophs are microbial types whose survival seem dependent on a nutrient supply that is typically 100 times higher than that found in oligotrophic habitats, but oligotrophy has not been distinguished from copiotrophy (Poindexter, 1981a). The present designation of a given isolate as an oligotroph is usually based solely on its ability to produce visible growth on media containing only 5 to 10 mg of C/l (Poindexter, 1981a). Therefore, the operational definition of a copiotroph suffers because carbon may not always be biologically available as in a peat bog.

Kjelleberg et al. (1982) hypothesized that marine bacteria are mainly copiotrophic, requiring relatively high nutrient concentrations for growth. As a result, they exist as dormant cells adapted specifically to frequently and rapidly changing environments, taking advantage of nutrients that tend to accumulate at interfaces.

Better definitions are needed to describe the terms "copiotroph," "low-nutrient bacteria," "oligotrophic bacteria (facultative and obligate)," and "starved bacteria," because all are involved in oligotrophic environments. In view of the questions concerning oligotrophic and eutrophic bacteria raised by Martin and MacLeod (1984), are low-nutrient bacteria, copiotrophic bacteria, and oligotrophic bacteria merely starved bacteria? Are the oligotrophic, copiotrophic bacteria starved cells in which the oligotrophs need some method of adaptation or resuscitation before good growth occurs on media, whereas the copiotrophs do not need adaptation or resuscitation? Good arguments can be put forward that most, if not all, of the above named forms are mainly starved bacteria in nature. These arguments can be listed as:

1. Nearly all ecosystems are oligotrophic and the bioavailability of the organic matter is very limited (see Chapter 3) and in most cases the organic matter present is not sufficient to maintain the indigenous microflora (see Chapter 10).

2. Nearly all organisms in oligotrophic environments are ultramicrocells so that the surface/volume ratio favors the capture of substrate or, in the case of the prosthecate bacteria, the prosthecae increases as the nutrient concentration decreases (see Chapter 5).

3. Starved bacteria also have the ability to use low concentrations of nutrients. Morita [1984b] grew Ant-300 at a glutamate concentration of 10^{-12} M but the value may be much lower. (Ant-300 is a psychrophile isolated from Antarctic waters and is classified as a species of *Moritella* [Moyer et al., 1996].)

4. The proper substrate or combination of substrates (including concentrations) have not been provided for the organisms and/or there is lack of proper chemical and physical environment.

According to the literature there are present in oligotrophic environments copiotrophic, low-nutrient bacteria, facultative and obligate oligotrophic bacteria, and starved bacteria. The question as to the occurrence of these organisms in these oligotrophic environments becomes an intriguing problem. Thus, when the discussion of the starved bacteria comes into play, one must consider all the other bacteria in the oligotrophic environment. Laboratory studies, dealing with bacteria from various ecosystems where the nutrient concentration is very low, should start with starved cells where the genetic expression and mechanisms for the starvation state is already in place. Thus, we are looking for the mechanisms by which the organisms can adapt to a higher concentration of organic matter (MacDonell and Hood, 1982). Unfortunately, there is no good method to detect starved bacteria in a given environmental sample. Measurement of cell size increases when the cells are presented with the proper substrate, but this is not a universal criterion for their detection.

1.6 Concepts and Definitions

The concepts and definitions of survival and energy, death of microbes, growth and reproduction of microbes, viable or nonviable microbes, moribund cells, and dormancy in microbes are separated in the following discussion because the various scientists have not made each concept an entity in itself but have used one or more terms to mean the same thing or have discussed a concept in light of another concept.

"The phenomenon of bacterial survival is fundamental to nearly all studies in bacteriology. But in spite of this, very little direct attention seems to have been given to the subject; it is very much taken for granted" (Sykes, 1963). All species require energy; thus the survival of the species during periods of energy starvation must be taken into account when studying survival strategies. Survival is also an important aspect of ecology occurring in all ecosystems. The manipulation of culture media is a valuable tool in microbiological research, and not too many microbiologists think in terms of energy deprivation as a tool for studies of microbial physiology and ecology. In order to provide a pragmatic approach to the above, a definition has been provided (Morita, 1982): starvation-survival is a physiological state resulting from an insufficient amount of nutrients, especially energy, to permit growth (cell size increase) and/or reproduction. There are quite a few studies dealing with starvation in the microbial literature, but the survival part of starvation-survival has only been the subject of reseach in recent years. In order to arrive at the starvation-survival state, the cells undergo a process by which they prepare for survival under starvation conditions. In certain species, the number of cells may even increase as a first stage in this process. Long term starvation-survival results in a shutdown of metabolism since no measurable metabolism occurs if energy (including intracellular energy sources) is not available. However, there are various degrees of starvation, starting with cells that have just utilized the last amount of nutrients for growth and metabolism to cells that have been deprived of nutrients for long periods of time. Survival can take place if the cell either retains its metabolic functions intact or is able to bypass or bridge in some way any gaps creating a proper metabolic cycle and no time factor is involved (Sykes, 1963).

In laboratory studies, it is easy to eliminate the energy source for the growth of a pure culture of heterotrophs. On the other hand, natural ecosystems have organic matter present to serve as an energy source, even when the organic matter is in extremely short supply. Furthermore, the organic matter may not be the correct substrate for any single species within the microbial community. Thus, within a microbial community, certain species may have the correct substrate(s) while others do not. Although we can apply the definition previously given for starvation-survival of a species, how can we apply it to various ecosystems? Should a bacterium be considered starved when it cannot reproduce at its maximum potential under ideal conditions but at a very slow rate such as once per day, once per month, once per year, etc., because there have been reports of bacteria multiplying at such low rates.

Death of a bacterial cell occurs because the cell is unable to follow its proper metabolic cycle, which results from a disruption of its enzymes and protein synthesizing structure. Death by senescence means simply that one or another of functions has been destroyed by an oxidation or other process (Sykes, 1963). Death of bacteria in nature occurs also by phage activity, bacteriovory, and by natural mortality. Natural mortality (loss of functional and morphological integrity, including lysis and destruction of genetic material) in the aquatic environment was determined experimentally to be at the rate of 0.010 to 0.030/hour (Servais et al., 1985). Within any given population, there are weaklings as well as "Amazons" and long-term starvation-survival involves these "Amazons." Resistance implies an active state of opposition to an attack on the cell by outside agents; whereas survival carries the idea of a more passive state of resignation induced by a natural turn of events (Sykes, 1963). The variability and individuality of bacteria is well documented by Koch (1987).

One should also recognize that growth and reproduction also require other nutrients besides energy. The term *nutrient limitation* has been used quite loosely and can mean (1) the limitation of the growth rate of phytoplankton populations currently present in a body of water, (2) the limitation of the potential rate of net primary production, allowing for possible shifts in the composition of phytoplankton species, and (3) the limitation of net ecosystem production (Howarth, 1988). Although Howarth (1988) applied the foregoing to phytoplankton, it also applies to other organisms. Generally, nitrogen and/or phosphorus are the primary limiting factors in seawater; some investigators, however, will argue that nutrients are not limiting at all in marine ecosystems, including highly oligotrophic waters (Howarth, 1988).

According to Williams (1985), it is possible to regard *oligotrophic* and *autochthonous* as synonymous and likewise with *copiotrophic* and *zymogenous*. Williams (1985) further points out that both concepts are not always supported by experimental evidence and that these arbitiary categories are not necessarily mutually exclusive.

Nearly all the Earth's biosphere is oligotrophic, especially when one considers the vast volume of the oceans. Although soil is generally not placed in this category, soil microbiologists recognize its oligotrophic nature. It is rare that we see truly eutrophic environments, but even within eutrophic environments, there are some physiological types of bacteria that do not find available sources of nutrients (energy) or find nutrients only in limited supply. Nutrient deprivation can occur for some physiological types of bacteria within eutrophic environments, mainly because the species-specific substrate(s) are not present. Thus, starvation can occur in the midst of plenty. It should also be noted that in environments where energy is limited, slow growth may take place. Hence, we are in a "fast or famine" (Poindexter, 1981b) or "feast or famine" (Koch, 1971) situation. However, in nature, it is probably "fast and famine" with an occasional feast (Morita, 1984a). The dynamics of microbial activities in any ecosystem depends on the quality and quantity of the available nutrients, especially energy. One is then forced to look carefully at the available energy in ecosystems. Microbial activities, including growth, must be "in tune" with the available energy supply. Laboratory growth studies conducted under optimum or nearly optimum conditions do not represent growth as it occurs in nature, mainly because growth in nature relies on syntrophy. Because the amount of energy sources is very seldom optimal in nature, one is then faced with the problem of where to draw the line in terms of time between growth, slow growth, and no growth. The same can be said for microbial activity in any ecosystem. Naturally, nearly all microbiologists like to think that microbial activity in ecosystems proceeds at a reasonable rate—a rate suitable for experimental verification by laboratory or field experiments. Many biologists seem intrinsically unable to imagine survival in the absence of growth, in spite of the extreme longevity of some bacterial spores, plant seeds, and other "survival structures." The fallacy "If it is not growing, it must be dying" has a strong grip on most humans. This had led to the misconceptions as to maintenance energy, employing definitions suitable for growth but not necessarily survival, and to a tendency to regard as miraculous the survival for decennia of ordinary, "run of the mill" bacterial isolates in unsupplemented seawater, distilled water, or soil (see Chapter 2). Should we consider the proper time frame for survival to be in seconds, hours, days, months, years, or geological time? Nonetheless, general slow growth may be the result of an insufficient amount of energy to provide for good growth of any organism, and the formation of new ultramicrocells from parental ultramicrocells may be a manifestation of starvation.

The physiological state in which bacteria are present in any ecosystem is difficult to determine. To complicate matters, we are bothered by definitions and methodology. In microbiology, the distinction between life (viable) and nonlife (nonviable) is difficult. Viable bacteria are defined as those capable of dividing to form progeny, usually determined by colony forming units (CFUs), but this definition is restricted to microbes; whereas there are many individual macroorganisms, including humans, viable but not capable of reproduction (Postgate, 1967, 1969). This is a restatement from Gay (1936), who states that "it is very probable that many cells may be unable to multiply but are still able to carry on other metabolic functions." When CFUs are not formed, the cells are considered dead. Indeed, this is an oversimplification. When organisms do not form a CFU on a medium, what terms should we use? Inactive, dead, nonproliferating, surviving, viable but nonculturable? In order to take care of microbes having many of the characteristics of normal, living organisms, but unable to divide and form a colony and committed to death, Postgate (1967) employed the term *moribund*. Bretz (1962) also used the term "moribund" to describe cells that only swelled during the same time interval in which other cells were able to divide in a slide culture method. Eventually these swollen moribund cells, when incubated longer, produced microcolonies and were considered to be viable. Kaprelyants and Kell (1993) state "if the ability to form a colony is the sole criterion of whether a cell is alive, it is reasonable to suggest that dormancy is likely to be far more common than death in stationary-phase cultures." Cells that do not form colonies on nutrient media but that will give a viable direct count can be referred to as "pseudosenescent" (Postgate, 1976) or "somnicells" (Roszak and Colwell, 1987a; Barcina et al., 1990). Some cells, that have been injured by some type of stress, may be resuscitated by preincubation in nutrient media before plating out on agar media (Ray, 1979; McFeters, 1990).

According to Valentine and Bradfield (1954), "viable" was used to describe cells capable of multiplying and forming colonies, but they suggested that "live" be used for cells showing other signs of viability, such as respiration, even if the cells were unable to divide under the prevailing conditions. Thus, viable cells would then be considered not alive. The term *viable,* on the other hand, is applied to an organism capable of multiplying to form two or more progeny under conditions that are "optimal" for the species concerned. Again, optimal conditions may not prevail during sampling. To further complicate the picture, bacteria can lose their ability to multiply but remain biologically functional as individuals (Postgate, 1976). In the sea, most of the metabolic activity is attributed to cells that do not form colonies on nutrient agar plates (Hoppe, 1976).

Dormancy has been defined as a "rest period" or reversible interruption of the phenotypic development of an organism (Sussman and Halvorson, 1966). Mason et al. (1986) classified bacteria on a physiological basis as dead microbes; nonviable, active microbes; dormant microbes; and viable, active microbes. Mason et al. (1986) go on further to state that dead bacteria, based on the inability to reproduce, are false and lead to embarrassing and dangerous misinterpretations of data; nonviable microbes can carry out substrate transformations when they possess appropriate enzymes. Examples of the latter are where, at superoptimal growth temperatures, substrate energy dissipation can be mediated by nonviable cells (van Uden and Madeira-Lopes, 1976; Weimer and Morita, 1974) or the use of enzymatic conversion for immobilized cell biocatalysts (Chibata and Tosa, 1977). The dormant cells differ functionally, leading to either active microbes or to death or lysis (Koch, 1971).

Recognizing the difficulties concerning the physiological state of dormant bacteria, Kaprelyants and Kell (1992) suggested that all the cell types could be reduced to three groups, as follows: *viable* to refer to a cell that can form a colony on an agar plate, *vital* to refer to one that can only do so after resuscitation, and *nonviable* to refer to a cell that cannot do so under any tested conditions. Hence, dormant cells are vital. Unfortunately, many species of bacteria will not grow on an agar surface including many endosymbionts. On the other hand, Kaprelyants and Kell (1993) also suggest that the phrases "starvation" or "starving cells" refer to the environment under which cells are incubated, rather than a physiological state. I would disagree with the latter since a starved organism does have a different physiological state than one that is in the log growth phase.

Postgate (1967) discusses the difficulties associated with defining "viable" cells. Methods used to assess viability, such as staining and dye-uptake, optical tests, direct counts, ATP content, and respiration or other enzymatic activity are generally unreliable as estimates of viability. Roszak and Colwell (1987a) also discuss the concept of viability in microbiology, especially in relation to the public health area. Roszak and Colwell (1987a) list 47 different microscopic methods to distinguish between living and dead bacteria, but none of the methods is satisfactory. However, we now have another method that employs nalidixic acid (0.002%) and yeast extract (0.025%) (Kogure et al., 1979). They found that the mean viable cell size in seawater was 0.6 μm but on incubation with the yeast extract and nalidixic acid for 6 hours, it increased to 3 or 4 μm. This method showed three orders of magnitude higher counts than the direct viable counts in seawater. Torrella (pers. communic.) noted that when marine water samples were incubated with yeast extract 2 hours before the addition of nalidixic acid, higher numbers of viable cells were found. This preincubation with yeast extract permits the starved bacteria to resuscitate themselves before the addition of nalidixic acid. Previous investigators have noted that cells do not immediately respond when substrates are added (Vaccaro, 1969; Kirchman, 1990; Williams and Gray, 1970; Miyamoto and Seki, 1992; and Preyer and Oliver, 1993). This delay may be as long as 36 hours. However, yeast extract is not a panacea for the growth of all bacteria in the environment.

Still to be further investigated is the subject of "viable but nonculturable" bacteria. Nevertheless, Kaprelyants et al. (1993) consider the phrases "direct viable count" and "viable but non-culturable" as misnomers, because the terms do not fit their criteria of physiological states mentioned previously. As a result of certain stresses, the ability of bacteria to multiply can be lost but they can remain completely biologically functional as individuals (Postgate, 1976). According to Hoppe (1976), the fraction of very small heterotrophic cells that cannot be cultured on nutrient media is responsible for the continuous breakdown of organic matter in off-shore regions of the sea.

Within a population, some cells have the ability to survive longer than others when stressed. In nature, the most natural and mildest stress is starvation (Postgate, 1989). Death in microbes is difficult to ascertain. Fissile, vegetative microbes do not grow old and die, they vanish (Postgate, 1976). Bacteria do not have "natural" senescence (Postgate, 1967). A senescent cell is effectively nonviable but it may retain all the other definitive properties of a living organism (Postgate and Calcott, 1985). Two equally young individuals replace the parent. Since an "old" cell can divide to form two cells when the physical and chemical conditions become right, it is difficult to state definitively what an "old" cell is. Death results only from some environmental stress (Postgate, 1976). The term "death" is used synonymously with "nonviable on a rich medium" (Postgate and Hunter, 1963). However, Steinhaus and Birkeland (1939) consider when an organism is grown in culture medium, there is a senescent phase (log death phase) of the culture, but that during this phase there are small rises and falls in the curve, indicating spurts in multiplication.

The concept of mortality of bacteria in nature must also be addressed. The subject has received very little attention except in terms of grazing by protozoa. Death of bacterial cells, other than by predation, parasitism, and environmental factors, is not well understood—mainly because very little research has been done on the subject. For a brief discussion on this subject, Pace (1988) should be consulted. When various bacteria were placed in lake water (15-day experiments), their decline in numbers was due mainly to predation by protozoa (Gurijala and

Alexander, 1990). Predators are also the main cause of decline of the microbial population in soil (Casida, 1980). In addition to grazing, bacterial mortality occurs in nature by phage lysis, bacteriolysis, and probably by autolysis.

The philosophical concept of "age" of a bacterial cell must still be answered. This subject is discussed in detail by Yanagita (1977), but not in relation to starvation-survival. The phenomenon of "ageing" does not appear to be confined to cells in exhausted media but also occurs during the lag phase in a medium (Midgley and Hinshelwood, 1961). The concept of physiological youth of bacteria, as discussed by Sherman and Albus (1923) must also be addressed. Physiologically young bacteria are more sensitive to adverse conditions than old (Sherman and Albus, 1923). However, the most stress encountered by microbes in any ecosystem is the lack of nutrients, especially energy (Postgate, 1976). If some other process arrests multiplication, death does occur. "The basic difference between fissile microbes and organisms with more complex mechanisms of multiplication is that, in the former, death only results from some environmental stress" (Postgate, 1976). There are no shortcuts, according to Postgate (1967, 1969), that would permit assessment of the moment of death. Death in microbiology is defined as the inability to reproduce, but this inability to reproduce may be the result of using the wrong medium and incubation conditions. Is this "true" death, because some microbes can be resuscitated? It is well known that higher bacterial counts are obtained when dilute media, rather than conventional full-strength media, are employed (Stark and McCoy, 1938; Taylor and Collins, 1940; Collins and Willoughby, 1962; Melchiorri-Santolini and Cafarelli, 1967; Fonden, 1968; Carlucci and Shimp, 1974; Ishida and Kadota, 1974, 1977; Akagi et al., 1977). Is the use of dilute media a form of resuscitation? It may be that many of the starved bacteria cannot handle full-strength media immediately. Could full-strength media result in the inability of the starved microbes to physiologically handle the sudden input of nutrients? Resuscitation has been studied mainly by the microbiologists who have devised different media to determine the presence of microbes or injured microbial cells that could not be cultivated by routine media (Andrew and Russell, 1984). Harris (1963) discusses the influence of rich and poor media and incubation temperature on the survival of damaged bacteria. It appears that recovery on rich or poor media depends on the organism in question as well as factors that produced the damaged bacterium. When taking environmental samples, the questions that must be asked, according to Postgate (1989), are (1) were the organisms dead when the samples were taken, (2) did they die afterwards, (3) did they die on the medium or would they have died on a different medium? There is no reliable method to determine any of the forgoing (Postgate, 1989). In addition, the type and brand of agar used can influence the recovery of bacteria from environmental samples (see Chapter 4).

Heinmets (1953) and Heinmets et al. (1953, 1954a,b) considered the "killed" state of bacteria to be relative and represent only a loss of cellular viability in terms of the inability to multiply, but not necessarily in the termination of synthetic and metabolic activity. If the cells were "guided" to function specifically to reestablish the specific synthetic capacity necessary for cellular division, then viability was restored. Cells can recover from irradiation "killed" injury, heat "killed," chlorine "killed," Zephiran chloride "killed," hydrogen peroxide "killed," or ethyl alcohol "killed" cells by incubation with the proper metabolites of the tricarboxylic acid cycle.

> The term "survival," the maintenance of viability in adverse circumstances, need not suffer from this linguistic restriction. Survival of individual microbes has a clear meaning, distinct from (and usually incompatible with) survival of a clone of population. In many areas of microbiology, only clonal survival is normally susceptible to experimentation. (Postgate, 1976)

The concept of the "resting" stage of bacteria should also be considered (Questel and Whetham, 1924; Kendall et al., 1930). Dormancy has been defined by Sussman and Halvorson (1966) as "any rest period or reversible interruption of the phenotypic development of an organism." This dormancy, according to Sussman and Halvorson (1966), can be constitutive and exogenous, where constitutive dormancy involves the formation of spores, cysts, etc., and exogenous involves an unfavorable chemical or physical environment. The term "dormant" population describes that part of the living microbial biomass that survives in a reduced state of metabolic activity (Gray and Williams, 1971). Naturally, starvation-survival is an example of exogenous dormancy. Dormancy was redefined by Kaprelyants et al. (1993) as a reversible state of low metabolic activity, in which cells can persist for an extended period without division. Resting cells, according to Sudo and Dworkin (1973), include only those cells in which division does not occur, endogenous respiration is absent or much decreased, and in which formation of the resting state is part of the natural life cycle of the organism. "The resting cell is usually, but not necessarily, morphologically distinct from the growing or vegetative form, and most types are more resistant than vegetative cells to adverse conditions" (Sudo and Dworkin, 1973). Starvation-survival could be a type of dormancy or resting state due to the lack of energy. During the initial period of starvation, however, the cell is not resting but preparing itself for survival. Depending on the ecosystem in question, it may be a matter of minutes to years. Much activity can occur in bacteria even when they are not growing, especially during the initial period of starvation-survival. For instance, in the study of microbial physiology, especially during the period when Warburgs were employed, we used to utilize the term "resting cells" but now we employ the term "washed cells." In the classical Warburg studies, investigators used manometric technique with cells that were not growing but were still active. Growth did not occur in the Warburg vessel because either a nitrogen source or some other limiting compound(s) was not present in the reaction vessel. What little energy there is in oligotrophic ecosystems may permit the bacteria to metabolize but there is an insufficient amount for growth and/or reproduction.

The starvation-survival state can be readily broken provided energy and nutrients are introduced into the system. Each species enters into the starvation-survival period in its own way, but certain generalities can be made from the studies performed to date. If the cells are deprived of an exogenous source of energy for a long period of time, however, they may enter a state of metabolic arrest. Metabolic arrest is a stage that may be analogous to a microbial spore in terms of metabolic activity. In essence, the process of starvation-survival is survival of the species due to the lack of energy, and not due to a lack of phosphate, nitrogen, specific amino acid(s), vitamins, etc. As I have stated earlier (Morita, 1982), only one viable cell per stressed environment is required for the continuation of the species, not a few cells per milliliter, liter, etc. The only example in the literature that points in this direction is the research of Rodrigues-Valera et al. (1979), who demonstrated that extreme halophiles could be recovered from seawater when it was previously thought that they were incapable of surviving in less than 10% NaCl. It took 5 l of seawater to demonstrate the presence of 10 to 35 viable extreme halophiles.

There is also a time scale that must be imposed on the starvation-survival process and the time scale may be different for different species of organisms. How long can a species survive without an energy source? All ecosystems are changeable, hence the available energy for any one species may be transient in that, when it occurs, the organism utilizes whatever amount it can until it is gone and reenters the starvation state. Furthermore, the introduction of specific energy sources in any ecosystem may be sporadic and may result in the sporadic growth of the microorganisms. In nature the steady state is not all that steady (Wright, 1988a). Steady state is not realistic (Jannasch, 1974; Krambeck, 1979). This situation is probably dominant in oligotrophic ecosystems. In general, short-term starvation experiments of several hours, days, or a few weeks will not answer the true question of the survival of the species—mainly because energy input into any ecosystem may take a long time to occur.

Many different organisms have evolved on Earth due to evolutionary processes. Microorganisms have been present much longer than any other forms of life. If organisms have evolved to cope with changes in the environment, then microorganisms must also have evolved so that survival of the species can occur in face of the temporary lack of energy. The spore state is a mechanism for the spore-formers to survive periods with the lack of energy. What then is the mechanism for the heterotrophic non-spore-formers or the chemolithotrophic non-spore-formers to survive the periods of lack of energy in nature? In this book, only the non-spore-forming bacteria will be considered. Fungi will also not be included but there are sufficient data that indicate that these organisms, through their spores or other specialized structures, can survive long periods of energy deprivation. Lockwood (1981) provides further information on survival of fungi subjected to energy deprivation and the response of fungi to nutrient competition.

All organisms need energy to grow. A competitive advantage is given to those species that find their energy supply in their immediate environment. To a large degree, this is the basis of the enrichment technique. In the microbial world, certain organisms must have a higher amount of energy and nutrients compared to others. Nevertheless, these species continue to be present on Earth in the face of competition with other organisms. The oligotrophic nature of the various ecosystems are the result of microorganisms utilizing the energy and nutrients faster than they are replenished. Utilizable organic matter enters the microbial world mainly by photosynthesis with a small amount due to chemolithotrophic mechanisms. Water and carbon dioxide are not the only products that result from the decomposition of plant material, but also a large number of secondary products with varying degrees of resistance to further decomposition. Included in the decomposition process is the humification of some of the plant material. As a result, most organisms must rely on syntrophy for their energy source. Ayanaba and Alexander (1972) noted changes in nutritional types with bacterial successions. In that study, the fastidious population developed only after an initial growth by bacteria with simple nutritional requirements. Thus, growth of bacteria in nature is sporadic. To illustrate this point, McCarthy and Goldman (1979) noted that nutrient regeneration is not a process that occurs uniformly through a volume of water, but rather occurs at discrete points for short time periods. Locally, a high concentration of nutrients may exist where animals happen to be when they excrete metabolic products. McCarthy and Goldman (1979) further suggested that short-term, high-speed nutrient uptake by phytoplankton within a few micrometers of zooplankton excretion would not be observed in the liter-size water samples that are usually collected. Nevertheless this short-term, high-speed uptake could supply the nutrient needs for the growth of phytoplankton in oligotrophic environments of the ocean. Each species of bacteria has its own spectrum of energy and nutrient sources as well as its own spectrum of chemical and physical factors. Even in a eutrophic environment, certain physiological types of bacteria may be undergoing starvation-survival due to the lack of their specific substrate. During early evolution, organic matter was created by means other than photosynthesis. It is doubtful that, through the billions of years that microbes have evolved on Earth, all potential microbial energy sources were plentiful at all times. Some mechanisms must have developed to permit the non-spore-forming bacteria to survive long periods of energy deprivation. Although Gest (1987) only considers the spore-formers (mainly *Bacillus, Clostridium,* and *Thermoctinomyces*) as hardy survivors in the microbial kingdom, it is inconceivable to this author that non-spore-forming microbes have not evolved methods to survive in the face of the lack of energy. There must have been periods during their evolution, especially before green plant photosynthesis evolved, where energy (organic matter) for heterotrophs was absent. There were many fluctuations of environmental extremes during the evolution of Earth and the starvation-survival process aids in the ability of the heterotrophs to withstand these environmental extremes.

There is a genetic base for starvation-survival in that new proteins are synthesized during starvation-survival and that the genes responsible for cell size reduction are located on the chromosome (Smigielski et al., 1990). These genetic studies come under the heading of "life after log." However, in the various ecosystems, the log phase of growth seldom occurs. Also the stationary phase of growth is common due to a lack of nutrients and suboptimal microenvironments that cannot support active growth (buildup of metabolic end products, a change in pH, etc.). The deprivation of energy probably was the driving force for the evolution of certain microbes to form spores, cysts, and other resting structures. Whatever the case may be, the organic matter produced by photosynthetic and chemolithotrophic processes does not keep up with the demand on a global scale. Thus, oligotrophic environments dominate on Earth. It is also the reason why I have coined the term "starvation-survival" so that we can take into consideration the general lack of energy in various ecosystems. I believe starvation-survival to be very important in the ecology of organisms. Liebig's Law of the Minimum states that under steady conditions the essential material available in amounts most closely approaching the critical minimum used by a given organism will tend to be the limiting one. However, the growth or activity of any organism may be limited by either too low a level of a given factor or too high a level of the same factor; each has an ecological minimum and maximum with a range that represents the limits of tolerance (Odum, 1971). Nevertheless, Liebig's Law of the Minimum also applies when energy is a factor in ecosystems as well as nutrients, ions, trace elements, etc. Recognizing that the normal state of most microorganisms on Earth is some degree of starvation, growth of microorganisms in the laboratory is an idealized artificial situation. The size of many species of bacteria originally isolated from various ecosystems does not reflect their size in nature, but represents "giant cells" that result from growth under optimum conditions (physical and chemical factors plus an abundant supply of nutrients). From an evolutionary viewpoint, the starved cells may have adapted to a nutrient-rich environment. Thus, in starvation-survival we are dealing with the survival of the species and not the survival of a clone or population in a test tube. Unfortunately, clonal survival is the subject most studied and deals with "gluttonous" bacteria with all their reserve energy supplies, which are probably not found in bacteria growing in an oligotrophic environment. (Jose Amador proposed the term "gluttonous" bacteria to refer to bacteria grown on rich media under optimal conditions.)

Starvation-survival of cells does not mean that the cells are dormant. They are only dormant in the sense that very little or no metabolism is taking place due to the lack of proper substrates. One must recognize however that much endogenous metabolism takes place during the initial stages of starvation before it enters survival stage, hereafter referred to as the metabolic arrest stage. To break the starvation state, one needs to add the right nutrients (mainly energy) and provide other conditions to the system for replication of the cells. However, there is an exception to this when one considers the "viable but nonculturable" microorganisms.

Of all the environmental factors an organism encounters, availability of energy is the most important. Energy is needed by organisms to help ward off other environmental factors that may stress them. For instance, under the stress of heat, within limits, the presence of energy can aid the survival of the organism better than the absence of energy (Fig. 1.3). To a large degree, the microorganism brings about the depletion of energy in any ecosystem, making it oligotrophic (Morita, 1986). This is due to the large number of physiological types of bacteria capable of using organic matter under different environmental factors. This is well evidenced in many processes (in sewage treatment plants, sewage lagoons, etc.) such as those studied by Naganuma and Seki (1988). Occurring simultaneously with making the ecosystem more oligotrophic, the remaining organic matter becomes more recalcitrant to degradation. An environmental stress may not be lethal if energy is available to the organism, helping to ward off the effects of the stress. Thus, starvation-survival becomes an important concept in ecology because it involves the survival of the species under unfavorable conditions, especially lack of energy. Growth and survival are important in microbial evolution. The lack of energy is generally not a subject taken into consideration in textbooks of ecology.

Within any ecosystem we find a diversity of microorganisms. If we take into consideration the competition theory, based on laboratory models, then one species would eventually displace all others (Pielou, 1974). An analogy to this situation is the use of the enrichment culture technique or chemostat technique used to isolate specific physiological types of bacteria. However, in nature, due to the lack of energy, this does not happen and permits many species to coexist. This situation preserves the diversity of microorganisms in ecosystems so that all types of microbial associations can function when the systems again receive sufficient energy-yielding compounds. For any given species in the environment, unbalanced growth is probably the rule. To compete with other organisms, energy is needed. The concept of competition (mechanism or consequences) among microorganisms (especially bacteria) is generally ignored. For more definite details among fungi, Lockwood (1981) should be consulted. "Competition is an active demand in excess of the immediate supply of a material or condition on the part of two or more organisms" (Clark, 1965), and this competition can be divided into exploitation and interference. According to McNaughton and Wolf (1973), exploitation is restricted to the depletion of resources by one organism or population without reducing the access of another organism or population to the resource pool, whereas interference deals with behavioral or chemical mechanisms by which access to a resource is influenced by the presence of a competitor. Food

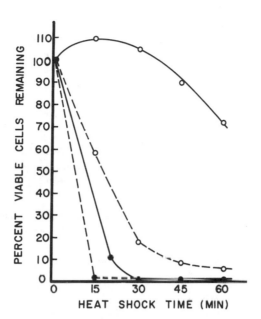

Figure 1.3 Effect of nutrients on viability retention of *Vibrio marinus* MP-1. Symbols: ○, heated at 25°C with nutrients (solid line) and without nutrients (dashed line); ●, heated at 31°C with nutrients (solid line) and without nutrients (dashed line). Reprinted by permission from *J. Bacteriology* 92:1388–1393, R. D. Haight and R. Y. Morita, 1966; copyright 1966, American Society for Microbiology.

is the component most likely to be in short supply and is the prime target of competition (Clark, 1965). The extreme importance of substrate (carbon source) as the primary object of competition has been emphasized in terrestrial systems (and also applies to aquatic systems) by various investigators dealing with energy requirements for turnover and maintenance of measured populations to the annual input of substrate (Babiuk and Paul, 1970; Clark and Paul, 1970; Gray and Williams, 1971; Shields et al., 1973; Gray et al., 1974). Because there is very little energy in oligotrophic environments, competition among microorganisms becomes nil or nearly nil so that many species as well as a high population density can coexist, mainly because competition requires energy. If energy were present, competition for the energy source would be fierce between species and between individuals of the same species. Thus, starvation-survival becomes an important factor in the coexistence of microorganisms in various ecosystems.

Unfortunately, there is no good means to detect starved cells in any given environmental sample other than a fatty acid profile on extracted samples (see Chapter 7). Naturally this procedure involves the entire microbial population present in the sample. Starved cells will take up substrate as well as thymidine. What criteria do we use to determine whether bacteria are in the starved state? Measurement of cell size (which increases when these cells are presented with the proper substrate) could be used, but is not a panacea for their detection. Conversely, could we use cell size decrease to indicate a starved cell? Unfortunately, some cells do not decrease in size when starved. As considered previously, is a bacterial cell in the starved state if it only reproduces once a day, once every two days, once a week, once a month, or once a year as an ultramicrobacterium and not up to its potential in a richer medium? Some organisms are naturally slow growers. Are various bacterial species in the starved state when the substrate becomes limiting in the environment so that rapid multiplication does not take place?

Starvation-survival of bacteria in oligotrophic environments is made more difficult by the following questions:

1. How does one define and determine viability of microorganisms, especially in a natural ecosystem?
2. How does one define growth over the long span of time? (In other biological sciences it is defined as an orderly increase of biomass of an organism and not the number of cells/ml as in microbiology.)

This book reviews the literature and discusses the listed questions in more detail. I have also retained the term oligotrophic bacteria, especially in Chapter 5, to minimize confusion in relation to the published literature.

1.7 Importance of Starvation-Survival

As I have stated previously (Morita, 1982), microbes are the principal catalysts in the ocean. This statement can be extended to all other environments. Thus, for the biological cycles of matter to operate, the preservation of the microbial species is essential. When the ecosystem is appropriate, the starved cell will express itself. Starvation-survival becomes important in agriculture (especially plant pathology), freshwater and marine systems, food microbiology, pathogenic microbiology, bioremediation (Kong et al., 1992; Truex et al., 1992; Kong and Johnstone, 1994) and finally the electronic and pharmaceutical industries (Gough et al., 1986; Mittelman and Geesey, 1987). Although high-purity water is needed in the pharmaceutical and electronic industries, virtually all industrial waters will support the growth of microorganisms, regardless of the efforts directed toward maintaining aseptic conditions or sterility (Geesey, 1987). Ultrapure water is essential for semiconductor manufacture where particulate contamination of silicon wafer surfaces can cause significant lowering in the performance of semiconductor devices and any contamination must be removed by frequent rinse procedures (Brumbach, 1972; Taubenest and Ubersax, 1980). Starved bacteria can form ultramicrocells and are then able to pass through low-porosity cartridges and thus come into contact with semiconductor devices undergoing "ultrapure" water rinses during manufacture (Gough et al., 1986). When a pathogen leaves its host, there is a period until it finds another host when the pathogen is faced with many stresses, including the lack of an energy source. Starvation is apparently responsible for triggering the synthesis of virulence factors in some pathogenic bacteria (e.g., *Yersinia enterocolitica*) (Bolin et al., 1982; Brown and Williams, 1985). Because starvation can also trigger the formation of certain resistant proteins (see Chapter 7), the question is how it affects the ability of pathogens to survive between their transmission from one host to another.

Lappin-Scott et al. (1988a,b) and MacLeod et al. (1988) demonstrated the use of starved cells in petroleum recovery. Essentially, starved cells penetrate the core strata due to their small size and then are resuscitated by the use of sodium citrate as an energy source permitting the cells to become large again. Furthermore, these small cells do not have any glycocalyx. In these experiments, *Klebsiella pneumoniae* (0.25 × 5 µm) was starved and injected into cores, followed by the addition of sodium citrate. This energy source permitted the resuscitated cells to become larger (up to 1.7 µm). This cell size increase brings about selective plugging of core strata. They theorized that starved cells could be used to selectively plug high-permeability rock strata already drained of oil so that further recovery techniques can focus on the low-permeability strata that still contain oil. In bioremediation of subsurface material, the penetration of the inoculum is extremely important. It has also been found that these starved small cells have the ability to mineralize various subsurface contaminants.

Starved cells of *Pseudomonas syringae* release an ultraviolet-absorbing material (free of purine-bound ribose and inorganic phosphate) that has the ability to bind with Cu^{2+} (Cabral, 1992). This binding substance may also be found in the supernatant of other starved bacterial species. Starved cells are not subject to metal stress and can release strong Cu^{2+}-binding substances. This could be the cause of pitting of copper pipes. Co^{2+}, Ni^{2+}, Cd^{2+}, Zn^{2+}, and Hg^{2+} also have high affinity to most of the molecules released by starved cells (Cabral, 1992). Almost all products of endogenous metabolism of starved cells are strong metal-binding molecules (Sillén and Martell, 1964, 1971; Hughes and Poole, 1989).

1.8 Conclusions

I firmly believe that the terms "oligotrophic bacteria" and "copiotrophic bacteria" should be dropped from the literature, mainly because the experimental evidence does not support their existence. Until there is rigorous experimental analysis of oligotrophic bacteria (especially obligate oligotrophs), the term cannot be accepted and oligotrophic should refer to the environment. If true oligotrophs are isolated in the future, then the law of priority in science should be applied and the organism should be termed "oligocarbophile."

In order for a species to survive, only one organism of that species per environment needs to survive. If and when the proper environmental conditions become right, the organism will express itself in terms of growth and reproduction and take its proper place in the ecosystem (Morita, 1982). In addition, the survival of any species against various environmental factors, including the lack of energy, must also be taken into account when the evolution of the species is addressed. If the organism cannot survive, then further evolution comes to a standstill. Thus, growth and survival go hand in hand in evolutionary processes. The degree to which starvation-survival plays a role in the evolution of microbes remains to be investigated, but energy (food) certainly had a role in the survival and evolution of higher forms.

Every organism has an ecological right to exist and each organism possesses a specific biochemical trait or group of traits that account for its existence (Alexander, 1971). "Bacteria deal with physical and chemical alterations of their surroundings by exhibiting a different set of metabolic activities. In other words, they are capable of existing in a variety of physiological states which can be quite different from one another. In addition, bacteria shift from one such physiological state to another in a rapid and efficient manner" (Schaechter, 1968). This apparently occurs when bacteria enter the starvation-survival state.

During the evolution of microbes, the bacteria faced many different types of stresses. However, I believe the most important stress during evolution was the lack of organic matter as an energy source for the bacteria. With sufficient energy sources, the bacteria evolved mechanisms to resist the stress. The adage "stress builds character," used by Koch (1991) when writing about the evolutionary development of bacteria, certainly fits the concept of starvation-survival. With the ability to survive starvation, our currently known species of bacteria evolved. Starvation-survival is also survival of the species. In nature, the species only needs to survive until conditions become favorable (Morita, 1982).

Current research by molecular biologists concerning "life after log" (Siegele and Kolter, 1992) and "life and death in the stationary phase" (Kolter, 1992) emphasizes the importance of starvation in the study of microbiology in general. However, for those molecular biologists studying the problem, I feel that I must emphasize that microbial life in nature is rarely in the log phase of growth.

Chemical, physical, and biological factors steadily challenge the capacity of bacteria to survive. The survival potential exhibited in most laboratory microcosm experiments are of little value other than academic interest (McCarthy, 1971; Hoff, 1989; Rose et al., 1990b; Austin and Austin, 1987). Nevertheless, survival of the species is an important doctrine in biology and survival due to the lack of energy must be addressed if we are to fathom the microbe's role in ecology. The bacteria have already met the challenge to survive in oligotrophic environments but exactly how they are capable of doing so remains to be investigated in much greater depth than that discussed in this book.

2

Survival of Bacteria in Energy-Deficient Systems

The ways of bacteria are odd
They're understood only by God
Any man who proclaims
That he's fathomed their aims
Is either a fool or a fraud.
—F. M. Harold, 1986

2.1 Introduction

The concept of "ancient" bacteria from various geological samples has been debated in the past and will continue to be debated in the future. Nevertheless, the mechanisms by which survival of bacteria occur in natural samples are of great interest. Most of these samples contain extremely small amounts of energy, if any, available to microorganisms. Thus, an oligotrophic environment and starvation-survival become involved. This chapter reviews the known data concerning these "ancient" bacteria, early laboratory studies of bacteria in nutrient-deficient situations, as well as bacteria from various geological samples. One should take into consideration that the older studies never employed resuscitation methods to revive the organisms; it takes only one organism of a species per environment for survival of that species, hence the distribution of the organism in natural material will probably be extremely sporadic; and even if contamination of the geological structure has taken place after the laying down of the geological structure, the main question is how long the contaminating organism survived without a readily available exogenous energy source.

"It is well established that some micro-organisms can, under certain conditions, be deprived of all visible signs of life and yet these organisms are not dead, for, when their original conditions are restored, they can return to normal life" (Keilin, 1959). In the literature, this state of an organism is referred to as visible lifelessness, suspended animation, viability, latent life, or the not very suitable but widely used term anabiosis (latent life). (Keilen [1959] addressed not only bacteria but higher forms.) The concept of anabiosis originated in 1702 from Leeuwenhoek's observations (see Keilin, 1959), but the term *anabiosis* (return to life; *Wieder-belebung*) was introduced by Preyer in 1872 and 1891 (see Keilin, 1959) for the resuscitation or resurrection of completely lifeless but viable organisms and later extended to the state of viable lifelessness. In microbiology, the term "quiescence" was used to describe the cryptobiotic state (Burke and Wiley, 1937; Lewis and Gattie, 1991; Lewis, 1991). In the modern plant science literature, quiescence is mainly used to describe plant dormancy (Lang et al., 1987).

Although Keilin's (1959) Leewenhoek Lecture presented a fascinating discussion concerning anabiosis, it deals mainly with anahydrobiosis (due to dehydration), cryobiosis (due to cooling), anoxybiosis (lack of oxygen), and osmobiosis (due to high salt concentration). Keilin (1959) proposed the term "cryptobiosis" (latent life) for the state of an organism when it shows no visible signs of life and when its metabolic activity becomes hardly measurable or comes to a standstill. Cryptobiosis also requires an intact structure, but if the structure is damaged or destroyed, death results. Today, we would refer to this as the starvation-survival state of microbes, provided it was brought on by the lack of nutrients. Keilin (1959) did not discuss cryptobiosis brought about by the lack of nutrients. Hinton (1968), dealing mainly with macroorganisms, states that the capacity of organisms to tolerate a total suspension of metabolism (cryptobiosis) is a primitive characteristic of protoplasm. In the cryptobiotic state, all chemical reactions responsible for the maintenance or both the maintenance and growth of the organism cease and if any chemical reactions occur they are purely adventitious and would occur if the organism were dead. The point is, if higher life forms enter into the state of cryptobiosis, microbes can certainly do the same. Therefore, one should not be surprised to find microbes in ancient material as described in this chapter. The detection and subsequent growth of microbes from ancient material indicates that there is a mechanism for survival. Starvation-survival is the mechanism by which these organisms have survived over the years.

The survival of bacterial spores is not detailed in this book. However, Sneath (1962) does deal with the survival of spores from various materials. Sneath (1962) was able to isolate *Bacillus subtilis* from 200–320-year-old sediment samples. It is interesting to note that desiccation is the most detrimental factor to the survival of bacterial spores, fungi, and streptomyces. Kennedy et al. (1994) list the survival of spore-forming bacteria, non-spore-forming bacteria, yeasts, fungi, green algae, and viruses that have persisted long periods of time in various natural samples. It is my opinion that many of these microorganisms were in the starvation-survival state mainly because the environment from which they were isolated is energy poor.

2.2 Fossil Bacteria

The concentration of available organic matter is minimal. In all probability, the energy for the organisms was exhausted before geological processes took place.

Thus, the cells were probably in the starved state before fossilization took place because sedimentary material, due to diagenesis, has very little organic matter. Also, it appears that cell wall material is resistant to decomposition (Moore, 1969) and there would be few microorganisms without an adequate supply of energy to produce the enzyme(s) necessary for the decomposition of the cell wall of other bacteria. In the starved state, lysis also would not take place. The exceptions to lack of organic matter in geological formations are substances such as coal and petroleum, which have been bypassed by the carbon cycle due to their accumulation under conditions that preclude the optimal functioning of the microbial decomposition processes. Neither the depth of the geological structure (sediments included) at which microbial activities cease nor the depth at which microbes are no longer found is unknown. Both these latter situations probably depend on the location and type of geological structure.

Walcott (1915) reported the existence of bacteria with algal deposits in thin sections of Newland limestone, a formation of the Beltian series of Algonkian rocks in central Montana. Fossil bacteria have been identified in thin sections of Precambrian iron ores from Jurassic and Cretaceous rocks in Great Britain (Ellis, 1915) and from the Lake Superior area (gunflint chert) (Gruner, 1917; Barghoorn and Tyler, 1963; Cloud, 1965; Schopf et al., 1965). They have also been reported in coal (Renault, 1900), pyrite formations in Scottish oil shales (Love, 1957), in shale (Love and Zimmerman, 1961), and in phosphate rocks (Oppenheimer, 1958). Numerous investigators (see Kuznetsov et al., 1963; Moore, 1969) have found fossil bacteria as well as other fossil microorganisms (some identified as to genera) in materials from different geological periods dating back to the Precambrian. Bacteria have been recorded in thin sections of geological material as old as the Onverwacht Series (3.2 billion years old) (Barghoorn and Tyler, 1963; Barghoorn and Schopf, 1966; Engel et al., 1968). Unmineralized fossil bacteria have been detected in lake beds of the Newark Canyon Formation, Eureka, Nevada (early Cretaceous) (Bradley, 1963). Fossil bacteria have also been found in the Nevada Cretaceous rocks in nearly black, unlaminated limestone and occluded originally in calcite crystals. Vallentyne (in Bradley, 1963) also found bacteria occluded in the limey mud of Green Lake near Fayetteville, New York. This could result in the microbial formation of calcite (Morita, 1980a) because bacteria can be found in calcite buttons formed by microbial action. One may ask what was the physiological condition of the microorganism in order to become fossilized? Before fossilization took place, the organisms were probably in the starved state so that their shape remained intact during the fossilization process. In this light, the presence of "exostosial growths" on skeletons of mesozoic animals is common, which indicates that pathogenic forms were involved (Moodie, 1916). They have also been identified in coprolites of fishes and in the coal of the Autun Basin.

The geomicrobial processes on organic matter in sedimentary material leading to organic matter preservation is well discussed by Moore (1969), who states that, after completion of the original diagenetic processes, time as such appears to have little effect, as indicated by the preservation of bacterial, fungal, and algal remains in Precambrian rocks and the presence of viable bacteria in Permian salt deposits. The distribution of fossil microorganisms in the marine and nonmarine environments occurs throughout geological time. Fossilization takes place by either petrification or opalization and takes only 20 to 30 years under certain conditions (Smith, 1926; see Leechman, 1961, p. 146). For more complete fascinating details on the preservation and occurrence of fossil bacteria, consult Knoll and Awramik (1983).

2.3   Survival of Bacteria in Rocks and Coal

The amount of organic matter for heterotropic bacteria in sedimentary material is scarce, hence if microorganisms are to survive in rock structures they must have undergone starvation first. On the other hand there is much organic carbon in coal and peat (precursor for coal), but it is not bioavailable to most heterotrophic bacteria. Thus, starvation also results.

Galle (1910–11) first reported the presence of living bacteria in coal and attributed these organisms to the production of mine gases such as methane and carbon dioxide. (The early researchers used the term "living," since they were able to culture the bacteria. However, the terms "viable" or "cryptobiotic" bacteria would more adequately describe the "living" bacteria isolated. I have retained the terms employed by the researchers.) Although Schröeder (1914) substantiated Galle's finding of microorganisms in coal, he claimed that they did not produce these gases. Schröeder (1914) believed the microorganisms infiltrated from the surface. Lieske and Hoffman (1929) found bacteria in a coal mine at 1089 meters below the crust of the earth. They found gram-positive spore formers and an occasional gram-positive coccus in coal but the latter were not the same as the 43 colonies on a petri dish exposed to air in the mine (Lieske and Hoffmann, 1929).

Living bacteria in ancient rocks were reported by Lipman (1931). These organisms were isolated from a sample of Precambrian rocks from the Algonkian in Canada and in one from the Grand Canyon in the United States. Other types of bacteria were also isolated from Pliocene rock. Lipman (1935) also reported the presence of filamentous bacteria from travertine rock from Yellowstone Park. Although Lipman (1931) employed drastic sterilization measures, he acknowledges that some of the organisms doubtlessly were derived from the free air, which had momentary access to the rock in the process of the techniques employed. On certain plates, some organisms occurred that were strikingly different from any usually found in soils or rocks. They grew very sparsely on media that supported excellent growth of other organisms. Many types of rocks were studied, especially rocks obtained from great depths where surface contact could have no part in the results obtained. The organisms from the Pliocene were different than those from the Precambrian.

Later, Lipman (1930) reported living bacteria from anthracite coal from mines in Wales and in Pennsylvania. In 1932, Lipman performed all his research on anthracite coal obtained at a depth of 1800 feet in a mine near Pottsville, Pennsylvania. Depending on the method of dating the coal, it was laid down either 15 million years ago or from one to 200 million years ago. Elaborate measures were taken to ensure sterilization of the coal surface. Most of the isolates were short bacilli, cocco-bacilli, and egg-shaped coccus forms, varying considerably in size and shape. No algae, molds, or yeasts could be detected. Lipman (1931) acknowledges the studies of Schroeder et al. (see Lipman, 1931; paper has no literature citation section) who also found organisms in coal. However, these investigators suggest that the organisms they found are nothing more than modern bacteria that have come into the coal from the outside in very recent times. Lieske (see Lipman, 1931) considered the organisms he found in coal to be *Bacterium liquefaciens-fluorescens* (*Pseudonomas fluorescens*) and perhaps one other form and did not consider them to be representative of any ancient bacteria laid down when the coal was formed. Furthermore, Lieske (in Lipman, 1931) expressed the view that the organisms that he found in anthracite coal were washed down into the coal from the surface of the grounds which cover the coal by seepage water and other surface waters percolating downward. Lipman (1931) countered by stating that particles as large as a coccus are too large to penetrate the coal, either through crevices or through microscopic pores. Lipman (1931) strongly believed that the bacteria he isolated from rocks and coal were actually survivors, imprisoned at the time the formations were created.

Farrell and Turner (1932) obtained coal from the same place as Lipman but could not demonstrate the growth of bacteria. However, the handling of the samples was different in that the coal was transported back to the laboratory in large cans of formalin. Other pieces of coal were taken back to the laboratory in an untreated condition and served as controls. One-half-inch squares were cut out of the center of large coal pieces by means of sterile instruments in an inoculation chamber. Seventeen different media were employed. Since their first results were negative, they sent away for more samples. They were not from the identical site, but from a new mine. When the coal samples arrived at the laboratory they noted that they contained a layer of bone and pyrite and were filled with minute cracks. Contrary to the first samples, growth occurred in 24 hours in all samples with the exception of the controls. They isolated a gram-positive coccus and a gram-negative spore-forming rod similar to *Bacillus megatherium,* except that it did not utilize lactose or reduce nitrates. A series of tests showed conclusively the relation of cracks to the presence of bacteria. They later stated that bacteria isolated from the mines, surface soil, and water were all identical, but they did find iridescent bacteria belonging to the genus *Pseudomonas* in the mine and surface soil, but not in coal. The latter is one of the forms reported by Lipman (1931) in anthracite coal. They concluded that (1) Pennsylvania anthracite from the Primrose vein contains no bacteria other than common living forms that have found ingress through fracture cracks communicating with surface and air, and (2) the two forms found in the coal, a *Staphylococcus* and a gram-positive spore-forming rod, are very plentiful in the mine and surface waters and soils. Turner (1932) also throws doubt on the finding of Lipman (1931).

Burke and Wiley (1937) also became interested in the problem of living bacteria in coal. The similarity of organisms in the coal and at the surface, according to Burke and Wiley (1937), is to be expected because coal is brought to the surface of the Earth, and that similarity and dissimilarity between organisms in buried geological formations and those of the surface of the Earth have no bearing on the question of longevity. Of special interest is that Burke and Wiley were able to obtain filterable forms of Lipman's coccus. Lipman's coccus survived in a dried state in test tubes and on coal for four years and a number of common non-spore-forming organisms, such as *Staphylococcus aureus, Sarcina lutea,* and *P. aeruginosa* survived the same conditions. They also demonstrated that passage of air and water is possible through anthracite and bituminous coal, marine sandstone of Miocene age, recent sedimentary spring deposit, Oligocene sandstone, and Eocene basalt under pressure of 15 to 20 psi. These geological specimens did not have any fissures visible to the unaided eye. Experiments were also undertaken to determine if bacteria could penetrate coal. Ten pieces of sterile bituminous and 10 pieces of sterile anthracite coal were placed in cultures of Lipman's coccus and left for 3 weeks or longer. Six of the 20 pieces of coal placed in broth produced growth, but the organisms recovered were spore-forming contaminants. The remaining 14 pieces of coal were removed and surface sterilized, ground up, and then placed in fresh broth. Pure cultures of the experimental organisms developed in 6 of the 14 flasks, and the rest remained sterile. Other studies dealt with penetration of microorganisms into bricks, which are more porous. Nevertheless, Burke and Wiley (1937) stated that if penetration can be ruled out, there still remains the problem of whether organisms have longevity in a quiescent condition or whether they survive only in those environments for carrying on their physiological activities. According to Burke and Wiley (1937), "incredible penetration is no more difficult to accept than incredible longevity, and incredible heat resistance and it requires no more support from assumptions."

In a rebuttal to Farrell and Turner (1932) and Burke (1936), Lipman (1937) stated the penetration of bacteria into coal does not take into consideration the thickness of the coal vertically and horizontally. On the other hand, Sneath (1962) doubted the longevity of bacteria in coal. Penetration of bacteria into rocks and coal could take place and the penetration of the starved ultramicrocells of bacteria can occur more easily (Lappin-Scott and Costerton, 1990). Penetration was reported by Myers and McCready (1966) in Berea sandstone and different types of limestone employing the *Serratia marcescens*. Each type of geological material has its own porosity characteristics and the time necessary to penetrate various types of stones varies. It should be noted that penetration and migration of *Serratia marcescens* did not occur under atmospheric pressure (Myers and McCready, 1966). The presence of bacteria in rocks, as well as deep soils, was noted by Beloin et al. (1988).

I do not believe that the bacteria isolated from coal represents organisms laid down at the time of deposition, mainly because coal has too many fissures and is quite porous compared to many other types of rocks. If one accepts the concept that the bacteria in coal and rocks are the result of penetration, then the questions that must be asked are: (1) when did it take place, since it takes time for the penetration process to occur (how old are the microbial cells that penetrated the material), and (2) what was the physiological state of the organisms penetrating the rock of coal? In the latter situation, soil and subsurface strata are grossly oligotrophic and the organisms are probably already in the starved state when they contact rock and coal. These organisms in the starved state are, in all probability, ultramicrocells of bacteria and it is also possible that these starved bacteria, themselves, have persisted in the starvation-survival state for a long time.

Living bacteria were also cultured from the interior of adobe bricks obtained from California missions (112 to 150 years old) (Lipman, 1934). Bacteria were also isolated from bricks obtained from Arizona pueblos, which are definitely known to be at least 600 years old. Bricks from pre-Inca pyramids near Lima, Peru, (1000 to 1400 years old) and from Aztec pyramids (no less than 800 to 1000 years old) in Mexico also had living bacteria present. Bricks from the missions were protected from water during their entire history (details not given, but presumably they were bricks from the interior of the missions). In all cases, the bricks were found to be extremely desiccated.

Gallipe (1920) and Gallipe and Souffland (1921) isolated living organisms from crushed amber, meteorites (21 samples), quartz, basalts, etc. Viable microorganisms were found in amber from 14 different geographic areas, with generally more than one sample from each geographic locale, and negative results were found in amber from 3 different locales (Gallipe, 1920). However, in their short papers they do not present sufficient details concerning their methodology. As a result, Keilen (1959) suggests that their results may have been due to faulty techniques. It should be mentioned that Cano et al. (1993, 1994) were able to isolate DNA from the abdominal tissue of four extinct stingless bees embedded in Dominican amber. Later Cano and Boruchi (1995) reported the isolation of bacterial spores from 25- to 40-million-year-old Dominican amber. These spores were revived. They offered several lines of evidence indicating that the isolated bacterium was of ancient origin and not an extant contaminant. Rigorous aseptic techniques were employed. The isolated ancient bacterium was characterized enzymatically, biochemically, and by 16S ribosomal DNA profiles, indicating that the organisms were related, but not identical, to extant *Bacillus sphaericus*. All the tests conducted on the ancient bacteria indicate that it is unique. Thus, by use of all the modern molecular biological techniques, an "ancient" species was cultured. From a microbiological viewpoint, an excellent example of cryptobiosis was presented.

Weirich and Schweisfurth (1985) studied the microbiology of sandstone, employing sterile techniques and a special type of drill to obtain samples from the interior of the new red sandstone obtained from an abandoned quarry near Kaiserslautern, Germany. The rock fragments were about 3 years old. The number of revolutions of the drill during the drilling process influences the number of microorganisms recovered, mainly due to the heat generated. A drilling temperature should not exceed 40°C (Schwartz and Muller, 1958). A minimum of 10^3 heterotrophs per gram of rock were found at 12–14 cm from the rock surface, but below 10 m, no more heterotrophs could be detected (Weirich and Schweisfurth, 1985).

Decay in stones of various types, especially in castles in England, led Paine et al. (1933) to do an extensive study of the bacteria on and in various building stones obtained from numerous castles located in England. Bacteria were always found on the surfaces but not always beneath the rock surfaces. Although he did make finely powdered stone, which was suspended in water, he noted that shaking of the sample was not an effective way to separate bacteria from the stone particles.

Three rock (ashfall tuff) samples and one water sample were taken from several active and inactive mines (called tunnels) of the Nevada test site, Rainier Mesa by Amy et al. (1992) and tested for the presence of bacteria. These tunnels were used for underground nuclear testing and are located 350 to 450 m below the surface. These samples were taken by hand chipping into the existing tunnel walls after several centimeters of rock had been removed. Water was taken from a free-flowing seep emanating from a fault that crosscuts the tunnel. Microbial counts ranged from 10^2 to 10^3/g (dry weight) of rock sampled and 10^2/ml of water. Many of the isolates were small (<1 µm) when viewed in the rock matrix and remained small when cultured. The isolates were characterized by use of three identification systems, API-NFT (Bio Meriux Vitek, Inc.) strips, BIOLOG (Biolog, Inc.), and Microbial Identification System (MIDI). Each system indentified only a small percentage of the total isolates. *Pseudomonas* was the most commonly identified genera. The water isolates were considerably different from the endolithic isolates. The data presented suggest that these populations are different from those currently known. The scenario proposed by the authors is that colonization of rock occurred between depositional events, or through the movement of water from the surface or in lateral fluxes. Surface water can reach subsurface sites via three methods: fracture flow, matrix flow, or a combination of the two. The travel time for fracture flow recharge water in the Rainier Mesa has been estimated between

6 and 30 years by ^{3}H dating (Clebsch, 1960). If one uses calculations based on Thordarson's (1965) matrix flow data, the minimum age of these microbes is calculated at 250,000 years. According to Amy et al. (1992) the actual age is probably greater because others have shown that bacteria do not travel well through soils of low matrix potential and lateral water flow is essentially negligible in the sample area. Whatever the case, the living cells have survived periods of time (years to thousands of years) with little or no nutrients since the diagenesis of sedimentary material into rock.

Haldeman and Amy (1993b) classified some isolates from the ashfall tuff rock as *Moraxella phenylpyruvica* and *Flavobacterium indologenes*. No plasmids were detected in any of the isolates. Later, Haldeman et al. (1993) defined the distribution of subsurface bacteria on a 21-m^3 section of rock from a 400-m-deep subsurface tunnel (U12 tunnel system at Rainier Mesa, Nevada test site) and determined how related groups of bacteria were distributed through the rock section. They found direct counts were several orders of magnitude higher than the viable counts and no definitive pattern of distribution. The isolates were analyzed for fatty acid methyl esters (FAME) using MIDI. Two genera, containing 16 isolates, were unmatched to known organisms within the MIDI database and clustered with other isolates at a Euclidean distance greater than 50. Twenty-nine genera (Euclidean distance of ≤25) were found within the rock structure, while 28 of the 210 bacterial types isolated were nonculturable under the growth regime required for cluster analysis. Isolates mainly clustered at the genus level with *Arthrobacter, Gordona,* and *Acinetobacter*. The larger cells from the rock usually had a higher metabolic activity and there appears to be no correlation between moisture and the types of organisms (gram-negative/gram-positive) recovered. Gram-positive organisms were predominately recovered from subsurface sites (vadose zone volcanic tuffs in New Mexico, vadose zone basalt/sediment interface, and intersedimentary samples) (Colwell, 1989) and a shallow water aquifer in Oklahoma (Bone and Bilkwill, 1988); whereas gram-negative bacteria from the saturated sediments in the Savannah River (Balkwill, 1989), unsaturated palesols at Hanford, Washington (Brockman et al., 1992), an aquifer in Germany, and most of the Nevada test site tunnel system (Haldeman and Amy, 1993a).

Six subsurface endolithic isolates from the 450-m-depth rock matrix of the Rainier Mesa, Nevada, test site were subjected to starvation studies in artificial pore water (Amy et al. (1992) formulated to mimic the in situ conditions of the nearly saturated rock. All isolates formed ultramicrocells when starved. Isolates NO5R4 and NO5R5A were identified as *Pseudomonas vesicularis,* by API-rapid NFT method, to a good and acceptable level of confidence but the same isolates, when identified by BIOLOG, were identified as *Sphingobacterium multivorum*-like, with good to poor certainty values. MIDI recognized them as *S. multivorum,* but to low certainty values. Isolate NO5R9 was subsequently found to be related to *Arthrobacter* after 16sDNA sequencing. No other isolates were identified by any of the three systems. All isolates demonstrated an initial increase in cell numbers during the first portion of the starvation experiments. All but one isolate displayed a peak in viable count followed by a rapid loss in viability, reaching a plateau level. NO5R7 displayed a constant decline in viable count from the peak viable count value throughout the 100-day starvation period. For all isolates, 60 days appears to be the critical time, after which the viable count either held steady or began a second decline. Two isolates (NO5R8 and NO5R9) retained higher viability after 100 days of starvation-survival and this high survival correlated with sustained respiration, as measured by iodonitrotetrazolium-formazan production during starvation.

Later, the results obtained by Russell et al. (1994) indicate that the culturable microbial community size and composition exhibit random spatial variability within the geological/geochemically homogeneous rock section. The pore water concentration of nitrate correlated with numbers of bacteria testing positive for nitrate reductase and indicates that these bacteria probably exist in the dormant form in situ.

Viable bacteria were isolated from deep volcanic rock formations (50–450 m) within the Rainier Mesa, Nevada, test site (Haldeman et al., 1994). Heterogeneity in microbial communities, as measured by diversity and evenness indices and by comparison of morphological and physiological tests of representative isolates, was demonstrated and provided evidence against the hypothesis for recent bacterial transport from the surface.

Krumholz and Suflita (1996; pers. communic.) obtained consolidated rock cores from a depth of 150 to 250 m at a site in central New Mexico. These cores were maintained under anoxic conditions at all times and sampling was conducted in an anerobic glovebox. The activity of sulfate reducers was determined by incubating fresh core faces with $Na^{35}SO_4$. These core faces were overlaid with silver oxide coated with silver foils to trap $^{35}S^{2-}$. As a result, the contaminated zones on the outer rock could be distinguished from endogenous rock activity. The incubation period was from 4 to 6 weeks. Sulfate reduction was noted and the greatest activity (100- to 1000-fold greater) occurred at the sandstone-shale interface core material. Their data demonstrated that the organically rich shale can slowly deliver organic matter to the active microbial communities living at the sandstone-shale interfaces. Acetogenic bacteria could also be isolated from this material. It appears that the mixing of shale (which possesses a greater amount of organic matter) provides to the microorganisms in sandstone sufficient energy for a slow rate of metabolism. Greater enzymatic activity of the indigenous heterotrophic bacteria was noted when sediment slurries made from adjacent sandstone and shale lithologies were combined than in slurries made from only sandstone or shale (R. P. Griffiths, pers. communic.). The lithologies must be adjacent.

Because of the large effort on the part of Department of Energy (DOE) Deep Microbiology Subsurface program, there will be much scientific literature in the future concerning bacteria isolated from various subsurface environments. All of the foregoing data on the occurrence of viable bacteria in coal and rocks is indicative of a survival mechanism, which, in all probability, is initiated by the lack of organic matter as the energy source in the original source material giving rise to the process of starvation-survival.

2.4   Survival of Bacteria in Solar Salts

Solar salt deposits can be very old and the amount of organic matter present is very low. During the formation of salt deposits, saltwater was exposed to a drying process, mainly due to sun rays. Before the drying process, the organic matter content of the water is low and further decrease in the organic content results from the indigenous bacteria in the saltwater, creating a very oligotrophic condition. In addition, the halobacteria will begin to use concentrated organic matter resulting from dessication. Again, we are dealing with a survival mechanism(s) of the bacteria isolated from solar salts. One of these mechanisms is starvation-survival, but it is not known if the high salt content or drying may provide other mechanisms for survival. All mechanisms may or may not complement each other. In the early 1930s, it was postulated that perfectly dry bacteria could live practically indefinitely and that death occurred as a result of the drying method, rather than dryness itself. Stark and Herrington (1931) examined this postulate. Various species of bacteria were dried under vacuum in the presence of $CaCl_2$ and P_2O_5; apparent dryness was achieved in less than 3 minutes. After 97 days under dry conditions, approximately two-thirds of *Streptococcus parachtrovorus* remained viable; whereas only 2–3% of the *Streptococcus lactis, Saccharomyces cerevisiae, S. aureus, Staphylococcus albus, Bacterium coli,* and *Lactobacillus acidophilus* remained viable. This process could be the forerunner for the lypholization technique used today. Thus, in the creation of solar salts or salt deposits, both starvation and drying (in nature, it is over a much longer period of time) are probable mechanisms that complement each other in the survival of certain microorganisms.

Red halobacterium was found to remain viable in crude solar salts for periods of months after the salt had been harvested (Harrison and Kennedy, 1922); whereas, Dussault (1958) established that bacteria in crude solar salt could survive 20–60 days. Bain et al. (1968) recovered viable "pink bacteria" from solar salt samples that have been stored for 4 years.

Crystals of solar salt contain water droplets. From 6 out of 13 water droplets in crystalline Permian salt (225 million years old), Reiser and Tasch (1960) isolated living bacteria (diplococcus) under conditions they believed to eliminate contamination. Evidence is presented by Tasch (1960) that the bacteria were laid down at the time of salt deposition and that they were not the result of water intrusion at a later geological period. [*Author's note*: I had the opportunity to visit

Dr. Tasch at Wichita State University in the summer of 1958. I found his laboratory setup to be excellent in terms of providing sterile conditions for examining the salt samples. He used ultraviolet lights to sterilize the air in his isolation chambers.]

Dombrowski (1963) also reported the isolation of bacteria (*Pseudomonas halocrenaea* and *Bacillus circulans*) from Zechstein rock salt samples (deposited 200 million years ago, Permian Age). The rigorous technique employed is detailed in Dombrowski (1963). He discarded all samples that came from near faults or the upper salt layers. The salt was primary Zechstein salt. Petrographic thin sections of the salt were made and it was found that the bacteria were embedded in the crystalline structure and not in capillary crevices. Pollen grains make it easy to establish the age of the bacteria. *P. halocrenaea* was indentified in 61 of the 128 rock salt samples, 41 samples contained fossil bacteria (nonviable), and 36 samples had no bacteria. In order to demonstrate the isolates could withstand the high osmotic pressure created by the salt, he cultivated them in a nutrient solution and permitted the solution to become saturated by adding 1 gm of salt per week. Once saturation occurred, the culture medium was slowly permitted to dehydrate and crystalize until only crystalline salt remained. The same bacteria could be recovered from the salt crystals. In all likelihood, the bacteria undergo starvation due to the long period of time needed for the system to become saturated with salt. If the inoculated culture was permitted to slowly dehydrate, the bacteria died; thus indicating it was not possible to reproduce in a laboratory the probable sequence that occurred geologically. *P. halocrenaea* isolated had an optimum temperature between 34°C and 55°C. A geologist confirmed this was the temperature that was present when the Zechstein sea was slowly drying up. Dowbrowski (1963) was also able to isolate six different species of bacteria from the Middle-Devonian salts from Saskatchewan; three different species from Silurian salts (from Meyers, New York) and two species from Precambrian salt (650 million years old) specimens from Irkutsk.

P. halocrenaea, isolated by Dombrowski (1963), was studied by De Ley et al. (1966). The DNA base composition variance of the compositional distribution of the DNA molecules of the organism was compared to *P. aeruginosa* and a Xanthomonad. DNA hybridization between the organisms was undertaken. From these studies, De Ley et al. (1966) concluded that *P. halocrenaea* was not a living fossil.

From the description of the organism that De Ley et al. (1966) received from Dombroski, it appears that they received a contaminant. Was this the same organism that Dombrowski (1963) first isolated? If *P. halocrenaea* was described as a true halophile, I would have agreed *P. halocrenaea* was, indeed, the organism deposited when the salt was laid down.

During crystallization of a halophilic medium containing pure culture of *Halobacterium salinarium* strain 1, the organisms were encased within the salt crystal formed. These organisms were recovered from the salt crystal 30 years later (Ian Dundas, pers. communic.).

One of the keys to understanding survival of halobacteria in salt is in the salt's mineralogy—mainly its large number of fluid inclusions (Norton and Grant, 1988). When sodium chloride crystallizes, the living halobacteria are entrapped with the fluid inclusions. Trapped cells have been observed in the marine salterns from Lake Magadi, Kenya, and Puerto Rico as well as in salt crystals grown under controlled conditions from solutions containing pure culture suspensions of halobacteria. Representative strains of each major grouping of the *Halobacteriaceae* except *Halococcus* survive entrapment within salt (Norton and Grant, 1988). Later, Norton et al. (1993) were able to isolate halobacteria and obligately halophilic eubacteria from the Winsford salt mines (Triassic Period; 195–225 million years ago) in Cheshire, England, and from the Boulby potash mine (Permian, 225–270 million years ago) in Cleveland, England. The halobacteria were characterized by chemotaxonomic methods and most but not all were shown to be very similar but not identical to those halobacterial types that dominate in highly concentrated surface brines. Norton (1992) applied the term "living fossil" to the halobacteria isolated from salt mines. It should also be pointed out that these organisms belong to the *Archaea* and many organisms classified as *Archaea* are found in extreme environments. Furthermore, the *Archaea* are considered by some researchers to be ancient prokaryotes (see Potts, 1995).

2.5 Survival of Bacteria in Frozen Material

Freezing is another mechanism for the preservation of microbial cells (not 100% of the frozen cells survive the freezing process). Geological material, except in rare situations, is oligotrophic. Therefore, bacteria in frozen material in both polar regions represents bacteria of geological origin or from the oligotrophic atmosphere. Both freezing (cold temperature and dehydration) and starvation may complement each other in the survival of bacteria in naturally frozen material. Experimentally, this subject has never been studied. Although I discuss the presence of bacteria in frozen organic material, the specific substrate(s) may not be present for the isolated organism(s). The freezing of material can also make organic matter unavailable to the microorganisms (water is no longer liquid). As I envision the process, starvation takes place first followed by the freezing process.

During the first half of the twentieth century, Russian microbiologists established the presence of viable bacteria in frozen Holocene sedimentary deposits (see Khlebnikova et al., 1990). R. H. McBee (in Sneath, 1962) found 10^3 viable coliform bacilli/g of poly feces from Scott's Antarctic Expedition of 1911, which had been continuously frozen for nearly 50 years. ("Poly" is not defined by McBee, but I believe it refers to feces orginating from humans and dogs.) Previously mentioned was the fact that *Escherichia coli,* in the intestinal tract, only reproduces one or two generations a day. Thus, the coliforms were close to a starved state when defecated and then frozen.

During the second Byrd Antarctic Expedition, Darling and Siple (1941) isolated many common species of spore-forming and non-spore-forming bacteria, as well as yeast and molds from snow, ice, plant debris, mud, and rock fragments. Because the bacteria were so widely distributed, they surmised that the bacteria were carried by air currents into the Antarctic. Becker and Volkman (1961) recovered viable bacteria from permafrost cores near Fairbanks, Alaska. The shallowest core appeared to be about 20,000 years old whereas the deepest core was about 70,000 years old. Likewise Boyd and Boyd (1964) found numerous bacteria in permafrost near Barrow, Alaska. Although the mesophilic bacterial count decreased with core depth, the thermophile count remained more or less constant. Boyd and Boyd (1964) were also able to sample a peat core drilled from the permafrost and the same bacterial profiles were obtained. The peat was radiocarbon dated and found to be $10,525 \pm 280$ years old. Cameron and Morelli (1974) reported viable microbes in Antarctic permafrost.

Viable bacteria were found in permafrost, ancient buried soils, and sedimentary rocks of various genesis (Khlebnikova et al., 1990). No correlation between the organic content and the number of bacteria in permafrost could be noted. The general distribution of viable cells in relation to the age of the deposit was established, which showed a decreasing number with increase in age of the permafrost. About 10^3 to 10^5 cells/g in the Late Pleistocene (Edom) deposits, 10^3 to 10^4 cells/g in the Late Pleiocene-Early Pleistocene (Oler), and 10^2 to 10^4 cells/g in the Pliocene (Tomus-Jara, 3 to 5 million years ago) were found. No yeasts or fungi were present.

Issatschenko and Simakove (in James and Sutherland, 1942), Kazansky (in James and Sutherland, 1942), Levinskaya and Mamincheva (in James and Sutherland, 1942), and Kapterev (in James and Sutherland, 1942) all reported finding bacteria in frozen Arctic soil.

In permanently frozen sedimentary rocks and buried soils of the tundra zone in the Kolyma lowlands, frozen for about 1 million years (Pleistocene), 10^3 to 10^4 viable bacteria per g were found (Zvyagintsev et al., 1985a,b). In this case, there was a correlation between the organic content and ice content of the samples to the number of organisms. Again, only procaryotes were found. In terms of enzyme activity, no catalase or dehydrogenase activity could be demonstrated but invertase activity was detected.

There was a decrease in eukaryote microorganisms (fungi and yeasts) in a microbe complex, decreases in morphological variability of prokaryotes, decrease in their numbers and an increase in the number of samples without microorganisms in Pleistocene rock compared to Holocene (10,000 years) rocks (Zvyagintsev et al., 1990). This finding indicates that even in the survival state, death of cells continues with time. Denitrifiers were found in 76% of the samples and cellulose decomposers in 28%. The Russian literature should be consulted for a thorough review of bacteria in permanent frozen material.

In the Ross Desert of Antarctica, the main forms of terrestrial life are lichen dominated or cyanobacterium-dominated crytoendolithic microbial communities. According to Johnson and Vestal (1991), these organisms do not show signs of nutrient limitation and, thus, may be the slowest growing communities on earth.

However, James and Sutherland (1942) were not able to isolate any bacteria in permanently frozen samples obtained aseptically near Churchill, Manitoba, Canada. These authors report that a negative finding could mean that the sample tested did not contain any bacteria at the time of freezing or could be due to the conditions of the experiment.

One could easily come to the conclusion that survival of bacteria in frozen natural material is the result of freezing. Again, I wish to emphasize that the bacteria in the air, seawater, or soil are generally in the starved state. The long-term survival of bacteria in permanent frozen geological material in both polar regions are the result of the combination of starvation and freezing. The latter also involves dessication. Thus, cold temperature, dessication, and starvation all contribute the the survival state.

2.6 Survival of Bacteria in Soil

Outside of the rhizosphere, soil generally lacks organic matter, which decreases with depth. It was first published by Waksman (1916) and is now well known that the number of bacteria decreases with soil depth. The ability of bacteria to survive in soil is well documented by soil scientists. Many soil bacteria do not produce resting cells that are obviously different in gross structure from the vegetative form and probably exist in soil much of the time as vegetative cells in a reduced state of metabolism (Gray and Williams, 1971).

The preservation of microorganisms in air-dried soil is well known and has previously been an accepted method for preserving some bacteria (McLaren and Skujins, 1968). These bacteria kept on bacteriological maintenance media for the same period of time would expire. Generally a 100-fold decrease in the number of viable organisms (i.e., from about 10^7 to 10^5/g soil) was noted by McLaren and Skujins (1968) in air-dried soil. Bacteria preserved in this manner probably also experience a lack of energy before the moisture content becomes equilibrated with the soil moisture. Unavailability of organic matter may also result from the lack of moisture in the system. Then the cell faces two harsh environmental stresses.

Gibbs (1919) and Winogradsky (1949; research published in 1931) were able to obtain nitrifying bacteria from soil maintained in the laboratory for 7 years.

Sneath (1962) found living bacteria in soil specimens up to 320-year samples. Nitrifiers were still present in air-dried soil after two years (Wilson, 1928). Lipman (1934) kept soils in sealed bottles up to 65 years. In all the samples, living bacteria were numerous and in great variety. Bollen and Byers (unpublished report, see Bollen, 1977) isolated species of *Bacillus* soil coryneforms and *Streptomyces* from soil samples from the 70-year-old collection of E. W. Hilgard. When ammonium sulfate was added to dry soil samples held in bottles for more than 15 years, nitrification was demonstrated (Fraps and Sterges, 1932). Soil samples kept at 20°C to 30°C for 2 years conserved their nitrifying power (Greaves and Jones, 1944). No decrease in the number of nitrifying bacteria could be detected in air-dried litter maintained at a normal temperature for 1 year (Hale and Halversen, 1940). Nitrifying bacteria can therefore survive for long periods of time in anoxic environments (Painter, 1970).

Garbosky and Giambiagi (1962) took 25 different soils from a depth of 0–40 cm from Argentine Patagonia and placed them in 5-l sterilized bottles. These were maintained for 5 years in the dark. Bacterial nitrifiers were found in nearly all samples. Soil temperature and soil type as well as the joint action of lime and potassium in quantities not less than 200 mg% CaO and 30 mg% K_2O respectively were shown to be important factors in the survival of the nitrifiers. The content of organic matter, organic N, and pH of the soils employed did not influence the survival.

Bosco (1960) kept 65 strains of known species of non-spore-forming bacteria in soil in the dark at room temperature and, depending on the species, they survived for 6 to 48 years. Jensen (1961) was able to show that *Rhizobium* could survive in soil for 30 to 45 years. Six soil samples, collected in 1921 from various areas in Oregon, were stored by Bollen (1977) and tested periodically for 54 years. These soil samples were airdried and stored in jars with clamped but unsealed lids and subjected to temperatures as low as 12°C in winter and as high as 40°C in summer. Sulfur, as well as the presence of molds, bacteria, streptomyces, and endospores oxidation, could still be found to occur in all samples.

The survival of bacteria in soils is influenced by a variety of chemical, physical, and biological factors, hence the effect of clay minerals in modifying bacteria survival involves many different mechanisms. Marshall (1975) reviewed the data of clay mineralogy in relation to survival of soil bacteria and found the following effects:

1. Absorbtion of antibiotics in the clay minerals.
2. Protection from desiccation—the formation of an envelope around bacterial cells as a consequence of electrostatic attraction between charged groups on the bacterial cell and clay surfaces.
3. Modification of the host-parasite relationship. Soil may form aggregates, especially when drying.

Hattori (1973) presented a simplified aggregate model to emphasize the advantages of the inner portion of the aggregate as a site for survival. The inner portion helps prevent desiccation of the system.

The survival of bacteria in soil is influenced by many factors, two of which are starvation and dessication. Because soil is oligotrophic, starvation takes place first, followed by dessication. This is true in laboratory studies because the bacteria are placed in the sterile soil in a liquid medium (generally sterile water). Thus, starvation and dessication are the two main mechanisms for the long term survival of bacteria in soil.

2.7   Survival of Bacteria in Subsurface Sediment and Water

Subsurface environments contain so little organic matter because the indigenous bacteria have had a long time to reduce the organic matter content. Furthermore, water containing organic matter, when percolating through the sediment, comes in contact with the sedimentary bacteria, which remove most of the organic matter. Thus, the systems are oligotrophic.

Issatchenko (1940), studying the bacteria in oil wells, demonstrated the occurrence of sulfate-reducing bacteria, denitrifying bacteria, and purple sulfur bacteria from depths of 1400 to 1700 m. (Although oil is energy rich, most bacteria cannot use it as a substrate.) He stated that the depth limit where bacteria can be found extends to 2000 m. Just how far down in the biosphere bacteria do exist becomes a fascinating question. Other studies dealing with the occurrence of sulfate-reducing bacteria in oil brines from depths exceeding 1000 m have been reported in the early days by Gahl and Anderson (1928), Bastin (1926), Bastin and Greer (1930), Ginter (1930), Ginsburg-Karagitscheva (1933), and others. The recovery of bacteria from oil-bearing and sulfur deposits is reviewed by Beerstecker (1954), ZoBell, (1958), Davis (1967), Kuznetsov et al. (1963), and Atlas (1984). Tiedje verbally reported, at the Fifth International Microbial Ecology Meetings in Kyoto, Japan, in 1989, that an organism capable of denitrification in Earth strata as old as 45 to 90 million years was recently isolated.

Higher organic matter content in sedimentary material is generally associated with the small size fractions. In aquifer sediment samples, Albrechtsen (1994) found the opposite to be true. In his studies, 99.9–100% of the viable bacteria were in the 1.2–100-μm sediment fraction, 40–96% in the 1.2–55-μm fraction, and only 0.01–5.4% in the 0.2-μm fraction. Greater heterotrophic activity was also shown in the larger fractions.

Starvation-survival appears to be a requisite trait for existence of bacteria in subsurface systems (Kieft et al., 1990). The geochemical conditions in subsurface environments suggest that the indigenous microorganisms are under starvation conditions as evidenced by the extremely low concentrations of organic matter and the low adenylate energy charges of the subsurface microbial population (Kieft and Rosacker, 1991). When the ratios of total organic carbon to substrate-induced respiration values are compared between surface soils and subsurface samples, the ratios are considerably lower in subsurface samples. This indicates that the subsurface microbial population is extremely old and stable and survives amid predominantly stable recalcitrant organic carbon (Kieft et al., 1990).

The occurrence of bacteria in subsurface material is well documented in Chapelle's (1993) book. However, the reason(s) for the occurrence and survival of microbes in subsurface samples is not addressed adequately. Again, starvation-survival plays a major role in the occurrence of bacteria in these subsurface environments where the energy for heterotrophic bacteria is in very low concentration.

2.8   Survival of Bacteria in Marine Environments

In the marine environment, one must consider not only the residence time of the water masses, which can exceed 1000 years, but also the depth and the low organic matter content. In the euphotic zone, photosynthesis does take place actively, but the rate is dependent on other factors (nitrogen, phosphate, iron, temperature, etc.). Because the thermocline (discontinuity layer) acts as an imperfect barrier, much of the organic matter produced in the euphotic zone does not enter the deep sea. Bacteria below the euphotic zone appear to be in an inactive metabolic stage (Wirsen and Jannasch, 1975). The bacteria in nearshore sediments have been studied quite extensively because this habitat contains more organic matter as an energy source than open ocean sediments. There is a decrease in organic matter as depth of the sediment increases and as a result, in the deeper portions of sedimentary material, anaerobic bacteria dominate. When dealing with the benthic division, scientists do not know the lower limits of the biosphere.

Because sedimentation of material in the aquatic environment requires time, finding of bacteria in the deeper layers of sediments indicates that bacteria do survive for extremely long periods in this habitat. In many cases, the sedimentary material has little energy associated with it. The organic content of the sediment decreases with depth (age). Nearshore sediments have organic matter from less than 1% to 10% (Trask, 1939); whereas open ocean sediment (red clay and globigerina ooze) contains less than 1% organic matter (Revelle, 1944). In all sediments, water is present. Although organic matter is present in small amounts, mostly recalcitrant, the metabolic rate of bacteria in sedimentary material is probably not measurable. In sediments, there is a continuous liquid phase, but diffusion of organic matter from the upper layers to the lower layers of sediment is always possible.

Finding bacteria in the deep strata of sedimentary material is an indication that the bacteria were laid down at the time of sedimentary deposition thereby representing old bacteria. Arguments exist against bacteria having been deposited in the lower strata. Thus, the bacteria in the deep strata of sediments may not be as old as the time of deposition. These arguments include the migratory ability of bacteria and the possibility of contamination during sampling and cultural operations. Bacteria in sediments generally migrate toward a more favorable environment, and the lower portions of the core do not, in all probability, represent a more favorable environment—mainly due to a decrease in an energy source for heterotrophs. Counterarguments are that bacteria display a random direction of motion and the plastic nature of sediment (especially red clay) would discourage bacterial migration. Migration by reproduction would probably not result in any great movement, because adsorption of clay particles and lack of organic matter would tend to hinder this type of movement. In addition, the metabolic rate, retarded by lack of bioavailable energy, would tend to curtail multiplication.

Bacteria have been isolated from marine sediments. The earliest report is by Certes (1884), who reported the presence of bacteria in sediments collected during the Talisman Expedition to the central Atlantic Ocean. Bacteria were found in all but 4 of the 100 samples analyzed. The quantitative results are of little significance because the sediments were not processed immediately. Other early investigators (Russell, 1891; Fischer, 1894; Drew, 1912; Thiel, 1928) have reported finding bacteria in marine sedimentary material. Coring devices did not exist at that time so that bacteria in the deeper portions of the sedimentary material were not sampled. Deeper portions have less organic matter because the microbes in the upper surface have utilized most of it for their own metabolism. From her investigations on the bacterial population of mud from the Clyde Sea area, Lloyd (1931) concluded:

1. The numbers of bacteria decrease from the mud surface downward.

2. The number of bacteria fluctuate in the top mud layers and there is evidence of bacterial zonation.

3. In the deeper mud layers, the bacterial content is fairly constant for any given station.

4. The predominant organisms were found to be water bacteria of the *Achromobacter* and *Chromobacter* types and the large spore-forming bacilli similar to common soil bacteria.

5. The factors affecting the bacterial content of the muds are oxygen and temperature, which, in turn, affect the rate of multiplication; H-ion concentration, which decreases with the depth of the mud; and the food supply.

It is of interest to note that she demonstrated the presence of bacteria in layers as deep as 76 cm. Lloyd (1931) did not concern herself as to the metabolic state of the bacteria found in her studies, nor did most investigators in the early studies dealing with the occurrence of bacteria in sediments. As soon as you get below the surface-water interface, a big decrease in the organic matter occurs, which results in the starvation-survival state.

Reuszer (1933) observed a relationship between the heterotrophic bacterial population and the carbon content in the deepest layers of the sediment obtained from around Cape Cod. According to Reuszer (1933), the numbers of bacteria decrease with depth in the mud, the most marked decrease being in the first 2.5 cm. The observations by Waksman et al. (1933) established a definite parallel between the number of bacteria and the abundance of humus in mud. They also demonstrated the ability of bacteria in sedimentary material to oxidize ammonia to nitrite as well as nitrate reducers. Working off the coast of Southern California, ZoBell and Anderson (1936) demonstrated numerous physiological types of bacteria present in the sedimentary material. They also found that the bacterial distribution in sediment is more or less independent of the depth of the overlying water, the temperature of the ocean floor, and the distance from land. They further stated that the bacterial population is more dependent upon the organic content of the sediment than upon any other factor, and the irregularities in bacterial population counts are due to the unevenness of organic particle distribution and the tendency of bacteria to colonize. ZoBell (1938) postulated the occurrence of strict aerobes at great depths in sediment lacking oxygen as indicating that such organisms have been buried there in a dormant state for a long time. Rittenberg (1940) reported the presence of aerobes and anaerobes in cores as long as 355 cm. ZoBell and Feltham (1942) and ZoBell (1942) found both the total numbers and kinds of bacteria decreased with core depth. The longest nearshore core examined for microbes was 752 cm and microbes were found (Morita, 1954). Calculations based on a deposition rate of 25 cm per 1000 years, as reported by Revelle and Shepard (1939), indicates that the 752-cm layer was deposited 30,000 years ago. It should be mentioned that the lower layers of the cores all were in a reduced state. From these data, there is no evidence to show the lower limit of the biosphere in which bacteria occur in nearshore sediment.

In the Channel Region off the coast of California, Bartholomew and Paik (1966) isolated many obligate thermophiles in all core samples (except one) ranging from 5 to 900 per g/wet wt sediment and from core strata as deep 175 cm. Thermophiles isolated from the 150-cm core strata represent an age of 7800 years. In this case, the temperature is about 3–5°C but also the amount of organic matter for their growth is limited. These thermophiles could not have originated in these sediments. Although one can question the original source for the thermophiles, there is no question that they were transported first by the wind and then by sedimenting through the seawater. Both air and seawater are oligotrophic. In addition the environment from which the thermophiles originate was also oligotrophic. Thus, I believe this is an example of starvation-survival of thermophiles that takes place in nature. Another example of survival of thermophiles is their isolation in Arctic soil and water samples (McBee and McBee, 1956).

Open ocean sediments beyond the continental shelf are mainly red clay and globigerina oozes. Red clay is a pelagic deposit characterized by the absence of terrestrial mineral grains larger than colloidal size (Kuenen, 1950). It is low in calcium carbonate; soft, plastic, and greasy to the touch; and low in organic matter (Shepard, 1948). Found in the deep-sea regions, globigerina ooze is a calcareous ooze that results from the skeletal remains of this important planktonic foraminifera settling to the bottom. Fischer (1894) noted the growth of only one or two colonies on plates of seawater gelatin medium inoculated with one-gram sediment samples obtained from depths of 1523 m and 1406 m in the Gulf of Mexico. Waksman (1934) declared that true oceanic bottom deposits in the Atlantic contained few bacteria, but he gave no data to substantiate this conclusion.

During the Mid-Pacific Expedition of 1950, Morita and ZoBell (1955) recovered bacteria from various core depths of red clay and globigerina ooze. These core samples were taken with a Kullenberg piston corer and all cores were taken beyond the continental shelf. This piston corer permitted us to take samples up to 10 m into the sediment. The Mid-Pacific Expedition of 1950 was the first oceanographic expedition after World War II. All layers of all cores taken were in the oxidizing state (Eh values ranging from $+150$ to $+450$ mV and pH values ranging from 6.58 to 7.97 in red clay; Eh values ranging from $+150$ to $+350$ mV with pH values ranging from 7.47 to 7.55 in globigerina ooze). The Minimum Dilution Method, employing different media, was used to determine the titer of bacteria. Incubation periods for these open ocean samples were at least 21 days indicating a long lag period occurred first. In all probability, this long lag period permitted the starved cells to become resuscitated. According to Arrhenius (1952), red clays are deposited in the Pacific Ocean at an average rate of 2.3 mm/1000 years. At this rate, some of the lower strata in core MP3-1 and MP35-1, in which a minimum of 10^3 bacteria g were found, were deposited more than a million years ago. These investigators raised the questions as to what depth and in what age of sediment these bacterial activities continue.

Deeps and trenches of the ocean, on the other hand, contain more organic matter than open ocean sediments beyond the continental shelf—mainly because all deeps and trenches are near island arcs and as a result, terrestrial debris can be observed.

Bacteria have been found in various deeps and trenches (ZoBell and Morita, 1957, 1959). Shore plants are known to be transported into these waters (Wolff, 1976; George and Huggins, 1979). The Puerto Rico Trench sediments are rather unusual in that low redox potentials of $+40$ to -30 mV have been recorded (George and Huggins, 1979), indicating sufficient organic matter as energy for microbes to lower the Eh. Generally, abyssal sediments have a high redox potential simply because insufficient organic matter sediments to the bottom so that microbes can bring about reducing conditions. As a rule of thumb, high redox potentials in sediments reflect a lack of organic matter indicating that the heterotrophs are in a starvation state; whereas, low redox potentials indicate sufficient organic matter present for the heterotrophs to continue their metabolic activity.

The bacteriology of the experimental Mohole drilled during March and April 1961 off Guadalupe Island, Mexico, was conducted by Rittenberg et al. (1963). The drilling penetrated 2.5 m of red clay followed by 172.5 m of hemipelagic ooze, both calcareous and siliceous. The sediment ranged back to the Middle Miocene Age. These sediments contained small concentrations of hydrocarbons, porphyrins, amino acids, and sugars. All the sediments had an Eh ranging from 250 to 400 mV. Bacteria were cultured from only a few samples (4 positive out of 100 samples), although a large variety of media were employed. These positive samples were from core depths of 44 m and 138 m. The bacteriology was conducted under the most aseptic techniques possible during the shipboard operations. However, due to the nature of the drilling operations, which involved drilling mud, the investigators could not rule out contamination in the few samples in which bacteria were isolated. All enrichment cultures of the drilling mud gave heavy growth except in media designed to elucidate cellulose digestors, nitrifiers, cellulose fermentators, and photosynthesizers. In one of the positive anaerobic samples, a pure culture of typical sulfate-reducing bacteria was found. An uneven distribution of colonies was observed on plates and there was no correlation between spread plates and pour plates. The media employed for growth of the usual heterotrophic bacteria used the usual concentrations of yeast extract and proteose peptone. No resuscitation media were employed, since the use of resuscitation media for starved organisms was not known when these field experiments were conducted. I believe that the negative results obtained in 96 samples speaks well for the aseptic techniques employed. Although Rittenberg et al. (1963) would not rule out contamination, I believe the organisms that were found were not contaminants. One should not expect to find positive results in every sample tested. Since the organic carbon for the positive cores was 0.41 and 3.8% (not all bioavailable, see Chapter 3) and the Eh in the positive range for the bacteria isolated, the bacteria found were in the starvation-survival mode.

Parkes et al. (1990) were able to demonstrate active bacteria in the Peru margin as deep as 80 m below the sediment surface. These strata represent sediments laid down approximately a million years ago. Bacteria were cultured from all strata of the sediment and acridine orange direct count (AODC) provides an accurate estimate of in situ distribution of bacteria in the sediment. The Frequency of Dividing Cells gives an index of growth and viability. Both aerobic and anaerobic bacteria were found at all strata. Figure 2.1 illustrates the anaerobic counts of different physiological types of bacteria with depth. Nearshore sediments are anoxic below sediment-water interface, but the depth at which sediment becomes anoxic depends on location and richness of organics in the sediment. Nevertheless,

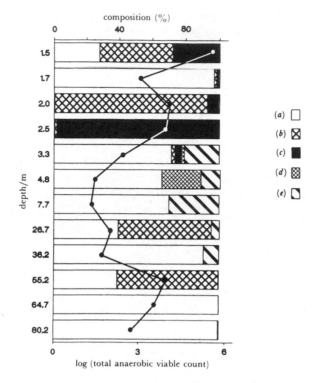

Figure 2.1   The distribution of different types of anaerobic viable bacteria and total anaerobic viable count with sediment depth. Heterotrophic bacteria (*a*), nitrate-reducing bacteria (*b*), sulphate-reducing bacteria (*c*), methanogenic bacteria (*d*), hexadecane-oxidizing bacteria (*e*). Reprinted by permission from *Philosophical Transactions of the Royal Society of London, Series A* 331:139–153, R. J. Parkes, B. A. Cragg, J. C. Fry, R. A. Herbert, and J. W. T. Winpenny, 1990; copyright 1990, The Royal Society.

they were able to isolate viable spores in the deepest strata. In the Sea of Japan (Oki Ridge) 518 m below the surface of the sediment, bacteria were also found (Parkes et al., 1993, 1994). The latter sediment depth represents sediment laid down more than 4 million years ago. A variety of modern techniques (MPN [most probable numbers], AODC, epifluorescent microscopy, and radioisotope methods) were employed to demonstrate the presence of bacteria. The data on bacterial profiles and activities were strongly correlated and were consistent with the chemical changes (e.g., sulfate reduction, methanogenesis) in the sediment. Recognizing that the AODC technique was initiated immediately on the core material, the number of bacteria are much too high to be due to contamination.

The depth of sediment and geological material at which bacterial activity ceases and where bacteria cease to be present is not known, but it will depend on the geological structure as well as the physical and chemical conditions when the geological structure was laid down.

Numerous investigations have shown the presence of viable bacteria in marine sedimentary material that was laid down years ago. I do not believe that the bacteria found in the deeper layers of sedimentary material should be considered contaminants. With the modern microbiological techniques, such as those employed by Parkes et al. (1993, 1994), contamination can be ruled out. Depending on the type of geological sedimentary material being examined, there will be a point where viable bacteria will cease to exist. In some geological material, some of the microbes will then become fossilized.

2.9 Preservation of Bacteria: Early Laboratory Studies

Microbiology requires the various species of bacteria be preserved for future studies. This has led to many stock culture collections. As a result, many early microbiologists looked for different ways to preserve their cultures, because periodic transfers of microbes on culture media requires an expenditure of time. Many of these studies (without knowing it) practiced starvation-survival as a means of preserving their stock cultures.

2.9.1 Bacteria in Water

There are quite a number of reports dealing with the preservation of microorganisms, especially plant pathogens, in distilled water (see Chapter 6). These studies are an extension of the starvation-survival problem in that there are nutrients added to the water. Because there is a lack of nutrients, preservation of some species of microorganisms occurs. In other words, we have a situation where starvation-survival has taken place. In the following studies reported, a colony is taken off the appropriate agar surface with a loop and then transferred to sterile distilled or tap water containing no amendments.

Kelman (1956) was probably the first to report the preservation of a microorganism, *Pseudomonas solanacearum,* in buffer, distilled, and tap water. After 220 days at 21°C in buffer, distilled, and tap water, less than 1% of the organisms remained viable. Nei (1977) reported that preservation of *Agrobacterium tumefaciens, Corynebacterium insidiosum, Pseudomonas syringae,* and *Xanthomonas albilineans* could be preserved in water. Many different species of yeast can be preserved in distilled water at room temperature and 5°C (Okuno and Kanai, 1981), but preservation was better at room temperature. The last test was after incubation for 48 months and it is not known how long yeasts will continue to survive in distilled water. Twelve different species of yeast were reported to survive over 48 months in distilled water (Emejuaiwe and Okafor, 1989).

Lynch (1990) starved *Enterobacter cloacae, Pseudomonas putida,* and *Azospirillum brasilenses* in sterile distilled water and found that the viability of the former dropped from only 10^8 to 10^6 in over 400 days and its cellular dimension did not change. *P. putida* and *A. brasilenses* were sustained for the 154 days.

The following are reports of bacterial growth in distilled water, seawater, or phosphate buffer. Unfortunately, growth is defined by the number of colony-forming units (CFUs) resulting from the incubation of the cells in distilled water, seawater, or phosphate buffer and not increase in biomass. Thus growth reported by the various scientists may be due to fragmentation, hence it is a manifestation of starvation-survival process (see below). I will retain the term "growth" in the studies reported.

Bacterial growth was shown to occur when washed cells of *Sarcina lutea, S. albus, S. aureus, Rhodococcus rosaceus, E. coli,* and *B. megatherium* were placed on a medium consisting of distilled water and 1.5% agar (pH 7.2) and incubated for 3 to 5 weeks (Burke, 1936). The water was distilled from a solution containing 200 mg of potassium permanganate and 750 mg of potassium hydroxide. Agar was thoroughly washed in distilled water and then dried before use. Variations in colony forms were noted with some species and pigmentation and in some cases were lost. Burke (1936) noted that double distilled water was very toxic and distilled water was slightly toxic to the organisms. Shearer (1917) found that physiological saline was more toxic than distilled water or 1.5% NaCl to meningococci and that calcium or potassium antagonized this toxicity. *B. coli* was found to survive longer in physiological saline than in distilled water (Winslow and Falk, 1923a,b). *B. coli* and *Bacterium typhosum* died more rapidly in distilled water or dilute buffer at elevated temperatures and high pH (Cohen, 1922). Washed buffered populations of *E. coli* maintained high viabilities at room temperature, compared with unwashed or unbuffered aqueous suspensions, and survival times were influenced by the prior growth conditions of their populations (Cook and Wills, 1958).

Growth of *E. coli* (24-hour culture grown on nutrient agar), washed three times with double distilled water, was noted in six different water-agar media in about 5 days (Burke and Taschner, 1936), but not in double distilled water-agar. The waters used to make up the different water-agars were river water filtered through filter paper, tap water from a deep well but stored in a reservoir, tap water filtered through a Berkefeld N filter, freshly distilled water from a Stokes steam still, tap water singly distilled with potassium permanganate and potassium hydroxide through a pyrex glass still, and double distilled water using potassium permanganate and potassium hydroxide in the still. Comparable experiments employing water without agar and making plate counts to determine growth indicate that *E. coli* was able to grow in filtered tap water and distilled water from a Stokes still. However, when placed in double distilled water, the cells were dead in 72 hours but no evidence was found that they died of starvation. In all flasks, growth took place when the oxygen consumption was 0.5087 ppm or more. Growth of coliform bacilli in various types of water was also noted by Bigger (1937), but no mention is made as to the organic content of these waters.

In distilled and double distilled water, the growth of bacteria was reported by Kalinenko (1957) when the flasks were cotton plugged and the energy for growth was absorbed from the atmosphere. Leifson (1962a,b) detected many gram-negative bacteria, including *P. aeruginosa,* in 36 samples of distilled water from ten different sources. The assumption was that these organisms were simply surviving. Favero et al. (1971) detected many gram-negative bacteria, including fecal coliforms and *P. aeruginosa,* during an environmental microbial contamination in the pediatric wards of two different hospitals' distilled water. The viable count in the distilled water from mist therapy unit reservoirs ranged from 1.7×10^1 to 2.7×10^7 (mean $= 2.4 \times 10^6$)/ml in one hospital whereas in the other, it ranged from 3.5×10^3 to 3.5×10^8 (mean $= 1.4 \times 10^7$) cells/ml with *P. aeruginosa* counts ranging from 1.5×10^0 to 4.3×10^4 (mean $= 7.5 \times 10^3$)/ml and 2.3×10^0 to 4.3×10^3 (mean $= 4.9 \times 10^2$/ml respectively. They found that naturally occurring *P. aeruginosa* reached a higher maximum cell population in poor quality distilled water that had been permitted to stand for a few days compared to fresh high quality distilled water. The survival and growth of naturally occurring and washed suspension of trypticase soy agar (TSA)-cultured *P. aeruginosa* were compared in mist therapy unit (MTU) distilled water and in buffered distilled water (BDW). After 72 hours, the TSA-cultured was not detectable, whereas the naturally occurring organisms actually increased in number and were still viable in high numbers after 168 hours. These investigators also noted that naturally occurring *P. aeruginosa* survive physical and chemical stresses better than cells of the same organism that have been subcultured on TSA.

The following organisms also have been shown to survive for a long period of time in sterile distilled water: *A. tumefaciens, Corynebacterium michiganense* pv. *insidiosum, C. michiganense* pv. *sepedonicum, Pseudomonas caryophylli, Pseudomonas marginalis* pv. *marginalis, Erwinia carotovora* subsp. *carotovora, X. albilineans, Xanthomonas campestris* pv. *oryzae, Xanthomonas campestris* pv. *campestris* (DeVay and Schnathorst, 1963; Tsuchiya and Wakimoto, 1971 [see Wakimoto et al., 1982], Perez and Cortes-Monllor, 1967; Wakimoto et al., 1982).

2.9.2 Phosphate Buffer

E. coli strain 28.D.10 and *P. aeruginosa* were found to be capable of growth in phosphate buffer (Garvie, 1955). These cells were first washed in phosphate buffer before inoculation into phosphate buffers (pH 7.0) made up with Analar chemicals and purified (recrystallized) Analar chemicals. When 60,000/ml *E. coli* cells were placed in the buffers and incubated 3 days, 330,000/ml were found in the phosphate buffer and 200,000/ml in the purified buffer. If 2,000,000/ml dead cells were placed in the buffers, then only 300,000 cells/ml were found in the regular buffer and 425,000 cells/ml in the purified buffer. More cells appeared in the purified buffer while there was a drop in the number of cells in the regular buffer. When 40,000/ml of *P. aeruginosa* cells were inoculated in the regular and purified buffers, 4,000,000/ml and 3,000,000/ml resulted. An increase in the number of bacteria when placed in buffers was also noted by Potter (1960).

King and Hurst (1963) presented data on the ability of some bacteria to survive in different diluents. Only one of the diluents contained added organic matter (0.1% peptone, w/v). In Table 2.1, it is interesting to note that some of the bacteria actually increased in counts during their presence in diluents that possessed no added organic matter. Is this increase in number due to fragmentation of the organism when placed in a starvation menstruum? The function of diluents is to enable a true assessment of the condition of the bacterial population and death and revival of the organisms should not take place during the dilution process (Jayne-Williams, 1963). Now, we must add that diluents may bring about an increase in the number of cells.

2.9.3 Seawater

Seawater, depending upon where it is taken, may have less organic matter than distilled water. The early studies did not place cultured bacteria into sterile seawater, but dealt with seawater samples taken and stored. If one seals a sample of seawater in a flask and leaves it for 2–5 days, tens and hundreds of thousands of bacteria per milliliter can be found. Storage of samples of water and mud is generally accompanied by a slight decrease in the number of bacteria, but is followed by a rapid increase (ZoBell, 1946). During the storage period, there is a detectable decrease in the number of species. According to Butterfield (1933), "the more interference with normal conditions, the greater the increase in bacterial numbers." The phenomenon of a sudden, rapid increase in heterotrophic bacteria in isolated samples of water or sediment has been noted by early researchers (Whipple, 1901; Fred et al., 1924; Gee, 1932; Waksman and Carey, 1935; ZoBell and Feltham, 1934; ZoBell and Anderson, 1936; Lloyd, 1937; Voroshilova and Dianova, 1937; Kriss and Markianovich, 1959). For example, Fred et al. (1924) observed an increase from 126 to 7,400,000 heterotrophic bacteria/ml in the water that had been kept in flasks. With the limited labile substrate, it is hard to account for these large increases in microbial populations when stored in flasks. Kuznetsov (1955, see Kriss et al., 1967) and Sorokin (1955, 1957) maintained that this large multiplication was due to chemoautotrophic bacteria but Kriss et al. (1967) pointed out that it was an erroneous assumption. In addition, one would not be able to see such a large increase in heterotrophic bacteria grow-

Table 2.1   The survival of four bacterial species in seven diluents.

Organism	Contact Time (min)	Tap Water	Charcoal-Treated Tap Water	Quarter-Strength Ringer's Solution	0–85% (w/v) NaCl	0–1% (w/v) $Na_2S_2O_3$ in Tap Water	Glass Distilled Water	0–1% (w/v) Peptone Water
					Mortality (%)[a] in			
Staphylococcus	30	40	36	66	38	+5	7	−36
aureus	60	70	36	99	67	−5	−3	−36
	90	80	38	99	95	−15	20	−36
	120	91	46	99	99	16	20	−39
Streptococcus	30	99	83	99	96	5	60	8
pyogenes	60	99	95	99	99	5	99	4
	90	99	99	99	99	5	99	4
	120	99	99	99	99	5	99	8
Escherichia	30	12	−9	−16	34	13	1	−7
coli	60	0	−5	1	98	2	−12	−3
	90	10	−26	25	94	0	−16	16
	120	10	−2	40	99	4	1	0
Salmonella	30	57	−5	−53	31	40	6	−3
typhimurium	60	60	−21	−43	58	34	30	−7
	90	80	−21	−30	84	40	36	
	120	70	−26	−20	91	14	30	

Source: Reprinted by permission from *Journal of Applied Bacteriology* 26:504–506, W. L. King and A. Hurst, 1963; copyright 1963, Academic Press, London.
[a] A minus sign indicates an increase in count.

ing on agar plates. Voroshilova and Dianova (1937) reported the large increases by plate count but could not verify this large increase by direct count. One wonders why the large increase when the amount of labile substrate is so limited and this large increase cannot be found by the direct count method. The probable answer to these questions are (1) the bacteria use all the labile compounds up immediately and then fragment to form a large number of cells, and (2) the reason the large numbers of bacteria cannot be seen by direct count is that the size is too small and, therefore, near the limits of the resolving power of the light microscope. This is also reflected in the research by Novitsky and Morita (1977), who demonstrated that 10 times the concentration of amino acids that occur naturally in seawater does not stop the fragmentation process when cells are starved. If one takes into consideration the increase of cell numbers when bacteria are placed in distilled water or buffers, one is faced with the dilemma of trying to determine if it is growth or a manifestation of the starvation-survival process. The Russians have given much thought to this problem of a large increase in numbers when natural waters are stored in flasks (Kriss et al., 1967). Is growth defined as an increase in the number of cells, an increase in biomass, or should some other definition be created? It is difficult to define growth in microbes.

The so-called growth of bacteria in distilled water, buffers, or menstruums without energy may be a manifestation of starvation-survival. It is not an increase in biomass but an increase in the number of cells. Nevertheless, when water samples are stored, it is difficult to distinguish if growth results from the bottle (surface and confinement effect), which is also discussed later in section 4.2.1.

2.10 Long-Term Survival of Bacteria in the Presence of Organic Matter

Survival also occurs in the presence of energy. It is discussed here in terms of starvation because of the possibility that the specific organic substrate(s) in the organic matter for the organism(s) in question have been utilized. If this be the case, then starvation can take place. In the natural environment this situation is real because syntrophy is the rule. However, in the laboratory other factors (change in pH, build up of metabolic endproducts, etc.) can bring about cession of growth.

In 1939, Steinhaus and Birkeland noted that *Sarcina lutea* and *Serratia marcescens,* when grown in nutrient liquid medium, did not display a log death phase. Instead, the organisms survived 2 years at concentrations of approximately 10^6 cells/ml. There was a drop of approximately one log from the stationary phase to the period they called the senescent phase. They were the first to suggest that if the cells were grown on a fermentable sugar, the cells did not survive long periods of time. Gay (1936) indicated that cells could carry on metabolic functions but were unable to reproduce in the senescent phase. Previously, Sherman and Albus (1923) noted that physiologically young cells (log growth cells) were more sensitive to adverse conditions (alkaline reactions, accumulation of waste products, and other detrimental environmental conditions) than old cells.

Rhizobium spp. have been reported to survive as long as 15–16 years in cultural media (Fred et al., 1932, in Jensen, 1961; Allen and Allen, 1958, in Jensen, 1961). Cultures of *Rhizobium meliloti* remained viable for 30 to 45 years when placed in a medium of loamy garden soil plus 0.5% mannitol and some calcium carbonate and placed in the dark at room temperature (Jensen, 1961). Nutman (1946) found that *Rhizobium trifolii* frequently produced nodulation-ineffective variants when grown in sterile soil. Jensen (1961) tested his old cultures (age range from 30 to 45 years, the majority being 30- to 40-year-old cultures) and all formed normal nodules with red pigment. However, one isolate (strain L.4, age 44 years), was ineffective immediately after inoculation, but was effective a couple of months later. According to Jensen (1961), "sterilized soil as culture medium thus seems to keep vegetative cells of *R. meliloti* alive and effective in a desiccated condition for periods approaching half a century; moreover, it preserves for more than 10 years the viability of several other microorganisms that have been claimed to form endospores." It appears that *Rhizobium* spp. can survive in nature for a very long time without its legume counterpart (Lowendorf, 1983). From soils that had wheat grown on them for 125 years, Nutman (1969) was able to demonstrate the presence of *Rhizobium leguminosarum* (28,000/gram dry soil; all effective nitrogen fixers), *R. trifolii, R. meliloti,* and *Rhizobium lupini.*

Viable bacteria, yeast, and molds were found in a 50-year bottle (labeled "Rising-up Yeast, Dauaerhefe and Levure Inalterable") left at New Camp Evans, Ross Island in the Antarctic by the Captain Robert Falcon Scott Expedition (Meyer et al., 1962). No mention was made as to whether it was frozen all the time or not.

P. aeruginosa (26 strains) stabbed into a nutrient agar column (salt free), obtained by B. Lanyi, survived for 14 years (Liu, 1984). These cultures had not dried out but were not overlaid with oil. He also reported that Dr. Lanyi had some cultures of *Proteus* spp. that are alive after 30 years in nutrient agar. English and McManus (1985) reported the longevity for *P. aeruginosa* to be 30 years in a medium that contained salt. Cash (1985) contended that salt does not aid survivability of microbes. Palleroni (1985) states that Lanyi's medium does contain salt. Survivability, according to Calhoun (1985), may result from the deprivation of a carbon source after endogenous reserves are depleted, thus, no high level of metabolic by-products accumulate. Bass (see Cash, 1985) believed that one of the conditions for long-term storage is the lack of any fermentable carbohydrates. Unfortunately, much of the foregoing is a matter of conjecture and no experimental approach has been made.

Palleroni (1985) was able to recover *Pseudomonas saccharophila* from liquid minimal medium cultures kept at room temperature that had dried. E. G. D. Murray sealed a large number of agar slant cultures of enteric bacteria between 1917 and 1954; some were tested for viability by his son (Murray, 1985). He found that some tubes of agar slants inoculated with *Shigella* between 1917 and 1924 and opened in 1967 were still viable and possessed the same phenotypic properties as when they were inoculated. The National Collection of Type Cultures (United Kingdom) determined the overall viability tested to be 89.2% (*Salmonella,* 95.2%; *Shigella,* 83.5%; *Proteus,* 90.5%; *Escherichia,* 91.7%; *Klebsiella,* 97%; and *"Bacillus alkalescens,"* 66.6%). For 106 cultures inoculated in 1918–1919 (61 to 62 years old), the overall viability was 70%. Unfortunately, there is no mention under what conditions the organisms were stored. Survival of the bacteria may have been due to a paucity of specific nutrients as proposed by Cohen and Barner (1954).

Battley (1987) cited some unpublished data from the Laboratorium voor Microbiologie of the Technische Hoogeschool in Delft, Holland. According to

J. van der Toorn (see Battley, 1987), either C. B. Van Niel or L. E. den Dooren de Kluyver in 1925–1926 inoculated luminescent bacteria into a series of peptone or fish extract water contained in glass ampoules. They were permitted to incubate for 2 days before the ampoules were hermetically sealed and kept at room temperature, which fluctuated considerably during winter and summer. When van der Toorn, Curator of the Culture Collection, opened the ampoules in 1979 he inoculated them into Difco Photobacterium broth. None of the cultures inoculated into the fish extract water grew. One culture in peptone water, *Photobacterium splendidum* isolated by M. W. Beijerinck in 1914, did grow at room temperature. During incubation for 2 days prior to being sealed, aerobic oxidation of the substrates took place and after being sealed, facultative growth also occurred. However, Battley (1987) surmised that, over a short period of time, the energy present must have been exhausted. Thus, the metabolic activities slowed down until the organism came into some state of energetic equilibrium with the environment within the ampoule and then stopped. When the organism was placed in new medium, the cellular machinery returned to an active metabolic state. Bacteria that are starved for a long time will immediately take up an introduced energy source (Novitsky and Morita, 1977).

Goldstein (1991) recovered *E. cloacae* from a well-preserved American mastodon (radiocarbon dated to 10,860 years ± 70 years B.P.). A discrete, cylindrical mass of plant material was found in association with the ribcage and interpreted to be intestinal contents of the mastodon. No *E. cloacae* was detected in a dozen soil samples near the mastodon bones (Folger, 1992). This could also be a starvation-survival situation since *E. cloacae* probably completely utilized its substrates but could not produce the necessary enzymes to degrade the mastodon's diet. *E. cloacae,* therefore, must rely on syntrophy for most of its energy sources when living in the gut of the mastodon. According to the paleontologists, the diet consisted of twigs and needles from spruce trees; but, in the case of this mastodon, its last meal consisted of wetland grasses, water lilies, and pondweed, due to the shrinking forest (Folger, 1992).

Although the above data concern long-term survival in the media, one hypothesizes the organism's growth mechanisms are shut off, providing another example of starvation-survival. The question that remains is how much of the increase in cell numbers is the result of the starvation process versus absorption of organic matter from the air.

2.11 Longevity of the Microbial Cell

The question of the longevity of bacteria cell begs for an answer. De Ley et al. (1966), citing the data of Abelson (1959) and Vallentyne (1965), who showed experimentally that serine and threonine decompose within 10^4 to 10^6 years, supported Sneath's (1964) conclusion that recovery of autochthonous bacteria from old strata is highly unlikely. Very recently, Poinar et al. (1996) brought up the question of amino acid racemization, where the L-amino acids racemize to the D form. Racemization depends on the presence of water, the temperature, and the chelation of certain metals to proteins (Bada, 1985) and the rate of DNA depurination is also dependent on the same factors (Poinar et al., 1996). The activation energy and rate constants over a wide temperature range at neutral pH for aspartic acid are similar to those of DNA depurination (Lindahl and Nyberg, 1972; Lindahl and Andersson, 1972). Because of these similarities, Poinar et al. (1996) determined that racemization of amino acids, especially aspartic acid, is a useful indicator of the extent of DNA degradation in ancient specimens. When the D/L aspartic acid ratio was lower than 0.08, no DNA sequences could be found in ancient samples. From the table presented by Poinar et al. (1996), the oldest sample with a higher D/L aspartic acid ratio than 0.08 was 20×10^3 years ago. An exception to the racemization of amino acids occurring in some insects preserved in amber may be due to anhydrous conditions (Poinar et al., 1996).

How does one resolve the difference between the longevity of microbes with the natural depurination of DNA and the racemization of amino acids? Just how much racemization of amino acids and depurination of DNA can a microbe withstand before it expires? Have the microbes found in ancient material evolved mechanisms to get around the natural rate of depurination of DNA and racemization of amino acids? One exception noted by Poinar et al. (1996) was the research of Cano and Boruchi (1995). What other scenarios could possibily permit microbes to exist in ancient material? All the microbes isolated from ancient material have organic matter present, albeit in extremely low concentration. Because they are living, they might use this organic matter at a very low rate for metabolism and/or growth. These rates would not be measurable under laboratory conditions and essentially are in the starvation state. In the case of microbes in solar salt, the organic matter is being concentrated along with numerous species of bacteria. Thus, cryptic growth and use of the organic matter might occur extremely slowly. Microbes in marine sediment are at low temperature, which decreases the depurination of DNA and racemization of amino acids quite sharply. Microbes in rocks (also at low temperature) may receive added organic matter seeping through the rock structure in addition to what is in the rock structure itself. The presence of metals (e.g., magnesium) also decreases the depurination of DNA and racemization of amino acids. In some cases, the cell may be dehydrated or the water in the cell is not free to act since it may be "bound" water. Low temperature, the presence of metals, and dehydration may act synergistically to help prevent the expiration of microbes from racemization of their cellular amino acids (in proteins) and the depurination of their DNA, but eventually, microbes will expire. Because there is a small amount of organic matter in rocks, salts, soil, seawater, etc., the organisms may metabolize extremely slowly. This metabolic rate would be just sufficient to take care of the racemization of amino acids and the depurination of their DNA. In other words, this metabolic rate would not be measurable by our current techniques.

2.12 Conclusions

Although one can question the ability of bacteria to survive in ancient material, the many publications indicate that it is possible. The evidence being accumulated by a large number of investigators in subsurface microbiology indicates that the presence of bacteria in ancient material is within the realm of possibility. Although one can never rule out contamination, it is difficult to determine when and how the contamination occurred. When high counts of bacteria are recovered from freshly extracted materials, contamination can be ruled out. Organic matter is present in all ecological materials and its utilization by microbes may not be measurable. Gaseous organics can diffuse into various environmental niches at an extremely slow rate so that the bacteria can also utilize it at an immeasurable rate, a situation analogous to that which occurs with the bacterial spore. Each publication must be judged on its own scientific merit. The question that remains is how long can a specific bacterium survive in various environmental conditions, since latent life may occur for at least 1000 years (Sneath, 1962)? The modern techniques employed by Parkes et al. (1990, 1993, 1994) and the investigators of subsurface microbiology strongly suggest that anabiosis in nature does take place, but for how long a period is still a microbiological mystery. One should never underestimate the resilience of the microbe.

3

Bioavailability of Organic Matter in the Environment

There are two limiting strategies for microorganisms to cope with the inevitability of Malthusian catastrophe. One is to consume resources rapidly and profligately and then find a new habitat. Such organisms need to become resistant to starvation and capable of rapid growth and dispersal of propagules to search for new habitat whether it be a newly fallen log or a bowl of milk. The other strategy is to adapt to habitat in which there is a continual renewal of resources and to use them efficiently, even when they are available at very low concentration. Such organisms need to have effective transport systems, large surface to volume ratios, and low maintenance energies.

Koch, 1979

3.1 Introduction

All organisms need energy and ecosystems are mainly oligotrophic. The quest for energy (food) in higher forms determines the survival of the species. Territorial instinct is another means by which many higher forms ensure themselves sufficient energy. Although microbiologists rarely address survival of the fittest or survival of the species, both take place in the microbial world and should also be addressed in terms of the energy available to microbes.

Before the evolution of green plant photosynthesis occurred, the heterotrophic bacteria must have experienced periods in which organic matter was in short supply. Organic matter produced by photosynthesis is the major source of organic matter for all ecosystems and this organic matter is mainly polymeric. The spore formers developed a mechanism to survive long periods of time without any energy source. Microbes have evolved longer than any other life forms, so in all probability, the non-spore-forming heterotrophic bacteria must have also developed mechanism(s) to survive long time periods when no energy and/or nutrients were available. Thus, the concept of starvation-survival is ecologically and evolutionarily important for microorganisms.

Although one can measure the organic matter in various environments, what you measure may not be bioavailable to microbes. The organic matter content is generally low, but to make matters worse, much of it is in the form of resistant molecules or complexed to resistant molecules or to clay minerals. Microbes are the reason why most ecosystems are oligotrophic. Thus, one is faced with the fact that many of the indigenous microbes are in the starvation mode. In any ecosystem, the bioavailability of organic matter for utilization by bacteria depends on various enzymes produced by the community of bacteria. Most of the organic matter in nature, other than the readily labile compounds such as free amino acids, free carbohydrates, and free fatty acids, is polymeric and high in molecular weight. Yet the term "free" may not be free but may be an artifact due to the chemical method of analysis, or these compounds may be complexed or adsorbed to clay or humic substances, respectively, rendering them unavailable. Microbiologists should not consider all the monomolecular compounds in environmental samples to be labile. Polymeric substances must first be enzymatically degraded before they can be transported into the bacterial cell and used as a source of energy. The synthesis of enzymes requires much energy on the part of the cell (see Chapter 5 and Chróst, 1991, for review of enzymes in aquatic environments). Where does the energy come from in an oligotrophic environment? It is recognized that there is a slow degradation of the polymeric and high molecular weight substances in nature, but not all of the various physiological types of microbes in the community are participating at any given time. Thus, syntrophy plays a key role in the activity of the microbial community and the diversity of different microorganisms creates different substrates. Certain bacteria while waiting for their substrate(s) to become available are in varying degrees in the starvation mode. As mentioned previously (Chapter 1.6, "Concepts and Definitions"), there are degrees of starvation.

In this chapter, I document, in sufficient detail, the dearth of organic matter in the major ecosystems for heterotrophic bacteria. With the dearth of organic matter, one begins to realize that the majority of ecosystems are oligotrophic and that starvation is the way of life for microbes in nature. In addition, I also point out the mechanisms that make the organic matter unavailable to microbes. Unfortunately, many microbiologists do not recognize the complexity and dearth of the organic matter in ecosystems, let alone the various components of the organic matter. Generally, most microbial ecologists do not even provide any data on the organic matter present in the ecosystem they are studying or they assume that the organic matter measured by chemical analysis is completely bioavailable to the microbes.

Chemical analysis of the organic carbon in any environmental sample certainly does not determine what portion of the organic carbon is actually available for use by the organisms present. When one discusses the organic matter in any ecosystem, one must remember that a distinction must be made between the living and dead (chemical analysis of soil and water samples include the microbes as part of the organic matter), dissolved and particulate, as well as a separation between labile and refractory organic matter. A labile compound is one that can be metabolized in a fairly short time, whereas a refractory compound exhibits resistance to breakdown, is usually complex, and has a large molecular weight. Labile organic compounds are not easily identified—mainly because they constitute a very small part of the total organic matter in the natural environment. Labile (bioavailable) forms of organic matter make up only a small part of the total organic matter in ecosystems (Kriss et al., 1967). Often, these labile compounds are removed as fast as supplied (Iturriaga and Zsolnay, 1981; Fuhrman, 1987). Some labile organic compounds, such as amino acids, can be measured at low levels but their flux cannot be directly related to the total turnover of labile substrates because simultaneous fluxes of other unmeasured organic compounds are not known (Gardner et al., 1989).

Flux data are often difficult to interpret because they involve conversion factors that may vary with location. Obtaining flux data may also involve physical manipulation such as perturbation by filtration (with the release of dissolved organic matter) (Fuhrman and Bell, 1985). To complicate matters, growth efficiencies are usually not accurately known and may vary within and among investigations (Gardner et al., 1989). Gardner et al. (1989) state that the relative degree of substrate limitation can be approximated by comparing the heterotrophic potential of microbes to ambient mineralization rates. Unfortunately, many investigators, especially those who employ models, assume that the total organic matter measured can be completely used within a reasonable time by the organisms. A low-nutrient environment measured as the standing crop of organic matter (OM), should be measured as the flux of nutrients across the ecosystem (Hirsch et al., 1979).

There is a difference between bioavailable and biodegradable. For example, biodegradable compounds are those that are fully, but very slowly, degraded and can be considered highly resistant to biodegradation (Pitter and Chudoba, 1990). Pitter and Chudoba (1990) classify biodegradable compounds into three categories: readily degradable, inherently (potentially) biodegradable, and persistent, resistant, refractory, and recalcitrant compounds. These latter compounds need not be "nondegradable" but their degradation requires prolonged acclimation, appropriate selection of microorganisms, and other suitable parameters. Unfortunately, the categories of biodegradable compounds depend greatly on the investigator and we are not consistent. On the other hand, bioavailable compounds that may be readily available to microbes in the laboratory may not be available in nature due to their adsorption or their being complexed to humic acid, phenols, and particles, or the physical environmental conditions may not be right. "What a species can do in a culture medium is not necessarily what it is doing in its natural habitat" (Brock, 1971). This becomes especially true when substrates are not available. In addition, Brock (1971) pointed out that most microbes live in microenvironments and that there is a large difference between the macroenvironment and the microenvironment and, thus, it is virtually impossible to measure microbial growth rates in nature. As a result, there are many different microenvironments in a macroenvironment. Do we destroy these microenvironments when we sample natural environments? Yes.

The utilization of organic matter in the environment depends on an interactive community of bacteria since there is a myriad of different organic compounds, each requiring different enzymes; no one bacterium is capable of producing all these different enzymes. Enzymes of anaerobes and aerobes must be examined because anaerobic conditions can occur as microniches in aerobic environments. Thus, consortia play an important part in the utilization of organic matter in nature.

Most ecosystems have gradients of bioavailable carbon with the highest availability near the input and the lowest levels of readily degradable organic carbon in areas further, in time and space, from the primary producers. This concept was formalized by Cummins (1974) and Cummins et al. (1984). The same situation has been observed vertically in marine water columns starting within the euphotic zone with a reasonable organic C content and ending in sediments of deep ocean basins with the least organic C (Menzel and Ryther, 1968). The same principle has been suggested for soils with the continuum starting in the litter layer and extending into the lower soil strata (Waring and Schlesinger, 1985). There are a few exceptions where there is a constant recharge of new carbon from surface aquifers, but the organic matter present in deep sediments represents the extreme end of the carbon quality continuum.

Substrates in ecosystems are discontinuous in time and space so that microbes must possess mechanism(s) for obtaining new substrates once existing ones are depleted. If no new utilizable substrates are found, then the microbes go into the starvation-survival state or die. Thus, it is possible for more than one species to occupy the same niche if their unique substrates are separated in time.

In order to make a comparison of the amount of organic carbon in various ecosystems and laboratory media, nutrient broth has 3400 mg/l of dissolved organic carbon (DOC) (Yanagita et al., 1978); whereas various environments have only a few mg/l in aquatic environments or a few mg/g in soil (see various figures and tables in this chapter). In reading the remainder of this chapter, a comparison should be kept in mind at all times between the oligotrophic environment's organic matter content and nutrient broth. The reader should also bear in mind that much of the organic matter present in any ecosystem cannot be used as an energy source for the heterotrophic bacteria. There are many research papers, reviews, and books dealing with the organic matter in soil and aquatic environments (e.g., Hood, 1970; Wangersky, 1965; Wagner, 1969; Duursma and Dawson, 1981; Joint and Morris, 1982; Thurman, 1985; Stevenson, 1986; Tate, 1987; Frimmel and Christman, 1988; Engel and Macko, 1993), hence the organic matter is addressed mainly in terms of its dearth in the various ecosystems and its recalcitrant nature to heterotrophic microorganisms. Bioavailability of organic matter is addressed mainly in terms of experimental time for its degradation. It is essential for microbiologists to know about the amount of organic matter in ecosystems, their bioavailability, their recalcitrant nature, etc., in order to grasp the true nature of the oligotrophic environment and why starvation is the main mode of life in these oligotrophic environments for microbes and many other higher forms.

3.2 Organic Matter in Soil

The major environmental stresses on bacteria that occur in soil include *unavailability and low rate of supply of nutrients,* unavailability of water or oxygen, and the development of acidity and toxic compounds. Readily utilizable carbon is apparently severely limited in soil (Clark, 1965; Lockwood, 1981; Lockwood and Filonow, 1981). The primary source of organic matter in soil originates from plants, but much of this organic matter is decomposed in the litter zone and not in the soil. Plants produce approximately 1 kg/m^2/y (dry weight) and most terrestrial and freshwater communities have similar primary production capabilities (McFadyen, 1970). Grazing by macroorganisms (cattle, worms, etc.) of plant material, along with microbial decomposition, can then be important in limiting primary production. Besides bacteria, fungi are the important colonizers of fresh newly fallen organic debris (Harley, 1971). The microbial biomass in the top 10 cm of grassland soil is estimated to be 70 g/m^2 (Clark and Paul, 1970). It is essential to have some knowledge of the stability of this microbial biomass in soil in order to assess its contribution to soil organic matter formation and nutrient cycling. Plant residues, on an average, contain 15% to 60% cellulose, 10% to 30% hemicellulose, 5% to 30% lignin, and 2% to 15% protein, plus minor amounts of phenols, sugars, and amino acids (Haidler, 1992) and therefore contribute signnificantly to the complex carbon in soils. Biomass C may also be a source of readily available C, especially when some of the biomass is dead (Soulides and Allison, 1961; Jenkinson, 1966a, 1976; Jenkinson and Powelson, 1976; Draycott and Last, 1971; Shields et al., 1973; Marumoto et al., 1977, 1982a,b; Brookes et al., 1984).

In soil, the scarcity of food, or lack of a suitable and available energy source, is the principal factor limiting bacterial growth (Clark, 1967). As a result, any addition of fresh energy-yielding material to soil will almost invariably elicit an increase in bacterial activity. Increase of soil biomass is theoretically controlled by the amount of C supply above that required for maintenance energy (Smith and Paul, 1988). Plant litter is undoubtedly a major source of energy input to soil in some ecosystems, but the litter layer is distinct from the soil system. The amount of litter in relation to its accumulation varies considerably in different climatic zones. Much of the litter is decomposed in the litter layer, hence the amount of readily utilizable energy that seeps into the soil is probably very little and is mainly humic material. For example, in a grassland prairie soil, the energy input was sufficient to allow bacteria to divide only a few times a year (Babiuk and Paul, 1970). The rhizosphere is not considered under this section, mainly because it is known that living roots release into the soil a variety of low molecular weight substances, including simple sugars, oligosaccharides, and amino acids (Sparling et al., 1982). All of these compounds have a great impact on the growth of organisms in the rhizosphere, but their presence cannot be demonstrated unless the system is bacteria-free (Hunt et al., 1985). This latter situation only demonstrates the fierce competition for substrate by the bacteria living in the rhizosphere. In addition to the availability of energy as organic carbon, nitrogen and phosphate may also be possible limiting factors for heterotrophic production in soil.

The energy budget of soil provides a solid basis for the concept that fungi and other microorganisms in soil exist in an environment characterized by insufficient resources for sustained growth (Lockwood, 1981). A change of 1 μg biomass C/g soil is equivalent to about 10^5–10^6 bacterial cells (Sparling et al., 1981). The main contribution of organic matter in soil to soil fertility is its capacity to supply inorganic nutrients for plant growth (Duxbury et al., 1989). The stability of organic matter in soil results from protective processes (complexing with clays, humus, etc.) occurring within soil rather than from the microbial creation through degradation of recalcitrant chemical structure (Duxbury et al., 1989).

Organic matter, mainly polymeric, in soils varies widely. The soil organic matter, is 10% to 20% is carbohydrates, primarily of microbial origin; 20% is nitrogen-containing constituents, such as amino sugars and amino acids; 10% to 20% is aliphatic fatty acids, alkanes, etc.; and the rest is aromatic carbon (Paul and Clark, 1989). Clark and Paul (1970) and Paul (1970) divide the organic input in soils into three divisions: (1) rapidly decomposing plant residues and the soil biomass, with a relatively short half-life of a year or less, (2) microbial metabolites and cell wall materials with a half-life of 5 to 25 years, and (3) the humic substances with much longer half-lives as measured by radiocarbon dating.

In measuring organic matter in soil, the various methods give different results. Extraction techniques are important in the identification of the various components of the soil organic matter. The 1 N sulfuric acid extract method gives the highest C values (500–1700 mg C/kg of soil); whereas the substrate-induced respiration method gives the lowest (1–5 mg glucose equivalent/kg soil). The anthrone method and a total organic-C analysis of extracts indicated that only a small portion of C was in carbohydrates (Sikora and McCoy, 1990). Thus, the organic content of soil depends on the method of analysis.

3.2.1 Refractive Nature and Bioavailability of Organic Matter

The main mechanism for the formation of refractory carbon compounds in soil is the degradation of plant material. It has been shown that bacteria can form refractory dissolved organic matter while in the process of degradating labile organic compounds. Refractory dissolved organic matter must also be considered in the total overall carbon cycle, for it may make more refractory material from simpler, bioutilizable compounds. Even in the older literature, there is ample evidence that proteins, polysaccharides, and related compounds combine with mineral soil colloids and lignins to form complexes and that complexing increases resistance of the compounds to microbial attack and enzyme action (Basaraba and Starkey, 1966).

Coal and oil, including methane, are examples of organic matter that essentially have been bypassed by the carbon cycle due to their accumulation under conditions that preclude microbial decomposition processes. It is well known that many organic compounds are found in nature that have resisted microbial degradation (peat, petroleum, amino acids in shale, etc.). The persistence of pesticides, common polymers, and paleobiochemicals in various environments is discussed by Alexander (1965, 1973) and attention should be directed to the persistence of paleobiochemicals and not the pesticides or common polymers. Among the most persistent components of natural organic residues in soil are phenols, lignins, and waxes.

Burns (1983) divides the soil organic matter into three physically and chemically discernible fractions: (1) a physically recognizable component, containing recently incorporated animal, plant, and microbial debris, in the early stages of disintegration and decay, (2) a biochemically recognizable, largely soluble and labile component (e.g., carbohydrates, peptides, organic acids) arising from the degradation of material contained in fraction 1, and (3) a brown polymeric component, largely phenolic but also containing polysaccharides and complexed and stable peptides, amino acids, etc. Fraction 3, the humic fraction of soil, is traditionally divided into humic acids, fulvic acids, and humins on the basis of their solubility in aqueous acids and bases. The humic polymers have a tertiary structure that permits them to expand upon hydration, thereby exposing an extensive internal surface area that may have significant influence on microbe-enzyme-substrate interactions. The half-lives of fraction 3 can be measured in terms of centuries. Organic material capable of supporting microbial growth occurs in a variety of forms and in different relationships with the soil fabric. These forms and relationships determine the availability of substrates, either directly to bacteria or to the extracellular enzymes. Thus, Burns (1983) points out that organic material may be a very suitable substrate in vitro but not in vivo. Labile substances, readily available in laboratory culture, may be recalcitrant in soil. This may be due to adsorption on clay particles (Pinck et al., 1954), formation of metal complexes (Alexander, 1965), or protection by recalcitrant substances such as tannins (Basaraba and Starkey, 1966). When labile organic matter is adsorbed or complexed to various material, it is possible that its reactive sites with the necessary enzymes are no longer available for reactivity. The mechanisms of adsorption include ion exchange, polyvalent ion bridging, van der Waals forces, hydrogen bonding, and complex formation (Theng, 1979; Burchill et al., 1981).

Approximately 10% of the organic matter in soil is carbohydrates. Acid releases many types of monosaccharides from soil. Hexoses, pentoses, methylpentoses, uronic acids, and hexosamines make up most of the monosaccharides in soil. Yet monosaccharides are generally very labile when microorganisms are present. In his study of the origin and stability of soil polysaccharides, Cheshire (1977) concluded that the stability of soil polysaccharides in their native state is not related to their chemical composition but to their unavailability. This inaccessibility is caused within undecomposed biological residues whose insolubility results in absorption on clay, the formation of metal complexes, or tanning of soil humic substrates. Hydrolytic enzymes may also be inhibited by tanning and metal complexes. Once the polysaccharide is removed from soil, it is water soluble and readily available to microorganisms and its inaccessibility while in soil is due to chemical combination with other soil constituents, or adsorption onto particle surfaces that are not accessible to water (Cheshire, 1977).

In the laboratory, the carbon becomes stabilized against degradation when 30% or less of the original carbon remains (McGill et al., 1973). Refractory organic matter in soils is old, and the age depends on the location, type of soil, etc. The mean residence time of C in soil is initially short, but approaches that of natural humus C within a relatively short time. The mean residence time of stable humus varies from several hundred to somewhat over a thousand years (Stevenson, 1982). The ages of the organic matter in various soils, obtained by radioactive dating, is reviewed by Stout et al. (1981). In general, the organic content of soil decreases with depth and the age increases with depth. The age of the organic matter in the O horizon is approximately 0–100 years, in the A horizon it is 100–500 years, and in the B horizon it is 500–1000 years (Thurman, 1985).

When [14]C- and ^{15}N-labeled plant materials were added to virgin and cultivated soil and permitted to undergo biological transformation for up to 3 years, the synthesized microbial metabolites resulting from the decomposition process were more resistant to decomposition than a significant portion of the initially added plant material (Shields and Paul, 1973; McGill et al., 1973). When ^{14}C-labeled barley straw was incubated in sandy soil under field conditions, 16% of the labeled C remained after 8 years and 9.3% remained after 20 years, making the half-life about 15 years (Sørensen, 1987). On the other hand, the soil contained 1.9% native C in organic matter and the rate of decomposition has a half-life of 91 years (Sørensen, 1986). An average of 15% of the labeled C in amino acids remained almost constant throughout the study. Laboratory experiments during a 3-month period increased the rate of decay, as measured by CO_2 production, by a factor of 1.2 and that of native organic matter by a factor of 4.3. When [14]C- and ^{15}N-labeled legume (*Medicago littoralis*) was added to soil, about one third of the legume ^{14}C disappeared after 4 weeks' decomposition and after 4 years about 18% of input plant ^{14}C remained and about 10% after 16 years (Ladd et al., 1985). Each type of soil appears to differ in the amount of labeled microbial biomass retained, but, nevertheless, after 4 + years, it still remains. This also indicates that microbial cells are surviving in soil. The turnover rates of ^{14}C-labeled plant material have been extensively investigated and reviewed by Führ and Sauerbeck (1968), Oberlander and Roth (1968), Jenkinson (1971), and Shields and Paul (1973). These studies reveal a very rapid loss (ca. 65%) of ^{14}C in the first year, followed by a less rapid rate in subsequent years, with about one-fifth of the added carbon remaining after 5 years. The proportion of added C retained by the soil is largely independent of the plant material and climatic conditions.

Intact plant globulin lost 45% of its carbon as $C^{14}O_2$ when incubated in soil for 30 days, but the stabilization of plant protein by sorption on soil colloids does occur (Simonart and Mayaudon, 1961). Under the same conditions, globulin hydrolysates lost 71% of their carbon as $C^{14}O_2$.

When ^{14}C-labeled substrates were incubated in soil, glucose was most readily decomposed, followed by hemicellulose, cellulose, maize straw, and barley straw, respectively (Sørensen, 1963). In the first 10 days of incubation 60% of the glucose-C left the soil as CO_2 compared to only 23% of the barley-C. The humified material, resulting from the addition of the various materials, remained in the soils after 3 months and decayed at almost the same rate, regardless of the origin of the organic matter (glucose, hemicellulose, cellulose, or straw) and the half-life of the labeled C ranged from 5 to 7 years, depending on the type of soil. The half-life of the native soil C estimated from CO_2 evolution ranged from 13 to 29 years. When glucose was employed, the amino acid C tended to be the largest and the amino acid C increased with the clay content of the soils. The influence of added organic residues on the rate of decomposition of native soil organic matter is not of short duration but apparently continues as long as the residue constitutes an important part of the decomposing mass (Hallum and Bartholomew, 1953; Broadbent, 1964).

When organic matter is added to any ecosystem, it serves as an energy source and activates a complex array of microorganisms. It should be emphasized that the utilization of organic matter in nature by one species alone is extremely rare; instead a consortia of microorganisms are required. The microorganisms use the added C and either respire it to CO_2 or transform it into microbial tissue, excrete it as metabolites, or find it resistant to decomposition. The resistant components of the added organic matter, a portion of the newly synthesized microbial tissue, and the stabilized metabolites contribute to the organic carbon in the ecosystem. According to Jenkinson (1971), the proportion of added plant carbon retained in the soil under different climatic conditions, using different plant materials and soils, is remarkably similar (excluding very acid soils) in that one third of the added plant carbon remains after 1 year, falling to about one-fifth after 5 years. Cellulose is the most abundant biogenic material on earth, probably followed by chitin. Definitely, not all microorganisms have the ability to utilize these forms of carbon energy, thus syntrophy must play a great role in microbial ecology. Even when readily decomposable substrates such as glucose, hemicellulose, or cellulose are decomposed in soil, the residual carbon is distributed in a pattern more similar to that of soil organic nitrogen and that of soil organic carbon (Mayaudon and Simonart, 1963, 1965). In fact, some of the glucose ends up in the humus fraction of the soil (Mayaudon and Simonart, 1958).

Decomposition rate constants per day for unprotected and protected decomposable organic matter were calculated by Paul and Voroney (1980) and found to be 8×10^{-2} and 8×10^{-4}, respectively. The equivalent rates for recalcitrant organic matter were 8×10^{-6} and 8×10^{-8}. The amount present in undisturbed (uncultivated) soil is governed by soil-forming factors of age (time), parent material, topography, vegetation, and climate (Stevenson, 1986). Organic matter entering the soil should be split into a stable fraction (ca. 40%) and a labile fraction (60%) (Sauerbeck and Gonzalez, 1977; Ladd et al., 1981, 1983, 1985). The residence time of organic matter in soil is initially short but approaches that of the native humus within a relatively short time and the residence time of stable humus varies from several hundred to over 1000 years (Campbell et al. 1967; Stevenson, 1978). Certain paleobiochemicals such as fatty acids or alkanes in shale are known to persist for 3.5×10^8 years (Alexander, 1973).

There is no correlation between microbial numbers in soil and activity as determined by the evolution of CO_2 (Colbert et al., 1993a). Nor is there a correlation between CO_2 production and soil organic matter content (Stotzky, 1965). When sodium salicylate (30 to 1500 µg/g soil) was added to soils, the metabolic activities of *Pseudomonas putida* PpG7 (natural host of plasmid NAH7, which encodes enzymes necessary for the utilization of salicylate in two operons) peaked between 18 and 42 hours with a concomitant population increase of approximately 10^1- to 10^5-fold. This peak population was maintained for the duration of the experiments (5 to 7 days). This elevated population of PpG7 thus represents inactive cells (Colbert et al., 1993b). Because soil is generally carbon limited, periods of great activity and competition of PpG7 will only take place when nutrients are available. The same holds true for other organisms in any ecosystem. For example, Jansson (1960) reported that over time, labeled microbial tissue decomposed at a remarkably decreasing rate. Humification of the mycelium of *Aspergillus* resulted in greater stability in soil than cell walls of *Azotobacter* (Mayaudon and Simonart, 1963). According to Parkinson et al. (1978), Anderson and Domsch (1975), and Bewley and Parkinson (1985), fungi comprise about 80% of the total biomass while bacteria comprise 20%, indicating fungi are more slowly biodegraded. There appears to be a regular sequence of fungal species during decomposition of organic matter. It is generally accepted that the rate of organic matter decomposition with attendant nutrient cycling is determined by the physicochemical nature of the organic matter (resource quality and availability), abiotic factors (temperature and moisture availability), and the available decomposer community (Parkinson et al., 1978).

Although many investigators have discussed the concept of a microbial microenvironment (niche), the microbial activity predominates on or in close proximity to the surfaces of soil particles, or within the lattice structure of silicate minerals. Stotzky (1974) points out that all compounds, both organic and inorganic, needed for nutrition of microbes are present in soil, albeit often in forms that are not readily available to them. Soil is a complicated three-phase (gas, liquid, solid) system presenting many chances for spatial and temporal variations in nutrient concentrations and environmental conditions and all microorganisms present in soil are not equal in their biomass, ATP content, metabolic activities, etc. (Williams, 1985). The aqueous phase surrounding soil particles in unsaturated soils is normally discontinuous and this restricts the movement of microbes and other nonfilamentous forms. This restriction results in local accumulation of nutrients and toxicants and allows cells to escape from grazing predators (Stotzky, 1986).

3.2.2 Bioavailability of Humus and Humus Complexes

Humus serves as a reservoir of N, P, and S for higher plants, it improves soil structure, drainage, and aeration, and increases water-holding capacity, buffering, and exchange capacity, but it is a poor source of energy for the growth and development of microorganisms (Stevenson, 1978). Thus, it is a blessing in disguise that microbes are not readily able to utilize humus. The average humus residence time (time the molecule remains) is 1000 years (Campbell et al., 1967; Barber and Lynch, 1977). Griffin and Roth (1979) argue that humic materials provide sources of energy for microbial growth in soil, but the question is to what degree. Since oligotrophic bacteria are supposed to grow in low concentrations of organic matter, are they capable of utilizing humus? Nikitin and Chumakov (1986) could not find evidence that oligotrophs used humic acid, fulvic acid, or their monomers and appear not to have any exohydrolytic activity. Apparently high molecular weight inorganic polyphosphate can be used in place of ATP in some reactions by oligotrophs. The oligotrophs appear to be able to use gases as well as C_1 compounds and could be classified as oligocarbophilic.

The amount of humus on land is much greater than that of all other organic matter combined. Humic material does not accumulate even though it is recalcitrant to microbial degradation. It is decomposed over time. Humic substances are a category of naturally occurring, biogenic, heterogenous organic substances that can generally be characterized as being yellow to black, of high molecular weight, and refractory to microbial decomposition. Three major fractions are defined in terms of their solubility: humin (not soluble in water at any pH), humic acid (not soluble in water under acidic conditions below pH 2.0, but becomes soluble at greater pH values), and fulvic acid (soluble in water under all pH conditions). These humic acid type substances appear as polymers of phenolic and other aromatic units complexed with amino acid compounds, polysaccharides, and possibly other organic constituents of biological origin (Broadbent, 1964; Burges et al., 1964; Kononova, 1966; Ladd and Brisbane, 1967; Ladd and Butler, 1966; Mathur, 1971; Stevenson and Butler, 1969; Swincer et al., 1968). Of the amino acids in Podzol illuvial soil 80% is present in the NaOH-insoluble residue plus the purified fulvic acid and the NH_4OH eluate (Sowden and Schnitzer, 1967); whereas, between 80% and 95% of the amino acids in volcanic soils was accounted for in the humic materials plus NaOH-insoluble organic residues (Sowden et al., 1976).

Humic materials have a large number of surface functional groups, which include carboxyl, carbonyl, amino, imidazole, phenylhydroxyl, sulphydryl, and sulphonic groups. In well-oxidized environments, the carbonyl, carboxyl, and phenylhydroxyl groups are the most significant, binding protons with increasing degrees of relative stability (Stevenson, 1982). Humic substances are known to be extremely resistant to microbial decomposition and this stability may be due either to toxic moieties in the humic structure, rearrangement of the organic matter making it nondegradable by the available enzymes, or the fact that utilizable organic matter is protected deep within the humic structure, the outer surfaces of which have been degraded and become resistant. The variability of its structure requires a variety of enzymatic activities to depolymerize or catabolize humic acid, and, it is extremely expensive in terms of energy for the microbes to synthesize all the necessary enzymes (Tate, 1987).

Some enzymes resist destruction when complexed with humic substances. Physical entrapment (Burns et al., 1972b), ionic, covalent, and hydrogen bonding (Ladd and Butler, 1975), and copolymerization with humic molecules (Rowell et al., 1973) have been suggested as the mechanisms for humic-enzyme complexes. Copolymerization with humic molecules is supported by the studies of amino acid-, peptide-, and protein-humic complexes (Verma et al., 1975). It is known that soil organic matter, especially humic acid fraction, can sorb certain constituents of both plants (Oberländer and Roth, 1968; Sauerbeck and Führ, 1968) and microorganisms (Mayaudon and Simonart, 1965). On the other hand, the occurrence of active, stable enzyme-humus complexes in soil has been demonstrated (Chalvignac and Mayaudon, 1971; Burns et al. 1972a,b; Ladd and Butler, 1975; Rowell et al., 1973). Enzyme activity in soil is hindered by limiting and unevenly distributed substrate, suboptimal and fluctuating pH, stationary incubation, and fluctuating temperature. Certain enzymes retain their catalytic activity when adsorbed to clay surfaces. Stability of soil enzymes is believed to be a function of their adsorbed state on clays (McLaren, 1973b), which, in turn, makes certain organic substrates unavailable for bacteria.

Serban and Nissenbaum (1986) suggested that humus-immobilized enzymes seem to retain their activity more consistently than those bound to clay and might be responsible for the persistence of peroxidase-like activity in sediments 7 million years old. Enzyme activities, produced mainly by microbes in soil, are faced with limiting and unevenly distributed substrate, suboptimal and fluctuating pH, stationary incubation, and fluctuating temperature. Certain enzymes retain their catalytic activity when adsorbed to clay surfaces. The proteolytic enzyme, pronase, is inhibited by soil humic acid (Ladd and Brisbane, 1967). The protective effect has been attributed to the inhibition through tanning of hydrolyzing enzymes and also to a direct effect involving the polysaccharide. For more information concerning the complexing of organic matter to soil as well as organic matter complexes in soil, consult Coleman et al. (1989).

To analyze soil for humus, soil chemists separate the organic matter from the clay matrix and sesquioxides by treatment with sodium hydroxide or sodium pyrophosphate. That portion not dispersible by the peptonizing action of sodium, the chelating action of pyrophosphate, and the hydrogen-bond-breaking activity of alkaline pH values is known as humin. The dispersable material precipitated at acidic values is known as humic acids and the material that stays in solution is fulvic acids (Paul and Clark, 1989). This treatment of soil is drastic and it certainly does not permit one to determine the available carbon for microbial activity, mainly because the soil-organic complex has been destroyed.

Bacteria in soil are the principal agents producing humus and its resistance to attack may result from its adsorption to clay minerals (Russell, 1950). The role of microorganisms in the formation of humic acids, especially the microbial transformation of phenolic constituents in plants and the microbial biosynthesis of phenolic compounds from aliphatic precursors and their transformation into humic acid are reviewed by Haider and Martin (1968). For further information on humus, consult Stevenson (1982, 1986).

3.2.3 Bioavailability of Clay-Humus Complexes

Soil structure and texture also have an effect on the activity and dynamic of the soil microbial population (Van Veen and Van Elsas, 1986). For instance, Roper and Marshall (1974) demonstrated that clay minerals, particularly montmorillonite, may even form clay "envelopes" around bacteria. Previously, Marshall (1964) had shown that colloidal clay permitted root nodule bacteria in the predominantly sandy soils of Western Australia to survive intense desiccation. For *Rhizobium trifolii,* immunofluorescent counts remained constant through a 60-day period in the presence of bentonite but decreased significantly in the control without bentonite (Van Veen and Van Elsas, 1986). The protective effect of clay minerals, as well as humic material, on microorganisms against the abiotic stresses of desiccation, heating, and heavy metals resides mainly on four characteristics of colloidal clay and humic material: (1) their high surface to volume ratio, (2) their cation exchange capacity, (3) their ability to retain water, and (4) their ionic properties. Thus, one is forced to consider the surface interactions between clay minerals, microorganisms, and organic material.

Clays, which are negatively charged, adsorb nutrients and bacteria do not have the ability to metabolize adsorbed nutrients (Dashman and Stotzky, 1986). High concentrations of clays may also interfere with oxygen diffusion by increasing the viscosity of a solution, thereby having a negative effect on aerobic microbial respiration (Stotzky, 1986). Adsorption of nutrients to soil does occur and it depends on the initial solution pH, not on the ionic strength, but the predominant mechanism was hydrophobic bonding (Weber et al., 1988). Weber et al. (1988) also found that the DOC adsorption increased with soil profile depth. This phenomenon is demonstrated in adsorption experiments with microorganisms, substrates, and soil components (Burns, 1983).

In clay-humate complexes (Fig. 3.1), the presence of many types of bonds, the three-dimensional configuration, the aromaticity, and the production of free radicals all slow decomposition rates. In addition the complexes can tie up peptides, sugars, etc. Thus, the complexes permit humus to persist and accumulate in aerobic surface soils for hundreds or thousands of years (Paul, 1992), making it the dominant form of organic carbon in nature. Clay-humate complexes are important phenomena in stabilizing the humic and fulvic acids. Guckert et al., (1975), Alexander (1965), and Stotzky (1986) provide a review that details adsorption, complexing, and encrustation as a means of protecting the organic molecules from degradation. Proteins, and a variety of polysaccharides (including cellulose and pectins), amino acids, ammonia, sugar phosphates, and phosphorus- and nitrogen-containing compounds, can be readily adsorbed or fixed on the external surface of clay particles, or within the lattice structure of silicate minerals.

Between 52% to 98% of the soil carbon is bound to clay minerals (Greenland, 1971). Polysaccharides (cellulose, starch, pectins, glycogens, alginates, dextrans) from different sources are adsorbed. Montmorillonite has a protective effect against the decomposition of polysaccharides (Lynch and Cotnoir, 1956). Some of the polysaccharides become bound within the interlamellar spaces of smectite clay, making them inaccessible to extracellular enzymes and removing them from direct contact with microorganisms (Burns, 1983). This inaccessibility has been studied by Burns and Audus (1970). X-ray diffraction analyses demonstrated that carbohydrate materials were adsorbed between the interplanar spacings of the montmorillonite clay, whereas infrared spectra analysis suggests that hydrogen bonding plays a role in the adsorptive mechanism between clay and carbohydrates (Lynch and Cotnoir, 1956). When relatively pure compounds such as glucose and cellulose are added to soil, some of their carbon enters the humic fraction (Sørensen, 1963).

In the study of the oligotrophic nature of the environment one must consider the bioavailability of the complexes of protein-clay, peptide-clay, amino acid-clay, etc. Clay was found to protect organic matter from microbial decay (Cabrera and Kissel, 1988). According to Sørensen (1981), the effect of clay in increasing the

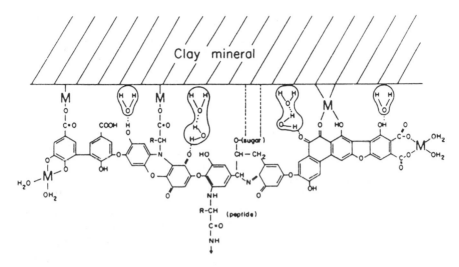

Figure 3.1 Schematic diagram of a clay-humate complex in soil. Reprinted by permission from *Micronutrients in Agriculture*, J. J. Mortvedt, P. M. Giordano, and W. L. Lindsay (eds.), F. J. Stevenson and M. S. Ardakani, pp. 79–114, 1972; copyright 1972, American Society for Agronomy.

content of organic matter in soil is possibly caused by newly synthesized matter, extracellular metabolites, as well as cellular material, forming biostable complexes and aggregates with clay; hence, the higher the concentration of clay, the more readily the interactions take place. Clay minerals apparently exert their primary influence on microbial activity by modifying the physicochemical characteristics of microbial habitats, and this either enhances or attenuates the growth and metabolic activities of individual microbial populations. This in turn influences the growth and activities of other populations (Stotzky, 1971, 1974, 1980; Filip, 1973, 1975; Hattori, 1973; Hattori and Hattori, 1976; Stotzky and Burns, 1982).

Many of the more labile compounds (such as polysaccharides, amino acids, etc.) may be complexed to lignins and phenolic compounds or absorbed to clay minerals (Stotzky, 1986). Hydrolytic enzymes may be inhibited by tanning and metal complexes (Cheshire, 1977), whereas polysaccharide complexes with Fe, Cu, and Zn have been shown to protect some polysaccharides from decomposition (Martin et al., 1966). These metals may inhibit enzymes or affect the solubility of the polysaccharides. Baker et al. (1967) concluded that the polysaccharides were present as insoluble salts in the Ca^{2+} and Mg^{2+} form or bonded to such humic acid salts. Likewise, in very dry soil, the polysaccharides are not available because water is not present for their solution. Even a change in pH may make organic matter inaccessible to microorganisms (Stewart and Wetzel, 1981).

According to Stotzky (1986), ". . . one of the assumptions is that in soils and other environments with low levels of organic substrates, particulates—especially surface-active ones, such as clay minerals—concentrate these substrates at the solid-liquid interface, where they presumably reach levels high enough to support the growth of microbes which adhere to these surfaces in response to the nutrient enrichment." The evidence available does not support this assumption. Samples of mined montmorillonite, attapulgite (polygorskite), and kaolinite that contain 0.24%, 0.24%, and 0.15% organic carbon (Stotzky, 1986) did not support microbial growth in the absence of exogenous substrates (Stotzky and Rem, 1966; Stotzky, 1974; Dashman and Stotzky, 1986). When proteins, peptides, amino acids, polysaccharides, nucleic acids, and nucleotides have been adsorbed to clay minerals, their bioavailability has been reduced. On the other hand, enhancement has also been noted. This subject was reviewed by Stotzky (1986). Thus, a variety of additions to soil may not be able to support microbial growth.

Adsorption has been implicated in stimulation as well as retardation of the degradation rates of organic matter (protein, dextran, diquat, alkylamine, and 2,4 D) in various ecosystems (ZoBell, 1943; Eastermann and McLaren, 1959; Eastermann et al., 1959; Vallentyne, 1962; Olness and Clapp, 1972; Gerard and Stotzky, 1975; Wszolek and Alexander, 1979; Gordon et al., 1983; Marshman and Marshall, 1981; Orgam et al., 1985; Stotzky, 1980; Subba-Rao and Alexander, 1982; and Weber and Colbe, 1968). For some compounds, adsorption decreases only the rate of biodegradation while, in other cases, adsorbed molecules become totally protected from biodegradation (Marshman and Marshall, 1981; Weber and Colbe, 1968). Adsorption has been postulated to be the mechanism for the preservation of organic matter in soil and sediment (Vallentyne, 1962). When clay minerals are present, the distinction between adsorption and intercalation (insertion between clay particles) cannot be made (Gordon and Millero, 1985). Intercalation can also make organic matter unavailable to microbes. Marshman and Marshall (1981) suggested that clays such as montmorillonite could affect microbial growth through interactions with the organisms, their substrates, individual enzymes, and growth factors. Because complexes between different amino acids and peptides and montmorillonite are differentially available to microorganisms, cysteine was not utilized when bound to montmorillonite or kaolinite, whereas proline and arginine were (Dashman and Stotzky, 1986).

Thus, the presence of clay makes much of the organic matter intrinsically recalcitrant. If organic matter is intrinsically recalcitrant, the soil becomes more oligotrophic and the starvation-survival process comes into play. For further information concerning the clay-protein interactions, consult Theng (1979) and Stotzky (1986).

3.2.4 Bioavailability of Clay-Protein Complexes

Enzymes, especially exoenzymes, are necessary to act on organic matter polymers so that cells can use the organic matter as a source of energy. Enzymes are known to be adsorbed to anionic clay particles in the soil as well as to be intimately associated with soil colloidal organic matter and humic materials (Burns, 1983). This property is due to three important properties of soil colloids as listed by Burns (1983): (1) They have high surface to volume ratios and, when separated from whole soil, tend to have a significant proportion of the total microflora associated with them, (2) they have ionic properties such that, as a general rule, cations are attracted to surfaces and anions are repulsed, and (3) they have a high affinity for water molecules. As a result, microorganisms, substrates, enzymes, metabolites, and inorganic nutrients tend to be concentrated at clay-water and humic-water interfaces. If the right substrates, enzymes, microorganisms, etc. are present, microbial growth will be enhanced in this microniche. McLaren (1960) concluded that most enzymes are complexed, or immobilized, with clay minerals and/or humus. However, some of the nitrogen is present in the form of active cell-free enzymes (Skujins, 1976). These cell-free enzymes are very stable in soil (Mulvaney and Bremner, 1981). For example, dead cells with membrane-bound systems of enzymes can carry out nitrification (Vela, 1974). Many carbon and nitrogen hydrolases that are cytoplasmic, such as ß-glucosidases, ß-galactosidases, and urease, still remain functional in dead cells, cell debris, and in the soil aqueous phase (Burns, 1983).

Eastermann et al. (1959) investigated microbial activity on solid-liquid interfaces. They found that monolayers of lysozyme adsorbed on kaolinite and bentonite could be digested by chymotrypsin, mixed soil cultures, and pure cultures of *Bacillus subtilis, Bacillus mycoides,* and *Pseudomonas* sp. On the other hand, lignin-protein complexes were extremely resistant to decomposition. Montmorillonite-protein monolayer and kaolinite-protein formation retard the rate of enzymatic proteolysis, but most could be hydrolyzed under optimum conditions in one day. Dried, rewetted montmorillonite-protein complexes are more resistant to decomposition than freshly prepared complexes. Silica-protein complexes are comparable to lignin-protein complexes.

Many enzymes persist as active moieties in soil for very long periods of time and have been detected in soils stored for decades (Skujins and McLaren, 1968, 1969). From the study of Burns et al. (1972a,b), it was found that the indigenous urease was resistant to proteolysis by pronase; this resistance was attributed to its association with the soil organic matter and not with the clay colloids, whereas urease added to soil may be subjected to pronase activity. The addition of lignin to a bentonite-urease complex affords protection to urease from pronase attack (Eastermann et al., 1959). McLaren (1963) suggests that the enzymes are situated within the organic matter per se. These enzymes could be liberated during decomposition of the plant roots, microorganisms, etc., and become internally complexed with organic matter during humification (Kononova, 1966).

Catalase and pepsin, as protein substrates, can be utilized by microbes but when complexed to clays they are not utilized. On the other hand casein, chymotrypsin, and lactoglobulin are utilized to a lesser extent when not complexed (Stotzky and Burns, 1982). The binding may result in a structural change of the protein (enzyme), resulting in less activity, depending on the protein (enzyme).

An unusual ecological role for the stable colloid-enzyme has been proposed by Burns (1980, 1982). In this scenario, the stable enzyme complex serves as a detector of the microorganism's substrate within a microhabitat, freeing the microorganism from the need to continually produce extracellular enzyme. In addition, the free enzyme would have a poor chance to persist for sufficient time to react with a proper substrate. The organism remains enzymologically inactive until such time as it receives a signal (i.e., product released by accumulated enzyme activity) that induces the microorganism to produce additional enzymes of the same type to degrade the exogenous substrate. Thus, the initial step(s) in the extracellular degradation of certain substrates may involve a process independent of microbial growth and may be induced by the activity of enzymes contained within the humic-enzyme complexes with subsequent degradation occurring through direct microbial activity. This would then be an energy-efficient strategy in the utilization of exogenous substrates in an oligotrophic environment. Without the proper enzyme(s) function, complex substrates cannot be hydrolyzed to simpler compounds for use by the microbes as an energy source.

Amino acids are an excellent source of energy for microbes but 40 to 50% of the amino acids are absorbed to the inorganic colloids (McGill and Paul, 1976). As a result, microbes have a difficult time using adsorbed amino acids. Amino acid utilization depends on the amino acid–clay complex in question (Stotzky and Burns, 1982). For example, cysteine or aspartic acids when complexed are not utilized, whereas complexed arginine and proline are. Again, the microbes are faced with less energy than would be indicated by a chemical analysis due to the formation of the clay-protein complex. Peptide binding also occurs (Stotzky, 1980; Stotzky and Burns, 1982). Nearly all the noncellular soil proteins are associated with clay minerals and humus fractions, with more than 90% of the soil nitrogen in organic form in the surface layers (Boyd and Mortland, 1990). Up to 50% of the nitrogen can be identified as amino acids derived from proteins. This acellular proteinaceous material in soil is resistant to decomposition and this resistance to decomposition is demonstrated by the fact that less than 5% of it is decomposed annually (Stevenson, 1982).

The silt plus clay fraction in soil ensures stabilization of amino acid metabolites produced during the time of intense biological activity brought about by the addition of decomposable energy sources; the amount of stabilization increased with increasing concentrations of silt plus clay, but during the latter stages of decomposition the stabilization was independent of the silt plus clay concentration (Sø-rensen, 1975). A half-life of about 8 years for amino acid metabolites formed in a loam soil during decomposition of ^{14}C-labeled cellulose was noted by Sørensen and Paul (1971).

Hedges and Hare (1987) studied the amino acid adsorption (15 amino acids at a concentration of approximately 10 μM in distilled water) onto organic-free kaolinite and montmorillonite clay minerals (1 wt% suspension). The systems came into a steady state within 2 hours at room temperature. Of the basic (positively charged) amino acids, 40 to 80% were strongly adsorbed by both clay minerals. Of the neutral (uncharged amino acids) 10 to 15% were taken up by montmorillonite, but little, if any (<5%), by kaolinite. Of the acidic negatively charged amino acids, 20 to 35% were adsorbed only by kaolinite. The adsorption patterns appeared to be related in part to electrostatic interactions between the clay mineral surfaces and the different types of amino acids. When microbiologists add radioactive compounds to an environmental sample to determine microbial activity, the radioactive compound is in the free state. The radioactive compound(s) should be allowed to equilibrate with a sterile environmental sample and then be added to the test sample in future studies.

Protons generated by metabolism could displace the adsorbed organic matter on clay. However, since most ecosystems are poor in readily available energy this does not take place easily. During low rates of metabolism, very few protons are generated and hence low growth rates are dependent on protons displacing the adsorbed material.

It is assumed that the bound amino acids in the soil are in the form of peptides or proteins (Bremmer, 1965). Sowden (1969) demonstrated that the mixture of peptides and amino acids released during partial acid hydrolysis of soil is similar to the mixture released when proteinaceous material is treated in the same way. Amino acids in soil have existed for a long time but the nature of the compounds in which they are located is unknown. The presence of these amino acids was demonstrated in a complex set of experiments by Sørensen (1967). It was postulated that the amino acids could be in intact living or dead cells (Jenkinson, 1966b) or that they could be free proteins or that peptides are established in the soils in such a way that decomposition by soil microorganisms is hindered or retarded. Clay minerals can be a protective mechanism for amino acids. When ^{14}C-labeled barley is added to soil, the ^{14}C-labeled amino acids are liberated during decomposition because of their reaction with soil organic matter (Sørensen, 1963).

Verma et al. (1975) found the linkage of proteins, peptides, and individual amino acids in a model phenolic polymer was slightly more susceptible to degradation than were the aromatic portions. They also found that all amino acid units of 2- and 3-unit peptides linked to form model humic acid-type polymers were readily stable to biological degradation in soil and in pure culture; the N-terminal unit, which was linked to the phenolic moiety, was little more stable than the COOH-terminal amino acid unit. Dissolved free amino acids (DFAA) is a misnomer in the microbiological sense because the amino acids are not freely available to microbes.

The half-life of ^{14}C–amino acids, produced from the transformation of added ^{14}C-acetate, exceeded 2000 days in soil, but when ^{14}C-amino acids were added directly to soil, their half-life was only a few days (Sørensen and Paul, 1971). This finding suggests that the turnover of organic materials in the soil is not only controlled by their chemical nature, but also by their physical location. ^{14}C-labeled metabolic products are retained with increasing clay content (Sørensen, 1975, 1981), which suggests that the effect of clays is to form clay-organic matter complexes with microbial metabolites. Sørensen (1967) observed in laboratory experiments that the half-life of amino acids metabolites formed in a loam soil during the decomposition of ^{14}C-labeled cellulose was about 8 years. Later, Sørensen and Paul (1971) calculated the half-lives of the metabolites left in soil after 90 days to be 800 days for carbohydrates, 1500 days for the insoluble residue, and 1600 days for amino acids. Although amino acids may be present in soil, the clay fraction of soil complexes with the amino acids making the amino acids unavailable to the microorganisms. This is another reason why bacteria enter the starvation-survival mode.

For a further review on this subject, consult Burns (1978, 1983).

3.2.5 Effect of Tannins

Plant tannins, water soluble phenolic compounds having molecular weights between 500 and 3000, have the ability to precipitate alkaloids, gelatin, and other proteins (Bate-Smith and Swain, 1962). Tannins in soil inactivate dehydrogenases, decarboxylases, amylase, invertase, cellulase, pectolytic enzymes, phosphatases, ß-glucosidase, aldolase, oxidase, polyphenoloxidase, catalase, lipase, urease, pepsin, trypsin, and chymotrypsin (Benoit and Starkey, 1968), but other enzymes are not affected because the specificity of the tannin is due to the peptide bond (Rhoades and Cates, 1976; Feeny, 1976). Tanning reactions can also preserve polysaccharides and these tanning reactions, favored at low pH, occur between polyphenols, proteins, or other polymers such as the cellulose of pectin (Swain, 1965). The preservative effects of tannins on polysaccharides added to soil has been studied by Benoit and Starkey (1968) and Griffiths and Burns (1972). The ability of soil bacteria to obtain energy from their immediate environment is hindered by the tannin-inactivated enzymes. Again, we have another reason why organic matter is rendered unavailable to bacteria.

For more information concerning tannins in soil, consult Zucker (1983).

3.2.6 Effect of Lignins

Lignins are present in soil as part of the organic carbon. Lignin interacts with proteinaceous enzymes and other enzymes by its reaction with the polyphenols, thus inhibiting further action with the substrate(s). Lignin, a product of dehydrogenative polymerization of cinnamyl alcohols, does not appear to be enzymatically controlled (Zeikus, 1981). Perhaps it is not readily degraded because no enzymes are involved. Fungi are responsible for most of the lignin decomposition (Bremner, 1955; Rashid, 1972).

Protein in soil is protected from decomposition by the protein-lignin complex (Lynch and Lynch, 1958), because lignin retards the action of the proteolytic enzymes. Although there are lignin-clay complexes that help account for the enzymic resistance, the protein-lignin complex resistance to decomposition persists in soils that have low clay content (Lynch and Lynch, 1958). Lignin, a product of dehydrogenative polymerization of cinnamyl alcohols does not appear to be an enzymatically controlled process (Zeikus, 1981).

Once again it can been seen that organic matter is protected from microbial activity. In this case, the protein is not available as a source of nitrogen and energy for the metabolisms of the indigenous soil population. It is a blessing in disguise because the above mechanisms protect the organic matter in soil from complete decomposition. Complete decomposition by microorganisms in the soil would be detrimental to plant life, including the water-holding capacity of soil. The non-availability of organic matter for microbes helps initiate the starvation-survival process in soil.

3.3   Organic Matter in Subsurface Environments

Over time, interactions between different gradients in soil provide fundamentally different environments for microbial growth but the most distinct factor is the lack of organic matter in the deeper strata of sediments. Leenheer et al. (1974) examined the DOC in groundwater from 100 sites in 27 states and found that the DOC concentrations ranged from less than 0.1 mg/l (limit of detection) to 15 mg/l. They also found that the DOC concentration was directly correlated with specific conductance and alkalinity, but not with pH. The DOC mean concentration was 1.2 mg/l but the DOC median value was lower at 0.7 mg/l. The difference in DOC concentration for various types of consolidated rock aquifers was slight (ranging from 0.5 mg/l to 0.7 mg/l for sandstone, limestone, and crystalline rock aquifers). No difference was found in the medium DOC concentration between shallow (<61 m) and deep (>61 m) sand and gravel aquifers.

Surface water, as it filters through soil and rock strata, undergoes changes in dissolved and particulate organic carbon as well as in its inorganic constituents due to cation exchange, adsorption onto soil and rock surfaces, and the metabolic activity of the indigenous microflora. Thus, polluted water may become purified provided it moves through a sufficient depth (vadose zone, or zone of aeration) and does not encounter cracks and fissures, provided that the vadose zone is not already saturated with water (Ehrlich, 1990). Thus, subsurface water contains very little organic matter for heterotrophic bacteria to use as an energy source. In addition, the measured DOC may be complexed, making the amount of DOC less bioavailable to microbes. Thus, microbes in the subsurface environment are mainly in the starvation-survival state. For further details, consult Chapelle (1993).

3.4 Organic Matter in Aquatic Environments

For aquatic systems, the organic matter is divided into the DOC and the particulate organic carbon (POC). By definition, the DOC is the organic carbon that has the ability to pass though a Whatman GF/C glass-fiber filter, whereas the POC is retained on the filter. The DOC is determined by oxidation to carbon dioxide and measuring the carbon dioxide by infrared spectrometry (Van Hall et al. 1963; Menzal and Vaccaro, 1964). The total organic carbon (TOC) is the sum of the POC and DOC. The chemists generally employ the Whatman GF/C (retention capacity of 1.2 μm) but others may use smaller pore size filters such as the Whatman GF/F glass-fiber filter (retention capacity = 0.7 μm). Carlson et al. (1985a) determined the molecular weight distribution of dissolved organic matter in seawater by ultrafiltration and found that only 34% had a molecular weight greater than 1000 while 6% had a molecular weight greater than 30,000, hence most had a molecular weight less than 1000. However, one has to question whether the term "dissolved" is actually dissolved, especially when Koike et al. (1990) noted the presence of submicrometer particles (0.38 to 1 μm) that were nonliving and composed of organic matter.

Approximately 20% of the bacteria in water from Yaquina Bay, Oregon, will pass through a GF/C filter (Lott and Morita, unpublished data). With the retention of bacteria on the GF/C filter, part of the POC is the result of the oxidation of microbial carbon. It is now known that DOC passing through a GF/F filter contains bacteria, eukaryotic microorganisms (Fuhrman and McManus, 1984; Chisholm et al., 1988), viruses (Borsheim et al., 1990), and submicron detrital particles (Koike et al., 1990). As a result, part of the DOC is also the result of the oxidation of microbial carbon to carbon dioxide. Toggweiler (1990) discusses which pore size filter should be used to distinguish between POC and DOC in reference to the results obtained by Koike et al. (1990). Because there is a lack of debris in deep water to help hold back the bacteria on the filter, a greater percentage of the bacterial population will pass through the filter. Now that we know that many bacteria are smaller than 0.2 μm, one has to wonder just how much of the DOC comes from microbial cells. This, then, makes the earlier measurements of the DOC too large due to including microbes that pass through the Whatman GF/C filter in the measurement. To microbiologists, cells and viruses that pass through a GF/C filter are not dissolved organic matter. Thus, POC and DOC are operational definitions.

Figure 3.2 shows the "average" concentration of POC and DOC in various systems. Note that the amount of POC is much smaller than the DOC. Seawater has the lowest DOC median concentration of 0.5 mg/l, whereas groundwater has a median concentration of 0.7 mg/l. None of these values comes close to the DOC of nutrient broth. These figures definitely establish the oligotrophic nature of aquatic environments. With so many microbes competing for this scarce commodity, part of the microbial population with the ability to utilize the POC and DOC are on a starvation diet. Another part of the microbial population cannot compete for this energy source and end up in a starvation-survival state.

Detritus itself can serve as a microhabitat, hosting redox gradients, altered microbial metabolism, and associated biochemical nutrient transformation processes both qualitatively and quantitatively (Paerl, 1984).

3.4.1 Organic Matter in Freshwater Environments

The organic content of many lakes in America was analyzed by Birge and Juday (1934). They found that the total dissolved and suspended organic matter ranged from 1 to 26 mg C/l, with an average of 1.6 mg C for the POM (particulate organic matter) and 15.24 mg C/l for the DOM (dissolved organic matter). The average DOC in North American lakes is approximately 15 mg C/l (Kuznetsov et al., 1979). The flux of DOC in eutrophic lakes is about 5 mg C/l (similar to the value for seawater) and about 0.1 mg C/l for oligotrophic lakes. The average organic carbon in rivers is 7.0 mg/l (Thurman, 1985).

Although measurements of POC and DOC are routinely made in the study of microbial ecology, these values do not provide the reader a measure of how much of the POC or DOC is bioavailable to the microbial population. Furthermore, each species has its own spectrum of nutrients as well as other physical and chemical parameters. Organic matter in the POC and DOC are so chemically complex that no general statement can be made about any fraction of these organic carbon compounds. What may be bioavailable to one species may not be to another species under identical physical and chemical conditions. Biochemical oxygen demand (BOD) is probably one of the best methods to measure the bioavailability of the indigenous POC and DOC to the indigenous microbial population, but this method does lack sensitivity.

Van der Kooij et al. (1982) proposed a method in which the water samples are sterilized at 60°C for 30 minutes, inoculated with a pure strain of bacteria (*Pseudomonas fluorescens* or *Spirillum* sp. NOX) and the bacterial growth monitored by plate counts. After 4 to 8 days of incubation, the maximum number of bacteria is taken as an index of assimilable organic carbon and this index can be converted to a carbon equivalent by reference to calibration obtained by measuring the

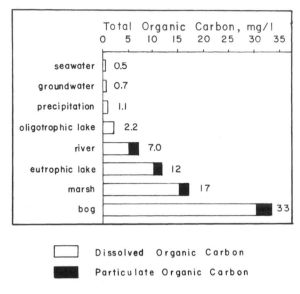

Figure 3.2 Approximate concentrations of dissolved and particulate organic carbon in natural waters. Redrawn and printed by permission from *Organic Geochemistry of Natural Waters*, E. M. Thurman, 1985; copyright 1985, Martinus Nijhoff/Dr. W. Junk Publisher. Reprinted by permission of Kluwer Academic Publishers.

growth of the same strain on a range of acetate concentration. Unfortunately, this method does not give the amount of organic matter available to a community of microbes in a sample. The use of heat in this procedure may cause unavailable organic matter to become available organic matter. The method proposed by Servais et al. (1987) also uses heat, but the sample is inoculated with a natural community of bacteria. The method was modified by Servais et al. (1989) in which the water (200 ml) sample was sterilized by filtration through a 0.2-µm Nuclepore membrane and inoculated with 2 ml of the autochthonous bacteria. These samples were incubated at 20°C for 4 weeks in the dark. The bioavailable DOC was oxidized by the UV-promoted persulfate oxidation to carbon dioxide, and the carbon dioxide measured by infrared spectrometry at time 0 and after 2 weeks of incubation of the organic matter. The value obtained for Seine river water upstream from Paris was approximately 0.70 mg/liter of bioavailable DOM; whereas in three Belgian rivers the values varied from 0.7 to 6.1 mg/liter (Servais et al., 1989). The bioavailable DOC was thus calculated as the difference between the mean values of the initial and final DOC. For further details, consult Servais et al. (1989).

The flux of the DOM in streams and lakes is controlled by the bacterial utilization rates of organic compounds (Wetzel, 1971). When dry hickory and maple leaves were added to a recirculating stream, enclosed in a nylon mesh bag for 30 hours, the DOC levels increased 10-fold (Wetzel and Manny, 1972). The bacterial population also developed rapidly, decomposing the labile organic carbon and nitrogen compounds in the leaf leachate within 72 hours. Bacteriologically labile and refractory DOC fractions, with $T_{1/2}$ decomposition rates of 2 and 80 days respectively, were present in the leaf leachate. The refractory dissolved organic nitrogen (DON) persisted unmodified at least for 24 days.

The quantity and composition of DOC released by phytoplankton vary with time and the changing algal population (Chróst, 1981). The DOC is dominated by MW (molecular weight) fractions less than 500, MW fractions 10,000–30,000, and MW fractions greater than 300,000; bacteria can utilize a significant portion of this released DOC, especially the fraction less than MW 10,000.

BIOAVAILABILITY OF ORGANIC MATTER

Dissolved organic matter is considered to be "microbially refractory" or "relatively microbially refractory," but, according to Lock and Ford (1986), there is no definition of the term refractory nor is the statement usually supported by references to specific studies. But "refractility" is time related in that even refractile molecules such as lignins may be ultimately degradable by microorganisms and their energy content usable over prolonged periods (Newell, 1987). Organic detritus is a complex mixture of labile, semirefractile, and refractile molecules and the overall decay rate, while initially fast, decreases exponentially with time as the labile components are more rapidly metabolized and removed (Godschalk and Wetzel, 1978). The amount of DOC in relation to depth is given in Table 3.1 for Lake Mergozzo (Ishida et al., 1977). This table includes low molecular weight fractions less than 10,000 with bacterial counts and the effect of the microorganisms on the DOC at two different depths. The extent to which Lake Mergozzo is polluted is not mentioned, because pollution can add to the DOC as well as the microbial population. When the microbial count goes up over an order of magnitude after incubation of the water samples for a day, the amount of total DOC or the lower molecular fraction does not seem to decrease (Table 3.1). If there is no decrease, the question as to where the energy for the production of new cells is coming from must be answered. There is a decrease in the DOC and the lower molecular weight fraction after 6 days of incubation of water samples but the largest increase is in the first day. There is an increase in cell numbers but not biomass. This may be due to a starvation-survival process or to the organic matter in the air supplying the energy.

The ability of microorganisms in rivers to use the DOC is discussed by Lock and Ford (1986). However, it is of interest that these investigators could not determine the in situ rates of uptake or turnover of the bulk DOC/DOM and direct comparison with the monomeric microbially labile compounds was not possible.

Table 3.1 Decrease of DOC in the low molecular weight fraction of dissolved organic matter, in consequence of growth of bacterial population, in water samples of Lake Mergozzo during incubation at 20°C (Nov. 25, 1975).

Water Samples of

	1 m depth			20 m depth		
Incubation (days)	Bacterial Number[a] (per ml)	DOC (mgC/l) Total	LMF[b]	Bacterial Number[a] (per ml)	DOC (mgC/l) Total	LMF[b]
0	1.2×10^3	1.6	1.5	7.5×10^2	1.5	1.4
1	4.5×10^3	1.6	1.6	1.3×10^3	1.5	1.4
2	6.9×10^4	1.7	1.6	2.4×10^4	1.6	1.4
3	8.4×10^4	1.6	1.5	2.9×10^4	1.5	1.4
6	4.6×10^4	1.5	1.4	2.7×10^4	1.4	1.3

Source: Reprinted by permission from *Bulletin of the Japanese Society of Scientific Fisheries* 43:885–892, Y. Ishida, A. Uchida, and H. Kadota, 1977; copyright 1977, Japanese Society of Fisheries Science.

[a]Plate counts of bacteria were made by use of F/5 medium.

[b]The low molecular weight (lower than 10,000) fraction.

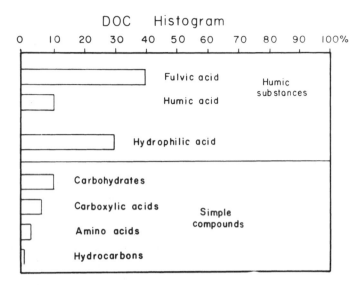

Figure 3.3   Dissolved organic carbon histogram for an average river water with a DOC of 5 mg/l. Redrawn and printed by permission from *Organic Geochemistry of Natural Waters*, E. M. Thurman, 1985; copyright 1985, Martinus Nijhoff/Dr. W. Junk Publisher. Reprinted by permission of Kluwer Academic Publishers.

They also postulated that there was a link between the refractory and labile organic matter pools. If a DOC concentration in the water was 5 mg/l, a degradation rate of 0.001% per day could result in 5 μg of labile substance/l/day. The defined bacterial inoculum grew best on DOC from the headwaters of the Ogeechee River, thus indicating that the bioavailability of DOC declined downstream (Leff and Meyer, 1991). Meyer et al. (1987) found that bacterial growth and amount of DOC was the greatest in low molecular weight fractions (MW <1000) of river water and the least in the intermediate MW (1000–10,000) fractions; while a high MW fraction (>10,000) supported more growth than did the intermediate MW fraction. The high MW fraction contained much humic material that was resistant to microbial degradation. In order to explain their results, the authors postulated that the low MW compounds complexed with high MW fraction but still remained available to the microorganisms.

Ittekkot (1988) reports that the labile fraction in riverine organic matter can be "oxidized" or lost within the rivers, their estuaries and in the marine environment. From his data, he extrapolates that 35% (81 × 10^{12} g C/yr) of the global carbon belongs to the labile fraction and the rest (150 × 10^{12} g C/yr) appears to be highly resistant to degradation. Thus, much of the organic carbon is recalcitrant to further microbial activity.

Table 3.2   Concentration of humic substances in natural waters.

Water Type	Concentration (mg C/1)
Groundwater	0.03–0.10
Seawater	0.06–0.60
Lake	0.5–4.0
River	0.5–4.0
Wetlands	10–30

Source: Reprinted by permission from *Organic Geochemistry of Natural Waters*, E. M. Thurman, 1985; copyright 1985, Martinus Nijhoff/Dr. W. Junk Publisher. Reprinted by permission of Kluwer Academic Publishers.

The DOM in lake water consists mainly of refractory, high molecular weight substances. With pure cultures, Stabel et al. (1979) were able to show that up to 30% of the DOM fraction was mineralized. Geller (1983a) demonstrated that 50% to 70% of the DOC was persistent, regardless of the season and community structure. Ryhaenen (1968) attributed the slow biodegradation of humic substances in lakes to a deficiency in inorganic nutrients.

Gardner et al. (1986, 1989) and Scavia and Laird (1987) demonstrated in Lake Michigan that the bacterial population size may be controlled by grazers, whereas growth rates are more likely to be limited by organic substrate availability. Bacterial biomass production is complicated by an incomplete understanding of the composition, availability, and supply rates of the various organic compounds (Cole et al., 1984).

BIOAVAILABILITY OF HUMUS AND HUMUS COMPLEXES

The DOC in river water can be divided into various fractions (Fig. 3.3). The organic carbon in rivers is 7.0 mg/l (Fig. 3.2) and the unavailable humic and fluvic acids make up more than 50% of the total organic carbon (Fig. 3.3). The concentration of humic substances in natural waters and in different types of lakes are shown in Tables 3.2 and 3.3 respectively. In all types of water, the concentration of humic substances is around a few mgC/l, except in wetlands where the concentration reaches 10–30 mgC/l. In oligotrophic lakes, it can be seen that the concentration of humic substances is between 0.5 to 1.0 mg/l (Table 3.3). The distribution of fulvic acid, humic acid, hydrophilic acids, and identifiable compounds in various natural waters is shown in Figure 3.4. The hydrophilic acids (resistant to microbial degradation) are thought to be made up of volatile fatty acids, hydroxyacids, and complex polyelectrolyte acids that probably contain many hydrolyl and carbonyl functional groups (Thurman, 1985). Simple compounds such as carbohydrates, carboxylic acids, amino acids, and hydrocarbons

Table 3.3   Concentration of humic substances in various types of lakes.

Lake	Concentration (mg C/1)
Oligotrophic lake	0.50 to 1.00
Mesotrophic lake	1.00 to 1.50
Eutrophic lake	1.50 to 5.00
Dystrophic lake	10.0 to 30.0

Source: Reprinted by permission from *Organic Geochemistry of Natural Waters,* E. M. Thurman, 1985; copyright 1985, Martin Nijhoff/Dr. W. Junk Publisher. Reprinted by permission of Kluwer Academic Publishers.

will support microbial growth, but they may not be bioavailable if they are complexed to humic substances or clay minerals. Since approximately 50% of the total DOC is made up of fulvic and humic acids, it further illustrates that very little of the organic matter is available to the microorganisms. Both riverine and deep-sea DOC are relatively resistant to biodegradation (Thurman, 1985). This is probably the reason for a succession of various species of bacteria with time in any environmental sample. Furthermore, certain fractions of the organic matter are recalcitrant to some of the bacterial species present in the environment sample. Lock and Ford (1986) explored the nature of DOC because of the magnitude of analytical problems in determining all of its constituents.

Humic material is defined differently when it occurs in soil rather than in aquatic systems. Soil humic substances, which are the colored, polyelectrolytic acids, are operationally defined by their isolation from soil with 0.1N NaOH; whereas, aquatic humic substances are operationally defined as colored polyelectrolytic acids isolated from water by sorption onto XAD resins, weak-base ion exchange resins, or a comparable procedure (Thurman, 1985). The major functional groups include carboxylic acids, phenolic and alcoholic hydroxyl groups, and keto functional groups. Humic substances that are precipitated by acid are known as humic acid and those that remain in solution are fulvic acid. Hydrophilic acids are not retained by the XAD resin at pH 2.0 and make up about 30% of the DOC. Due to the different methodologies involved in the analysis of humic material from soil and aquatic environments, the soil and aquatic humic material are probably different.

From a study of humic and clear water lakes in Germany, Tranvik and Höfle (1987) found that 15 to 20% of the total DOC was consumed by the heterotrophic bacteria. Later, Tranvik (1990) investigated the ability of bacteria to utilize different MW fractions of the DOC obtained from 10 different lakes of differing humic content. High MW DOC ($>$10,000) became an increasing fraction of the total DOC as the total DOC and humic content of the lakes increased. On an

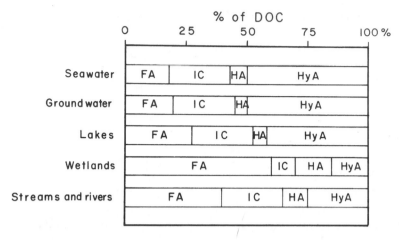

Figure 3.4   Distribution of humic substances (fluvic acid [FA], humic acid [HA], hydrophilic acids [HyA], and identifiable compounds [IC]) in natural waters. Redrawn and printed by permission from *Organic Geochemistry of Natural Waters*, E. M. Thurman, 1985; copyright 1985, Martinus Nijhoff/Dr. Junk Publisher. Reprinted by permission of Kluwer Academic Publishers.

average, 48% and 22% of the bacterial growth occurred at the expense of DOC of <10,000 MW and <1000 MW, respectively, with no significant correlation to humic content and DOC concentration of lakes. The size fraction (<10^3 dalton fraction) appears to support most of the bacterial activity (Meyer et al., 1987; Sundh, 1992) and the more complex higher molecular weight fractions are only utilized after the small molecular weight fractions are consumed. Connolly et al. (1992) divided the DOC into three pools—the L1 pool of labile compounds readily used by bacteria, the L2 pool of more complex compounds used at a slower rate, and the R (refractory) pool that is generally not metabolized by bacteria—but did not assign any MW to the various pools. Geller (1983b) found that 90% of the DOC in Mundelsee consisted of macromolecular (MW ≥ 1500) and oligomeric (MW 400–800) components, regardless of the season in which they were isolated. These macromolecules and oligomers repressed the growth of the natural bacterial population of the lake and exhibited resistance to microbial attack.

Mineralization of humic matter is an extremely slow process (Fenchel and Blackburn, 1979; Wetzel, 1983). Many of the subsurface bacteria are undoubtedly forced to subsist on recalcitrant polymeric compounds such as humic and fulvic acids (Kieft et al., 1990). Bactericidal action of humic acid on *Staphylococcus aureus* and *Serratia marcescens* has been reported (Hasset et al., 1987). However, growth occurred when flocculated humic matter was added to cultures of natural bacterial populations (Tranvik and Sieburth, 1989). However, the humic matter may have been rendered more available to the bacteria due to the acid and base treatments of the humic material used in these experiments.

"Free" sugars and amino acids may be complexed with metallic ions or adsorbed to clay particles (Gocke et al. 1981). Organic material complexed to aggregates cannot be used fully by bacteria (Gocke et al., 1981; Burnison and Morita, 1974; Carlson et al., 1985a,b). During relatively short experimental incubation times of experiments, the added ^{14}C-labeled glucose may not be complexed to metallic ions or absorbed to clay particles. As a result, the added ^{14}C-labeled glucose is much more bioavailable than the "free" glucose absorbed or complexed (Gocke et al., 1981). The same can be said for amino acids (Dawson and Gocke, 1978). Unfortunately, we do not know enough about the different types of organic compounds present in the environment, their distribution in space and time, and their importance for the species composition and ecological dynamics of microorganisms.

In rivers, the DOM is primarily allochthonous in nature and is composed mainly of humic substances, carbohydrates, and proteinaceous matter. In the Williamson River (Oregon), about 96% of the dissolved amino acids were associated with humus and amino acids comprised ~1% of humic carbon (Lytle and Perdue, 1981). From the humus-amino acid complex the most abundant amino acids, in order, were glycine, aspartic acid, alanine, serine, and glutaric acid. The concentration of total amino acids and humic carbon correlated significantly with each other. The principal source of the humic carbon and amino acids appeared to be surface runoff.

Brisbane and Ladd (1968) reported that under sterile conditions, the observed concentrations of released amino acids from the humus-amino acid complex either decreased markedly or remained comparatively stable on further incubation. The foregoing is related to the humic acid substrate and its effectiveness as a bactericidal agent; thereby, inhibiting the growth of the indigenous microbes.

A histogram of the carbohydrates in freshwater is given in Figure 3.5. The polysaccharides make up only 5% of the DOC and the monosaccharides make up less than 2% of the DOC. One wonders how much of the polysaccharides are truly "free" to be used as substrates by the microorganisms. The concentration of volatile fatty acids in different natural waters is given in Figure 3.6. For rivers and lakes the concentration is only 100 µg/l, and less in the ocean and in rain water. Volatile substances, such as formic, acetic, propionic, and butyric acids, have been found in water samples from three different types of lakes (Table 3.4). Again, the total concentration of volatile fatty acids in oligotrophic lakes is only 60 µg/l. However, these substances may be complexed to humus and not readily available to the microorganisms.

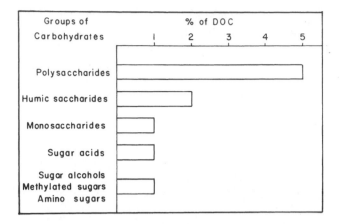

Figure 3.5   Histogram for carbohydrates in freshwater. Redrawn and printed by permission from *Organic Geochemistry of Natural Waters*, E. M. Thurman, 1985; copyright 1985, Martinus Nijhoff/Dr. Junk Publisher. Reprinted by permission of Kluwer Academic Publishers.

3.4.2   Organic Matter in Marine Environments

ORGANIC MATTER IN THE WATER COLUMN

In 1934, Krogh put forth the hypothesis that the DOM, and possibly the POM, is chronologically old, chemically and biochemically inert, and is not significant in the marine food chain in the deep sea. DOM, as a source of food for aquatic animals, is reviewed by Jørgensen (1976). This review includes Pütter's theory as well as Krogh's data on the subject. Both Jørgensen (1976) and Krogh (1934a,b) concluded that the amount of DOM is insufficient to support marine life. The significance of the organic components of seawater in the distribution and growth of marine algae, bacteria, protozoa and larvae is summarized by Lucas (1955). The terms "good" and "bad" ocean waters have been used in the past literature (Hood and Loder, 1973), where "good" waters support heavy growth and "bad" waters are nearly sterile. However, early studies indicate little difference between "good" and "bad" ocean waters (Hood and Loder, 1973). In the deep sea, Krogh (1934a,b) found the OM to be partly humus and resistant to bacterial degradation. The low values of the oxygen demand (OD) of the OM in deep waters only points to the resistance of the OM, which constitutes the almost constant values of ^{13}C (Williams, 1968; Willliams and Gordon, 1970). Much of the organic matter in the ocean exists as aggregates of organic detritus, microorganisms, and clay minerals (Alldredge and Silver, 1988). Approximately 5% of the DOM is utilizable by bacteria (Ammerman et al., 1984) and only a small fraction can be directly

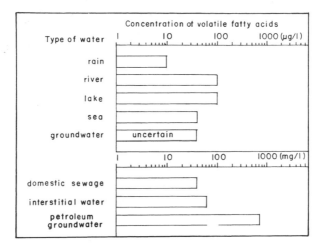

Figure 3.6 Approximate concentration of volatile fatty acids in different natural waters. Redrawn and printed by permission from *Organic Geochemistry of Natural Waters*, E. M. Thurman, 1985; copyright 1985, Martinus Nijhoff/Dr. Junk Publisher. Reprinted by permission of Kluwer Academic Publishers.

taken up by bacteria since a major portion consists of proteins and polysaccharides that must first be hydrolyzed. With the total organic carbon at 0.5 mg/l (Fig. 3.2), the microbes in the sea are faced with a starvation diet.

Seki (1982) discusses the turnover rates of organic matter and divides the organic matter into three broad categories, depending on its biodegradability. These categories are: (1) constituents easily metabolized by most microorganisms such as amino acids, monosaccharides, and organic acids, (2) constituents moderately resistant to biochemical breakdown such as cellulose and chitin, and (3) refractory constituents highly resistant to biochemical breakdown such as aquatic humus. The turnover times of the first category are put at less than a few days in hypereutrophic waters, a few days in eutrophic waters, between a few days and several tens of days in mesotrophic waters, and several tens of days in the surface layers of oligotrophic waters. The moderately resistant category's turnover time is several days and several tens of days in hypereutrophic waters, between several tens of days and several months in eutrophic water, several months in mesotrophic waters, and a few years in oligotrophic waters. The refractory category has been estimated to be between half a year and one year in hypereutrophic waters, several years in eutrophic waters, and several tens of years in mesotrophic waters, several tens of years and hundreds of years in the surface layer of oligotrophic waters, and several thousands of years in the deeper layer of oligotrophic waters. Some of these calculations are based on data while others are speculation.

Table 3.4   Concentration (in µg/l) of volatile fatty acids in lake water by type of lake.

Acid	Eutrophic	Mesotrophic	Oligotrophic
Formic	15	3	23
Acetic	123	63	27
Propionic	10	19	5
Butyric	10	39	5
Total	162	125	60

Source: Reprinted by permission from *Organic Geochemistry of Natural Waters,* E. M. Thurman, 1985; copyright, 1985, Martin Nijhoff/Dr. W. Junk Publisher. Reprinted by permission of Kluwer Academic Publishers.

DOC, POC, and POM in the Marine Environment

The organic content of seawater is low. Dissolved organic matter makes up the largest pool of organic matter in the sea (Mopper and Degens, 1979), but unfortunately more than 60% of the organic matter remains uncharacterized (Degens, 1970; Skopintsev, 1971; Williams, 1971, 1975; Hedges, 1987). Joint and Morris (1982) state that "it is obviously easier to study compounds which are present at relatively high concentrations but, almost by definition, such compounds are unlikely to be important in the cycling of organic matter because their high concentrations are the result of non-utilization by bacteria." In the marine environment, the low concentration in which identifiable compounds occur in seawater indicates that most biochemical organic carbon is transformed or removed at rates comparable to production (Gagosian and Lee, 1981). The values for POC and DOC are between 3 and 10 µg C/l and between 0.35 and 70 mg C/l respectively (Menzel and Ryther, 1970). DOM can be transformed into POM by physicochemical actions such as bubbling, adsorption, and coprecipitation (Baylor and Sutcliffe, 1963; Riley, 1963, 1970; Barber, 1966; Khailov and Finenko, 1968) and can occur at depths in the ocean (Riley et al., 1965; Riley, 1963; Wangersky, 1965).

It is recognized that the DOC in seawater is a global reservoir of carbon (Hedges, 1987). The rate at which this global reservoir of DOC is recycled to CO_2 is important in any discussion of carbon flux in the ocean (Toggweiller, 1988) as well as globally. The major contributor to the recycling process is bacterial respiration, mainly because of the large number of bacteria in the marine environment and the diverse physiological types (Van Es and Meyer-Reil, 1982). Most of the DOC cycling is accomplished via the turnover of just a small fraction of the large DOC pool, while the rest of the pool remains static on the time scales of hours to weeks observed by most experimental biologists (Ducklow and Carlson, 1992). The small fraction is composed of labile substrates. This time scale could be much longer if one includes all the organic compounds in the system.

The euphotic zone is relatively rich in nutrients but it is only a minor part of the entire ocean. The fraction of the net primary production released as DOM is estimated from 5% to more than 50% of the net primary production (Hellebust, 1965). Below this euphotic zone at depths greater than 200–300 m, the DOC concentration ranges from 0.35 to 70 mg/l and the particulate organic carbon ranges from 3 to 10 µg/l (Menzel and Ryther, 1970). Because we cannot separate the bacteria from either the POC or DOC, the old values given by Menzel and Ryther (1970) are probably too large. This, then, makes one wonder what the true values for the DOC and POC are. The major input of organic carbon in the ocean appears to be primary production in the surface waters and associated heterotrophic activity is mainly responsible for the recycling of the labile organic compounds produced by primary producers (Mopper and Degens, 1979). The efficiency rate in the surface waters is probably >90% and the remaining 10% is distributed into deep-sea pools of refractory dissolved and particulate organic carbon (Mopper and Degens, 1979).

The main body of the ocean has a nearly constant temperature, salinity, and pH as well as a low organic content. It is covered by a thin layer (photo zone) where photosynthetic processes produce organic matter. The margins (within the continental shelf) of the oceans are where there are extremes in heterogeneity. Thus, in the margins, bacteria must have developed strategies to overcome the rapid fluctuations in the physical and chemical environments.

The continental shelf is the main site of mineralization of organic matter in the sea bottom and around 20–50% of the local phytoplankton production in the shelf areas is deposited on the sediment (Jørgensen, 1983). For the deep sea, the figure is 1% or less; hence 83% of all remineralization in the ocean bottom takes place on the shelf, which is only 8.6% of the total bottom area (Jørgensen, 1983). The sedimented material, although only a small fraction of the total primary productivity of the ocean, does play an important role in controlling the chemical composition of the seawater when one considers the time range between 10 and 100,000 years (Jørgensen, 1983).

Due to pollution by drainage from land, coastal seawater has higher concentrations of DOC than the open ocean (Fredericks and Sackett, 1970; Foster and Morris, 1971b; Maurer and Parker, 1972; Ogura, 1972) and seasonal DOC variations also occur (Duursma, 1963; Holmes et al., 1967; Strickland et al., 1970; Morris and Foster, 1971; Banoub and Williams, 1973; and Ogura et al., 1975). In the latter situation, maximum DOC concentrations are found in the spring to autumn. Spectroscopic analysis of DOM indicates distinct differences in the DOM from the open ocean, estuarine waters, and sediment (Kerr and Quinn, 1975; Stuermer and Harvey, 1977). Prahl and Meulhausen (1989), using plant biomarkers, estimated that 20% of the abyssal sedimentary organic carbon is terrestrial in origin, although this has not been confirmed (Ittekkot and Haake, 1990; Pocklington et al., 1991).

In the open ocean, the DOC is around 1–2 mg C/l at the surface and in deep water, it decreases to 0.4–0.8 mgC/l (Duursma, 1961; Menzel, 1964; Menzel and Ryther, 1968; Ogura, 1970c; and Ogura and Hanya, 1967). Williams (1971) gives the average concentration of DOC from 0 to 300 m to be 1.0 mg C/l (range: 0.3 to 1.0), from 300 to 3800 m, 0.5 mg C/l (range: 0.2 to 0.8); and for the POC from 0 to 300 m to be 0.1 mg C/l (range: 0.03 to 0.3 mg C/l), from 300 to 3800 m to be 0.01 mg C/l (range: 0.005 to 0.03). Hence, the surface water has 10 times more DOC than POC and in deep water, it is 50 times higher. The transition depth of 300 m may vary from 100 to 400 m, depending on the surface production of organic matter and the physical mixing processes. Although the concentrations of DOC and POC may vary widely above the transition depth, they remain relatively constant below it to the sea floor. According to Menzel and Ryther (1970), the DOC in subsurface waters is not related to its concentration at the surface or to the organic production in the euphotic zone. Of the total POC in the bathypelagic region, 20% is enzymatically hydrolyzable and the remaining 80% is refractory and theoretically unsuitable for organisms to utilize (Gordon, 1970).

Approximately 0.4% of the carbon fixed by photosynthesis enters the deep sea DOC (Williams, 1971) and approximately 0.01% of the total POC produced by photosynthesis passes into the sediment and the turnover time of the POC is 3000 years (Bryuevich, 1962 [see Williams, 1971]; Romankevisch, 1968 [see Williams, 1971]). The flow of POC to the deep sea has been estimated to be in the range of 20–800 mg/m^2 per year (Honjo, 1978; Hinga et al., 1979; Brewer et al., 1980) with a much higher estimate (2–6 g/m^2/year) by Rowe and Gardner (1979). How much of this material reaches the various depths still remains unclear since falling through the water column takes time and decomposition processes are taking place throughout this time period, although there are reports that fecal material and chitinous skeletons do fall at a rather fast rate. There is a tendency for DOC to concentrate at interfaces; therefore, its absorption onto particles suspended in the seawater changes their charge and surface properties (Hedges, 1987).

The downward vertical flux of POM in the open ocean exhibits a nonlinear decrease with depth (Knauer et al., 1979; Suess, 1980; Betzer et al., 1984; Pace et al., 1987; and Martin et al., 1987). Greater than 75% of the net POM loss occurs in the upper 500 m of the water column (Martin et al., 1987). Sinking particles contain viable metabolically active microorganisms (Caron et al., 1982; Ducklow et al., 1982, 1985; Silver et al., 1984; Karl and Knauer, 1984; and Taylor et al., 1986); there is a selective loss of biochemical labile compounds from sinking particles (Wakeham et al., 1980; De Baar et al., 1983; Wakeham et al., 1980; Lee and Cronin, 1984). Thus, the microbial decomposition process is an important mechanism controlling the POM flux. Karl et al. (1988) concluded that large sinking particles are generally poor habitats for bacterial growth and are not active remineralization sites of organic matter. Karl et al. (1988) suggested that emphasis of particle decomposition be shifted to suspended (nonsinking) particulate matter. Large food falls into the deep sea are rare events, but when they happen they may affect areas of the benthos for long periods of time (Stockton and DeLuca, 1987).

The rain of particulate matter to the deep-sea floor has been suggested as the primary means by which organic matter is supplied to organisms inhabiting the sea floor (Menzies, 1962). The sinking rates for various types of particulate matter have been measured. For instance, euphausiid carapaces sink at a rate of 288 m/day (Lasker, 1966) and fecal pellets at 36–376 m/day (Osterberg et al., 1963; Smayda, 1969; Fowler and Small, 1972). These materials are generally missed when samples are taken by routine means rather than by sediment traps. Wiebe et al. (1976) estimated that 14% of the total carbon in the deep sea could be supplied by rapidly sinking particulate matter. The pools of deep-sea POC and DOC are probably in a "quasi-steady" state as the result of losses of POC to the sediment and deep-sea biological oxidation of POC and DOC in the water column and at the sediment-water interface. Fecal pellets are altered chemically by microbial degradation originating either internally or externally (Johannes and Satomi, 1966; Honjo and Roman, 1978) but this process may be slow when temperatures are low, around 3–5°C (Honjo and Roman, 1978). Soluble organic matter within fecal pellets diffuses rapidly into the water as the fecal pellets fall, resulting in the soluble portion remaining in the upper mixed layer (Jumars et al., 1989). The labile DOC in the deep ocean may be an important source of food to all heterotrophic organisms (Hedges, 1987). Although the large fecal pellets may fall to the bottom of the deep ocean, many fecal pellets are too small and have too low density to sink to the bottom (Pattenhöfer and Knowles, 1979). According to Honjo (1990), fecal pellets of adult copepods are mostly utilized by microorganisms in the warm mixed layer such as the Sargasso Sea. Thus, through turbulence in the upper mixed layer, much of the particulate detritus remains suspended in the upper layer of the ocean or becomes trapped in the pycnocline (Lande and Wood, 1978). Since all biological processes do not hasten organic matter sinking into the deeper portions of the ocean, less than 5% of the total primary production in the euphotic zone is estimated to reach the benthos (Pomeroy and Wiebe, 1993).

POM that falls from the euphotic zone is the main food source for organisms in the deep sea. As POM falls through the water column, it is altered by grazing, microbial degradation, and chemical processes (Karl et al., 1988). Thus, the more labile components (sugars, amino acids, readily degradable protein) are degraded leaving the more recalcitrant polymeric material, which is not very nutritive. The nutritive value decreases with prolonged sinking times (Banse, 1990). Up to 90% of the total benthic biomass is microbial (Rowe et al., 1991; Tietjen, 1992; Pfannkuche, 1992). Thus, extracellular enzymes acting on polymeric compounds play a key role in the degradation of POM, which in turn provides the monomeric and oligomeric molecules necessary as an energy source for bacterial metabolism. Although some papers (Helmke and Weyland, 1991; Meyer-Reil and Köster,

1992; Lochte, 1992; Boetius and Lochte, 1994) refer to the enzymes being ba-
rotolerant, the depth (hydrostatic pressure) must be considered. Thus, in shallow
environments, enzyme barotolerance is not necessary but it is in the hadal environment of the oceans. There is much literature that definitely indicates that high
pressure inhibits enzyme activity.

Rowe et al. (1990) state that the consumption of POM in the benthos is very
nearly equal to inputs, but in the abyssal depths, sediment trap data show the
respiratory rates of organisms increased, indicating an excess of POM input. They
attribute this to the refractory nature of the deep-sea detritus and not to physiological limits imposed by hydrostatic pressure or low temperature. They further
stress that caution must be used when interpreting the relationship between total
organic carbon fluxes measured with sediment traps at great depths and rates of
metabolic processes. The shortest detrital carbon turnover found by Rowe et al.
(1990) was just over 8 years in a sandy continental shelf environment versus 609
years at 5.3-km depth on the Hatteras Abyssal Plain. For more information concerning the interaction of particulate detritus and bacteria, consult Kirchman
(1993).

In the aquatic environment, substrate supply and grazing are the factors with
the greatest potential for short-term control of planktonic bacteria density and
productivity (Wright, 1988a). Substrate concentration will change under conditions of changing bacterial densities, and when grazing occurs, it will raise the
substrate concentration and begin to drive bacterial growth so that the steady state
will be equal to the grazing rate (Wright, 1988b). This steady state may apply to
bacteria over a short term, but shifts in the mode of control will occur and temporarily disturb the steady state (Wright, 1998b).

Analysis of DOM, DOC, POC, and DON

The difficulties in analyzing the organic matter in seawater are discussed by
Hedges (1987). These difficulties include that (1) few methods are sensitive
enough for direct quantification of organic matter in the low concentrations (ppm
to ppb) found in seawater, (2) no technique has yet been devised to concentrate
DOM from the mass of sea salt, (3) the most successful isolation method for
DOM involves selective adsorption onto synthetic resins, which recovers less than
10% of the DOM, (4) some DOC concentrates at interfaces, and (5) some DOC
components chemically combine with dissolved metals to form complexes that
affect the chemical properties of seawater and phytoplankton production.

Jeffrey and Hood (1958) evaluated various means for the isolation of the organic matter in seawater and concluded that quantitative isolation of soluble organic compounds from seawater without alteration of one or more constituents is
difficult because:

(1) many compounds are easily destroyed by bacterial action, (2) the ratio of inorganic to organic material in seawater is extremely high, (3) destruction by heat or slight change in chemical conditions is difficult to prevent in the case of many compounds of biological origin, and (4) organic matter readily adsorbs to surfaces where it becomes difficult to desorb.

This situation still remains today. If definite conclusions are to be drawn concerning the nature and quantity of the organic constituents present in the soluble organic compounds in seawater, extreme care must be exercised in the isolation procedure.

Currently there is a controversy concerning the amount of DOC and DON in the sea. Sugimura and Suzuki (1988) and Suzuki et al. (1985) reported that the DOC and DON are 50–400% higher than previously reported for oceanic surface and deep water. These results were obtained by using a high-temperature catalytic oxidation method. The high values for DOC and DON by Sugimura and Suzuki (1988) would help overthrow Krogh's (1934a,b) concept that there was insufficient organic matter in the ocean to support life processes. The vertical profile of the amounts of DOC reported by different methods is shown in Figure 3.7. This can be compared to the data of Sugimura and Suzuki (1988) and Suzuki et al. (1985) in Figure 3.8. However, in the chromic acid oxidation method used by Plunkett and Rakestraw (1955) the chloride ion was removed from dried seawater by precipitation with thallium (Fig. 3.7) and showed an increased DOC value. Different methods were used by Suzuki et al. (1985) for DON and DOC and Keil and Kirchman (1991a,b) for DCAA (dissolved combined amino acids). The latter used vapor phase hydrolysis instead of the traditional acid hydrolysis method. Eventually, a standard method(s) must be employed by various investigators so that data can be compared (Mopper, 1977). If extraction is difficult, how difficult is it for the microbes to synthesize the necessary enzymes needed to utilize the various components of the organic matter? The formation of new enzymes by the microbes takes energy, which is in short supply in many environments. Hydrolysis by chemical means may be very harsh compared to enzymatic hydrolysis, thereby indicating that these substances are not readily available to the microorganisms. Many of the extraction techniques call for drastic hydrolysis of the organic matter.

According to Williams and Druffel (1988), the amounts reported by the Japanese researchers would help explain the discrepancy between the potentiometric and manometric measurement of total CO_2 in surface seawater and the imbalance between reduced carbon exported out of the euphotic zone (as measured by sediment traps) as well as the higher amounts expected from in situ production inferred from seasonal oxygen zone. According to Williams and Druffel (1988), the various methods used to detect DOC show little or no change in values below the upper 300–500 m. The values reported by Sugimura and Suzuki (1988) could

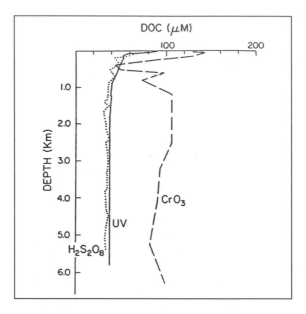

Figure 3.7　Profiles of DOC (μM) in the North Pacific using different oxidative techniques: UV radiation (–, 31°N, 159°W, Williams and Druffel, 1987), peroxodisulfuric acid (···, 31°N, 158°W, L. Gordon, pers. communic., Geosecs Stn. 204), and chromic acid (——, 32°N, 142°E, Plunkett and Rakestraw, 1955). Reprinted by permission from *Oceanography* 1:14–17, P. M. Williams and E. R. M. Druffel, 1988; copyright 1988, The Oceanography Society.

not be repeated on Suzuki's catalysts (Toggweiler, 1988). Unfortunately, Suzuki (1993) could not repeat the values previously reported by Sugimura and Suzuki (1988). This discrepancy appears to come from instrumental carbon contamination, or "blanks" (Toggweller, 1992). This has ended in a plea for marine chemists to develop a standard and precise method for accurate measurement for DOC and DON analysis (Hedges et al., 1993). Even if Sugimura and Suzuki (1988) were correct, the question that remains is whether the extra DOC is bioavailable to microbes, especially in light of the harsh treatment of the DOC for analysis. Some of the new DOC revealed by high temperature catalytic oxidation is biologically labile (Kepkay and Wells, 1992). A significant fraction of the DOC can be respired to CO_2 in the upper ocean before more resistant fractions are incorporated into the longer-term consumption of DOC at greater depths (Kepkay and Wells, 1992; Kirchman et al., 1991).

Most of the DOC oxidized by seawater using bulk oxidation techniques is chemically unidentifiable and is often referred to as marine humus (Toggweiler, 1988). DOC provides the main substance for bacterial growth. The distribution

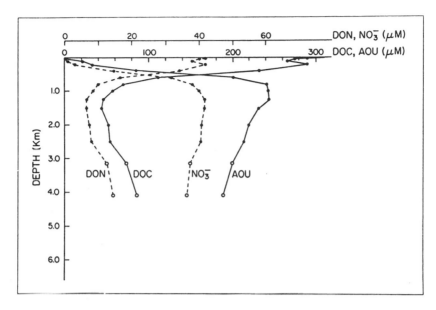

Figure 3.8 Profiles of DOC, DON, apparent oxygen utilization (AOU), and NO_3^- as measured by Sugimura and Suzuki (1988) at 20°N, 137°E (•) and 5°N, 135°E (–○–). The depth of the O_2 minimum at 20°N, 137°E is between 800 and 1000 m. Notice the similarity between this DOC profile and that of Plunkett and Rakestraw (Fig. 3.7). Reprinted by permission from *Oceanography* 1:14–17, P. M. Williams and E. R. M. Druffel, 1988; copyright 1988, The Oceanography Society.

of readily characterized biochemical compounds such as amino acids, carbohydrates, and lipids has been reviewed by Duursma (1965), Wangersky (1965), and Wagner (1969). This relatively stable and complex fraction has been variously termed *Gelbstoff* (Kalle, 1966), aquatic humic material (Skopintsev, 1972) and refractory organic matter (Ogura, 1972). DOC exudates of phytoplankton and the main food sources for the bacteria are identifiable low-molecular weight metabolites and polymeric carbohydrates that are produced and consumed within hours or days rather than years.

Butman (1986) and Butman et al. (1986) state that any direct approach of analyzing POC recovered in bottom-moored sediment traps is arguably compromised by problems associated with the design and deployment of the traps themselves. The thigmotactic (attachment) ability of bacteria, even starved bacteria, must also be considered when working the sediment traps.

Although the total amount of DOC and POC are of great interest to marine chemists and geochemists, what is important to marine biologists is the amount of the DOC and POC that can be used as an energy source for the organisms in the sea.

Bioavailability of DOC, DOM, POC, and POM

Most of the DOC in the ocean must be considered biologically refractory and unavailable for microbial growth (Jørgensen, 1976). The organic matter of phytoplankton is not completely converted into the mineral components that initially served as nutrients, despite all the transformations that occur in the sea (Skopintsev, 1981). Skopintsev (1981) reviews most of the studies on the decomposition of plankton and notes that the labile portion is oxidized within 1 or 2 months but about 25% remains recalcitrant and that "stable" organic matter increases with depth. The vertical flux of organic matter in the ocean is reviewed by Joint and Morris (1982).

Menzel (1974) posed the question, "How are energy requirements of the deep sea met?" This still remains an unsolved problem, not only for bacteria but for all biological forms. In dealing with higher forms of deep-sea marine life, various mechanisms have been proposed: (1) sinking of large particles (fecal pellets, fecal material, microscopic aggregates, detrital macrophytes, remains of small organisms, or carcasses of large animals), (2) migrations of zooplankton or micronekton and overlapping distributions of predators and prey, (3) adsorption of dissolved organic matter on particles, (4) use of bacteria as food by filter feeders, and (5) transport of organic matter from continental shelves via turbidity currents and suspended particles. The primary production due to chemolithotrophic bacteria in and near hydrothermal vents as well as seeps of methane and hydrocarbons could also be included. The results of Pearcy and Stuiver (1983) indicate that rapidly sinking particles from surface waters, such as fecal pellets, are not the major source of organic carbon for the deep-sea fishes and large benthic invertebrates. If surface water is permitted to incubate with the indigenous microflora, the DOC levels are reduced to the DOC concentrations in deep water (Barber, 1968), suggesting that the baseline fraction of DOC is only very slowly utilized by bacteria.

Sheldon et al. (1967) determined that less than 5% of the DOC in seawater is utilized by bacterial growth. There has been controversy over the exact values of DOC in the deep ocean where the values are stable on a time scale of centuries (Mopper and Degans, 1979). It is recognized that particles in the ocean can sink rapidly (hundreds to thousands of meters per day) (McCave, 1984). Most of the readily (labile) utilizable compounds have either already been used by the associated bacteria in and/or on the particles, leaving the particles more recalcitrant, or the labile compounds are complexed so that they are not bioavailable to the bacteria. Microbes must degrade polymeric compounds before their subunits can be actively or passively transported into the cell. Of the energy the cell uses 20%

is for active transport of substrates into the cell (Stouthamer, 1973). There must be sufficient energy in the system to permit the bacteria to biosynthesize the exoenzyme necessary to degrade the polymer. Since an enzyme is composed of many amino acids, it takes a large amount of energy to synthesize the enzyme(s). Where does this energy come from?

The breakdown of organic matter in the ocean is due primarily to microbes. Deming and Baross (1993) gave four reasons why bacteria are the primary agents for the early diagenesis of organic matter; these reasons also explain the decomposition of organic matter in the entire marine environment. These reasons are

> (1) they [bacteria] can occur abundantly and universally throughout all marine sediments; (2) they can respire, reproduce, and, therefore, use organic matter more rapidly than other organisms; (3) they possess enzymes and enzyme systems (in many cases, unique to the prokaryotic kingdom) that make them extraordinarily versatile in their nutritional requirements and have abilities to alter a wide variety of particulate and dissolved organic (and inorganic) material; and (4) they can readily enter into complex associations with each other and with higher organisms in ways that produce powerful degradative capabilities beyond those of a single organism in isolation.

Menzel (1964) reached the conclusion that the DOC and POC in the ocean was extremely stable and subject to limited change by biological action. Barber (1968) showed that about 50% of the DOM of the surface of the Atlantic Ocean is oxidized in 1 to 2 months, but there was no change in the DOM for deep waters. Sorokin (1972) postulated that the low oxygen demand in the deep oceanic waters was not due to the resistance of OM but to the negative effect of low temperature and increased hydrostatic pressure. Pomeroy and Deibel (1986) and Pomeroy et al. (1991) also present evidence for the negative effect of low temperature. The deep sea is characterized by Gooday and Turley (1990) as "a food-limited environment in which the abundance and biomass of the benthic organisms is related directly to the amount of food reaching the sediment surface." This may include bacteria as food. The patchy distribution and intermittent or seasonal pulsed distribution of organic matter into the deep sea probably play a major role in the structuring of deep-sea benthic ecosystems (Gooday and Turley, 1990).

The poor nutritional quality of the DOC and POC (especially if the microbes are not included) is reflected by the fact that it is resistant to decay (Menzel, 1967, 1970; Menzel and Goering, 1966; Menzel and Ryther, 1970). The DOC does not significantly decrease with depth from 400 m to 6000 m. After concentrating the DOC from surface and deep water by pressure dialysis, Barber (1968) found that about one-half of the DOC from surface water is decomposed by the indigenous microflora in 1 month, whereas there was no change in the DOC from deep water after incubation up to 2 months. Viable bacteria were still present after the incubation period. From his research, Barber (1968) concluded that the DOC from deep water is a class of compounds or complexes that are relatively resistant to degradation by the bacteria. In order for the POC to be more useful to microbes, it must be hydrolyzed chemically (Seki et al., 1968) or enzymatically (Gordon, 1970). These hydrolyzed products represent good substrates for the bacteria because they are probably proteins that are refractory but slowly being converted to living material by the bacteria (Gordon, 1970). In assessing the quality of the POC and DOC of the deep sea as nutrients, Morita (1979) posed the following questions: "(1) What portion of the DOC and POC can be used as energy sources for the organisms inhabiting the deep sea? (2) What portion of the DOC and POC can be used to synthesize new cellular material? and (3) Do deep-sea organisms require ectocrine compounds for growth?" If the DOC was subject to appreciable microbial decomposition, then its concentration would decline with depth. Thus, analysis has shown that there is no significant decrease in the DOC below the photo zone with depth. Hedges (1987) maintains that the DOC must be more reactive to maintain a balance between organic carbon and CO_2 inputs to the surface and deep oceans. Kepkay (1994) provides a partial answer, stating that DOC degradation in the mixed layer near the ocean surface is enhanced by coupled physical and microbial processes.

The turnover times of DOC, determined by isotopically labeled solutes, increased in a seaward direction (Williams, 1975) by 1 day in estuarine waters, 1 to 10 days in coastal waters, and 10 to 100 days in oceanic waters. In terms of depth, surface water showed turnover rates of 20–50% per day, depths of 400 m showed >1% per day, and at 2000 m, the turnover rate was undetectable. The slow rate of microbial metabolism in the deep sea was termed a "blessing in disguise" by Morita (1984a) because a rapid metabolic rate would degrade all the organic matter in the deep sea and result in no energy being left for the other forms of life. This becomes more important when one takes the residence time of water masses into account. Because it is hard to date microbes, except those that are fossil bacteria, an example of slow growth in the deep sea is the slow growth of a clam taken from a depth of 3800 m. This clam, by 228Ra dating, took 100 years to reach a size of 8 mm and 50 to 60 years for gonad development (Turekian et al., 1975).

Another explanation for the decomposition of POM is that the POM is ingested by various organisms and the decomposition takes place in the intestinal tract by psychrophilic and barophilic bacteria (Alldredge, et al., 1986; Deming et al., 1981). When macroorganisms ingest POM, the adhering bacteria may become the most important nutritive material ingested because they contain the essential amino acids, especially methionine, needed for protein synthesis and vitamins. This is discussed in more detail by Morita (1980b). Recognizing that food supply in the deep sea is limited, Smith and Baldwin (1982) put forth a metabolic strategy for deep-sea scavengers so that the deep-sea organisms have a selective advantage to optimally utilize the available food energy. The strategy is for the deep-sea scavenger (1) to withstand long periods of starvation, (2) to respond rapidly to an organic fall, (3) to locate rapidly an organic fall, (4) to maximize the rate of consumption and quality of the organic fall ingested, and (5) to utilize the consumed food efficiently. They further state that starvation involves a state of temporary "torpor"; the respiration rate would be lowered while the amphipod subsisted on a high energy reserve such as a lipid. Growth does not have to be continuous. It can occur sporadically, using the bioavailable substrates immediately as they become available, followed by metabolic arrest until sufficient nutrients (including energy) become available again, possibly through syntrophy for bacteria.

Coffin et al. (1993) compared the bacterial production and respiration with total DOC concentration and suggested that approximately 1–3% of the pool supports growth. Thus, there must be a tight coupling between bacteria and sources of substrate. It is well known that the chemical speciation of substrate affects bacterial production (Connolly et al. 1992), with the readily utilizable compounds such as amino acids giving bacterial growth efficiency values as high as 70% and bulk DOC giving values of 20% or less (Meyer et al. 1987; Hopkinson et al., 1989; Griffith et al., 1990; Connolly et al., 1992). In water samples taken from the Santa Rosa Sound between January 31 and April 10, 1989, Coffin et al. (1993) found that the total degradable organic matter was similar for all sample dates, but the oxygen consumption was markedly different. Thus, the substrate composition is important and when labile compounds are present, much greater activity occurs. Radioactive experiments by Coffin (1989), Fuhrman (1990), and Suttle et al. (1991) examined the turnover of specific pools of DOC and demonstrated that small labile fractions of the DOM may be turned over remarkably fast. Although the dissolved amino acid pool is but a small fraction of the DOM, it may supply 20–65% of the bacterial carbon and nitrogen requirements for bacteria (Fuhrman, 1990; Suttle et al., 1991).

The stoichiometry of the substrate also influences the bacterial growth efficiency (Goldman et al., 1987b; Hopkinson et al., 1989). Because of this situation, bacterial growth efficiency estimates vary widely. The high range for growth efficiencies (based on change in bacterial biomass and oxygen consumption) is 42% for the Riskilde Fjord, Denmark (Jensen et al., 1990), and 40% for the Southern Ocean (Bjørnsen and Kuparinen, 1991). In the North Atlantic, it ranges from 1.6 to 9.0 (Kirchman et al., 1991) and from 11 to 2% from estuarine to open ocean across the Georgia continental shelf (Griffth et al., 1990). Likewise, growth efficiencies will vary over the diel cycle and consideration must be given to large temporal and spatial variations in the dynamics of bacterial growth efficiencies (Coffin et al., 1993).

Incubation experiments done by Ogura (1970) suggested that the DOM in surface water is composed of an oxidizable half and a refractory half, the latter not easily utilized by microorganisms. Later Ogura (1972a,b) divided the oxidizable half into two fractions. The F_I (easily utilizable by microorganisms) fraction ranged from 10% to 20% of the total DOM and is utilized within the first 50 days; whereas, the F_{II} (utilizable by microorganisms) represents 30–40% of the DOM. The F_{III} (50–60%) of the DOM is refractory and not easily available to microorganisms. He estimated that the rate constant for the decomposition for the total DOM was 0.0052/day and 0.33/day for the F_I fraction. The half-life for the total DOM was calculated to be 130 days and 20 days for the F_I fraction. He suggested that the low molecular DOM is first utilized by microorganisms. After 6 to 10 months, the proportion of low molecular weight DOM decreased and that of the high molecular weight DOM increased. Unfortunately, none of the studies by Ogura takes into consideration the bottle effect.

The doubling time of 260 hours for deep-sea bacteria would result in a maximum O_2 consumption rate of 1.4 ml/m^3/y (Williams and Carlucci, 1976). It should also be noted that the number of bacteria in deep waters ranges from 10^6 to 10^7/l. Carlucci and Williams (1978) estimated the utilization of DOC in the central North Pacific to be 0.3 μg C/m^3/d from simulated in situ growth rates. In situ utilization rates of one component of DOC, glutamic acid, by bacteria in the benthic boundary layer in the eutrophic Panama Basin (2800 to 3850 m), and extrapolated to carbon utilization, ranged from 17.4 to 174.8 μg C/m^3/d (Smith et al., 1986). Based on these estimates, the bacteria in the Panama Basin could utilize up to 0.04% of the available DOC per day, but the rate of supply is unknown. DOC may be supplied through vertical diffusion from the sediment to the overlying water column (Williams et al., 1980). Yet, on the other hand, the high C:N ratio of DOM implies that it is old and refractory. It has been suggested that there may be a less refractory component of the DOM that has a rapid turnover rate and is never "free" in detectable quantities.

Oxygen uptake is a measure of the recalcitrant nature of the organic matter as well as the ability of aerobic organisms to metabolize. Knowledge of the temperature and availability of oxygen and phosphate over a wide area of the north Atlantic Ocean led Riley (1951) to attempt to estimate the respiration rates in the deep Atlantic Ocean. Using various assumptions, he arrived at a figure of around 1 μl O_2/l/year. With better data and theory, Munk (1966) and Arons and Stommel (1967) estimated rates about an order of magnitude higher for depths up to 2000 m. Naturally this includes respiration from all forms of life in these waters. Packard (1969) used electron transport system (ETS) activity to estimate several hundred μl O_2/l/year at the surface of the ocean, about 7 μl O_2/l/year at 500 m, and about 5 μl O_2/l/year or less at 3000 m. Strickland (1971) estimated the respiration rates at 200 m in the Atlantic Ocean to be 100 μl O_2/l/year using Riley's (1951) procedure, 140 μl O_2/l/year by the concentration-respiration method, 30 μl O_2/l/year by the ETS activity method, and 125 μl O_2/l/year by the ATP method. The concentration-respiration method was found to be faulty by Griffiths et al. (1974). The foregoing figures include the use of oxygen by all life forms. With the oxygen consumption figures being in terms of a few microliters per liter per year, many of the microorganisms must be in the starvation state.

The oxygen demand values calculated by various investigators are shown in Table 3.5. These values would be greater if the organic matter present in deep water were more labile and in greater quantity. These oxygen demand values are another line of evidence that the organic matter in the deep sea is recalcitrant and in short supply. If the rates of oxygen consumption were greater than 0.6 ml O_2/ m^3/y (0.006 ml O_2/l/yr), Williams and Carlucci (1976) indicate that vast areas of the deep sea would be anoxic and point out that the calculations based on hydrophysical (vertifical diffusion) models are more valid. Unfortunately, we cannot yet partition oxygen utilization between DOC and POC oxidation (Toggweiler, 1988).

At depths >2000 m, Arons and Stommel (1967) calculated that 2.0 to 2.5 $\times$ 10^{-3} ml of oxygen/l/year was utilized. The residence times (a reflection of the recalcitrant nature of the organic matter) in the deeper portions of the ocean range from 230 to 950 years (Broecker, 1963). The residence time of total organic detrital carbon, based on concentration in the top 15 cm of sediment divided by "estimated remineralization" increased from 11 years on the shelf to a high of 756 years on the abyssal plain, whereas the biomass appeared to turn over on time scales of months (Rowe et al., 1991). The average residence times for the Pacific Ocean and the Atlantic Ocean were calculated to be 500 and 275 years respectively (Stuiver et al., 1983). Radioactive dating of dissolved organic matter in two water samples taken in the Northwest Pacific at depths of 1980 and 1920 m was found to be 2740 ±300 years (Williams et al., 1969). Skopintsev (1972) agrees with this residence time calculated by Williams et al. (1969). The apparent age of surface water in the central Pacific Ocean was estimated to be 1300 years at the surface and 6000 years from 900 to 5720 m, and the DOC is recycled within the ocean on a >10^3 to 10^4 year time scale (Williams and Druffel, 1988; Hedges, 1987). The age of 6000 years indicates that the deep-ocean DOC is more refractory than previously thought and that an as yet undiscovered process exists by which the sea eventually recycles this huge reservoir of recalcitrant organic molecules back to inorganic carbon (Hedges, 1987).

The oxidation of DOC was calculated to be 0.1–0.2%/year, or 0.004 ml O_2/l/ year (Craig, 1971). A generation time of 210 hours (8.87 days) was calculated by Carlucci and Williams (1978) for bacteria. The turnover time of DOM was estimated to be 3300 years by Menzel (1974). There are discrepancies among data. Williams and Carlucci (1976) calculated the DOC residence time to be 738 years, POC to be 4.9 years, and the O_2 to be 1214 years. The DOC residence time is thus three to five times lower than the value of 2600 to 3500 years reported by Williams et al. (1969). The long residence time of OM in the ocean is also evidence of its resistance (Skopintsev, 1971; Wangersky, 1978). DOC appears to

Table 3.5 Oxygen demand values of seawater at various locations by various means.

Location and Depth	$O_2/l/yr$	Reference
Based on hydrophysical parameters		
North Atlantic, 1000 m to bottom	0.001 (average)	Wyrtki (1962)
North Atlantic	0.002 to 0.0025	Arons and Stommel (1967)
Pacific Ocean, 1000 m +	0.002 to 0.0025	Munk (1966)
Pacific Ocean, 1000 m +	0.0027 to 0.0053	Craig (1971)
Pacific Ocean, 3000 m +	0.0015	Tsunogai (1972)
Based on ^{14}C-labeled substrates		
Northeast Atlantic[a]	0.0046 to 0.0296	Williams and Carlucci (1976)
Western Pacific[b]	0.0256 to 0.0767	Williams and Carlucci (1976)
Based on electron transport system		
Eastern tropical Pacific[c]	0.00009	Williams and Carlucci (1976)
Based on the ATP content of the cells		
Northeast Pacific[d]	0.0014	Williams and Carlucci (1976)

[a]Calculated from the data obtained by Jannash et al. (1971).
[b]Calculated from the data obtained by Sorokin (1972).
[c]Calculated from the data obtained by Packard et al. (1971).
[d]Calculated from the data obtained by Holm-Hansen and Paerl (1972).

be largely unreactive over time spans on the order of decades to hundreds of years (Williams and Druffel, 1988; Bauer et al., 1992).

The dwarfism that occurs among the higher organisms emphasizes the lack of organic energy in the deep sea (Allen, 1979). Because of the lack of energy, the average size of ascidians decreased from more than 1 cm in the neritic zone to 2.5 mm in the abyssal plains (Monniot, 1979). This, by volume, is a decrease of 25 to 1. As a further indication of the lack of energy in the deep sea, it has been found that a bivalve (*Astarte borealis*) required 50–60 years for gonad development and 100 years to reach 8 mm in size.

For the best reviews dealing with organic matter in the ocean and its relationship to microbes, consult Joint and Morris (1982) and Moriarty (1988).

Colloidal and Aggregate Nature of Organic Matter

How much of the colloidal organic carbon (COC) can be used by microorganisms in the aquatic environment? One of the largest reservoirs of organic carbon on the planet is COC (Cauwet, 1978; Hedges, 1987; and Druffel et al., 1992), outweighing phytoplankton or bacteria by a large margin (Cauwet, 1978; Druffel et al., 1992). This abundance of colloid particles between 1 nm and a few μm can reach densities as high as 10^7 to 10^9/ml (Koike et al., 1990; Wells and Goldberg, 1991, 1992, 1993, 1994). The COC is not readily accessible to bacteria, especially the smaller colloids, but when 0.2 to 2.0 μm in diameter and incorporated into microaggregates, the COC is broken down as the aggregates become bioreactors for organic material (Kepkay, 1994). An important item to be noted is that when aggregation of colloids and bacteria takes place by surface coagulation, it triggers a brief 2- to 4-h episode of bacterial respiration and the OM released by this activity remains largely uncoupled from bacterial growth (Kepkay, 1994). How much of this activity is reflected when marine samples are taken for study? If bacteria are in the starved state, where does the energy come from in order to produce the enzymes necessary to metabolize the COC?

Batoosingh et al. (1969) determined that colloids in the range of 0.2–1.2 μm size were the primary components of the organic matter removed from seawater by surface coagulation onto bubbles. The organic matter in a submicrometer size from 0.38 to 1.0 μm was found to be an important and labile fraction of the DOC in the upper ocean (Koike et al., 1990; Toggweiler, 1990). Johnson and Wangersky (1986), Kepkay and Johnson (1988, 1989), and Kepkay et al. (1990a,b) found that bacterial respiration and cell numbers both increase when the production of microaggregates by surface coagulation bring colloidal DOC and bacteria in close contact, confirming the fact that the colloidal DOC is labile; it does not, however, appear to be accessible to bacteria when dispersed throughout seawater (Kepkay and Johnson, 1989). The colloids ranging in size from 0.001 to 1.0 μm escape bacterial degradation by virtue of their particle size characteristics (Johnson and Kepkay, 1992). It has been suggested that colloidal DOC, referred to as "submicrometer" or "high molecular weight" carbon, is a predominant fraction of the DOC in the upper ocean (Sugimura and Suzuki, 1988; Koike et al., 1990, Wells and Goldberg, 1991; and Toggweiler, 1990, 1992). On the other hand, Benner et al. (1992) and Ogawa and Ogura (1992) suggest that low molecular weight material (<0.001 µm) is the predominant fraction of the DOC, making up to 65–80% of the total DOC. Colloidal DOC, according to Benner et al. (1992), includes substantial carbohydrate content, making it the most likely candidate for labile material for the heterotrophic bacteria present. This limited access is evidenced by the studies dealing with free-swimming bacteria and the cells associated with aggregates.

Marine snow exists in the oceans as aggregates of organic detritus, microorganisms, and inorganic particles. They may range from 500 µm to many centimeters and their abundance may range from 1 to 100 aggregates/l in surface water to less than 0.1 aggregates/l in the deep sea (Alldredge, 1989). They are a means by which organic matter can be transported to the deeper portions of the oceans. These aggregates provide the surfaces where microbes can attach and grow rapidly. Hodson et al. (1981b) found that bacterial populations are more numerous and more active in marine macroaggregates than in the surrounding water. A significant mass-transfer advantage is associated with bacteria on these aggregates, but only when their metabolism is supplied by transport of larger colloids (Logan and Hunt, 1987). A substantial amount of DOC may exist as biologically labile colloids in seawater, but the way in which bacteria gain access to this globally significant reservoir of carbon remains unknown. Thus, according to Johnson and Kepkay (1992), models of collision efficiency and the transport of colloids to bacteria by Brownian movement, turbulent shear, and bacterial swimming suggest a significant fraction of the DOC could escape degradation by virtue of its particle size characteristics. On the other hand, coagulation of DOC on bubble surfaces (concentrating colloidal material into larger aggregates) appears to enhance the microbial respiration of the DOC (Kepkay and Johnson, 1988, 1989; Kepkay et al., 1990a,b). By this means, approximately 5–15% of the ocean's recalcitrant DOC reservoir could be recycled back to CO_2. This would be an effect of physical perturbation on the DOC, making the DOC more readily bioavailable to the microorganisms. With depth, the quality of the organic matter gets lower and lower, mainly because microbes have the time to work over the organic matter as it sinks. This concept indicates the importance of coagulation and the formation of bacterial-colloid aggregates before larger colloids can be utilized efficiently by the bacteria and recycled to CO_2. The question that Johnson and Kepkay (1992) put forth is "what processes are most important in the coagulation of colloid-size particles into aggregates?" These possible processes include surface coagulation of colloidal organics on bubbles, especially at the air-sea surface interface (Kepkay and Johnson, 1989), interaction of large colloids with preexisting macroaggregates (McCave, 1984; Logan and Hunt, 1987; Jackson, 1990), and the Browning pumping of small particles to larger aggregates (Honeyman and Santschi, 1989). For more details concerning all aspects of colloids in the marine environment, consult Kepkay's (1994) review.

Humus and Humus Complexes

Marine humus is different from soil humus, mainly because of its origin. Soil humus is derived from plant remains, whereas marine humus is derived from phytoplankton remains and fecal residues (Jackson, 1975; Moore, 1969). The abundance and nature of humic substances in natural waters and sediments show marked geographical variability (Jackson, 1975). Humus in sediments is concentrated in clay- and silt-size sediments in relatively nearshore sediments and in the central regions of lakes and inland seas. Marine humus obtained from three northern Pacific sediments contains from 0.14 to 0.35% organic matter, including 20 to 1145 ppm of amino acids, and the remainder is kerogen (Palacas et al., 1966). The relatively nonpolar fraction of the marine humic material consists primarily of acid-rich aliphatic polymers with molecular weights of 500–1000 that bear little resemblance to any known biochemical or humic substances in soil (Hedges, 1987). According to Ehrlich et al. (1972), the organic matter of the deep sea sediments contains a fraction which, although refractory to microbial attack in situ, is readily attacked by microbes when brought to the surface. Presumably, the hydrostatic pressure and low temperature deter the microbial activity in situ.

Apparently the most successful method to isolate dissolved organic matter from seawater is to use selective adsorption onto synthetic resin, which recovers only less than 10% (Hedges, 1987). Also fulvic acid from seawater differs from soil and marine sediment fulvic acid (Stuermer and Harvey, 1977). For instance, *Gelbstoff* (marine water humus) is undetectable in large areas of the world ocean (Kalle, 1966), hence most DOM cannot be called *Gelbstoff*. *Gelbstoff* can be distinguished from its terrigenous counterpart by its greater fluorescence in ultraviolet light and its lower degree of condensation (Kalle, 1966). *Gelbstoff* was isolated from DOC from waters off the coast of Trondheim, Norway (6 km), and found to be phenolic in nature (Sieburth and Jensen, 1968). The exudate of brown seaweeds is phenolic in nature and forms polyphenols in the alkaline seawater. In turn, the polyphenols react rapidly with proteinaceous and carbohydrate material of either algal or other origin (Sieburth and Jensen, 1969).

Base-soluble organic matter (humic substances) has been demonstrated to be an important fraction of the uncharacterized organic matter in seawater (Jeffrey and Hood, 1958; Williams, 1961; Khaylov, 1968; Sieburth and Jensen, 1968, 1969; Stuermer and Harvey, 1976; Kerr and Quinn, 1975). Alcohol-soluble organic substances are also present in important quantities (Stuermer and Harvey, 1976).

The $\delta^{13}C$ for DOC is close to that for the so-called cellulose and lignin fractions of the marine phytoplankton and zooplankton (Degens et al., 1968).

Characterization of humic substances isolated from open ocean environments indicates that it is primarily from an autochthonous source (Harvey and Boran, 1985), whereas humic substances in estuaries indicate that about 50% of the organic DOC originates in freshwater and terrestrial environments (Mantoura and Woodward, 1983)

Gagosian and Stuermer (1977) and Gagosian and Lee (1981) present a hypothetical structure of a typical marine humic substance, which has amino acids, sugars, amino sugars, and fatty acid moieties incorporated into its structure (Fig. 3.9). Other molecules (carotenoids, chlorophyll pigments, hydrocarbons, and phenols) may also be incorporated into the structure. It can be seen that if labile substrates (amino acids, amino sugars, etc.) are incorporated into the humic substance (Fig. 3.9), it would be difficult for enzymes to react on them, mainly because the reactive sites are not free. Recognizing that labile compounds are complexed to the humic substances, the reporting of DFAA, dissolved carbohydrates, etc., will depend on the method of hydrolysis. One wonders if these are truly free and bioavailable to microorganisms. The formation of marine humic substances is discussed by Gagosian and Lee (1981). Kerr and Quinn (1975) and Stuermer and Harvey (1977) should be consulted concerning the origin and chemical and physical properties of humus.

Thus far, the labile compounds known to be transformed into higher molecular weight refractory compounds microbially are palmitic acid, alanine, valine (Carlson et al., 1985a), glycine (Carlson et al., 1985b; Iturriaga and Zsolnay, 1981) glutamic acid (Iturriaga and Zsolnay, 1981; Geller 1985b), glucose (Brophy and Carlson, 1989; Shields et al., 1973), and leucine (Brophy and Carlson, 1989). Radioactive leucine and glucose added at natural concentrations to seawater were transformed to high molecular weight dissolved organic materials that persisted through a 6-month incubation period (Brophy and Carlson, 1989). At the end of the 6-month incubation period, only 1–17% of the high molecular weight materials were respired when reincubated with the microbial population of seawater, whereas 40–70% of the monomers were respired over the same period of time. In situ transformations of biologically available carbon, according to the investigators, may be an important source of refractory dissolved organic carbon in the ocean. This process has been shown to occur in the ocean as well as in lakes. Although the magnitude of this process in nature may be slight, it is a mechanism to make ecosystems energy poor for heterotrophic bacteria.

Rice (1982) proposes the following steps for the formation of complexes:

> 1. After initial leaching of soluble material from fresh detritus, microbial enzymes depolymerize the detritus substrate producing reactive carbohydrates, phenols, small peptides, and amino acids. 2. The reactive carbohy-

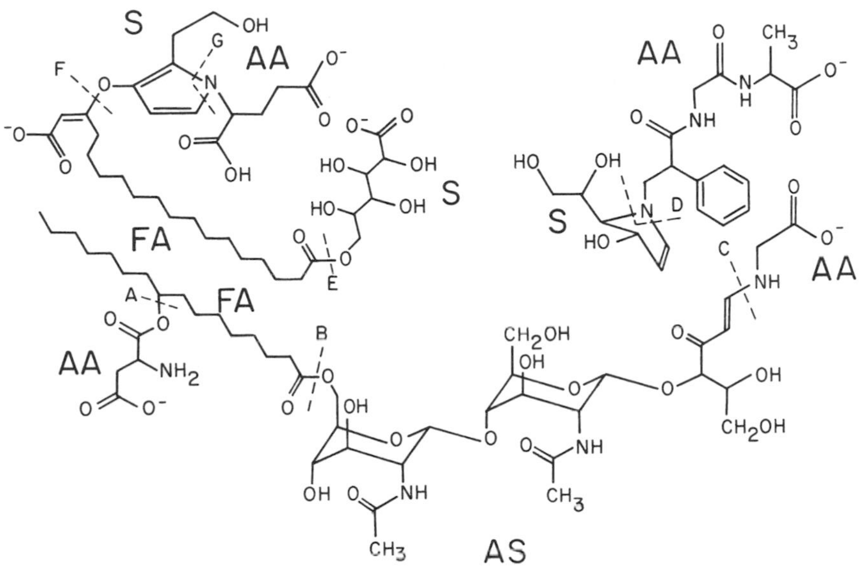

Figure 3.9   Hypothetical structure of seawater humic substances with amino acid (AA), sugar (S), amino sugar (AS), and fatty acid (FA) moieties incorporated. The dashed lines A–G represent sites of bond formation of these compounds. Reprinted by permission from *Marine Organic Chemistry*, E. K. Duursma and R. Dawson (eds.), R. B. Gagosian and C. Lee, pp. 91–123, 1981; copyright 1981, Elsevier.

drates and phenols condense with polypeptides and amino acids to form nitrogenous geopolymers. 3. Microbes associated with detritus assimilate the dissolved nitrogen from seawater for protein (including exoenzymes). 4. Exoenzymes and other proteinaceous materials exuded by microbes condense with reactive carbohydrates and phenols in the humifying detritus. If the condensation products are not easily solubilized in seawater, the detritus particle becomes enriched with humic nitrogen. 5. As humification proceeds from relatively labile amino-sugars, amino-phenols, and polypeptide forms into chemical recalcitrant (and progressively less assimilable) heterocyclic aromatic forms typical of mature sedimentary humic material.

Proteins and Amino Acids

In the ocean, monomeric organic compounds (amino acids, sugars, and fatty acids) have been found to bind abiologically to dissolved macromolecular material in particle-free seawater at natural substrate concentrations (Carlson et al., 1985b). As in the soil, metal complexes are also known to occur. Individual amino acids were usually found at concentrations of 1–10 µg/l and the total dissolved amino acids pool ranged from 5 to 90 µg/l (Williams, 1975). A total free amino acid concentration of 60 µg/l (25 µg C/l) has been reported in marine surface waters (Siegel and Degans, 1966). Concentrations are not much higher in coastal or estuarine water (Wright, 1984). Although the dissolved free amino acids are a very small portion of the total DOC, all evidence points to their importance in bacterial growth; hence, there is a tight coupling between production and utilization (Wright, 1984).

In the marine environment, the dissolved free amino acid (DFAA) pool, in all probability, plays an essential role in the cycle of nitrogen (Coffin, 1989; Mayer et al., 1986; Burleigh and Martens, 1988). DFAA represents about 15% of the dissolved nitrogen in the sea. DCAA (dissolved combined amino acids) may contain many types of bound amino acids including proteins and oligopeptides (Lee and Bada, 1975). DFAA was found to be an important carbon source to marine bacteria. In the Santa Rosa Sound and Flax Pond (shallow coastal environments) cultures, the DFAA sustained from 32 to 91% of the bacterial carbon demand (Jørgensen et al., 1993).

The concentration of dissolved free amino acids in the deep sea ranges from 10^8 to 10^9 M, depending on the amino acid in question (Menzel and Goering, 1966; Lee and Bada, 1975). The bulk of the amino acids in Pacific waters are a constituent of a dissolved component. In deep water, alanine accounts for ~6–70% of the combined amino acids, whereas in surface water, alanine makes up only ~30% of the combined amino acid fraction. Serine is present in the combined fraction in surface waters, but is essentially absent in deep water (Bada et al., 1982). Bada and Lee (1975) point out that the amino acids detected in seawater (including DFAA) may not really be dissolved compounds but may actually be constituents of intact bacterial cells that simply pass through the glass-fiber filter used during the initial water processing.

The amino acid content of aquatic and soil humic and fulvic acids is shown in Figure 3.10, contrasting the values with that of soil humic and fulvic acid. The DFAA or dissolved compounds, analyzed by chemical techniques, are not truly dissolved, otherwise drastic steps (hydrolysis, generally by different acids) would not be needed. Bada and Lee (1977) pointed out that amino acids detected in seawater, including combined compounds, may not really be dissolved compounds but actually the constituents of intact bacterial cells that pass through the glass-fiber filter used during the initial water processing. As mentioned elsewhere, Barber (1968) had to hydrolyze the DOC with acid before the bacteria could utilize it. Current analytical procedure calls for the adsorption of DFAA onto resins and then derivatization. Does derivatization free amino acids from clay particles or from cells? Does the use of resins adsorb off the amino acids complexed to clay minerals? It should be noted that in the literature concerning the amino acid pool of microbes, no distinction has made between those amino acids

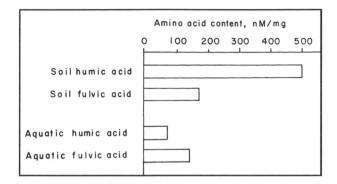

Figure 3.10   Histogram of amino-acid content in aquatic humic substances with soil humic substances shown for comparison. Redrawn and printed by permission from *Organic Geochemistry in Natural Waters*, E. M. Thurman, 1985; copyright 1985, Martinus Nijhoff/ Dr. Junk Publisher. Reprinted by permission of Kluwer Academic Publishers.

on the surface of cells and those inside the cell, mainly because the technique to separate them was too difficult. (For more information, consult Griffiths et al., 1974.) It then becomes conceivable that resins might also adsorb the amino acids off the surface of bacterial cells. The fact that amino acids bioavailable to micro-organisms are much less than the amino acids measured chemically has also been shown by Burnison and Morita (1974), Dawson and Gocke (1978), and Christensen and Blackburn (1980, 1982). This discrepancy also exists for glucose (Gocke et al., 1981).

Vaccaro (1969) pointed out for the first time the phenomenon of inactivity of microbes in the open ocean with respect to the utilization of a labile substrate; in this case the heterotrophic potential method was employed using ^{14}C amino acids. An uptake response occurred within 24 to 36 hours followed by Michaelis-Menton kinetics. Williams and Gray (1970) also showed the same results with different amino acids. Thus, it appears that there was a time delay, which was attributed to the time it takes for enzyme induction of the existing population. However, the synthesis of enzymes takes much energy and the energy is not readily available in the original samples. Does the addition of a labile substrate act as priming mechanism? Cells in the starvation-survival state do take time to recovery (see Chapter 7).

Adsorption to Clay Minerals

The formation of organic polymers on minerals (Degans and Matheja, 1971), selective adsorption of organics on inorganic surfaces (Gustafson and Paleos,

1971), principles of complex formation (Martell, 1971), metal-organic interactions in the marine environment (Siegel, 1971), and metal-organic interactions in soils and waters (Schnitzer, 1971) are all discussed in Faust and Hunter's (1971) book. Amino acids can be adsorbed to clays or other materials (Carlson et al., 1985b; Hedges and Hare, 1987).

The main question concerning adsorption of organic matter to clay particles is the bioavailability of these organic compounds to the microbes, recognizing the fact that biodegradation rates decrease when compounds are in the presence of complex-forming minerals (Gordon and Millero, 1985). This decrease in metabolic activity of bacteria on the organic matter adsorbed to clay particles may be a mechanism to keep the bacteria from a rapid utilization of their energy source so that it will be available for a longer period of time. Chelation of DOM with metals also occurs (Lewis et al., 1971; Foster and Morris, 1971a). Recognizing that DOM can be adsorbed to carbonate minerals, the deposition of these minerals naturally will carry the adsorbed organic matter to the sediments. The interaction of organic matter with various calcium carbonate particles (skeletal fragments, precipitated aragonite needles, or ooid) also occurs in sediments (Suess, 1973).

By use of microcalorimetry and radiorespirometry, Gordon et al. (1983) could not enhance the effect of hydroxyapatite particles on the metabolism of glucose or glutamic acid when using *Vibrio alginolyticus* as the test organism, but did find that bacteria attached to the hydroxyapatite particles were inhibited compared to free-living cells. The respiration of glutamic acid was seven times higher for unattached bacteria than attached bacteria. Their results are in agreement with the observations of Hattori and Furusaka (1960, 1961).

All surface waters are supersaturated with respect to calcite (Wattenberg, 1933; Weyl, 1961; Schmalz and Chave, 1963; MacIntyre and Platford, 1968; Pytkowicz, 1965; Lyakin, 1968; and Broecker et al., 1968). Chave (1965) suggested and Chave and Suess (1967) experimentally showed that dissolved organic compounds in seawater interact with carbonate mineral surfaces to form organo-carbonate associations, as coatings or as monomolecular layers. This organic coating on calcite, aragonite, and a variety of magnesium calcites completely inhibits the equilibrium of these minerals with surface waters of the Caribbean Sea, Sargasso Sea, and the Gulf of Mexico. In seawater, 14% of the total DOC is removed from solution by adsorption processes onto carbonate minerals and these mineral surfaces are saturated with organic carbon after adsorbing 0.1 to 1.5 mg C/m^2, which is the amount of organic matter needed to form a monomolecular layer (Suess, 1970). Calcite selectively adsorbs 30–75% of the initial concentrations of lipoid material, phospholipids, and amino acid-containing substrates when added to water. Thus, between 10 and 14% of the total dissolved organic carbon in seawater interacts with calcite surfaces. Glucose, acetic acid, succinic acid, glutamic acid, and citric acid showed adsorption values ranging from 0 to 94% from a 2-μM solution and changes in both respiration and assimilation of the substrates in the presence of hydroxyapatite were inversely correlated with adsorption (Gordon and Millero, 1985). When peptone is added to water overlying sediments in tanks, a large portion of the peptone is unavailable to bacteria, probably due to adsorption mechanisms of the solid phase (Daumas and Bianchi, 1984).

Bioavailability of Humus and Proteins

Although we recognize that organic matter can be made bioavailable by experimental means, the concept of free enzymes should not be discounted. Free enzymes have been demonstrated by Kim and ZoBell (1974) and Meyer-Reil (1981). The degree to which free enzymes help make the more complex dissolved and particulate organic matter more bioavailable is not known. These complexes cannot be absorbed by microbial cells unless hydrolyzed (Rodgers, 1961). Where the organic matter is complex, exoenzymes are definitely needed to hydrolyze the material so that some of the resulting end products can be transported into the microbial cells. Thus, the exoenzyme of microbes in contact with particulate matter ensures that the exoenzyme is not diluted by the surrounding menstruum and can act locally on the particulate matter. Furthermore, some exoenzymes are bound to the external side of microbial membranes (Pollock, 1962). By adding aminoacyl-ß-naphthylamide (substrate for proteolytic enzymes), which gives rise to a fluorescent product after hydrolysis of the peptide bond, into unfiltered and filtered (0.8- and 0.2-μm pore size) seawater, Somville and Billen (1983) were able to demonstrate exoproteolytic activity in both the filtered and unfiltered seawater. However, it is now recognized that filtration can release organic matter from phytoplankton (Ferguson et al., 1984). Clay particles probably adsorb free enzymes; the question is whether the filtration process would release the exoenzymes so that they will be capable of reacting with the substrate.

Proteins undergo slow degradation in the marine environment, mainly because they are probably complexed to lignins and polyphenols (Odum et al., 1979; Rice, 1982). This fact becomes especially important in nearshore environments because lignins and polyphenols come in with terrestrial runoff. As much as 30% of the nitrogen of aged detritus may exist in the form of nonprotein nitrogen (Odum et al., 1979). Odum et al. (1979) suggest that nitrogen-containing compounds such as amino sugars (glucosamine and chitin), phenol-protein, protein-lignin, protein-chitin, complexes of inorganic clays and amino groups, and nitrogen-containing humic acids are resistant to chemical degradation and therefore not readily digested and assimilated by many detritivores. They question the usefulness of the detrital C:N ratios as food value indicators. Because the nitrogen complexes are not readily chemically degraded, the bacterial degradation processes are also insufficient to enzymatically degrade these complexes. During decomposition, detritus becomes richer in phenolic and carbohydrate groups, which may form condensation products with amino acids, yielding precursors to complex nitrogenous humic geopolymers, and biological availability of the humic nitrogen probably depends upon the extent to which proteinoid subunits are retained in the humic macromolecular structure (Rice, 1982).

Bioavailability of Other Organic Compounds

A glucose concentration of >2 to 78 µg C/l has been reported in the euphotic zone of the Atlantic Ocean (Vaccaro et al., 1968). Many low molecular weight organic acids and sugars are adsorbed onto hydroxyapatite surfaces and there is a negative correlation between adsorption of low molecular weight organic acids and sugars onto a hydroxyapatite surface and biodegradation rates of the compounds in the presence of the mineral (Gordon and Millero, 1985). It does not matter if the organics are equilibrated with the surface of the hydroxyapatite prior to the addition of the organisms or that organisms are preattached to the surface. From 2-µM solutions of glucose, acetic acid, succinic acid, glutamic acid, and citric acid, the adsorption values range from 0 to 95%. Hence, the respiration and assimilation of the substrates in the presence of hydroxyapatite were inversely correlated with adsorption. It is noteworthy that glucose is not adsorbed by hydroxyapatite (Gordon et al., 1983).

During a phytoplankton bloom, Ittekkot et al. (1981) found large amounts of carbohydrates were released into seawater in the northern North Sea, the major portion of this release occurring toward the end of the bloom. The combined dissolved carbohydrate (CHO) is determined by subtracting the free dissolved CHO from the total CHO. The total dissolved (free and combined) during the bloom varied from 40 to 400 µg/l and the dissolved combined CHO comprised more than 75% of the total dissolved combined, ranging from 180 to 360 µg/l. More than 60% of the dissolved combined CHO was glucose, followed by mannose, galactose, and xylose. These latter two were present at ca. 20 µg/l throughout the study. These authors suggest that large amounts of CHO are rapidly released from stressed cells, from lysing cells, and from planktonic debris during a phytoplankton bloom. Because large amounts of CHO were not released during the phytoplankton bloom, Ittekkot et al. (1981) concluded that they are not released in quantity by excretion or exudation; but Wright (1984) suggested that the dissolved CHO is used as rapidly as it is produced, giving a tight coupling between bacterial use and the production of low molecular weight CHO.

About 20% of the total DOC consists of chloroform-extractable compounds, which include fatty acids and triglycerides (10%), mono- and diglycerides (25%), and phospholipids (15%) (Jeffrey, 1968). The extraction process removes lipoid material but does not extract materials such as proteins, free amino acids, and carbohydrates. Nevertheless, phospholipids, amino acid-containing substances, and probably free fatty acids exhibit strong preferential adsorption at the calcite-seawater interface (Garrett, 1967). Other short chain fatty acids, besides acetate, in anoxic sediments have also been studied. However, it is difficult to extract the tightly bound fatty acids (Balba and Nedwell, 1982). When tracer amounts of [^{14}C]propionate and [^{14}C]butyrate were injected into cores, there was an exchange of the tracer between the extractable and tightly bound fatty acid pools. Nevertheless, both propionate and butyrate were utilized by the indigenous bacteria.

The concentration of cyclic-3′,5′-adenosine monophosphate (cAMP) in coastal seawater ranged from 1 to 35 × 10^{-12} M, being greatest at the surface (Ammerman and Azam, 1981). The presumed source of the dissolved cAMP were planktonic organisms (bacteria, algae, zooplankton, etc.). Uptake of the cAMP by marine bacteria may possibly play a role in their metabolic regulation. When substrates are introduced to measure the metabolic activity, the introduction procedure can induce artificially high levels of activity, with a possible disturbance artifact (Tunlid and White, 1992).

BIOAVAILABILITY OF ORGANIC MATTER IN MARINE SEDIMENTS

The organic carbon decreases with distance from land so that, in abyssal marine sediment, the organic carbon content is very low. Because there is such a low organic carbon content in sediments, it is only natural that the oxygen uptake in sedimentary material is also low. Thus, when the organic flux is recalculated relative to the local phytoplankton productivity, the fraction that reaches the sediments is mainly a function of water depth. According to Suess (1988), 30–50% of the flux between 100 and 300 m was living biomass. At a depth of 5000–6000 m, the fraction is about 1%. Consult Ducklow and Carlson (1992) for further information on the flux of organic matter into the deeper portions of the ocean. The environmental constraints on the diagenesis of POM in marine sediments is made difficult by the fact that the breakdown of most polymeric carbohydrates requires multiple enzymes and the activities of multiple species of bacteria (Deming and Baross, 1993). Other members of the consortium can produce proteases that will degrade enzymes. However, the biosynthesis of enzymes is expensive in terms of the energy costs, and the ecosystem, even microniches, is continually being depleted of energy source(s).

In the nearshore environment, the amount of organic matter deposited on the bottom depends on the depth, the phytoplankton productivity in the waters above, and the oxidation-reduction potential. Naturally, if the rate of organic matter production and sedimentation is greater than the oxygen needed to degrade the organic matter, anaerobic and anoxic conditions will result. Nevertheless, anaerobic decomposition also reduces the organic content. What remains is then buried with the sedimentary material and the indigenous microbes will also act on it. Naturally, the greatest microbial activity takes place at the sediment-water interface. Thus, from the surface of the water column to the sediment depths, all the processes are degrading the organic matter so that there is less and less organic matter in the ecosystem with time unless there is a constant input of organic matter. In many locations, the input of photosynthetic organic matter is seasonal and certain ecosystems become anoxic only during the warmer periods of the year. The process to complex the organic compounds, adsorption and absorption of the organic matter, takes place so that not all the organic matter is bioavailable to the various microbes in the system. According to Smith et al. (1987), Sayles and Curry (1988), and McCorkle and Emerson (1988), the pore water model's predicted rates of OM degradation conflict with measured OM fluxes on the sea floor; OM degradation rates calculated from them can only be considered approximations. For anaerobic degradation of organic matter in anaerobic marine sediments, consult Parkes (1987) and Deming and Baross (1993).

Total organic carbon in a salt marsh pan sediment increased with depth due to compaction, but the available carbon varied from 0.5 to 1% of the total carbon in the surface 0–1 cm to <0.2% at depths greater than 10 cm (Nedwell, 1987). Approximately 10 to 50% of the humic substances in recent sediment and soils is acid hydrolyzable and consists largely of proteins and other nitrogenous compounds. Nearshore sediments have organic matter from less than 1% to 10% (Trask, 1939); whereas open ocean sediment (red clay and globigerina ooze) contains less than 1% organic matter (Revelle, 1944). Rowe and Deming (1985) estimated that less than 10% of the organic carbon reaching the bottom is buried (not consumed biologically). Based on laboratory experiments they attributed at least 13 to 30% of the total inferred biological consumption of organic carbon to microbial utilization. Recycling and new production are coupled by nutrient cycles of long duration in deep waters (Eppley et al., 1983). Yet Goldman et al. (1987a) say that the "rate of nutrients cycle in the microbial loop is set by the maximum growth rate of phytoplankton." There is no phytoplankton photosynthesis below the photic zone.

There has been renewed interest in the organic matter in deep-sea sediments. It is now realized that the flux of phytodetritus to the benthic sediment is not uniform throughout the year and seasonal variations occur, with the maximum in the abyssal northeastern Atlantic taking place in July and August (Pfannkuche, 1992). The abyssal environment is also influenced by abyssal storms and deep-reaching vortices (Klein, 1987). Of the phytodetritus that does reach the deep-sea benthos, rapid colonization, growth, and decomposition are noted in sediment community oxygen consumption along with seasonal differences (Pfannkuche, 1992). The quality of sedimenting particles is extremely variable in origin and size as well as in time and space (Alldredge and Silver, 1988). Although models of POC diagenesis in the deep-sea sediments typically describe degradation as a direct function of the concentrations of various types of organic matter, Smith et al. (1992) state that such a formulation may be mechanistically misleading because the degradation rate of specific detrital components can vary solely due to changes in the community structure of the mineralizing organisms. After statistical analysis of the measurements of bacterial biomass, utilization of dissolved amino acids, and supply of organic carbon to the sea floor, Deming and Yager (1992) came to the conclusion that the depth of the ocean was a weak predictor of bacterial parameters whereas the magnitude of the organic carbon flux to the sea floor was a strong one. Deming and Yager (1992) came to the conclusions that: (1) the missing factor in our understanding of bacterial activities in the deep-sea sediments is the quality or hydrolytic potential of the organic carbon supply to the sea floor, and (2) high-latitude basins, from which the highest benthic bacterial biomass and activity rates were recorded, should provide ideal sites for testing hypotheses on the bacterial fate of organic carbon in the deep sea. Furthermore, Deming and Yager (1992) state that inference of bacterial activity from biomass measurements should be avoided, especially in relation to bacteria, since that portion of the total population active on a given substrate at a given time is almost always unknown and unpredictable. For more details concerning the organic matter in deep-sea sediment, consult Rowe and Pariente (1992).

Dead cells can also add to the amount of organic matter in any environmental system. Novitsky (1986) labeled indigenous bacteria in marine beach sand with [^{14}C]glutamic acid, [^{3}H]adenine, and [^{3}H]thymidine and then killed the microbes with chloroform. This sterile beach sand was inoculated with seawater, more beach sand, or both. Labeled RNA from dead bacteria was degraded more quickly than labeled DNA after 10 days of incubation, but both to the same extent of around 60 to 70%. Of the total macromolecular ^{14}C, 32% was respired in the 10-day period, irrespective of the inoculum. Respiration was complete in 3 days but nucleic acid degradation continued throughout the 10-day incubation period. However, in nature, one has to wonder how many of the cells are dead when one can view many more organisms in samples than can be determined by viable counts. Are those that are not viable subject to degradation by other microbial cells?

Gibbon et al. (1989) obtained data indicating that the acetate in pore water of sediments was used solely by the sulfate-reducing bacteria. Two pools of acetate apparently exist in pore water, each with different biological availabilities. Not all the acetate in pore water of marine sediments is degraded by biological processes (Ansback and Blackburn, 1980; Balba and Nedwell, 1982; Christensen and Blackburn, 1982; Fenchel and Blackburn, 1979; Gibbon et al., 1989; Parkes et al., 1984; Sansome, 1988) and the use of ^{14}C-acetate greatly overestimates the acetate turnover, the true rate in sediments (Parkes et. al., 1984). Acetate availabilities differ from one site to another, but the mechanism resulting in different availabilities of acetate await further investigation. ^{14}C-acetate, when added to sediments, was found to be distributed in a porewater pool and two adsorbed pools (Christensen and Blackburn, 1982). One of the adsorbed pools could be displaced by excess acetate while the other could not be displaced. As a result of their studies, they reported that about 84% is unavailable to microbes. Acetate is an important metabolic intermediate in anoxic degradation. Using *Desulfobacter* sp., Parkes et al. (1984) also demonstrated that there were two pools of acetate in marine sediment porewater, much of it unavailable to microbes. When employing radiolabeled acetate to measure in situ acetate turnover rates in sediments, the rates are overestimated (Ansbaek and Blackburn, 1980). Of alanine in porewater, 97% is unavailable to microorganisms (Christensen and Blackburn, 1980). King (1992) used an enzymatic method to detect acetate in sediment pools and was unable to measure the acetate in the sorbed pools. In Loch Etive, only 19% of the chemically measured acetate was available to microorganisms with corresponding values of 48% and 65% for Lock Eil and Tay Estuary respectively (Gibbon et al., 1989).

Christensen and Blackburn (1980) found that 30–40% of the added radioactive alanine was rapidly adsorbed to sediment particles over a wide range (36–230 nM) and that adsorbed alanine decomposed more slowly than dissolved alanine. Christensen and Blackburn (1980) hypothesized that alanine, like acetate, could be complexed by macromolecules in porewater, rendering it unavailable to microorganisms. Similiar results were obtained by Parkes et al. (1984). Thus, amino acids may be associated with particulate organic matter such as clay as exchangeable amino acids or irreversible bound amino acids (Cheng, 1975; Rosenfeld, 1979), as biologically resistant organic matter (in humic substances) (Rashid, 1972), and as intracellular amino acids in sediment-inhabiting organisms such as bacteria (Stanley and Brown, 1976), algae (Fowden, 1962), and invertebrates (Gilles, 1975). Furthermore, Thompson and Nedwell (1985) also demonstrated that much of the acetate and propionate is not available for use by microorganisms. Thus, three pools for these compounds can be designated as: (1) a free pool available to microorganisms for metabolism, (2) a slowly metabolizable pool, and (3) a refractile pool not available for metabolism. Glucose can also fit into this category (Gocke et al., 1981).

The most abundant amino acids found in sediments were glutamic acid, serine, glycine, alanine, and leucine, and glutamic acid and ß-aminoglutaric acid in anoxic sediments (Jørgensen et al., 1981). Glutamic acid has been found to be the most dominant free intracellular amino acid in marine bacteria (Stanley and Brown, 1976; Henrich, 1980) and is dominant in sediment extracts. Jørgensen et al. (1981) suggested that bacteria leak glutamic acid during centrifugation but the amount of intracellular glutamic acid would increase the amino acid concentration about 25 nM. Nevertheless, the length of centrifugation of sediments does play an important role in the amount of amino acids that can be detected (Jørgensen et al., 1980, 1981). Sugai and Henrichs (1992) found that free amino acids were lost from the dissolved pool by both bacterial uptake and adsorption to sediment particles and that adsorption was the dominant process for the basic amino acid lysine. Half of the glutamic acid and alanine was removed from solution. Sugai and Henrichs (1992) also found that serine and glycine were also adsorbed. According to Jørgensen (1977), a microscale environment in undisturbed sediments can support bacterial reactions in conflict with the larger scale oxidizing or re-ducing state of the sediment. In general, Höfle (1988) demonstrated that the tax-onomic structure of pelagic microbial communities responds rapidly to changes in nutrient supply but is rather inert to massive introduction of nonindigenous bacteria. At any one given moment, the rate of microbial activity in sediments is usually limited by the small amount (pool size) of available organic substances (Nedwell and Abram, 1979).

Although free amino acids constitute less than 1% of the total dissolved ex-tractable nitrogen in sediment interstitial water (Kemp and Mudrochova, 1973), they are biologically important in marine sediment. Amino acid turnover times from a few minutes (Christensen and Blackburn, 1980) to several hours (Hanson and Gardner, 1978) have been reported. Jørgensen et al. (1980) suggest that amino acids, as well as other low molecular weight organic compounds, diffuse into the overlying water.

In reference to amino acids in sediments, the term "free" amino acids is an operational definition, because a harsh treatment of sediments may be necessary to obtain the amino acids, resulting in overestimates of their numbers (Jørgensen et al., 1981). Many of the early analyses for amino acids in sediments usually acid hydrolyzed the sediment prior to the analysis for the amino acids (e.g., Ste-venson and Cheung, 1970). This procedure does not take into consideration the dead and living bacteria cellular protein that may be hydrolyzed to amino acids. The extraction of amino acids from sediments is a difficult problem but early investigators believed that the use of acids (HCl and HF) released amino acids and other nitrogenous compounds from inaccessible sites of clay minerals (Ste-venson and Tilo, 1969). Various extraction procedures for amino acids in sedi-ments that have been reported in the literature have been listed by Jørgensen et al. (1981). These are:

1. Pressure filtration with nitrogen gas (Jørgensen, 1979)

2. Centrifugation of sediment samples (Hanson and Gardner, 1978; Jørgen-sen et al., 1980)

3. Hydraulic squeezing (Henrichs and Farrington, 1979)

4. Dialysis membranes placed directly into the sediment (North, 1975)

5. Filtration of sediment in situ through aeration stones (Clark et al., 1972)

5. Extraction with 80% acidified ethanol (Starikova and Korzhikova, 1968)

6. Extraction with water (Brinkhurst et al., 1971)

7. Extraction with ammonium acetate solutions (Whelan, 1977) and

8. Direct derivatization with dansyl-Cl of amino acids in sediment slurries (Litchfield et al., 1974).

Each extraction procedure has certain advantages and disadvantages.

Carbonate sediments may contain OM as POM. This may be directly associated with mineral grains, either as an adsorbed fraction on grain surfaces or as an internal skeletal matrix. Amino acids were found to comprise 15% to 35% by weight of humic substances extracted from marine carbonate and noncarbonate sediments (Carter and Mitterer, 1978). Amino acids laid down during the Oligocene (30 million years ago) (Erdman et al., 1956) have survived. Humic and fulvic acids extracted from carbonate sediments were characterized by an amino acid composition consisting primarily of the acidic amino acids, aspartic and glutamic acids. Humic acid extracted by noncarbonate sediments had a distinctly different amino acid composition consisting primarily of glycine and alanine. According to the authors, carbonate surfaces appear to selectively adsorb aspartic acid-enriched organic matter while noncarbonates do not have this property. Carbonate grains have been shown by Suess (1970, 1973) to preferentially adsorb surface-active organic matter from seawater. Ooids have been artifically precipitated in the presence of humic extracts from sediments (Suess and Fütterer, 1972). The natural ooids also contain organic matter that has been shown to contain proteinaceous material and is similar in composition to skeletal carbonate with a predominance of aspartic acid (Mitterer, 1968; Trichet, 1968). It is surprising to note in the data of Carter and Mitterer (1978) that the amino acids methionine (needed for protein synthesis) and tyrosine are one to two orders of magnitude lower than other amino acids, and in some grain size fractions, methionine is missing. Using fluvic acid with calcite and quartz, Carter (1978) demonstrated that carbonate surfaces are able to selectively adsorb aspartic acid-rich organic matter, whereas quartz did not but had a preference for aspartic acid-poor organic matter.

Deming and Baross (1993) point out that bacterial degradation of refractory organic matter proceeds most efficiently and effectively through the actions of bacterial consortia that develop and persist in anaerobic (or microaerophilic) microenvironments. However, these consortia are not limited to anaerobic microenvironments. Much of the deep sediments beyond the continental shelf are oxygenated, but extremely poor in organic matter, as evidenced by the positive Eh values obtained during the Mid-Pacific Expedition of 1950 on red clay and globigerina ooze (Morita and ZoBell, 1955).

BIOAVAILABILITY OF ORGANIC MATTER IN DEEPS AND TRENCHES

Marine organic matter comes mainly from land drainage and photosynthesis by phytoplankton in the surface waters. Land drainage adds quite a bit of organic matter to nearshore environments, including trenches and deeps that are located near land masses (island arcs). Organic matter in trenches from terrestrial sources is well documented by Wolff (1970, 1976) and George and Huggins (1979).

Coconut husks, branches from trees, etc., have been identified. However, degradation of these materials is slow, mainly because there is a lack of utilizable nitrogen and phosphorus for the cellulose digestors. A similar situation occurs on land where many logs, even in streams, remain in the forest for long periods of time without degrading to any great extent, mainly due to the lack of nitrogen and phosphorus (Aumen et al., 1983, 1985). An analogous situation occurs in some marine environments such as in the trenches.

Shore plants are known to be transported into trenches and deeps (Wolff, 1976; George and Huggins, 1979). The Puerto Rico Trench sediments are rather unusual in that a low redox potential of $+40$ to -30 mv have been recorded (George and Huggins, 1979), indicating sufficient organic matter as energy for microbes to lower the Eh. Generally, abyssal sediments have a high redox potential simply because insufficient organic matter sediments to the bottom so that microbes can bring about reducing conditions. Bacteria have been demonstrated in various deeps and trenches (ZoBell and Morita, 1957) but exceptionally deep strata of the cores have not been taken. The depth of sediment and geological material at which bacterial activity ceases and where bacteria cease to be present is not known.

The availability and kind of food for higher forms are the main factors in structuring the organisms, that is, their number, genus, and species. These principles are manifested in the low diversity in the abyssal regions of oceans, as seen in opportunistic species' behavior, small size and weight, and organisms' internal structures (stomach volume, reduction of intestinal tract size, simplification of certain organs, etc.) (Gaill, 1979; Monniot, 1979; Thiel, 1979).

Smith and Teal (1973) found that sediment benthic communities at 1830 m (average depth of the ocean is 3800 m) showed oxygen uptake two orders of magnitude less than the uptake in sediment communities taken from shallow shelf depths. Another indication of the lack of organic material in deep sea sediments is the redox range ($+200$ to $+300$ mv) (Morita and ZoBell, 1955). If sufficient organic material rained down from the surface, the redox readings would be in the negative range.

According to Deming and Baross (1993), most long-term bacterial residents of the deep-sea sediment may have evolved substrate kinetics that do not allow them to take advantage of elevated levels of OM and they may be oligotrophic bacteria. However, no barophiles have been obtained, thus far, in culture because all known barophiles have been isolated in organic-rich media. Sediment community oxygen consumption and POC flux in the benthic boundary layer were concurrently measured at five deep-sea stations on a transect across the eastern and central North Pacific at different times of the year by Smith (1987). He found no direct coupling between the sediment community oxygen consumption and POC flux, which he attributed to the limitations of the measurements or other organic carbon sources. Although there are numerous studies dealing with the sediment community oxygen consumption, one has to wonder about the inherent errors in the measurement as discussed by Smith and Hinga (1983). On the other hand, the measured supply of POC is adequate to meet the estimated sediment community oxygen consumption in the Atlantic Ocean (Smith, 1978; Hinga et al., 1979; Rowe and Gardner, 1979; Smith and Hinga, 1983).

3.5   Conclusions

From the foregoing, it can be seen that the major ecosystems of the world are oligotrophic. The little organic matter present in these ecosystems is not completely bioavailable to the heterotrophic bacteria, mainly because the organic matter is inaccessible or tied up as a complex with humus or clay. The enzymes necessary to degrade polymers of organic matter so that the microbes can metabolize them are also complexed to humus or clay. The long residence times of organic matter and oxygen consumption in these ecosystems are indications that the organic matter is not available for the microbes to use as an energy source. In some environments, the primary source of the organic matter in time and space is a great distance away. Thus, one is forced to recognize that not all microbes within any oligotrophic environment have sufficient energy to carry on all the necessary functions for cell growth. With the major input of organic matter being the result of photosynthesis, polymers like cellulose are not utilized by all the indigenous microbes; thus syntrophy is the rule. This results in a situation in which some of the microbial cells lack sufficient energy and starvation sets in. If microbes are denied energy for a sufficient time, starvation-survival takes place. As a result, the lifestyle of most organisms in oligotrophic environments is starvation-survival, and this survival time may be short (minutes, hours, days), medium (weeks, months), or long term (years). Starvation begins as soon as the organism is denied an energy source.

4

Estimating Nutrient Bioavailability for Microbes

Each chemical and biological method has its advantages and disadvantages, and each equation or set of equations has its unique advantages and disadvantages. Both the activity and biomass measurements can be artifacts of the experimental methods employed.

—Hobbie and Ford, 1993

Our understanding of microbial processes in the environment is only as good as our methods.

—Kemp, 1994

4.1 Introduction

Microbial ecology is currently methods-oriented and these methods are the driving force in accumulating our current database. Nevertheless, it is desirable to step back and analyze the methods being used in microbial ecology so that we will have some idea as to the limitations of the methods as well as possible underestimates or overestimates of the methods being used in microbial ecology. Methods should incorporate the bioavailability of energy in ecosystems so that future studies can be more realistic. In this chapter various items are examined so that each investigator can take the advantages and disadvantages of the specific method employed in any given study as well as to help design methods that may be more realistic for environmental studies. In all probability, the measurements made on environmental samples may be overestimations of the microbial activity in any ecosystems.

4.2 Processes Making Organic Matter More Bioavailable

Any manipulation of an ecological sample that makes organic matter bioavailable to microbes leads to an overestimation of their activities. The decomposition of organic matter in soil involves aerobic and anaerobic processes that can be severely impacted by storage, mechanical manipulation, and moisture content (Greenwood, 1968). In very dry soil, the polysaccharides are not available because water is not available for its solution. Certain other organic compounds may also be recalcitrant under one set of conditions, but not in others.

The ability to utilize organic compounds may be a function of the nature of the compound or due to one of several environmental factors. For example, denitrifying bacteria may use a variety of organic acids, carbohydrates, and other organic compounds as carbon and energy sources when growing under aerobic conditions, but under anaerobic denitrifying conditions, these organisms are restricted to a few carbon sources (Beauchamp et al., 1989). *Pseudomonas stutzeri* metabolizes cysteine, isoleucine, leucine, and valine when grown aerobically, but under anaerobic conditions, the same amino acids are not used as carbon sources (Bryan, 1981). Koike and Hattori (1974) found that under anaerobic conditions, *Pseudomonas denitrificans* assimilates 60% of the substrate it does under aerobic conditions. Until we know the energy sources, their quantity, and their bioavailability under various environmental conditions for the various species of bacteria in any ecosystem, it will be difficult to fully understand the role in nature of individual bacterial species. For heterotrophic bacteria, their ecological success depends on both the bioavailability as well as metabolizability of carbon compounds needed to generate ATP and provide for the carbon skeletons for growth.

4.2.1   The Bottle (Surface or Confinement) Effect

Placing samples into containers terminates the exchange of cells, nutrients, and metabolites with the in situ surrounding environment. Although experimental manipulation of samples for microbiological analysis facilitates experimental studies, the results can be biased due to the bottle effect. The rapid multiplication of heterotrophic bacteria in flasks containing water from aquatic environments has been documented by many investigators (e.g., Whipple, 1901; Fred et al., 1924; Gee, 1932; Waksman and Carey, 1935; ZoBell and Anderson, 1936; Voroshilova and Dianova, 1937; Lloyd, 1937; Kriss and Markianovich, 1959). A high increase in cell numbers (126 to 7,400,000) due to the bottle effect was reported by Fred et al. (1924) and a similar increase is reported by ZoBell and Anderson (1936). It would be of interest to know how much of this increase in cell numbers accompanies starvation of the heterotrophic bacteria after it has exhausted the supply of available organics.

ZoBell and Anderson (1936) described the bottle (surface) effect (originally called the volume effect) when they observed that both the number of bacteria and their metabolic activity were directly proportional to the ratio of the surface area to the volume of the container in which the seawater was stored. ZoBell (1943) demonstrated that bacteria in seawater, when stored in clean glass bottles, had the ability to multiply and their ability to respire the organic matter present increased. The explanation for this is that nutrients present in low concentration are adsorbed and concentrated onto the surface and, thus, can be more utilizable by the bacteria. This was later verified by Waksman and Carey (1935) and Heukelekian and Heller (1940). This increase in cell numbers also takes place when underground water, surface water, seawater and sewage are placed in a container (Heukelekian and Heller, 1940). Stark et al. (1938) found that measurable amounts of organic matter accumulated over a period of an hour on the surface of chemically clean glass slides suspended in sterile lake water. This indicates that the accumulation process of organic matter is independent of and precedes bacterial growth. Container surface adsorption of organic matter is the basis for the adhesion of bacteria to solid surfaces as demonstrated in both the aquatic environment and in the laboratory (Corpe, 1980; Dempsey, 1981; Fletcher, 1979a,b; Fletcher and Loeb, 1979; Gordon et al., 1981; Kaneko and Colwell, 1975; Paerl, 1975; Sieburth, 1975, 1981) and because of the increased concentration, the nutrients are more available (Duursma, 1965; Fletcher, 1979a,b; Shilo, 1980), especially in oligotrophic environments (Jannasch and Pritchard, 1972; Filip, 1978). It is also possible that the increased adsorption of nutrients is due to the excretion of extracellular polymers resulting from the attaching bacteria (Geesey, 1982; Lange, 1976) or that the refractive organic material in seawater becomes biodegradable by interaction at the surface (Kriss, 1963). Some scientists claim that, in natural environments, the major microbial activity is attributed to attached bacteria (Goulder, 1977; Kirchman and Mitchell, 1982) but others (Azam and Hodson,1977a; Bell and Albright, 1982; Hodson et al., 1981a) attributed it to the free-living bacteria. Since a volume effect has been reported, the major portion of the microbial activity lies with the attached bacteria. It should also be mentioned that photosynthetic rates are influenced by the chemical composition of the incubation container and sampler and other items that come in contact with the seawater samples (Fitzwater et al., 1982).

The greater the surface area in relation to the volume of water, the more rapidly multiplication of bacteria takes place, hence small receptacles provide considerably more surface area (Lloyd, 1937; ZoBell, 1943). For survival, many marine bacteria are attached to detritus or solid surfaces, which adsorb organic substances to higher concentrations than those in the surrounding waters (ZoBell, 1943). On the other hand, Butterfield (1933) found that the size of the container does not appear to make any difference in the increase of bacteria in stored river water. According to Clark (1968), the volume effect has been known to soil microbiologists for a long time. Thus, comparison of decomposition rates are valid only when the same soil is used in all comparisons and when incubation quantities are equal. Different soils are likely to show the volume effect in differing degrees. When large numbers of bacteria are living in soil, the problem of spatial requirements may become critical (Clark, 1967, 1968). When a small amount of alfalfa (organic matter) is added to soil, decomposition is more rapid than when large amounts of alfalfa are added to small amounts of soil. The influence of the volume effect on the decomposition of organic matter, as measured by CO_2 evolution in soil, is shown in Table 4.1. Even after 8 weeks of incubation more alfalfa is decomposed when small amounts of soil are employed. When 10, 25, and 50 g

Table 4.1   Influence of differing soil volumes on residue decomposition.

2 g Alfalfa Meal Added to	Rate of Addition (%)	Weeks of Incubation				
		1	2	4	6	8
		% added C evolved as CO_2				
50 g soil	4	32.0	38.6	42.2	44.3	46.2
100 g soil	2	30.2	35.9	39.6	42.1	43.9
200 g soil	1	26.7	32.2	34.4	36.7	38.5

Source: Reprinted by permission from *The Ecology of Soil Bacteria,* T. R. G. Gray and D. Parkinson (eds.), F. E. Clark, pp. 441–457, 1968; copyright 1968, University of Toronto Press.

of soil are placed in small bottles, progressively more nitrification takes place with increasing amounts of soil, because less aeration takes place with the larger samples (Clark, 1968). Does this surface factor lead to an error in the Frequency of Dividing Cells method (see section 4.5.2), especially when coupled with the fact that this increase occurs within a few hours after sample collection?

Microbial growth on surfaces as well as the significance of surfaces in microbiology is well documented in Marshall (1976), Ellwood et al. (1980), and Bitton and Marshall (1980). It is now acknowledged that enrichment of bacteria occurs at interfaces (solid-liquid, gas-liquid) (Blanchard and Syzdek, 1970; Kjelleberg et al., 1979; Norkrans, 1980). Bacteria do not need to be firmly attached to a surface to benefit from the nutrients adsorbed to the surfaces, although some bacteria are capable of firm adhesion. Small (microcells) rod-shaped heterotrophic bacteria dominate during the early stages of attachment, followed by larger rods and then the prosthecate *Caulobacter* and the *Hyphomicrobium* after 1 or 2 days, followed by the *Cyanobacteria* (Marshall, 1979). Many of the prosthecate, stalked and budding bacteria, which are considered to be model oligotrophs, are known to produce adhesive polysaccharides for attachment.

Results with clay are different. It may be that when clay minerals are used, there is just too much surface area or that there is an intercalation with the clay minerals so that the organic matter is not used as efficiently.

Heterotrophic bacteria that are attached to particles have a higher metabolic activity than unattached bacteria. When bacteria are of the same size, the attached ones have been observed to show greater activity (Fletcher, 1986). This increased activity may, in part, be attributed to the larger size of the attached bacteria compared to the free-living ones (Kirchman, 1983; Alldredge et al., 1986; Simon, 1987) or to the utilization of the particle. Additionally, it may be due to fluid flow through aggregates or particles that the bacteria are attached to, so that they experience advective flow past their surfaces (Logan and Kirchman, 1991), whereas free-living bacteria must move with the bulk fluid. On a per cell basis, increased metabolic activity by attached bacteria has been demonstrated with glucose (Iriberri et al., 1987), glucose and glutamate (Kirchman and Mitchell, 1982), dissolved ATP (Hodson et al., 1981b), phosphate (Paerl and Merkel, 1982), thymidine (Jeffery and Paul, 1986), protein hydrolysate (Simon, 1985), and amino acids (Bright and Fletcher, 1983; Palumbo et al., 1984). Thus, when containers providing a surface are employed, the method may increase the metabolic activity, which may not be the actual rate in situ.

Waksman and Carey (1935) stated that the organic matter in seawater exists in a state of equilibrium between formation and decomposition, and this equilibrium is disrupted by confinement (bottle effect). This was reaffirmed by Jannasch (1967). Ferguson et al. (1984), working with seawater from Frying Pan Shoals, North Carolina, noted that the bacterioplankton community confined at 25°C changed significantly within 16 hours of collection, with an increase in colony-forming units (CFUs), total cell number (acridine orange direct count, or AODC), average cell volume of the bacterioplankton, and turnover rate of amino acids in seawater. Both the non-colony-forming cells and ultramicrocells increased in absolute numbers during confinement, but the CFU increased more slowly. Confinement in a bottle also brings about changes in culturable community species structure during incubation (Ferguson et al., 1984). The abundance of *Alcaligenes* sp., *Pseudomonas* sp. strain 2, *Flavobacterium* sp., and *Vibrio* sp. increased in absolute numbers during 0–16 hours and then continued to increase at a reduced rate through 32 hours. The percent increase in total bacterial cell numbers subsequent to confinement was 41% at 16 to 18 hours and 86% at 30 to 34 hours. This increase was due to bacteria that grew on the nutrients that were somewhat higher than that originally occurring in the water. The increase (76%) in organic matter was due to the filtration of the water sample through a 3.0-μm Nuclepore filter, where the filtration process probably releases organic matter (mainly amines) from the phytoplankton that are damaged during the filtration process and also removes predators. Because not all the bacteria in a sample are culturable, this suggests that confinement selects for bacteria that are culturable in complex media (Flynn, 1989). Köster et al. (1991) also noted the confinement effect in deep-sea sediments in untreated controls but used the term "incubation effect." If energy is added to the sample when confinement takes place, then it is difficult to distinguish between the confinement effect and the "feeding effect" (Köster et al., 1991). Because of the confinement effect, Ferguson et al. (1984) recommend caution in the interpretation of measured increases in bacterial abundance by direct cell counts of confined communities.

Some investigators (e.g., Marshall, 1980; Kjelleberg et al., 1982, 1983) have reported that attached bacteria are ultramicrocells, whereas others (e.g., Hodson et al., 1981b; Pedros-Alio and Brock, 1983) have reported that free-living cells are smaller than the attached cells. There also appears to be a temperature-dependent relationship in freshwater where the unattached cells are larger, when the percentage of attached bacteria and the number of attached cells per particle were also highest (Kirchman, 1983). In chemostat studies, Ellwood et al. (1982) demonstrated that bacteria attached to glass slides grew about twice as fast as free-living cells. Davis et al. (1993) reported that a solid substratum was necessary for expression of the alginate gene by *Pseudomonas aeruginosa* but not for the same organism grown planktonically.

Many investigators have been critical of extrapolating data obtained in bottle experiments to the natural environment. ZoBell and Anderson (1936) questioned the use of controlled bottle experiments to predict bacterial dynamics in the ocean, whereas Gunkel (1972), after noting that confinement of natural waters interfered with quantitative analysis of the relationship between bacterial dynamics and organic substrate concentrations, stated "it is still not possible to design experiments which would allow strict comparisons between results obtained in the laboratory and in the sea." According to Le Fèbre (1986), in situ bottle experiments can hardly be a satisfactory model of phytoplanktons in nature with respect to either the conditions experienced or the physiological activities observed. The same can be said for using in situ bottle methods for studying bacteria.

Samuelson and Kirchman (1990) found that surface energy (work of adhesion of water) determined the amount and availability of adsorbed protein and, consequently, the growth of attached bacteria. Initially growth rates were higher on hydrophilic glass than on hydrophilic polyethylene, but after 6 hours of incubation, growth rates increased with surface hydrophobicity because of increasing amounts of absorbed protein. Both protein adsorption and bacterial attachment decrease with increasing surface energy and availability of adsorbed protein. Thus, initial bacterial growth rates increased as surface energy increased.

4.2.2 Aggregate Formation

Amorphous aggregates that appear in natural aquatic environments contain various microorganisms (bacteria attached to zooplankton fecal pellets, bacteria attached to each other by long polymeric fibers, and bacteria, phytoplankton, and other suspended material attached to the cast houses of mucous netfeeders) (Silver et al., 1978; Alldredge, 1979; Prezelin and Allredge, 1983; Pomeroy, 1984; Amy et al., 1987). Rising bubbles through the water column can concentrate particles and bacteria (Weber et al., 1983; Stanley and Rose, 1967), which may aid in the formation of aggregates because particle collisions may result in the formation of aggregates (Baylor and Sutcliffe, 1963: Barber, 1966). Aggregates may form a microhabitat allowing interactions between microorganisms (mutualism, commensalism, parasitism, and the exchange of genetic material) (Alldredge, 1976; Calleja, 1984) and providing protection from some bacterivores (Curds, 1982). Aggregates can also serve as a food source to filter feeders (Baylor and Sutcliffe, 1963; Kuznetsova et al., 1984). Bacteria are exposed to an increase in nutrients when advective flow through high porous aggregates occurs (Logan 1987; Logan and Alldredge, 1989).

Bioflocculation is genetically controlled and may be expressed only in specific environments (Stewart and Russell, 1977; Johnston and Oberman, 1979). Biofloc-culation has been observed as substrates become depleted (Stanley and Rose, 1967; Bush and Stumm, 1968; Pavoni et al., 1972; Wu, 1978). Since the diffusion model predicts a decrease in substrate availability to cells within aggregates, this decrease increases starvation (Logan and Hunt, 1987). This aggregation is one means by which bacteria in starvation-survival exist. Many types of cells, including bacteria and yeast, aggregate when stressed for nutrients (Calleja, 1984).

4.2.3 Uptake of Substrates

The natural bacterial population in seawater takes up substrate rapidly when enriched with organic nutrients and is noted within 12 hours (Vaccaro and Jannasch, 1966). This uptake of substrate is considered by Wright (1973) to mean that some cells in the population were functionally switched off or dormant mainly because the 12-hour response seemed unlikely to allow time for substantial multiplication to occur. If the amount of substrate added to natural seawater is low, microbial oxidation takes place immediately, but when larger amounts are added, a longer time is needed for complete microbial oxidation to take place (Williams, 1970; Williams and Gray, 1970). Starved cells have the ability to take up substrate rapidly. Williams and Gray (1970) noted that there is an immediate uptake of amino acids by the marine population, followed by a second response after 20–36 hours. In term of labile substrates, bacteria are capable of removing free amino acids at concentrations of 1–10 nM (Kirchman et al., 1985; Morita, 1984b; Semenov, 1991), while glucose is utilized down to the nanomolar range (Semenov, 1991). According to Moran and Hodson (1989, 1990), refractory compounds (e.g., phenolic derivatives of lignocellulose) can also be used rapidly at nanomolar concentrations.

A high affinity uptake system comes into play when bacteria are exposed to low concentrations of substrate (Geesey and Morita, 1979; Azam and Hodson, 1981; Mården et al., 1987). The kinetic parameters for uptake differ when the cells are exposed to nM and µM quantities of arginine (Geesey and Morita, 1979). Thus, when the concentration of substrate is low the microbes have the capacity to bring into play a high affinity uptake system in order to utilize the low substrate concentration efficiently. This bimodal system is found to operate in natural assemblages of marine bacteria (Azam and Hodson, 1981; Davis and Robb, 1985).

Temperature is also known to influence substrate utilization, mainly due to the ability of the organism to maintain membrane fluidity. Rüger (1988) found that deep sea isolates (*Altermonas, Bacillus* and *Vibrio;* all psychrophiles) could utilize numerous substrates (cellobiose, D-galactose, D-glucose, maltose, D-mannose,

D-ribose, sucrose, trehalose, acetate, citrate, lactate, pyruvate, glycerol, and pu-trescine) at 4°C only and not at 20°C. All strains isolated by Rüger (1988) were capable of growth in complex seawater media at temperatures between 1 and 20°C. Depending on the strain tested, gluconate, fumarate, succinate, mannitol, L-arginine, L-aspartate, L-glutamate, and L-ornithine were used as sole sources of carbon and energy at either 4° or 20°C. Substrate enrichment was reported not to increase bacterial abundance, production, and specific growth rate within 24 hours when the temperature was <7°C, but at higher temperatures (>20°C), substrate enrichment effects occurred after 3 hours incubation (Shiah and Ducklow, 1994). This effect was not consistent across the study area investigated by Shiah and Ducklow (1994). Nevertheless, temperature-substrate interaction experiments showed that temperature was more effective than substrate in regulating bacterial production and specific growth rate when compared on the same time scale. This leads one to question the extent to which temperature affects the increased abun-dance and productivity during the period between obtaining the sample and in-cubation of the sample at in situ temperature. Noble et al. (1990) also reports that bacteria grown at low temperature and low nutrients are more nutritionally ver-satile. *Nitrosomonas cryotolerans* (psychrotroph) oxidizes ammonium better at 5° than at 25°C (Jones and Morita, 1985), which may be due to a difference in the lipid composition of the membrane when grown at two different temperatures (Jones and Prahl, 1985).

4.2.4 Utilization of Various Substrates by Starved Cells

Matin (1979) noted multiple substrate utilization by various microorganisms with decreasing dilution rates in a chemostat. Poindexter (1981b) proposed that bac-teria adapted to oligotrophic conditions might be expected to possess uptake sys-tems for a more diverse variety of substrates than bacteria adapted to copiotrophic conditions. Hence, oligotrophs would maximize simultaneous use of a variety of substrates, because energy is in short supply.

Harowitz et al. (1983) compared the characteristics and diversity of isolates from two different media: Bushnell-Haas agar (Difco), which contains mineral nutrients only (no organic carbon) and 2216 marine agar (Difco). They found a higher incidence of pleomorphism among the isolates from low-nutrient media than for the strains from high-nutrient media. Furthermore, they found that the isolates from low-nutrient media were nutritionally far more versatile than those isolated on high-nutrient media. The low-nutrient isolates could utilize two to three times the number of alcohol, carboxylic acid, amino acid, and hydrocarbon substrates as the high-nutrient isolates. The isolates on Bushnell-Haas medium were capable of growth over a wider range of temperature, pH, and salinity values than bacteria isolated on high-nutrient media (Harowitz et al., 1983). All the organisms employed in their study were psychrotrophs. Upton and Nedwell

(1989) showed that Antarctic bacteria behaved in a similar metabolic fashion as reported by Harowitz et al. (1983). In this study, the copiotrophic bacteria were grown on the casein-peptone-starch-agar (CPSA) medium of Wynn-Williams (1979) that incorporates casein hydrolysates, peptone, and glycerol, whereas the oligotrophic bacteria were grown on the oligotrophic enrichment (OEMS) medium based on Bold's (Nichols and Bold, 1965) medium, which contained 5 mg glucose and 5 mg arabitol/l, plus a small amount of vitamins B_6 and B_{12}. Upton and Nedwell (1989) showed that isolates placed on Bold's medium incorporating different substrates had the ability to use more different substrates than those isolates placed on the CPSA medium. The scientists applied a "nutritional flexibility" index, where the median flexibility index for the copiotrophs and oligotrophs was 0.222 and 0.889, respectively. Thus, the oligotrophic bacteria had more flexibility in terms of nutritional capabilities than the copiotrophs. At low substrate concentrations, nutritional versatility was also reported by van der Kooij and Hijnen (1983). The subsurface population as a whole may be capable of utilizing a wide range of organic compounds (Fredrickson et al., 1991). This nutritional flexibility may be due to the composition of fatty acids in their membranes, which, in turn, changes the membrane fluidity and the cell's ability to take up substrates (Cronan and Gelmann, 1975). Noble et al. (1990) demonstrated that bacteria, mainly *Vibrio, Pseudomonas, Flavobacterium,* and *Alteromonas* isolated from the pelagic zone and the benthic surface layer of the continental shelf (both oligotrophic environments) near Newfoundland and grown at low temperature and low nutrients are more versatile in terms of substrates used than when grown in rich media. However, some of the strains were fastidious. Most versatile were the *Pseudomonas* and *Flavobacterium.*

The differences in the induction response and the initial reactions of quinoline degradation between short-term (2 days) and long-term (60 to 80 days) starved cells of *Pseudomonas cepacia* were studied by Truex et al. (1992). With a concentration range of quinoline at 39 and 155 µM, long-term starved cells converted quinoline to degradation products more efficiently than did short-term starved cells. The induction process for quinoline degradation was equal or better in long-term starved cells than in short-term starved cells.

An environmental actinomycete studied by Herman and Costerton (1993), which was capable of utilizing *p*-nitrophenol as its sole carbon and nitrogen source, starved for 8 weeks did not shown any reduction in its ability to biodegrade *p*-nitrophenol. In fact, the mineralization of $[^{14}C]p$-nitrophenol was better for the first 8 days in starved actinomyces.

However, at low substrate concentrations (Standing et al., 1972) or at low dilution rates in chemostats (Harder and Dijkhuizen, 1982), there may be no catabolic repression or other inhibitory phenomena, such as inhibition of substrate uptake. This would enable the organism to metabolize more than one substrate simultaneously. Schmidt et al. (1985) states that it is likely that bacteria acting on low concentrations of organic molecules in natural ecosystems may simultaneously metabolize more than one organic substrate. Thus, it appears that bacteria growing in oligotrophic environments have adapted to utilization of a broader range of substrates and, therefore, have less specific uptake systems (Poindexter, 1981b). The ability to use a broader range of substrates by starved bacteria is also useful in bioremediation as well, permitting the microbes to use much of the indigenous organic matter as energy.

The reason for a greater ability to use a broader range of substrates probably lies in the fact that there are membrane changes when cells are starved. Permeability changes take place in *S. lactis* when it is starved in phosphate buffer (Thomas et al., 1969). Changes also occur in the phospholipids when cells are starved (Guckert et al., 1986). For more detail concerning changes in bacterial membranes, see Chapter 7.

4.2.5 Perturbation of Natural Material

Perturbation, when obtaining the samples as well as during laboratory manipulations, adds to the increase in the bioavailability of the energy in the sample. Kranck and Milligan (1988) warned that methods used to sample natural populations destroy delicate assemblages. We routinely shake a flask or bottle in order to break up clumps and to ensure that the sample represents an average component of the natural population. Although bacterial growth rates and production are measured by various techniques, the samples are generally filtered to remove predators (Wright and Coffin, 1984a; Landry et al., 1984). Filtration techniques also break up aggregates and flocs. The delicate nature of marine snow is an example of this effect. The perturbation caused by sampling, shaking, and filtration may cause organisms to be stressed, but this is seldom taken into consideration in field research as well as laboratory studies. It is impossible to duplicate the original environment, especially when samples must be taken. Therefore, precautions should be taken to minimize sample perturbation, storage time, and the temperature at which samples are transported. Routine laboratory procedures can bring about perturbation. This perturbation is caused during incubation of the samples accompanied by shaking, filtering the sample, incubation of sample at a different temperature rather than the in situ temperature, change in the physical makeup of suspended material, drying of soil samples, etc. In most cases, perturbation increases the activity (uptake of substrate or growth) of the bacteria being studied. This section details some of the perturbation effects that occur in soil and aquatic samples as well as perturbations caused by shaking, filtration, other organisms, etc.

PERTURBATION IN SOIL

Even after a rainfall, the size, shape, and biomass of bacteria change temporarily (Clarholm and Rosswall, 1980; Campbell and Biederbeck, 1976; Cutler and

Crump, 1920). Besides providing water, rain also washes down substrates from the canopy (Tukey, 1970) and leaches plant litter. Thus, in natural soil, some of the bacterial biomass increase may be due to the bioavailability of energy.

Drying and rewetting of soil brings about an increase in decomposition of organic matter until the organic matter reaches some relatively stable form (Birch and Friend, 1961). This is attributed to the effect of drying on water-soluble material (Birch, 1959) and on the microbial population (Stevenson, 1956). Plate counts by Lund and Goksoyr (1980) found that the bacterial population doubled 4–5 hours after dried soil was remoistened, but when microscopically counted it took 90 hours for the bacterial population to double. This rewetting was accompanied by a burst of respiratory activity of the bacterial population after 2–3 days. This drying and rewetting process brings about a rapid acceleration of the release of plant nutrients (Birch, 1960; Soulides and Allison, 1961; Harada and Hayashi, 1968; Marumoto and Yamada, 1977). The amount of nutrient mobilized in soils that had been dried or fumigated with $CHCl_3$ was closely related to quantities available in freshly killed biomass (Marumoto et al., 1982b). Air drying and rewetting every 30 days for 260 to 500 days of incubation caused an increase in the evolution of CO_2 ranging from 16 to 121% as compared to the controls kept continuously moist (Sorensen, 1974). Greater microbial activity when air-dried samples are rewetted was shown also by Lebedjantzev (1924), Stevenson (1956), and Birch (1958). In addition to killing microorganisms, the drying process causes a release of amino acids and other compounds from humic material (Stevenson, 1956; Soulides and Allison, 1961; Hayashi and Harada, 1969). The increase in microbial activity, when air-dried soil is remoistened, has been attributed to an increase in the amount of readily available soil organic matter (Lebedjantzev, 1924; Stevenson, 1956) and to the "physiological youth" of the microbial population (Birch, 1958). Some of this increase may result from the decomposition of dead microbial cells (Jenkinson, 1976; Anderson and Domsch, 1978; Marumoto et al., 1977), which make up a substantial portion of the organic matter released during the drying and wetting process. When ^{14}C- and ^{15}N-labeled dead microbial cells were added to soil, about 76% of the ^{15}N mineralized was attributed to the drying-rewetting process (Marumoto et al., 1982a). The sterilization of moist soil with ether or chloroform also resulted in an increase in decomposition of the soil organic matter, which was attributed to the development of a microbial population with the ability to utilize microorganisms killed by vapor treatment as substrates (Birch, 1959; Jenkinson, 1966a). Chapman and Gray (1986) have demonstrated that this could have an appreciable effect on microbial growth rates in sediments.

In some studies soil samples are frozen before they undergo microbiological and/or chemical analysis. Freezing and thawing of soil have been shown to increase the decomposition of organic matter (Soulides and Allison, 1961).

Perturbations, such as cultivation of fallow land, periodic flooding, clearcutting, freeze-thawing, chronic ionizing radiation, and fire and wind erosion, bring about disturbance of the soil, which, in turn, brings about changes in the activities of the microorganisms present. Conversion of land to agricultural purposes makes the labile pool of organic matter susceptible to mineralization, disrupts internal recycling of nutrients, and increases the potential for loss of nutrients from the system. Microbes also act as a sink for plant nutrients since mineralization-immobilization processes occur simultaneously. The most effective way to increase organic matter in cultivated soil is to stop tillage. Plowing improves the availability of organic substrate to microorganisms but in unperturbed soil, microbial growth can only be initiated if the soil is simultaneously amended with nitrogen fertilizer (Domsch, 1986). When virgin or fallow land is converted to a single crop production, a reduction in the diversity or resources and microhabitats as well as total carbon and nitrogen contents occur (Lemaire and Jovan, 1966; Chu and Stephen, 1967; Ayanaba et al., 1976) and the detrimental effects are outweighed by the benefits that accrue to the opportunistic fungal decomposers or bacteria. These opportunistic decomposers occur primarily as spores (Warcup, 1955; Christensen, 1969; Griffin, 1972), which germinate following crop growth. Perturbations in biomass can also occur when soils are disrupted by sampling, screening, mixing, and preincubation prior to biomass assay (Carter, 1986; McLaughlin et al., 1986; Sparling et al., 1985; and West et al., 1986).

Alexander (1979) points out that peat can be decomposed within a few years simply by draining (drying) the land area and that organic matter in virgin soil is destroyed by bacteria after ploughing or repeated cycles of drying and wetting. Thus, certain organic molecules are resistant to decomposition only under certain environmental conditions. Soil organic matter includes microbial biomass, an unprotected or labile pool that is protected in the absence of cultivation, and a pool that persists for very long periods (Duxbury et al., 1989). Even in laboratory experiments, mechanical disturbance of soil increased the rate of decomposition of the soil organic matter (Rovira and Greachen, 1957). There is a temporary flush of CO_2 evolution if soil is fumigated (Jenkinson, 1966a), and fumigation with chloroform is used to determine microbial biomass. There is a great increase in O_2 consumption in chloroform fumigated soil compared to untreated soil (Jenkinson and Ladd, 1981). Instead of using intact soil, it is common practice to collect samples from a soil profile and homogenize them after the removal of plant debris and macrofauna, sieving, and air-drying the sample before analysis takes place; this practice has been criticized by Burns (1986), mainly because of the gross amount of perturbations. There are many studies dealing with perturbed and nonperturbed soils. Sørensen (1971) compared the degradation of organic matter, as measured by CO_2 production, and found that ^{14}C organic matter (wheat straw) degraded 1.2 times faster in the laboratory than in field studies, whereas native organic matter increased 4.3 times faster in the laboratory than in field studies. Thus, perturbations during sampling and laboratory analysis will bring about increased activity of the indigenous microbes, giving an erroneous picture of the activity of microbes in situ.

Shield et al. (1973), using ^{14}C-glucose rather than plant material under field conditions on a clay soil, noted that microbial metabolites produced from the glucose were very stable. The soil was then partially sterilized with $CHCl_3$ followed by reinoculation and incubation. After 14 days of incubation 30% of the stabilized ^{14}C was released. Freezing and thawing and wetting and drying of soil released 16.2 and 7.9% of the glucose, respectively. Very little N was remineralized under these field conditions; under laboratory conditions 30% or more may be mineralized during the same incubation period.

N mineralization occurs more readily in disturbed than in undisturbed soil samples (Cabrera and Kissel, 1988), as can be see in Figure 4.1. Generally, models fitted to disturbed samples cannot be used to predict N mineralization in undisturbed samples. The method proposed by Stanford and Smith (1972) overpredicts the amount of N mineralized. Overprediction may be related to the drying and sieving of the sample before incubation and/or to the physical conditions of the samples before incubation (soil mixed with sand or vermiculite). Drying of soil samples can increase the amount of N mineralized after rewetting (Seneviratne and Wild, 1986) and disruption of soil aggregates can increase minerilization of organic N (Craswell and Waring, 1972; Hiura et al., 1976). Thus, N mineralized in disturbed soils was larger than in undisturbed soil samples. Large net N mineralization was found by Marion and Miller (1982) in air-dried and in field-moist intact samples of a tussock tundra soil incubated at 35°C and at 0.02 MPa. The appearance of a small pool of N mineralization that decomposes easily in disturbed soil samples may be the consequence of drying the soil, because Seneviratne and Wild (1986) have shown that drying of soil samples can cause a flux of N mineralization upon rewetting. After studying various homogenates of soil, Faegri et al. (1977) concluded that respiration is inhibited or suppressed in soil because soil homogenates are six to eight times higher than nonhomogenized soil. The microbial activity, not only in terms of respiration (CO_2) of organic matter, but also in terms of N mineralization is increased by mechanical perturbation of soil. Thus, in vitro data obtained should not be accepted as in situ data.

When deep subsurface volcanic rock samples were perturbed and stored for 1 week at 4°C, the abundance of viable bacteria recovered increased, while the diversity and evenness of recoverable heterotrophic bacteria communities generally decreased (Haldeman et al., 1994). Each distinct morphological type of colony was recovered before and after storage, purified and characterized by fatty acid methyl ester (FAME) and API-rapid-NFT (Analytab Products) strips. The composition of the recoverable microbial communities changed with storage of rock samples, as determined by Microbial Identification System (MIDI) cluster analysis. Some groups were recovered only before, only after, or at both sampling times. Those isolates recovered after storage generally had faster generation times

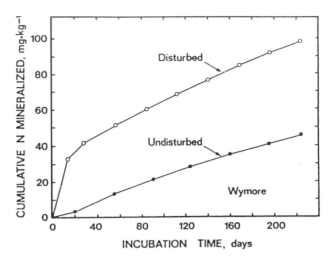

Figure 4.1   Cumulative N mineralization with time in disturbed and undisturbed samples of Wymore soil. Reprinted by permission from *Soil Science Society of America Journal*, 52:1010, M. L. Cabrera and D. E. Kissel, 1988; copyright 1988, Soil Science Society of America.

than isolates recovered only on initial plating. Since some new genera are recovered only after storage, the possibility remains that dormant bacterial types are resuscitated during sample perturbation and storage. The above situation was also noted with other subsurface samples (Amy et al., 1992; Haldeman and Amy, 1993b; Brockman et al., 1992; and Hirsch and Rades-Rohkohl, 1983).

When bacterial analysis is conducted on rocks, they are crushed first and then made into a slurry with water or some type of liquid medium and then analyzed. Nutrient-rich media, without exception, were unsuitable for rock investigations; soil extract agar (James, 1958) and PYGV agar (Staley, 1968) were found to be the best media for demonstrating growth of bacteria from rock material (Weirich and Schweisfurth, 1985). Further mechanical perturbation results when the slurry is shaken. The bacterial count rose steeply (approximately a little more than one order of magnitude) when compared with the time-zero samples, whereas in non-shaken samples the bacteria counts did not increase significantly (Weirich and Schweisfurth, 1985). This observation was noted as early as 1922 by Waksman and later corroborated by Stevenson (1958) and Hirte (1962). The optimum shaking duration depends on the shaking apparatus and shaking rate.

Storage of environmental samples is another form of perturbation of the sample that can lead to severe artifacts in a variety of parameters. Moreover, a gradual decrease in the aerobic bacteria and actinomyces was noted by Takai and Harada (1959) when water-logged paddy soil was stored. Stotzky et al. (1962) noted that the number of most bacteria decreased as a result of storage of soil samples; the percentage of distribution of several groups increased. Storage of soil samples for more than a few hours brings about changes in the sample's microbiology (Jensen, 1968). Storage of subsurface samples has also been shown to increase the AODC counts of bacteria (Ghiorse and Balkwill, 1983). Haldeman et al. (1994) noted that the abundance of viable bacteria from deep subsurface volcanic rock samples increased after rock perturbation and storage for 1 week at 4°C. The storage temperature makes a difference in the phospholipid profile of a soil microbial community, especially when stored at higher temperatures of around 25°C (Petersen and King, 1994), providing further proof that storage induces changes in microbial species.

Atlas (1991) found that disturbance of water samples resulted in lowered taxonomic and genetic diversities of microbial communities and that microbial populations within disturbed communities demonstrated capabilities for enhanced survival, indicating a capacity for generalized adaptations to ecological disturbances.

BIOPERTURBATION

Perturbation can also occur due to invertebrates (earthworms, mites, collembola, etc.) in soil. These invertebrates grind plant materials so that they are more accessible to bacteria and they also promote the formation of clay-organic matter complexes. This type of perturbation also mixes the detritus with the soil and spreads microorganisms throughout the soil. Predation also occurs and keeps the microbial population active by releasing nutrients that would be tied up in the microbial standing crop. Thus, we have to recognize that microbe-animal interactions play an important role in function and activity. For example in soil, the effects of gut passage by earthworms result in casts containing fewer fungal propagules and denitrifying organisms, higher bacterial total counts, and more cellulolytic, hemicellulolytic, amylolytic, and nitrifying bacteria than unworked soil (Loquet et al., 1977). In the total soil volume, 40% of all aerobic, free-living nitrogen-fixers, 13% of anaerobic nitrogen-fixers, and 16% of denitrifying bacteria were estimated to be located in a narrow zone a few millimeters around earthworm burrows (Bhatnagar, 1975). For more information concerning bioperturbation in soil, consult Anderson (1987, 1988).

PERTURBATION OF AQUATIC SAMPLES

Bacteria in aquatic environments may be exposed to a fluid mechanical environment (quiescence, advection, and fluid shear). It does not appear possible to eliminate these limitations in the laboratory. The difficulty with laboratory studies is simply the fact that, in general, the microbes are subjected to greater substrate concentration than found in nature. Furthermore, one has to question whether bacteria in batch or continuous culture behave as they do in nature.

Using *Escherichia coli* and *Zoogloea ramigera,* Confer and Logan (1991) demonstrated that increased uptake of macromolecules (bovine serum albumin and dextran) occurs under fluid shear. The uptake was measured by oxygen uptake employing biochemical oxygen demand (BOD) bottles and was found to be 2.3 and 2.9 times higher for bovine albumin and dextran, respectively, than in undisturbed samples. When radiolabile chemicals were employed, the amount of labeled bovine serum albumin and dextran retained in the cells in stirred bottles increased 12.6 and 6.2 times faster, respectively, than in unshaken bottles. The uptake of ^{3}H-dextran and the O_2 consumption by *E. coli* in stirred and still bottles is shown in Figures 4.2 and 4.3. However, the same effect cannot be demonstrated for leucine and glucose. Although the size distribution of macromolecules can vary in natural waters, they, nevertheless, make up a major portion of the dissolved organic matter (DOM). Of the total amino acids in the North Pacific Ocean survey 80% was in compounds found to have molecular weights of >1500 atomic weight mass (Sugimura and Suzuki, 1988). Although the molecular weight of bovine serum albumin is 68,000, products released during hydrolysis are much smaller. Confer and Logan (1991) consider any compound that must be cleaved into subunits before transport into the cell should be considered a macromolecule. In media containing macromolecules, these authors strongly suggest that the mixing conditions be reported, because mixing will influence the uptake of macromolecules. Employing *Z. ramigera,* Logan and Dettmer (1990) demonstrated that shear rates below 50/second did not increase leucine uptake, but when cells are fixed in a flow fluid of ~1 mm per second, leucine uptake is increased by 55 to 60% over uptake by suspended cells. Therefore, microbes that form aggregates attached to suspended particles, rising bubbles, and fixed surfaces can also be influenced by a flow field. As a consequence they have a better opportunity to obtain substrates in an oligotrophic environment.

Logan and Kirchman (1991) tested the effect of advective flow and fluid sheer on the uptake of leucine and glucose by natural assemblages of heterotrophic bacteria collected from Roosevelt Inlet, Delaware Bay (U.S.A.). They found that ^{3}H-leucine (ca. 1 nM) uptake by cells held in fluid moving at 20–70 m/day was eight times larger than uptake by cells at a velocity of 3 m/day. Flow rates of only 20 m/day may be sufficient to cause a severalfold increase in uptake rates of selected compounds by attached bacteria. This becomes especially important when sinking velocities of particles (fecal particles, dead small organisms, etc.) reach 200 m/day (Alldredge and Gotschalk, 1988). Attached bacteria may then have a greater ability to obtain nutrients in oligotrophic environments. Nevertheless, shaking of environmental samples may increase uptake of certain compounds.

When adding radioactive compounds to water samples (see section 4.5.5), it is

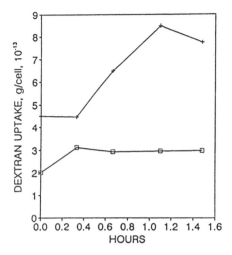

Figure 4.2.   Uptake of [³H]dextran by *E. coli* cultures in stirred (+) and still (□) bottles. Reprinted by permission from *Applied and Environmental Microbiology* 57:3093–3100, D. R. Confer and B. E. Logan, 1991; copyright 1991, American Society for Microbiology.

more than likely that the results will be much greater than what takes place in nature because of the perturbation of the sample. Thus, the aquatic environment is much more oligotrophic than past microbial analysis indicates. The ability of microbes to obtain energy when associated with sinking particles is aided by fluid movement. This, in turn, makes the particles more nutrient rich, aiding in the nutrition of particle-ingesting forms in the aquatic environment.

FILTRATION (PERTURBATION)-INDUCED RELEASE OF DOM

Freshly collected seawater was filtered through ordinary qualitative filter paper. This filtration process increased the apparent growth rate of culturable cells but the initial population was reduced by filtration (Waksman and Carey, 1935). Perturbation of a microbial community by filtration may result in the production and accumulation of compounds that would normally be metabolized by another component of the community that has been inhibited by the perturbation (Joint and Morris, 1982). Miyamoto and Seki (1992) state that it takes an incubation period of 6 to 12 hours for bacteria to respond to nutrient perturbation, a period necessary for acclimatization to the altered environment, mostly by the second generation bacterioplankton. Hermansson and Marshall (1985) state that bacteria may become detached from particles during filtration and may feed at surfaces while never attaching firmly in the first place (see section 4.2.1). They may attach and

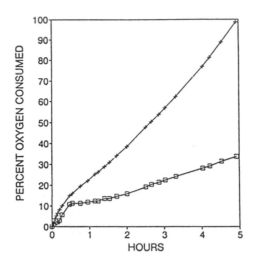

Figure 4.3   Oxygen consumption by *E. coli* growing on dextran in stirred (+) and still (□) samples. Reprinted by permission from *Applied and Environmental Microbiology*, 57:3093–3100, D. R. Confer and B. E. Logan, 1991; copyright 1991, American Society for Microbiology.

detach during periods of fluctuating nutrient sufficiency (Power and Marshall, 1988).

When seawater is used for microbiological analysis, various investigators have used filtered seawater (e.g., Höfle, 1984; Jannasch, 1967; Ammerman et al., 1984; Hagström et al., 1984; Landry et al., 1984; Li and Dickie, 1985b; Lee and Fuhrman, 1987; Simon and Azam, 1989), autoclaved seawater (Carlucci et al., 1986), ultraviolet-irradiated (UV) seawater (Hamilton and Carlucci, 1966), and filtered autoclaved seawater (Baxter and Sieburth, 1984; Hamilton et al., 1966). A consequence of perturbation is that a water sample becomes different than it was originally. After these manipulations of the sample, it has been found that during filtration and sterilization of seawater, a significant release of utilizable organic substrates may occur (Ferguson et al., 1984; Goldman and Dennett, 1985; Li and Dickie, 1985a). Also, absorption of organic particles in the air can easily occur during filtration. Generally the number of microbes in environmental samples is so low that membrane filtration, reverse filtration, centrifugation, etc., are used to concentrate the microbes. To further complicate the picture, microbes in environmental systems often exist in localized patches or are found in random or uniform distribution (Karl, 1986). Griffiths and Morita (1973) rejected the reverse flow filtration for concentrating microbes on membrane filters, because a significant and inconsistent number of microbes are lost on the membrane filter and there was a depression in the microbial uptake of radioactive glutamate. Thus, the filtered microbes were not the functional equivalent to those in the unfiltered water.

Attempts have been made to reduce the effects of filtration by use of very low vacuum. Very gentle filtration of exponentially growing *E. coli* did not affect the adenylate energy charge but did increase the adenosine/ATP ratio by 10-fold (Davis and White, 1988). Many laboratory protocols call for the filtration of the water samples. Several fragile species of phytoplankton have been found to be very susceptible to cell breakage when exposed to the air vacuum between filtration and rinsing steps in ^{14}C-fixation experiments. After 15-minute incubation, up to 60% of the fixed carbon was found in the rinse (Goldman and Dennett, 1985). These authors further noted that air exposure of phytoplankton caused extreme osmotic shock and rupture. Fuhrman and Bell (1985) noted an increase in dissolved free amino acids during filtration with some types of filters, indicating possible breakage of the planktonic organisms during the filtration process. It is also possible to release loosely bound amino acids by filtration from cells, mainly because the so-called amino acid pool is made up of internal and cell surface attached amino acids (Griffiths et al., 1974). Heterotrophic protozoans are also known to release DOM when filtered (Taylor and Lean, 1981; Taylor et al., 1985) as well as dissolved free amino acids (Andersson et al., 1986; Flynn and Fielder, 1989; Nagata and Kirchman, 1990). Nagata and Kirchman (1990) used Acrodisc filter with syringe filtration to collect dissolved free amino acid samples from a marine flagellate (*Paraphysomonas imperforata*) and a marine ciliate (*Urenema* sp.) and other microorganisms from the natural aquatic environment. Is it possible that laboratory handling can cause the extremely fragile phytoplankton and protozoans to release their DOM, including free amino acids? If samples are filtered for microbiological analysis, this added organic matter to the samples will cause an error in the data obtained.

Ducklow and Carlson (1992), in reviewing dissolved organic carbon (DOC) utilization, point out the disturbing possibility that enclosure or filtration somehow "turns on" DOC utilization. They further point out that experimental observations are consistent with observed short-term variations in DOC levels but more information is needed to define this phenomenon.

PERTURBATION OF MARINE SEDIMENTS

The value of experimental rate measurements depends on how well they reflect actual in situ rates since sampling and laboratory manipulations tend to affect microbial metabolism and thermodynamic equilibria (Jørgensen, 1983). For marine studies, this experimental problem increases in proportion to water depth, because a pressure decrease during sampling of deep-sea sediments may depress the in situ rate of bacterial metabolism by up to 100-fold (Jannasch and Wirsen,

1973) and may depress the oxygen uptake of undisturbed sediment cores as much as 10-fold (Smith, 1978).

Rhoads et al. (1978) postulated disturbances were mechanisms to keep parts of any ecosystem at a high rate of productivity. Macrofaunal burrowing in sediments stimulates microbial growth and metabolic activities (Hargrave, 1970; Aller and Yingst, 1978; Henriksen et al., 1980; Hylleberg and Hendriksen, 1980; Kristensen and Blackburn, 1987; Morrison and White, 1980; Hines and Jones, 1985; Hines et al., 1982; Kikuchi, 1986; Reichardt, 1986, 1988). Exceptions to the foregoing findings are also noted (Bianchi and Levinton, 1981; Hanson and Tenore, 1981; Pearson, 1982; Alongi, 1985; Alongi and Hanson, 1985). The exceptions focused primarily on microbial production and related food chain aspects. Bioperturbation by polychaetes and bivalves may account for three- to fivefold increases in the nutrient turnover that is mediated by bacteria (Hines and Jones, 1985). I believe that the effects of bioperturbation are real and must be taken into account in microbial ecology.

Disturbing sediments, either mechanically or by adding chelator, increased all major components of polymeric ß-hydroxyalkanoates (PHA) within microbes relative to the bacterial biomass, but a natural process, gardening of the sedimentary microbes by *Clymenella* sp., an annelid worm, induced decreases in PHA, with changes in the relative proportion of component beta-hydroxy acids (Findlay and White, 1983). Thus, the concentration of PHA relative to the bacterial biomass may reflect the recent metabolic status of the microbes.

When marine sediments are exposed to labeled substrates, a "disturbance artifact" is introduced into the measures of metabolic activity (Findlay et al., 1985). The artifact detected a quantitative measurement of the relative rates of incorporation of [^{14}C]acetate into phospholipid fatty acids (PLFA) and endogenous storage lipid (PHA). Mechanical disturbance, natural disturbance, and predation disturbance all affect the nutritional status, metabolic activity, and biomass in shallow marine sediments (White and Findlay, 1988). Later, Findlay et al. (1990a,b) expanded their studies on disturbances (sieving and bioperturbation) of sediment compared to ambient sediments and presented evidence that either biotic or abiotic disturbances may be important factors controlling the structure of microbial communities in sediments. Immediately following sieving (disturbance), the ratio of acetate incorporation into phospholipid versus acetate incorporation into PHA indicated the short-term metabolic status of microbes was shifted toward phospholipid synthesis. In addition, microbial growth rates were depressed and microbial biomass was unchanged. This was followed by a shift in 2 to 4 hours toward phospholipid synthesis and growth rates comparable to those in ambient sediments. Long-term measures of metabolic status (PHA, trans/cis ratios) indicated possible metabolic stress such as starvation or anoxia (Findlay and White, 1983; Guckert et al., 1985, 1986; Guckert and White, 1986).

The "disturbance artifact" was demonstrated by injecting [^{14}C]acetate samples to the core (Howarth and Teal, 1979; King, 1983) and by using the pore-water replacement or slurry technique to measure the activity. The PLFA/PHA ratio increase was observed in the cores when the injection technique was employed. Raking of the sediment prior to exposure to the label also increased the rate of PLFA synthesis. Changes seen after the disruption of intact sediment were due to stimulation of the microbial community and not to changes in the in situ acetate concentration or differences in the isotope distribution within the sediment. These sediments were tested prior to and after disruption. Increased PLFA production is also noted when the sediment is disturbed by a sand dollar. The "disturbance artifacts" were also noted by Martinez et al. (1983) and Revsbech and Jørgensen (1983). According to Findlay et al. (1985), a disturbance in sediment that results in the immediate burst of cellular growth is like a "shift up" to the faster growth rate seen in monocultures with given nutrients.

Sediment resuspension is an important factor in controlling the benthic microbial activity (Demers et al., 1987; Riemann and Hoffman, 1991; Troussellier et al., 1993). Storm-simulated conditions caused sediment resuspension leading to a dramatic short-term increase in enzymatic decomposition of organic matter, as measured by the aminopeptidase and ß-glucosidase activity, and bacterial secondary production (Chróst and Rieman, 1994). Four hours after the storm stimulation, the aminopeptidase and ß-glucosidase activity was 24 and 43% higher. These activities were short term and activity returned to the prestorm event within 24 hours.

Perturbation, resulting from any mechanical force, brings about increased microbial activity. Although the perturbation effect may be short lived, it is real and when samples are measured for microbial activity we may be measuring this perturbation effect. Just the handling of the microbial samples causes perturbation. The perturbation effect is not an artifact and probably results from exposure to oxygen and exposure to organic matter not previously bioavailable to microbes. Thus, when microbial activity is measured in environmental samples, one has to question the results in relation to the methodology employed.

4.2.6 *Priming of Environmental Samples*

When a change in the decomposition rate is caused by the addition of a small or large amount of fresh organic matter, it is described as a priming action (Jenkinson, 1966b). Any time substrate(s) are added to environmental samples and/or the samples are perturbed by shaking, filtration, or even the sampling process, one has to distinguish between "reality" and "artifacts." Although the priming action is short lived (Jenkinson, 1971), most of the data obtained are from ecological samples incubated for short periods. The degree of the priming effect differs according to the substrate(s) used, the soil sample, or the method used to demonstrate the priming effect. This priming effect was found to be significant in soil profusion studies due to the increased microbial activity (Szolnoki et al., 1963). Because most organisms in the environment are not growing under optimum conditions, the type of substrate becomes important as a priming agent.

Starved cells can take up energy more readily and may be more adept at taking up a greater variety of substrates than actively growing cells (Horowitz et al., 1983; Nedwell, 1987). Thus, in all probability, the addition of an energy source only means that the starved physiological state is awakened to produce an actively growing state in the microorganism(s) in any ecosystem. All methods that add an energy source to demonstrate productivity may be in error, because the amount of added energy may be excessive in relation to the amount present or the amount of energy added acts as a primer necessary for the organisms to show reproduction. Clark (1967) stated, "The nature of an added energy material influences both the immediacy and the duration of the rise in activity as well as the specificity of the responding flora." Added energy-yielding substances, when used in the field or in the laboratory, can, at times, bring about a dramatic stimulation of microbial activity.

Some energy-rich materials, including lignin and other recalcitrant substances, are not easily digested in soil without the addition of soluble organic substances. These substances may allow new enzyme synthesis (McLaren, 1973a,b). Mineralization of low concentrations of substrate may be enhanced by the addition of an easily metabolized compound (LaPat-Polasko et al., 1984; Law and Button, 1977). The addition of an easily metabolizeable compound may bring about co-metabolism or furnish just enough energy to synthesize the necessary enzyme(s). In all probability, the addition of an energy source in any ecosystem only means that the starved physiological state of the microorganism is awakened to produce an actively growing state of the microorganism(s) in any ecosystem. All methods that add an energy source to demonstrate productivity may be in error, mainly because the amount of energy may be excessive in relation to the amount present; or the amount of energy added, coupled with that present in the sample, may act as a primer necessary for the organisms to show reproduction. According to Davis et al. (1993), the soil microbial community in its nongrowing steady state appears to convert a much lower percentage of a radiocarbon substrate to $^{14}CO_2$ than a growing soil community that responds to a substrate addition. In their study, more than half of the net CO_2 production may represent the mineralization of biomass and soil organic matter.

When organic matter is added to any ecosystem, the microorganisms use the added carbon and either respire it to CO_2, transform it into microbial tissue, or excrete it as metabolites unless it is resistant to decomposition. The resistant components of the added organic matter, a portion of the newly synthesized microbial tissue, and the stabilized metabolites contribute to the organic carbon in the ecosystem. According to Jenkinson (1971), the proportion of added plant carbon retained in the soil under different climatic conditions, using different plant materials and soils, is remarkably similar (excluding very acid soils) in that one-third of the added plant carbon remains after 1 year, falling to about one-fifth after 5 years. Even when readily decomposable substrates such as glucose, hem-icellulose, or cellulose are decomposed in soil, the residual carbon is distributed in a pattern more similar to that of soil organic nitrogen and that of soil organic carbon, including the humus fraction (Mayoudon and Simonart, 1958, 1963, 1965). In the laboratory, the carbon became stabilized when 30% or less of the original carbon remained (McGill et al., 1973).

In soil, the scarcity of food or lack of a suitable and bioavailable energy source is the principal factor limiting bacterial growth (Clark, 1967). As a result, any addition of fresh energy-yielding material to soil will almost invariably elicit an increase in bacterial activity. Plant litter is undoubtedly a major source of input to soil, but in some ecosystems, the litter layer is distinct from the soil system. The amount of litter in relation to its accumulation varies considerably in different climatic zones (Jensen, 1974). Much of the litter is decomposed in the litter layer, hence the amount of readily utilizable energy that seeps into the soil is probably very little, but is mainly humic material.

When determining the growth of natural populations in oligotrophic environments, any substrate that can be metabolized must be used with extreme caution, mainly because starved cells can immediately utilize it. Any addition of energy or nutrient can act as a primer for growth and metabolism of the cells. Additionally, it may also be able to break constitutive dormancy of the organisms caused originally by the lack of energy. This primer effect has been noted in soil where addition of fresh organic residues or simple compounds such as glucose results in the decomposition of some of the native soil organic matter (Jansson, 1960; Macurak et al., 1963). Addition of small, frequent amounts of carbonaceous materials results in more pronounced decomposition of soil organic matter than large, less frequent additions (Hallum and Bartholomew, 1953; Macurak et al., 1963). Thus, Gray and Williams (1971), after reviewing the literature, came to the conclusion that it is possible to view the bulk of the soil population as being relatively inactive for a large portion of the time due to the lack of available nutrients. However, the idea of the addition of small amounts of labile organic matter in aquatic systems has never been viewed as a possible primer effect.

Figure 4.4 illustrates the priming effect when energy is added to the soil. Note that the zymogenous activity occurs first, followed by a long-term autochthonous activity. This effect in soil microbiology can be described as the tendency of an addition of fresh organic matter to soil to stimulate the decomposition of the native organic matter and can also be described as the greater loss of soil organic matter from a soil receiving an organic amendment than from untreated soil (Clark, 1968). Although Clark (1968) provides four possible explanations for the priming effect, he discards all of them because they ignore the fact that the substrate itself is changed, or more correctly, the conditions under which the substrate, the soil

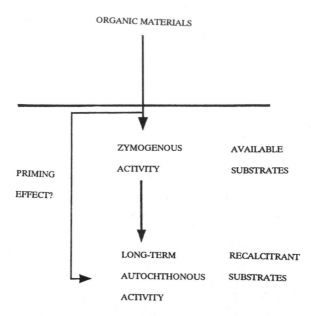

Figure. 4.4   The temporal distribution of energy substrates in soil. Reprinted by permission from *Bacteria in the Natural Environments: The Effect of Nutrient Conditions*, M. Fletcher and G. Floodgate (eds.), S. T. Williams, pp. 81–110, 1985; copyright 1985, Academic Press for Society for General Microbiology.

organic matter, is presented, are changed. This priming effect also is observed in greater mineralization of nitrogen when nitrogen fertilizer is added, than in unfertilized soil (Broadbent and Norman, 1947; Broadbent, 1965). ^{13}C-labeled Sudan grass (Broadbent and Norman, 1947), ^{13}C-labeled glucose (Chahal and Wagner, 1965), or ^{13}C-labeled plant material or its components (Mortenson, 1963; Sørensen, 1963; Sauerbeck, 1966) added to soil results in a priming action to some degree. The evolution of CO_2 after the addition of decomposable organic matter was found to be four to ten times greater than the evolution from the controls (Sørensen, 1974). The degree of priming effect differs according to the substrate(s) used, the soil sample, or the method used to demonstrate the priming effect. When ^{14}C plant residues are added to soil, the priming effect as well as the loss of radioactive CO_2 oocurs (Sørensen, 1974). The probable explanation for the priming action in soil, according to Stevenson (1978), is that there is a buildup of a large and vigorous population of microorganisms. When energy is added, these microorganisms subsequently produce enzymes that attack their native soil organic matter. The extent of organic matter decomposed depends on a variety of factors, such as the size and activity of the microflora.

The treatment of soil with ^{14}C-glucose, NH_4NO_3, or both resulted in a marked priming of the native C during the second and third days of incubation, with a second priming effect during the fifth day (Shields et al., 1974). When ^{14}C-glucose was amended with soil and incubated 2 months under field conditions, appreciable amounts of ^{14}C were mineralized following treatment known to partially sterilize soil. Freezing and thawing are more effective than wetting and drying, but less effective than $CHCl_3$ vapor in releasing stabilized ^{14}C materials (Shields et al., 1974).

When agricultural soil is amended with ^{14}C-glucose, there is an initial rapid increase in microbial activity and biomass during the first 7 days (Shields et al., 1973). It is estimated that about 60% of the added C was devoted to the synthesis of microbial biomass and metabolites. This is followed by a stabilization of the biomass C. When labeled rye grass is added to soil, the loss of organic matter usually represents a first-order process with exponential decay (Jenkinson, 1965). About two-thirds of the rye grass is lost in the first few weeks after addition. Ladd et al. (1981) showed that when plant residues are added to various soils, more than half the C is lost after 4 weeks. When ^{14}C-labeled barley straw was added to soil in the field and in the laboratory, Sørensen (1987) found that its degradation increased by a factor of 1.2 under laboratory conditions. On the other hand, Jenkinson (1971) produced conflicting evidence on the priming effect. No priming effect could be detected when fresh plant residues (Cerri and Jenkinson, 1981) or a small amount of glucose (Dallenberg and Jager, 1981) were added to soil.

A priming effect has been noted with lakewater isolates where addition of a labile compound increases the decomposition of refractory DOM by *Pseudomonas* sp. and *Erwinia* sp., but *Flavobacterium* sp. utilizes the original substances more effectively (Geller, 1986). Others (de Haan, 1977, 1983; Rifai and Bertru, 1980; Stabel et al., 1979) have reported that a pulse dose of organic matter enhances the degradation of refractory compounds. On the other hand, Steinberg and Herrmann (1981) and Strome and Miller (1978) could not show any enhancement when an additional substrate (cosubstrate) was added in cultures of natural bacterial communities, hence, these authors assigned the use of the term *cometabolism* to the addition of a cosubstrate. These researchers used glucose in their system but glucose by itself can bring about substrate accelerated death in starving bacteria (Novitsky, 1977). Aging also increases the ability of microorganisms to decompose refractory DOM. The minimum threshold of the cosubstrates necessary for this priming effect is not known, but it could be restricted to a period of enhanced algal productivity (Geller, 1986). This priming effect is never taken into consideration by the marine microbiologist when dealing with water samples where a labile compound is added to the system to measure activity. The priming effect should be considered when any environmental samples are spiked with a radioactive compound, especially a labile compound. Priming could activate the inactive cells and/or supply sufficient energy for the indigenous organisms to make the necessary enzyme to act on the indigenous substrates.

Substrate enrichment is known to stimulate rates of microbial metabolism in shallow water environments, but whether this holds true in the deep sea where nutrients are scarce is not known (Nealson, 1982). This is simply the fact that utilizable substrates in ecological samples are in extremely low concentrations and these utilizable substrates may be unavailable to the microbes (see Chapter 3). If the initial concentration of substrate added is low, microbial oxidation begins immediately, but with increasing concentration of substrate there is an increasing lag time observed for the oxidation of the substrate (Williams, 1970; Williams and Gray, 1970). Limitation of nutrients other than carbon is typical for resting cell suspensions in which the respiratory system is uncoupled from ATP production while oxygen consumption and CO_2 production increase. Nutrient limitation, other than carbon, for soil microbes may closely resemble that of resting cells, whereby the reducing power of substrate oxidation might be likewise "wasted" in a similar manner to that found in resting cell suspensions. This would explain how up to 90% of the added carbon is released as CO_2 while growth yields are very low (Schut et al., 1993). If radioactive glutamate is added to starved cells in extremely low concentrations, it is oxidized mainly to CO_2 (over 80% in 23 hours) (Morita, 1984b). The amount of CO_2 produced depends on the initial concentration of glutamate added.

Bacterial growth in the sea may be limited not only by the supply of DOM, but also its quality. The stimulation of bacterial growth is greater when dissolved free amino acids (DFAA) are added than when glucose and ammonium are added (Kirchman, 1990). From an energetics viewpoint, this appears logical because energy is needed to synthesize amino acids when glucose and ammonium are added.

In marine sediments, carbon-amended sediment slurries usually show a lag in metabolic activity for a period of 8 to 25 hours, but this lag period does not bring about an increase in bacterial proliferation. Instead, dormant bacteria are being "activated" by the added energy source (Gandy and Yoch, 1988).

Just how much substrate is needed to prime the organisms in any given environmental sample? Recognizing that starved Ant-300 cells can utilize glutamate at a concentration of 10^{-12} M (Morita, 1984b), the amount necessary may be less than this. Can the organics in the air, coupled with perturbation effects, be sufficient to prime cells in an environmental sample? How much priming effect takes place when radioactive labile compounds are added to aquatic samples, such as in the Heterotrophic Potential Method discussed in section 4.5.5.

Smith and Baldwin (1982) captured amphipods (*Paralicella caperesca*) from the western North Atlantic in a slurp gun respirometer. When bait (food) is available, the amphipods will swim into the overlying water in search of the food source. The respiration rate was monitored for 21 hours. During the first 6 to 7.25

hours, the respiration rate was elevated and then declined precipitously by one or two orders of magnitude to a constant rate for the duration of the incubation period. They interpreted these data as suggesting an adaptation to a low food-supply environment where these animals are capable of rapid response when food is available and can return to a state of dormancy when such fortuitous food sources are not available. This raises the question as to whether bacteria can do the same, being in a starvation-survival stage (dormant) and then display a greater activity when substrates (priming) are present or when perturbation takes place. The former appears to be the case because starved bacteria will immediately take up substrate when it is present.

Perturbation of the sample, coupled with the priming effect and the bottle effect, adds to the bacterial activity. It is impossible to duplicate an original environment, especially when samples are removed from that environment for the experiment.

4.2.7 Photodegradation of Organic Matter

Photodegradation of dissolved aquatic and terrestrial substances is a common phenomenon. Geller (1985a) studied the abiotic stability of refractory, high molecular DOM from lake water and found that these compounds were not stable in the dark and exposing them to daylight in the laboratory and to sunlight in quartz bottles enhanced conversion to microbial utilizable compounds. Soil humic material is also known to undergo photodegradation. Photolysis of refractory DOM, according to Wetzel (1983), may contribute to the relative constancy of the DOC in lakes. Smaller molecules, such as fatty acids found in humic substances, may then be released (Il'in and Orlov, 1973; see Geller, 1985b). Thus, the exposure time to sunlight of the refractory substances in environmental samples may again aid the growth of microorganisms because complexed substances such as carbohydrates, fatty acids, and amino acids may become bioavailable. Amador et al. (1989) prepared (2-^{14}C)glycine bound to humic acid and studied the effect of sunlight irradiation on the prepared substance. They found that microbial mineralization of these complexes in the dark increased inversely with the molecular weight of the complex molecules. Microbial mineralization increased with solar flux and was proportional to the loss of A_{330}. Their results indicated that photochemical processes (wavelengths below 380 nm) generate low molecular weight, readily biodegradable molecules from high molecular weight complexes of glycine with humic acid. Photochemical breakdown of humic substances has also been noted by Aiken (1976) and Kotzias et al. (1987).

Photochemical production of carbon monoxide from seawater samples were noted by Redden (1982) and Conrad and Seiler (1980). Redden (1982) found that radiation from 290 nm through at least 460 nm was the most effective in producing carbon monoxide, and that the addition of humic acid to seawater produced approximately three orders of magnitude more carbon monoxide/substrate weight than acetone or acetaldehyde. Seawater samples treated with acetone and acetaldehyde also produced methane. The possible mechanism for photochemical CO production is elimination of carbonyl functional groups from an undefined fraction of dissolved organic matter by carbon-carbonyl bond cleavage (Redden, 1982). He also noted that oxygen may photochemically oxidize organic matter, thereby increasing the number of photoreactive carbonyl groups. Mopper and Stahovec (1986) expanded the list of carbonyl compounds (especially formaldehyde, acetaldehyde, acetone, and methylglyoxal) formed by photochemical processes and stated that once formed, they appear to be readily consumed by the biota. Mopper and Stahovec (1986) speculated that photooxidative cleavage of biologically refractory dissolved organic matter in seawater yields low molecular weight organic fractions, such as carbonyl compounds. Photochemical processes in seawater have also produced reactive species, such as single oxygen (Zepp et al., 1985) and hydrogen peroxide (Zika et al., 1985). Carbonyl sulfide was also being formed (Ferek and Andreae, 1984).

Although Governal et al. (1992) demonstrated that both *P. fluorescens* P17 and *Spirillum* strain NOX can grow in ultrapure water, they did not mention that absorption of organics from the air could be the reason for growth. In addition, they also demonstrated that ultraviolet (UV)-185 caused a 144% increase in assimilable organic carbon and a 44% decrease in UV oxidizable carbon, which may be due to the UV breakdown of high molecular weight organics into low molecular weight organics.

4.3 Threshold Amounts of Energy

There are many papers dealing with growth of bacteria in low concentrations of substrate(s). However, the question is how low can the substrate concentration be before growth ceases for any particular species of bacteria. Naturally, when substrate becomes so low that it will not support metabolism and growth of microbes, then the starvation process sets in. The lowest substrate concentration that will support growth will probably depend upon the species in question. On the other hand, the threshold amount of energy for cellular metabolism only is much lower and probably is in line with diffusion (Koch, 1990). According to Poindexter (1981b), the overall nutrient balance, rather than the concentration of specific organic compounds, has the greatest effect on the growth of some low-nutrient bacteria. Threshold concentrations for bacterial growth have been shown to occur in the presence of a single substrate (van der Kooij et al., 1982a,b), presumably when a substrate concentration is too low to provide for both growth and maintenance of a metabolizing population (Pirt, 1975; Powell, 1967). Schmidt and Alexander (1985) and Schmidt et al. (1985) provide a theoretical basis for the inability of bacteria to grow at very low substrate concentrations. The kinetics of mineralization of organic compounds by *Salmonella typhimurium* in the presence of organic compounds too low to support growth was best described as pseudo first-order kinetics (Schmidt et al., 1985). Furthermore, these studies do not take into consideration the effects of surfaces, aggregate formation, perturbation of samples, the priming effect, or growth resulting from contaminants from the air, or the possibility that the number of cells growing in the absence of substrate might be the result of starvation. Below some finite energy concentration bacterial growth is thought not possible (Button, 1978; Jannasch, 1967). The reason for occurrence of low levels of many organic compounds in aquatic systems may be due to the existence of threshold concentrations for bacterial growth (Jannasch, 1968). However, metabolism, not growth, may further reduce the indigenous concentration of organic compounds in nature. As a result, the concentration of some organic substrates in nature will be below the threshold level for bacterial growth.

ZoBell and Grant (1942) reported that marine bacteria multiplied in mineral media when the concentration of glucose concentration was 0.1 mg/l (40 µg C/l). They subsequently reported that *E. coli, Staphylococcus citreus, Proteus vulgaris,* and *Lactobacillus lactis* could also grow at the same concentrations (ZoBell and Grant, 1943). Investigators studying the ability of bacteria to grow at low concentrations of organic matter are adding a readily utilizable compound and not naturally occurring substrates, many of which are resistant to degradation. Nevertheless, a review of the methodology indicates that the original inoculum of 10 to 100 bacteria was added to concentrations of peptone or glucose ranging from 0.1 to 100 mg/l. Unfortunately, no data concerning the number of cells and incubation times were reported. If one disregards the volume, the number of cells in relation to the added substrate is quite great. This is not taken into consideration by the investigators. The bacteria range from 10^4 to 10^7/ml in natural aquatic environments. The amount of substrate in relation to the number of cells present in the study of oligotrophic bacterial growth must be taken into consideration.

The effect of surfaces on the threshold amounts of energy needed is best illustrated by the early research of Heukelakian and Heller (1940), who found that growth of *E. coli* was restricted to flasks containing glass beads when the concentration of glucose and peptone was 0.5 to 2.5 ppm. The stimulatory effect of glass beads (bottle or surface effect) was observed at concentrations up to 25 ppm. Similar results were obtained by Jannasch (1958) and Corpe (1970a). Thus, when trying to determine the threshold amount of energy needed to support bacterial growth, the effect of surfaces must be taken into consideration. Again, we see that surfaces must be taken into consideration when environmental samples are analyzed.

A threshold of growth 0.21 mg glucose/l was obtained by Law and Button (1977) for a marine coryneform bacterium grown in a chemostat. However, a total concentration of 0.11 mg C/l would support normal growth, provided that the "right" mixture of substrates (mainly amino acids), including 0.3 µg glucose/l, was provided. Law and Button (1977) and Shehata and Marr (1971) provide evidence that bacteria are not able to grow solely at the expense of organic compounds that are present at low concentrations. Further evidence of threshold concentrations below which synthetic organic compounds fail to support the growth of microorganisms in pure culture is provided by Schmidt et al. (1985) and Boethling and Alexander (1979).

The growth of *P. aeruginosa* occurs in tap water at 25 μg of C/l supplied by any one of a number of compounds (van der Kooij et al., 1980) and *Aeromonas hydrophila* multiplies when glucose is added at 10 μg/l (van der Kooij et al., 1982b). Neither of these organisms could be considered oligotrophs. They cannot be considered low-nutrient bacteria or copiotrophs either. It should also be noted that concentrations of 1 to 15 mg C/l is several orders of magnitude greater than that found in seawater. Although van der Kooij et al. (1980) do not even mention that glucose could be a primer, data suggest that it does act as a primer. Organisms do not grow with yeast extract, D-glucosamine, and gluconate at an initial concentration of 1 mg/l. Slow growth does occur when 20 μg/l glycerol-C is added, but not at 10 μg/l. Growth of a *Flavobacterium* sp., that was found to be able to multiply in tap water without added substrates, was enhanced when starch of glucose in amounts as low as 1 μg of substrate C per liter was added (van der Kooij and Hijnen, 1981). Two strains of *Pseudomonas* were capable of growing in low concentrations of acetate (1 mg to 2.5 μg of acetate-C/l) (van der Kooij et al., 1982a). When labile substrates are added to tap water, they are readily bioavailable to the organisms and these substrates are not complexed to other material as they would be in nature. This might make the substrates unavailable to the cells.

An isolate of *A. hydrophila* (opportunistic fish pathogen) from tap water was found to be capable of growth in used tap water (2 to 3 mg DOC/l) when 2.5 μg/l of C in the form of glucose was added but there was no growth when no amendment was added, or when acetate, glutamate, or succinate was increased to 1 mg C/l (van der Kooij et al., 1980) How much the added glucose acts as a primer so that the DOC can be utilized is not known. Starch is not utilized unless glucose-C (10 μg/l) is added. No growth was noted when acetate or glutamate (10 μg/l) were added unless glucose was added.

In order to produce 10^7 cells/ml, it would require about 2.5 mg of C/l supplied as a useable carbon source (Davis et al., 1967). If one assumes that a colony on agar contains 10^7 cells, then the medium must contain 2.5 mg of C/l as a metabolizable carbon source, hence only 15 to 20 colonies could appear on a plate (Martin and MacLeod, 1984).

Unfortunately, all experimental studies dealing with threshold amounts of organic matter for growth are conducted with readily useable compounds instead of naturally occurring substrates. When compounds are found in environmental samples by chemical analysis, it is not known if those compounds are bioavailable to the organism(s) in question. Gunkel (1972) states that it is impossible to extrapolate threshold values obtained under laboratory conditions to the situation in oceans and coastal waters.

Organisms possibly are capable of utilizing the small amount of substrate (below the threshold level). The starvation-survival process may generate many cells through a fragmentation (reductive division) process. Essentially, it is a matter of how many cells are going to use the limited amount of substrate present. Unfortunately in microbiology, growth has been incorrectly defined as an increase in the number of viable cells. When taking all factors into consideration, it appears safe to say that we do not know the lowest concentration of substrate(s) needed for growth of a particular bacterium, especially in natural samples involving the indigenous substrates. Nevertheless, it would be nice to be able to discern the threshold amount of energy needed for growth of a species or for the growth of a consortia of bacteria in nature so that we would know when starvation of the species or consortia of bacteria begins.

4.4 Growth Resulting from Organics in Air

There are all types of volatile organic chemicals in the biosphere, including insect pheromones, flower attractants, vegetables and other foods, odors from fruits, putrefaction smells, perfumes, deodorants, and natural and anthropogenic air pollutants. The types, sources, and possible effects on soil microbes of volatile organics are presented by Stotzky and Schenk (1976). According to these authors, it is highly probable that volatiles are important in the activity, ecology, and population dynamics of microbes in natural habitats. The major external sources of DOM are the atmosphere and rivers and they are approximately equal in the amounts of DOM each contributes to the ocean—each being about 1% of the marine net primary production (Williams, 1975).

The amount of organic carbon in the atmosphere is 3.2×10^{15} g (Bolin and Cook, 1983). Bacteria capable of growth in mineral salt media with small amounts of volatile organic substances in the air have been isolated from water and soil (Hirsch, 1968; Moaledj, 1978; Witzel et al., 1982a,b). Agar plates, inoculated with lake water samples from the Plusssee, developed fewer colonies when plates were incubated totally free from organic carbon from the air and the investigators (Moaledj and Overbeck, 1982) used the term "oligocarbophilic" for the missing colonies. Moaledj and Overbeck (1982) also noted that the number of oligocarbophilic bacteria are generally twice as great as saprophytic bacteria when grown on mineral salts or dilute media. Growth of microbes in the presence of air has been noted by many investigators, but more lately by Jones and Rhodes-Roberts (1981) and Geller (1983a). Some DNB (dilute nutrient broth) microbes are capable of growth in distilled water (Hattori, 1984), which indicates that growth is supported by volatile organic compounds in the air. Gases in the air provide sufficient nutrients to distilled water for the growth of *Caulobacter* species, and *P. fluorescens* (Kushner, 1993). The latter usually grows in a rich medium.

Although carbon monoxide is not an organic volatile compound, I have included it in this section because it does occur in the atmosphere. Carbon monoxide can be used at trace concentrations (≤ 1 nM) by bacteria from the ocean, lake water, and soil, and permanent trace amounts of methane and carbon monoxide are in the Earth's atmosphere (Conrad and Seiler, 1982). Although the average mixing ratio of carbon monoxide in the lower atmosphere is just 0.1 ppm by volume (Seiler, 1974), atmospheric carbon monoxide can be rapidly utilized by microorganisms (Seiler, 1978; Conrad and Seiler, 1980a). According to Conrad and Seiler (1982), the mode of metabolism may be either specific (utilitarian, profitable), which means that the carbon monoxide-utilizing microorganisms gain an advantage from the oxidation of carbon monoxide for growth or maintenance; or the mode of metabolism may be nonspecific (adventitious, fortuitous) which means that carbon monoxide oxidation is of no use to the microorganism.

Geller (1983a) subjected a lake water bacterial community as well as 48 starved *Pseudomonas fluorescens* (German Collection of Microorganisms strain no. 50090) to a mineral salts medium containing trace elements (no organics). When different types of stoppers were placed on the containers of lakewater bacteria, growth was noted in all cases (Fig. 4.5). Growth of *P. fluorescens* also occurred in lakewater when stoppered with cotton (Fig. 4.6).

Unfortunately, the quality and quantity of airborne organic material is sorely lacking in the literature. Hahn (1980) reviewed the literature on the organic compounds found in natural aerosols. It is amazing that in the ether-extractable fraction of aerosol particles collected in the southern North Atlantic, there are more than 100 compounds, consisting of aliphatic and aromatic hydrocarbons, organic acids and bases, and high polar compounds. Organic compounds are also found in the troposphere, urban areas, etc. (Hahn, 1980). Williams (1971) estimated that precipitation brings organic matter equivalent to 1% of the total net production to the ocean. Spitzy (1988) analyzed for amino acids and combined amino acids in marine aerosols and in rain away from the coastal environment and found 0.47 and 1.13 nM/m^3 in their aerosol samples and 2.6 and 1.17 $\mu M/kb$ in rain water. These results are in the same order of magnitude found by Mopper and Zika (1987). Amino acids in continental and coastal rain samples have been reported by Fonselius (1954), Munczak (1960), Degens et al. (1964), Dean (1963), Sidle (1967), and Williams (1967). It can be assumed that rain gets its amino acids by aerosol scavenging. The laboratory air should have many more organics than natural aerosols. Organic volatiles probably serve as nutrients in soil, where their availability and utilization is influenced by the clay mineral composition (Stotzky, 1986). For more information on gases and organic volatiles on microbial activity, consult Stotzky (1986). Liss (1983) and Duce (1983), respectively, should be

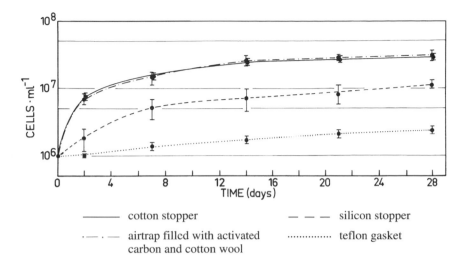

Figure 4.5 Growth of lake water bacteria at 20°C in mineral medium, using cotton stoppers (–), an air trap filled with activated carbon and cotton wool (—·—·—), silicon stopper (--), and screw caps with Teflon gaskets (· · ·). Data points represent means ± standard deviation from three parallel cultures. Reprinted by permission from *Applied and Environmental Microbiology* 46:1258–1262, A. Geller, 1983; copyright 1983, American Society for Microbiology.

consulted for more information concerning the exchange of biogeochemically important gases and aerosols across the air-sea interface.

Contamination by organics in the air must be guarded against when filtering environmental samples low in organic matter. Techniques such as the INT method (see 4.5.2 "Microscopic Methods") must take into consideration the possibility that metabolic activity might be due to the gaseous organic substances absorbed into the sample.

4.5 Problems with Currently Employed Methods

Probably one of the most difficult concepts to accept is that, in nature, microbial processes are not as dynamic as we would like. We are presently conditioned to expect that whenever we take a sample from the natural environment, bacterial growth and activity should occur. Growth may occur due to the processes mentioned above for making organic matter available but not all cells participate in growth. What we have done is to lump all physiological types of microbes to-

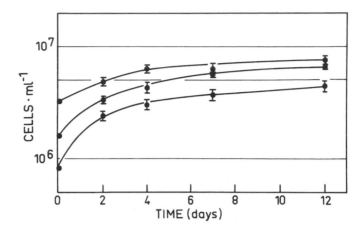

Figure 4.6 Growth of *P. fluorescens* in mineral medium at different initial cell concentrations (cotton-stoppered Erlenmeyer flasks, 20°C, means ± standard deviation from three parallel cultures). Reprinted by permission from *Applied and Environmental Microbiology* 46:1258–1262, A. Geller, 1983; copyright 1983, American Society for Microbiology.

gether for a given set of experimental procedures. In reality, syntrophy dominates the situation and growth of a species is sporadic, hence we are measuring only the active cells within the population. We have recognized that we have different physiological types of bacteria and each has its own chemical and physical factors in order to exhibit optimal growth. We have a tendency to forget that there are various fastidious bacteria present and growth may not result because the systems lack a specific amino acid(s), vitamin(s), trace element(s), specific Eh, etc. In other words, we are not dealing with pure cultures. What we end up with in our research is the response of certain bacteria that can grow under the set of conditions that we have employed. It is impossible to mimic field conditions in the laboratory. Throughout the book edited by Costerton and Colwell (1979), it is noted that the various methods that we employ in microbiology leave much to be desired.

Karl (1986) states that "our understanding of the integrated functioning of microorganisms in nature is methods limited." He further stresses that

> in nature, additional constraints may be placed on microbial populations, such that demonstration of growth per se, may not always yield the most reliable and meaningful ecological information. Measures of microbial biomass (i.e., standing stock of living organisms), metabolic activity (which may be uncoupled from growth and cell division), viability, dormancy, and death may also yield information on the effect of the environment on the microbiota and on microbiologically mediated environmental changes.

This is very true because most methods do not consider all or some of the various factors mentioned previously (e.g., bottle effect, perturbation of the sample, priming of the system with addition of organics, or the growth resulting from organics in the air). Our technical and analytical capabilities are still overwhelmed by the complexities of natural ecosystems and their resident microbial populations, yet the discipline must maintain a high level of self-criticism and protect its present rigorous standards for careful and thorough evaluation of new methodologies prior to general acceptance (Karl, 1986).

Much effort has been directed at the growth rate of bacteria, as well as biomass, biovolume, and productivity in the natural environment (Bell et al., 1983; Christian et al, 1982; Fuhrman and Azam, 1982; Hagström et al., 1979; Moriarty and Pollard, 1982; Newell and Fallon, 1982; Pedros-Alio and Brock, 1982; Kroer, 1994). These methods include the frequency of dividing cells (FDC), ^{3}H-thymidine uptake, isotope dilution, etc. Each of these methods suffers from the fact that samples are perturbated or that energy is added to the system. However, it is difficult to determine the effects of the foregoing because the size of the cells will differ under the various environmental conditions that are known to exist. The average cell size and composition in one environment may be different than the average cell size from another environment. The extrapolation of data into biomass, productivity, biovolume, etc. must be used with care.

Flynn (1988b) states: "Studies of bacterial production are adversely affected by ignorance of identification physiology and position in food webs. It could be argued that methods involving filter-fractionation and incubation in containers may at best adversely affect bacteria less than phytoplankton, and at worst actually stimulate bacterial activity." There are no currently available methods to study microbes and their reactions at the level of the microhabitat and extrapolation of in vitro observations to what occurs in the natural environment is usually an unreliable and misleading exercise (Nannipieri et al., 1990). Too often investigators have the tendency to extrapolate in vitro results to the in situ situation without verifying that the phenomena observed in vitro do occur essentially in the same manner in vivo (Stotzky, 1974).

The various methods (ATP, biochemical methods, phospholipids, lipopolysaccharides, muramic acid, D-alanine, diaminopimelic acid, chloroform fumigation incubation method, respiratory response method) for measuring biomass and metabolic activity are discussed in detail by Tunlid and White (1992) and Smith and Paul (1990). Tunlid and White (1992) proposed the concept of "signature compounds" for subsets of the microbial community. Signature compounds are compounds unique and always present in a subset of the microbial community, generally in low concentrations. All these measurements ignore the interaction of the many different individual microbial components that make up the entire system (Jenkinson and Ladd, 1981). Microbial biomass estimations are useful for comparing different ecosystems (Smith and Paul, 1990). Likewise, these estimates of microbial activity only measure those activities that are functioning at the given moment under the conditions of the measurement(s). One difficulty encountered is in extrapolating data obtained into biomass, especially when it is known that biovolumes will vary greatly (Sime-Ngando et al., 1991). Cell size will vary according to the environment and ecosystems are not homogenous. The advantages and disadvantages (mainly the latter) of the various methods (ATP, muramic acid, hexoamines other than muramic acid, nucleic acid, etc.) to determine bacterial biomass in soil are reviewed by Jenkinson and Ladd (1981).

Methods for estimating growth include the incorporation of thymidine into DNA, adenine into RNA and DNA, leucine into protein. These methods, as well as dilution culturing and measuring the frequency of dividing cells, share several characteristics that limit their usefulness, such as severely restricting the scope of questions that can be addressed (Kemp et al., 1993). Microbial growth estimates indicate indirectly the amount of organic matter available for energy and, thus, reflect the trophic nature of the sample. Numerous methods have been developed to measure growth and respiration of bacteria, but they remain controversial and rely on individual interpretation (Karl and Winn, 1986). The specific limitations of these methods are presented by the authors in *Native Aquatic Bacteria: Enumeration, Activity and Ecology* (Costerton and Colwell, 1979). For further discussion of these methods employed and their limitations, consult Karl (1986), van Es and Meyer-Reil (1982), and Costerton and Colwell (1979). Costerton (1979) states: "None of these methods has evolved to a point where it can be used in blind faith for all systems, but one must understand both the theoretical basis and the shortcomings of each method in making an intelligent choice for a particular study." One should always bear in mind the limitation of the method employed in any study. In field studies, the natural diversity, both in bacterial types and in metabolic states, is often overlooked (Kjelleberg and Hermansson, 1987). Both the activity and biomass measurements can be artifacts of the experimental methods employed (Hobbie and Ford, 1993). As Kemp (1994) states, our understanding of microbial processes in the environment is only as good as our methods for studying them. Because no one method can adequately measure microbial numbers, and biomass or activity, a battery of methods have been used by some investigators. A few of the techniques are described below in order to illustrate the limits of their application.

4.5.1 Plate Counts

The classical plate count method, which can only subject the sample to one type of medium incubated under one set of physical conditions, can only enumerate less than 1% of the organisms present in the environmental sample (Jannasch and Jones, 1959; Buck, 1979; Kogure et al., 1980; Button et al. 1993). The uncertainty and unreliability of plate counts on soil is discussed in detail by Jensen (1968).

Microbial biomass can be divided into active cells, inactive cells, and dead cells, but the plate count method largely recognizes the active cells, and then only provided the right conditions are met. In plate counts one always wonders whether the right medium was employed, the right physical and chemical conditions met, and whether resuscitation was employed. In addition, the following questions must be asked concerning the condition of the cells in the environmental sample:

1. Were some of the cells dead before the sample was taken?
2. Did some of the cells die during the sampling and/or storage process?
3. Did some of the cells die in or on the culture medium?
4. Are some of the cells viable but nonculturable?
5. What percent of the cells are moribund?

The balance between various bacterial species in the environment at any given time depends on the availability of compounds essential for growth of the organisms. Media are generally developed for environmental samples to demonstrate the highest number of CFUs per unit sample and not the highest different physiological types of bacteria. To illustrate this point, West and Lochhead (1950) isolated different bacteria from soil with the following requirements: bacteria that require cysteine, bacteria that require a mixture of amino acids, bacteria that require growth factors, and bacteria with undefined requirements. Unfortunately many microbiologists rely on peptone to satisfy nutritional requirements of bacteria in nature and peptone could not satisfy the nutritional requirements of all the different bacteria isolated by West and Lochhead (1950). In any given soil, an equilibrium exists between bacteria with simple requirements and bacteria with more complex ones; the balance may be modified by many factors (West and Lochhead, 1950).

The discrepancy between direct counts, membrane filter counts, and plate counts for marine samples was definitely established by Jannasch and Jones (1959). When low-nutrient medium is used for environmental samples, a significant increase in viable counts over the use of regular strength medium can be generally noted (Akagi et al., 1977; Buck, 1974; Carlucci et al., 1986; Eguchi and Ishida, 1990; Ishida et al., 1986; Yanagita et al., 1978). Extremely large numbers of bacteria are not cultivated as evidenced by the direct count of these samples. More bacteria are capable of growth when dilute media are used (Stark and McCoy, 1938; Collins and Willoughby, 1962; Melchiorri-Santolini and Cafarelli, 1967; Fonden, 1968; Carlucci and Shimp, 1974; Ishida and Kadota, 1974; Akagi et al., 1977; Hattori, 1980; MacDonell and Hood, 1982). Viable counts of soil heterotrophic bacteria were three to five times higher on low-nutrient media compared with those grown on a series of conventional agar media (Olsen and

Bakken, 1987). These viable counts only represent 2–4% of the total microscopic counts in different soils.

The use of agar for the isolating bacteria may cause flawed data. Agar contains soluble organic matter and the quality of agar differs according to the commercial brand used, which can cause inhibitory effects on the activity of microorganisms (Sands and Bennett, 1966). Different types of agar were compared in the isolation of bacteria from soil and found to give different viable counts (Hattori, 1980; Marshall et al., 1960). In general, agar media were found to be somewhat more inhibitory than broths (MacDonell and Hood, 1984). Furthermore, the selection of the type of agar used for solidifying media influences the number of DNB-microbes that can grow on plates (Hattori, 1980). The use of Bacto agar and Wako (first grade) agar gave the highest number of colonies from soil when regular strength nutrient agar was used, but when dilute nutrient agar was used, the highest colony counts were obtained when Noble (Difco), Purified Agar (Oxoid), and Purifed Agar (Difco) were used. The water content of the agar and incubation temperature are also important (Sieburth, 1967, 1968, and 1971; Hopton et al., 1972).

Counting colonies of bacteria on rich nutrient media as an approach to microbial ecology has dominated past research. However, the concept of substrate shock (too many nutrients) has not been addressed. If more microbes from environmental samples can grow on dilute media, why don't the others grow? This could be the result of substrate shock where too many nutrients are harmful to some of the starving cells in environmental samples. There is a hint that this could occur (Reichardt, 1979). If cells are in the starved state in the environment, substrate shock could be real. (In higher organisms substrate shock does occur, especially when a starving organism eats a big meal.) Winding et al. (1994) noted that, in soil, the highest number of microcolonies estimates of bacteria occur after 2 months' incubation on filters placed on the surface of nutrient-poor media. Winding et al. (1994) postulated that the long lag time was probably caused by activation of slowly growing or dormant bacteria during incubation. Nevertheless, the plate count method does have its place in microbial ecology.

There is a lack of logic in the methods used to enumerate and isolate bacteria from seawater, because routine microbiological methods require that the organisms be transferred to media of relatively high nutrient concentration (Jannasch, 1969). Plating on relatively rich medium favors those organisms that are probably atypical of low-nutrient habitats.

Recognizing that only a very small percentage of bacteria can be cultured by the plate count method points to the fact that no one medium can fulfill the requirements of all the different physiological types of bacteria in environmental samples. Thus, we are faced with the enormous task of trying to determine the chemical and physical requirements of those organisms that cannot be isolated with the media that we use today. For instance, when dwarf cells in soil did not grow on the medium provided, Bakken and Olsen (1986) suggested that their substrates may be the volatiles that occur in the soil.

Failure to grow has been ascribed to environmental fastidiousness, dormancy (Allen-Austin et al., 1984), inhibition by neighboring cells (Hopton et al., 1972), physicochemical differences between laboratory and natural conditions (Carlucci and Pramer, 1957), clumping (Parkinson et al., 1971), and to the fact that stressed and injured cells may have some difficulty in reproduction (Busta, 1978) because sublethal damage and inactivity are not distinguishable (Allen-Austin et al., 1984; Reichardt, 1979). For further discussion on agar-based cultivation, the reviews of Buck (1979) and Fry (1982) should be consulted. Van Es and Meyer-Reil (1982) summarize the reasons for the difference between total direct counts and viable counts. Some of these reasons are (1) a large number of cells are inactive, (2) bacteria may not be able to grow on the substrate of the media, (3) the substrate may be available, but in concentrations too high for the particular bacteria, (4) aquatic bacteria may not be able to grow on a solid surface, such as an agar plate, but may favor liquid media, (5) bacteria are inactivated by other bacterial colonies in their vicinity, (6) bacteria may tend to clump, resulting in only one viable colony originating from more than one cell, and (7) one bacterium may relate syntrophically with other bacteria.

Before an environmental sample is plated, there is no attempt to use any resuscitation process. A short incubation in the presence of a suitable substrate (e.g., 0.1% solution of pyruvate, acetate, or oxalacetic acid) prior to plating as shown by Heinmets (1953) and Heinmets et al. (1954a,b) permitted injured *E. coli* (which were considered to be nonviable on agar medium) to grow on agar medium.

Although Butkevich and Butkevich (1936) recognized that multiplication of marine bacteria depends on the composition of the medium and on the temperature of incubation, they concluded that a considerable portion of the bacteria in the sea must be present in a resting state.

Plate counts should not be used for environmental samples unless the technique can answer a specific question (i.e., coliforms in aquatic samples). Because of this situation, environmental microbiologists have gotten away from plate counts and have substituted other methods. Nevertheless, any microbial method used must be carefully analyzed for its advantages or disadvantages.

4.5.2 Microscopic Methods

DIRECT MICROSCOPY

It is imperative to know the total biomass so that it can be partitioned into major taxonomic groups such as bacteria, algae, and protozoans. This, coupled with other data, is needed to interpret the energy flux through microbial systems. Current procedure uses several microscopic methods. All are examined for efficiency. The current direct microscopy of choice is epifluorescent microscopy using ultraviolet fluorescence by adding fluorochromes such as AO, DAPI, FITC, or Hoechst 33258 to determine microbial biomass. The direct counting of viable cell routing uses an appropriate fluorochrome dye and counts under an epifluorescent microscope (Norris and Swain, 1971; Quesnel, 1971; Pettipher, 1983; Harris and Kell, 1985; Fry, 1990b; Hall et al., 1990; Herbert, 1990). Unfortunately the major drawback is the inability to distinguish between live and dead particles and is labor intensive. Even though this method has other drawbacks including a lack of a standard, relative insensitivity to detect single and ultramicrocells, precision and accuracy, etc., its biggest drawback is determining biovolume. The biovolume may be converted to biomass carbon. In making this conversion, a series of assumptions are made (Karl, 1993). These assumptions include the average bulk density of the cells, the average percentage dry weight, and the average dry-weight-to-C ratio. As a result, the biovolume-to-carbon conversion factors exhibit a wide range from 1.21×10^{-13} g C μm^{-13} (Watson et al., 1977) to 5.6×10^{-13} g C μm^3 (Bratbak, 1985). Karl (1993) concluded that a biomass-C estimate derived from direct microscopy of a field sample might be within only $\pm 100\%$ of the real value. During incubation, bacterial biovolume can double from $0.070 \pm 0.037 \ \mu m^3$ to $0.153 \pm 0.152 \ \mu m^3$ in the early stationary phase and bacteria C:N ratios ranged between 2.8 and 10.3 (mean 6.5), which were inversely correlated with cell volumes (Kroer, 1994)

After reviewing the literature, Austin (1988) concluded that fluorescing cells (as determined by epifluorescent microscopy) that do not produce colonies on agar-containing media should be judged to be intact and viable. Treatment of sediment samples with pyrophosphate and sonication yielded cell counts over a log higher than the nontreated sediment samples as evidenced by the AODC method (Velji and Albright, 1986) The difference between data obtained by direct counts compared to viable counts has shown that $<0.1\%$ of the bacteria from the marine environment formed colonies (Kogure et al. 1979, 1980; Simidu et al., 1983).

FREQUENCY OF DIVIDING CELLS

Hagström et al. (1979) proposed a simplified AODC method for estimating bacterial production in aquatic samples, which made use of the relationship between the frequency of dividing bacterial cells (FDC) and bacterial growth. This method required no incubation. Bacteria in aquatic samples are filtered out of samples by use of Nuclepore filter (0.02 μm) and stained acridine orange. The cells were counted microscopically recording the total number of bacteria/field and the number of bacteria that were dividing in the same field. Sufficient samples were counted to ensure statistical validity. These investigators also found a linear relationship between FDC and growth rate in the laboratory and in the coastal Baltic Sea and used this relationship to calculate environmental growth rates from the measured FDC. After reviewing the theoretical basis of the FDC method with the published literature and their own data, Newell and Christian (1981) expressed reservations concerning its validity as a predictor of environmental bacterial growth rates.

KOGURE METHOD

The method of Kogure et al. (1979) may not be totally reliable for determining the viability of cells during starvation or in the stationary phase (Preyer and Oliver, 1993). The Kogure method also uses filtration of the aquatic sample through a Nuclepore filter, that has been previously incubated in the presence of a small amount of yeast extract and nalidixic acid (prevents division of cells). Acridine orange is used to stain the bacteria and only those cells that show elongation (no division) are the viable cells.

INT METHOD

The number of respiring bacteria in an environmental sample can be determined by the INT [2-(p-indophenyl)-3-(p-nitrophenyl)-5-phenyl tetrazolium chloride] (an artificial electron acceptor) method (Zimmermann et al., 1978). If a microbe is respiring, it forms an INT-formazan precipitate that can be seen as red spots within the cell when observed microscopically. A sufficient amount of INT formazan must be deposited before it can be detected microscopically. The INT may also act as an inhibitor to respiration (Karl, 1986). When the cells are observed on a polycarbonate filter and epifluorescence (AODC) is applied to the INT-treated sample, it is possible to differentiate between respiring and apparently nonrespiring bacteria. The nonrespiring cells do not show any INT formazan within their cells. Initial use of this method indicates between 6% and 12% of samples taken from the coastal areas of the Baltic Sea and 5% and 36% of the samples taken from freshwater lakes and ponds are respiring (Zimmermann et al., 1978). INT cells between 1.6 and 2.4 μm dominated the freshwater samples and 0.4 μm INT cells dominated the Baltic Sea samples.

Later, others added a solution of NADH-NADPH to the INT method and considered the cells that formed INT-formazan to be active cells. This INT method, with the addition of NADH and NADPH, for measuring metabolic activity does not require that the organism have existing energy reserves, utilize a specific C source, or be in a growing state (Norton and Firestone, 1991). Nevertheless, the utilization of exogenous pyrimidine nucleotides has been demonstrated for mutants of *E. coli* and *Salmonella typhimurium* that are deficient in de novo synthesis of NAD (Foster et al., 1979).

The autoradiographic technique was developed by Brock (1967) but this technique does employ a radioactive substrate. Thus, if the right radioactive substrate is used, starved state receives energy making it possible for the cells to take up the radioactive substrate and the result is evident on the photographic emulsion.

Autoradiographic methods have been used to determine the fraction of metabolically active bacteria within a population (Brock, 1967; Hoppe, 1976, 1978; Meyer-Reil, 1978). All autoradiographic methods use radiolabeled compounds as an energy source for the organisms. Even with the addition of an energy source some of the microbes in nature are inactive, which may also indicate that the energy source is not the right substrate. Using labeled amino acids, Hoppe (1978) reports that up to 60% of the bacteria in seawater are actively metabolizing. This actively metabolizing fraction may be 10 to 1000 times more numerous than the viable population. These values indicate the presence within the natural population of a large fraction that is not capable of reproduction but that is metabolically active with regard to the uptake of exogenous substrate(s).

Meyer-Reil (1978) combines autoradiographic and epifluorescent microscopy for assessing the microbial population of natural water. For the autoradiographic method, he used tritiated glucose. By this method, 2.3 to 56.2% of the bacteria were found to be actively metabolizing. A significant correlation exists between the number of metabolizing cells and the observed glucose uptake rate, but not between the number of viable cells or the total number of cells and the uptake activity.

4.5.3 Chemostat Studies

As early as 1932, Kluyver and Baars pointed out that laboratory cultures of bacteria are, to some degree, laboratory artifacts. According to Postgate (1973), "this principle applies equally to laboratory models for microbial ecosystems: they are of necessity very artificial." Continuous culture selects for variants that have a high affinity for the limiting substrate; one may call these variants "mutants" as a matter of semantics (Postgate, 1965). However, the most natural microbial environments resemble a slow-growing, moribund chemostat. Postgate (1976) concluded that the most scientifically sound system for studying aging and death during starvation would utilize chemostats of known nutritional status running so slowly that the death rate of the cells made a significant contribution to the steady state kinetics of the system as a whole. The relationship between a low growth rate and a high death rate is not universally true since autochthonous microbes are often starvation resistant and tend to have a naturally slow growth rate (Dawes, 1976; Gray, 1976).

The early studies on *Klebsiella aerogenes* by Postgate and Hunter (1962) showed that only a proportion of the population was viable when the cells were grown in a carbon-limited continuous culture. No evidence for a shift in the starvation resistance of the population was noted after continuous culture for 3 months. This finding was corroborated and expanded by Tempest et al. (1967), who presented evidence that populations subjected to a growth rate (μ_{min}) of about 0.1/hour or less could not multiply. This decrease in viability in chemostats with decreasing dilution was also noted by Sinclair and Topiwala (1970) and Kaprelyants and Kell (1992). Kaprelyants and Kell (1992) have also shown that cells grown at a very low dilution rate are extremely heterogenous with respect to their ability to accumulate the lipophilic dye (rhodamine 123), and their viability and resuscitation could be well correlated with their ability to accumulate the dye.

Chemostats only operate properly as the culture conditions approach homogeneity and at least two characteristics distinguish microbial ecosystems from most homogenous laboratory cultures. According to Wimpenny (1981), these two characteristics are that (1) most habitats contain different species of microorganisms so that microbial ecology is generally concerned with the study of mixed cultures and (2) the most important property is the organization of the organisms within a particular habitat.

Chemostats are not a good tool for studying growth at low growth rates, especially when the D values are lower than 0.05/hour (van Verseveld et al., 1984a,b). At slow growth rates the viability of the cells is low (Tempest et al., 1967). It should also be pointed out that *A. aerogenes* harvested from a chemostat at dilution rates greater than 0.1/hour result in a viable population over 95%, indicating that death does occur in chemostats. This result indicates that lower dilution rates can be used. On the other hand, investigators may not consider growing bacteria at low dilution rates as low as 0.004/hour, which requires approximately 1000 hours for the steady state to be rewarding or they may just lack the patience (Gottschal, 1990).

Matin et al. (1979) noted the accumulation of poly-ß-hydroxybutyric acid in a *Spirillum* sp. during carbon-limited growth in a chemostat. This accumulation was highest at the lowest dilution rate (D = 0.03 to 0.05) examined. This reserve energy source permitted the organism to be more resistant to starvation. Its synthesis may result from adaptation to a low-nutrient environment. With this in mind, I would suggest that the organism physiologically senses the low nutrient conditions and resorts to accumulation of as much energy as possible in the cell, anticipating a complete lack of energy with time. This situation is analogous to some freshwater bacteria synthesizing polysaccharide material in low-nutrient water.

The continuous culture technique produces artificial growth conditions with the aim of providing a mathematically manageable system (Jannasch, 1977). The method appears to be mainly for pure cultures; with mixtures of cultures, such as occurs in an environmental sample, the method does not work. According to

Jannasch (1974), "the significance of 'average' growth and uptake constants for natural populations appears to be doubtful, since prolonged exposure to any nutrient concentration will result in unpredictable changes of the population and its uptake characteristics." Michaelis-Menton kinetics used for calculating potential rates of substrate uptake by natural populations relies on the apparent fit of the data (Wright and Hobbie, 1965). In addition to the above difficulties, washout occurs in a chemostat when the concentration of the organic matter in the system is equal to that of marine waters (Jannasch, 1967, 1969). The inability of microorganisms to grow in a chemostat below certain limiting concentrations of nutrients has been known since 1958 (Novick). Chemostats work poorly for defining the behavior of slowly growing bacteria with cultures having mass doubling times longer than 10–12 hours (Chesbro, 1988). Generation times of up to nearly 100 hours in chemostats have been reported by Postgate and Hunter (1962) and 60% of these cells were determined to be nonviable by a slide culture technique (Postgate et al., 1962). This naturally brings up the question as to the threshold level of energy necessary for growth as well as for reproduction. Chemostats, in a steady state mode, do not reflect natural environments (Dawes, 1989).

The ability of a population to scavenge glucose was increased as glucose carbon was replaced by amino acid carbon (Law and Button, 1977). With glucose as the sole carbon source, at a dilution rate of 0.008 to 0.05/hour, more than 200 μg/l of glucose remained unutilized. With the addition of 1, 2, or 20 amino acids, the residual glucose was progressively utilized. With 20 added amino acids the reduction of glucose utilization was at a minimum of 0.3 μg/l at a dilution rate of 0.02/hour. Jannasch (1967) reported that a threshold concentration of organic matter was necessary for marine bacterial growth in the laboratory and this threshold is higher than that found in seawater. In his chemostat studies, Jannasch (1967, 1968) found that one group of bacteria was adapted to marine environments as evidenced by its ability to grow at low substrate concentrations with low substrate affinity (K_s value) and the other group was inactive in natural seawater with high K_s values. The observed phenomenon appears to represent an expression of the suboptimal character of seawater as a base medium for growth of some heterotrophic bacteria (Jannasch, 1968). Specific growth rates as low as 0.005/hour (generation times of 20 to 200 hours) of aquatic bacteria in natural waters were calculated from significant differences between dilution rates and washout rates in a chemostat (Jannasch, 1969).

In a series of papers, a *Spirillum* sp. and a *Pseudomonas* sp., both isolated from the aquatic environment, were compared when grown in a chemostat at a low dilution rate. Chemostat studies dealing with the competition between an obligately psychrophilic *Pseudomonas* and a psychrotrophic *Spirillum* at low lactate concentrations at two different temperatures indicate that at 10°C the psychrophile was unable to compete with the psychrotroph but at 4°C, the opposite was true (Harder and Veldkamp, 1971). Their data suggest that, in the permanently cold ocean, mineralization is carried out by the psychrophiles. The physiological bases of the selective advantage of a *Spirillum* sp. and a *Pseudomonas* sp. in a chemostat study that was carbon limited appears to be mainly a function of the surface/volume ratio where the surface/volume ratio was higher in the case of the *Spirillum* (Matin and Veldkamp, 1978). In the *Pseudomonas* sp., NAD-dependent and NAD-independent L-lactate dehydrogenases, aconitase, isocitrate dehydrogenase, and glucose-6-phosphate dehydrogenase activities increased up to 10-fold when the dilution rate was decreased from 0.5 to 0.02/hour, regardless of whether the growth-limiting nutrient was carbon, ammonium, or phosphate, whereas the *Spirillum* exhibited an increase in the activity of several enzymes at low dilution values (Matin et al., 1976). Matin et al. (1976) suggest that increased enzyme syntheses at low dilution rates represent the normal physiological state for bacteria in aquatic environments where growth occurs slowly under nutrient limitations and such increases probably permit a more efficient utilization of nutrients at subsaturating concentrations. In the *Pseudomonas* sp., the NADPH:NADP(H) ratio was much higher than the NADH:NAD(H) ratio, averaging 55% in carbon-limited cells (Matin and Gottschal, 1976). The *Spirillum* sp. had the ability to accumulate poly-β-hydroxybutyric acid when grown under C-limited situations, and this permits the organism to enhance its survival capacity when it undergoes starvation in the natural environment (Matin et al. 1979).

Numerous researchers (Horan et al., 1981; Jones and Rhodes-Roberts, 1981; Zychinsky and Matin, 1983; Otto et al., 1985; Poolman et al., 1987) studied the decline in ATP, AEC (adenylate energy charge), and/or the ability to accumulate lipophilic cations in starving cells or in cells grown at low dilution rates. Generally, none of the bioenergetics parameters as judged by plate counts could be correlated with the loss of viability (Kaprelyants and Kell, 1992).

When the chemostat containing polluted water was charged with filtered, sterilized, polluted seawater (containing organic substances in high concentrations), the genus *Acinetobacter* was predominant; when the chemostat was charged with clean seawater (containing organic substances in low concentration) the genus *Vibrio* predominated (Ishida and Kadota, 1975). Thus, the growth kinetics of two different populations respond according to the concentration of organic matter in the chemostat. By using two different types of bacteria grown on Z1 (nutrient-rich) and Z/20-agar (nutrient-poor) plates, the washout rates of these bacteria could be demonstrated in chemostats that were fed with different concentrations of organic matter in seawater, including unpolluted seawater (Ishida and Kadota, 1975). The competition between the Z1 and Z/20 bacteria depended on the concentration of organic matter in the seawater. From these data, Ishida and Kadota (1975) concluded that the Z1 bacteria exhibited lower affinity toward organic substances at the low concentrations and the population of Z/20 bacteria displayed a higher affinity for organic substrates at the low concentrations—the same conclusion that Jannasch (1967) came to by using a mixed culture of aquatic bacteria.

Thus, both research groups came to the conclusion that structure of the bacterial community in the sea is intensely controlled by concentrations of organic substances in seawater.

The dormant state can be established in bacteria after long periods of cultivation at low dilution rates in a chemostat (Pirt, 1969). The use of flow cytometry and rhodamine 123 by Kaprelyants and Kell (1992) showed that 40% of *Micrococcus luteus* cells in a chemostat population (employing a low dilution rate) persisted as dormant cells but could be resuscitated to normal bacteria.

Chemostat studies cannot duplicate the growth and activity in the natural environment. Nevertheless, growing cells at very low dilution rates will furnish microbes for experimental purposes that are closer to the growth rate that occurs in nature. Unless dilution rates in a chemostat are very low, the data obtained should be used with extreme caution when extrapolated to the environment.

4.5.4 Growth on Nuclepore Membrane Filters

Because chemostats do not work well in growing bacteria in nutrient concentrations that exist in natural waters, the growth of organisms is best performed on Nuclepore membrane filters floating on top of the natural water sample (Meyer-Reil, 1975) (a modification of the original Kunicki-Goldfinger and Kunicki-Goldfinger [1972] method). Its main drawback is the exposure of the bacteria to the air. Growth can occur but one must be careful not to grow oligocarbophilic bacteria. This technique requires time and good aseptic technique and cannot therefore be used routinely for many ecological studies.

4.5.5 Heterotrophic Activity (Substrate-Induced Respiration)

The heterotrophic activity method for measuring microbial activity proposed by Wright and Hobbie (1965) for aquatic systems has been employed very extensively in field samples. This original method only counted the radioactivity associated with the cells. The correction, adding the radioactive CO_2 so that total amount of the added radioactive carbon compound metabolized to the method, was made by Harrison et al. (1989). The resulting mixed population of the environmental sample must be treated conceptually as a finite number of equivalent responsive metabolic sites that conform to a uniform set of kinetic variables. Because of faulty assumptions, this concept is not valid (Williams, 1973; Wright and Burnison, 1979; Azam and Hodson, 1981; Law, 1983; Li, 1983). Modifications to this method have been reviewed by Wright and Burnison (1979). This approach is limited by our lack of knowledge of the composition of the DOM (Thurman, 1985) and by the availability of radiolabeled compounds. Because most of the DOM is not characterized chemically, it is not clear if the monomers of radioactive substrates added are present in the field. If they are present, are they monomeric or complexed to other substances (such as humic material) or adsorbed to clay particles? The usefulness of the radiolabeled approach occurs when one has specific questions about specific substrates (Lock and Ford, 1986). The heterotrophic activity method has been used extensively in the marine environment and has also been adapted to measure microbial activity in surface and subsurface soil. In soil microbiology, this method is known as substrate-induced respiration, a term used by Anderson and Domsch (1978). Thus, any time a sample is amended with any type of organic matter, substrate-induced respiration will take place. Unfortunately, this concept has not been adapted by the marine microbiologists. Recognizing that added radioactive substrates are employed in the method, many marine investigators use the term *heterotrophic potential,* mainly to reflect the potential that indigenous microorganisms have to carry on the metabolic functions on the added labile substrate. Furthermore, the heterotrophic activity method with glucose and other substrates cannot be used below 1000 m in the ocean where uptake rates are 0.001 µg carbon/1/hour (Strickland, 1971). Williams and Carlucci (1976) compared the utilization of DOC and/or POC in deep waters and found, in terms of utilization and oxygen consumption, the values were much higher than those calculated from a vertical diffusion advection model.

The addition of radiolabeled compounds to soil has been used extensively. A labile substance added to soil is rapidly utilized. ^{14}C glucose disappeared in 2 days and a minimum generation time of 2 hours was observed from 5 to 10 hours after the addition of the glucose (Chahal and Wagner, 1965; Oades and Wagner, 1971; Behara and Wagner, 1974). On the first day 36% of the carbon was liberated and the rest was incorporated primarily into the bacterial cells, because fungal growth was noted when the bacterial numbers declined. Thus, it can be seen that the addition of a labile substrate in field experiments provides the potential for use of the substrate because the level of the substrate is not maintained in the system. The uptake potential for natural samples must definitely be higher in many cases than the actual uptake. There is a constant uptake potential in starving bacteria; this potential is maintained for long periods of time in the absence of any external substrates (Amy and Morita, 1983a; Kurath and Morita, 1983; Höfle, 1984; Güde, 1984). Any time utilizable organic matter is added to any sample, substrate-induced metabolism occurs. Starved cells can immediately take up substrate. Likewise, it is of interest to note that a greater number of bacteria can incorporate ^{3}H-glutamate than ^{3}H-thymidine, indicating that cells can incorporate carbon and energy substrates at a time when they are not synthesizing nucleic acids (Douglass et al., 1987). On the other hand, extremely long turnover times (≤ 500 years) for leucine assimilation by microbial assemblages in the Arctic Ocean were reported by Pomeroy et al. (1990). ^{14}C-leucine now appears to be the radioactive compound of choice to determine the heterotrophic activity in seawater samples, because it may give better results than other labile radioactive compounds. However, it should be pointed out that Yegian and Stent (1969) found that leucine tRNA appears during leucine starvation of *E. coli*.

The main drawback of the use of radiolabeled tracers is that they reveal the dynamics of the compound added, not the complete array of substrates actually being used at any given moment (Ducklow and Carlson, 1992). Furthermore, they produce substrate-induced respiration and may cause the priming effect.

4.5.6 ATP and Adenylate Energy Charge in Natural Samples

The ATP content of seven marine isolates ranged from 0.5 to 6.5^{-9} µg ATP/cell, corresponding to 0.5% of the cellular carbon, but in senescent cells after nutrient depletion, its content was one-fifth of the above values (Hamilton and Holm-Hansen, 1967). In a starving marine *Pseudomonas,* Hamilton and Holm-Hansen (1967) also showed a decrease in the ATP values per cell as the cell viability increased for the first 5 days of starvation. However, after the fifth day the cell viability began to decrease rapidly while the ATP content remained stable at 0.2 $\times$ 10^{-9} µg per cell. The ATP content of bacteria in exponential growth is fairly constant. However, when bacteria are subjected to different environmental conditions, the ATP changes. Karl (1980) reported a 7-fold variation in ATP content of a variety of bacteria, and Knowles (1977) indicated a 30-fold variation in ATP content for various growing bacteria. The ATP content of starving bacteria is reduced 3 to 16 times when either P or N is limited (Karl, 1980). Postgate (1976) used cellular ATP as a check on cell death and cryptic growth. A high ATP:C is maintained during growth.

While studying starvation in three soil isolates, Nelson and Parkinson (1978) found that the two isolates that survived longer showed a gradual decrease in ATP content with starvation. The third isolate, a *Bacillus,* lost viability rapidly when starved and showed a simultaneous rapid decrease in ATP content. Thus, it appears that the bacterial response to starvation is specific. The measured ATP values of the soil biomass strongly suggest that the concentration in a largely resting soil population is little different from that in actively growing organisms (Jenkinson and Ladd, 1981).

The adenylate energy charge (AEC) equation, proposed by Atkinson (1971), $(AEC = [ATP] + 1/2[ADP]/[ATP] + [ADP] + [AMP])$ provides a biochemical basis to assess the physiological and nutritional state of organisms. A theoretical value of 1.0 to 0 corresponds to the highest and lowest metabolic activity. AEC values in the range of 0.50 to 0.75 are considered to represent inactive or dormant cells; values below 0.4 are considered to indicate dead or dying cells (Karl, 1980). Maintenance, but not growth, in *E. coli* seems possible between 0.5 and 0.8 and loss of viability below 0.5 is not surprising (Chapman et al., 1971). Spores can have an AEC below 0.1 (Setlow and Kornberg, 1970).

Wilson et al. (1986) studied the relationship of ATP in subsurface samples to the rate of degradation of alkylbenzenes and chlorobenzene at the Lulu site in Oklahoma and the Conroe site in Texas. The typical ATP content of surface material was 44–500 ng/g. At the Lulu site, the ATP content ranged from 0.03 to 1.2 ng/g. The rate of toluene degradation decreased with decreasing ATP content but ceased when the ATP content was at 0.05 ng/g or less. For most subsurface bacteria, the AEC values indicate inactive or physiologically stressed bacteria. The respiration- and adenylate-based assays, according to Kieft et al. (1990), may be used for the viable nonculturable cells that comprise large populations of the microbial communities in soil and subsurface sediments, but their sensitivity is a disadvantage. Often, biomass and rates of activity are below detection.

Applying the AEC or the ATP values to environmental samples does have its limitations. The limitation, according to Karl (1980), are the occurrence of soluble (cell-free) ADP, ATP, and AMP, and the lack of validity of "community" AEC measurements. AEC values of 0.5 to 0.7 are indicative of either a senescent population or an association of actively growing cells (AEC >0.8) and dead or dying cells (AEC >0.5). The dead or dying cells may be cells that are in the starvation-survival state. However, Wiebe and Bancroft (1975) reported that environmental AEC values and community densities vary independently, such that the potential for microbial growth in the ocean is not necessarily correlated with biomass.

Other investigators have made measurements of ATP, AMP, and AEC on environmental samples (see Karl, 1980). Recognizing the above limitations of the AEC values, Karl (1980) states that the application of the AEC index may be in assessing the impact of pollution on selected components of an ecosystem.

4.5.7 Thymidine Method

The use of ^{3}H-thymidine (unless otherwise stated, thymidine in this section is tritiated thymidine) to determine the growth of bacteria in natural systems was initiated by Brock (1967) and further developed by Fuhrman and Azam (1982). This incorporation theoretically correlated with cell division and biomass production (Brock, 1971); thus estimates of bacterial production rates and specific growth rates could be determined. According to Brittain and Karl (1990), a uniform method has not been adapted by the various investigators using the thymidine method. From the growth response, an inference is made as to their activity in various ecosystems. The method has not been universally accepted and thymidine incorporation in certain marine environments may not be as universal as it was once assumed (Davis, 1989; Douglas et al., 1987; Jeffrey and Paul, 1988; Robarts et al., 1986). Over half of all published estimates of bacterial productivity have been made by the thymidine method, in spite of the method being controversial (Ducklow and Carlson, 1992). The assumptions made when one uses the thymidine method are: (1) all bacteria incorporate thymidine, (2) most of the TCA-precipitable radioactivity is found in the DNA fraction, and (3) there is dilution of the thymidine by internal or external DNA precursors (Karl, 1982; Moriarty and Pollard, 1981; Riemann et al., 1982). Item 2 was not completely verified (Güde, 1984). It appears to be more practical to relate thymidine incorporation measurements to bacterial production by using conversion numbers derived from independent direct growth estimates. This approach is justified by the remarkably good correlations usually found between thymidine uptake and direct growth measurements (Fuhrman and Azam, 1980, 1982; Kirchman et al., 1982; Bell et al., 1983). On the other hand, starved cells show low rates of thymidine incorporation, while taking up an appreciable number of amino acids (Nyström et al., 1986).

It is well known among microbial physiologists that thymidine is not used by all microorganisms. Some marine *Pseudomonas* strains incorporate thymidine (Pollard and Moriarity, 1984). Jeffrey and Paul (1990) found that 37 out of 41 isolates could not incorporate thymidine. Davis (1989) stated that one must distinguish between thymidine uptake and incorporation. Out of 17 isolates from upwelled waters she found that 6 out of 17 isolates could not incorporate tritiated thymidine. Some of the isolates employed that did not incorporate thymidine were *Vibrio, Cytophaga, Flavobacterium,* and *Serratia.* These organisms were isolated on high- and low-nutrient agar plates. There is also a diversity in the responses in organisms of the same genus and different locations. The data also show that thymidine uptake exceeds incorporation and varying portions of the transported nucleotide may be incorporated. The ability of freshly grown cells and 5-week-starved cells to take up tritiated thymidine is shown in Figures 4.7 and 4.8, respectively. Note the great difference between freshly grown cells and starved cells in their ability to incorporate thymidine. The average rate of uptake and incorporation was high in upwelled waters. The microbial population of upwelled water would naturally be in a different physiological state; that is, it would be in some degree of starvation. This uptake and incorporation takes place in the absence of any other added C or N sources. Davis (1989) cautioned that the ability of these cells to incorporate thymidine in the absence of cell division has implications for calculations of bacterial production on the basis of thymidine measurements. The physiological state of the organism plays a major role in the ability of the organism to incorporate tritiated thymidine. Davis (1989) also found that some isolates respond only to high concentrations of tritiated thymidine. Thus, the rate and incorporation by natural populations will depend on strain composition of the populations. Johnstone and Jones (1989) could not demonstrate thymidine incorporation in a chemolithotrophic organism. Others (Carlson et al., 1985a,b; Saito et al., 1985; Jeffrey and Paul, 1990; Güde, 1984; Novitsky, 1983a,b) found various species of bacteria unable to incorporate thymidine. Douglas et al. (1987) found that between 5.9 and 18.5% of the AODC count was able to incorporate thymidine, while 23.2 to 97% were able to utilize glutamate. Mården et al. (1988) found that marine bacteria undergoing starvation incorporated thymidine during the initial phase of starvation; even after 24 hours of starvation some incorporation

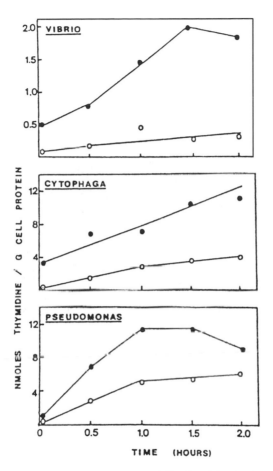

Figure 4.7 Uptake (●) and incorporation (○) of [³H]thymidine by freshly grown *Vibrio, Cytophaga,* and *Pseudomonas* isolates. The substrate concentration was 19 nM. Reprinted by permission from *Applied and Environmental Microbiology* 55:1267–1272, C. L. Davis, 1989; copyright 1989, American Society for Microbiology.

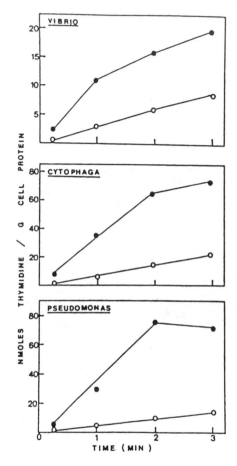

Figure 4.8   Uptake (●) and incorporation (○) of [³H]thymidine by starved (nongrowing) marine isolates. The *Vibrio, Cytophaga,* and *Pseudomonas* strains had been starved prior to the assay for 4, 3, and 5 weeks, respectively. The substrate concentration was 19 nM. Reprinted by permission from *Applied and Environmental Microbiology* 55:1267–1272, C. L. Davis, 1989; copyright 1989, American Society for Microbiology.

occurred. Bloem et al. (1989) reported significant thymidine incorporation in nongrowing marine and freshwater bacteria. Finally, after much research, Brittain and Karl (1990) concluded that the thymidine method is invalid in most oligotrophic habitats.

Working in lakewater, Chróst et al. (1988) reexamined the thymidine incorporation method and recommended (1) short incubation times (30–40 min) for thymidine incorporation, (2) the isotope dilution procedure for determining the degree of thymidine participation in DNA synthesis, or substrate-dependent thymidine uptake kinetics for estimating the required concentration for saturation, (3) parallel growth rate estimates from increasing bacterial abundance and thymidine incorporation, and (4) conversion factors derived from exponential rates of changes in thymidine incorporation and cell numbers during the growth of bacteria in undiluted water samples in the absence of grazers.

Although Moriarty (1988) expounds the virtues of the use of DNA for growth measurements, he also admits that much of the laboratory work from which conclusions are drawn and extrapolated to the natural environment has been undertaken on organisms such as *E. coli.* He states that applying the principles of bacterial growth processes to the natural environments, especially where the nutrient supply is small and variable, should be done with caution. Christian et al. (1982) have demonstrated the need for caution in their studies with mixed batch cultures from natural populations, especially when they found that the growth rate constants calculated from a variety of different measures of growth were variable. For more information concerning the use of thymidine in ecological studies, consult Moriarty (1988).

The inability to incorporate thymidine may be due to the lack of a mechanism to transport the thymidine into the cell and/or the lack of thymidine kinase. When catabolism occurs, nonspecific labeling of other macromolecules occur (Carmen et al., 1988b; Hollibough, 1988; Jeffrey and Paul, 1988; Robarts et al., 1986; Brittain and Karl, 1990). Thymidine is readily degraded within cells by the inducible enzyme thymidine phosphorylase. According to Jeffrey and Paul (1990), the ability to incorporate thymidine may also be a matter of the cell's physiological condition (not synthesizing DNA at the specific time as in starved cells). Incorporation of the radiolabile thymidine does not necessarily mean that it is in the DNA of the cell (Riemann et al., 1982; Robarts and Wicks, 1989; and Robarts et al., 1986) and thus a true picture of the growth of all bacteria in the various ecosystems cannot be precise. Furthermore, the addition of thymidine to cultural material also brings about another problem; it adds energy to a system that is void or nearly void of bioavailable energy sources. Robarts and Wicks (1989) stressed that thymidine uptake must be in excess of incorporation to exclude isotope dilution. In terms of energy needed for synthesis of bacterial growth, approximately 25% of DNA bases in the DNA do not have to be synthesized when tritiated thymidine is added to natural samples. "Degradation product formation, including that of both volatile and nonvolatile compounds, was much greater than the rate of incorporation of tritium in stable macromolecules" (Brittain and Karl, 1990). Hence, the catabolism of thymidine can add energy to the system (priming effect) making the cells more metabolically active. Douglas et al. (1987) reported that a greater number of bacterial cells in seawater were able to incorporate tritiated glutamate than thymidine, demonstrating that in natural populations, metabolically active bacteria are not all able to incorporate thymidine.

Carman et al. (1988) and Dobbs et al. (1989) noted that nonspecific labeling, catabolism, and recycling of [methyl]^{3}H- and [methyl]^{14}C-thymidine occur in marine sediments. Only 2% of the incorporated radioactivity was recovered in the DNA fraction of the TCA-insoluble material following time course incubations of marine samples for 1 to 300 minutes. The methyl group of the thymidine was extensively catabolized as evidenced by copious production of $^{14}CO_2$. According to the authors, the temporal patterns of ^{3}H:^{14}C ratios in macromolecular fractions indicated that the products of catabolism were recycled into the DNA fraction and the accuracy of the thymidine technique depends in large part on the degree to which such catabolism occurs. Nonspecific labeling of macromolecules during the assimilation of tritiated thymidine is now recognized to be a ubiquitous ecological phenomenon, and the conversion factor is a variable, habitat-dependent parameter (Karl, 1994).

The thymidine conversion factor (rate of incorporation of labeled thymidine in macromolecules into units of bacteria cell production per unit of time) as well as the carbon conversion factor (measured bacterial cell volumes into units of carbon biomass) are commonly employed in aquatic microbiology. The results obtained using the thymidine conversion factor can vary by nearly three orders of magnitude as documented in the literature (e.g., Fuhrman and Azam, 1982; Smits and Riemann, 1988), while the results obtained with the carbon conversion factor can vary over fivefold (Watson et al., 1977; Bratbak, 1985). A common assumption is that the thymidine conversion factor is constant in various systems (Fuhrman and Azam, 1980). However, the amount of DNA per cell varies during the cell cycle (Morse and Carter, 1949), increasing with growth rate in culture (Cho and Azam, 1988; Ingraham et al., 1983), and decreasing with nutrient limitation or starvation (Amy et al., 1983b; Hood et al., 1986). If the thymidine method is employed with sediments, thymidine adsorption onto sediment particles can occur (Moriarty and Pollard, 1990), resulting in lower activity.

In the Antarctic, the specific growth rates are low (μ = >0.1–0.3/day) and during the austral winter, growth may not occur at all and unbalanced and discontinuous growth may be the norm (Karl, 1993). Karl's (1993) opinion is that the four most crucial assumptions are that (1) the uptake of thymidine is restricted to, and occurs uniformly in, all heterotrophic bacteria, (2) the dilution of the exogenously added thymidine by ambient extracellular and intracellular pools and by de novo synthesis either is negligible or can be directly measured and corrected for, (3) catabolism of thymidine does not occur, and (4) there is a constant relationship between thymidine incorporation and biomass production. The first three assumptions appear to be invalid for microbial assemblages in nature (Law, 1983; Fallon and Newell, 1986; Douglass et al. 1987; Jeffrey and Paul, 1988; Davis, 1989; Brittain and Karl, 1990). Kogure et al. (1986) proved that there is no simple correction for isotope dilution of Antarctic assemblages. Also in the Antarctic, thymidine appears to be extensively metabolized (Karl et al., 1991). Finally, independently derived extrapolation factors used to convert thymidine uptake to bacterial production have varied by more than 100-fold (Fuhrman and Azam, 1980; Kirchman et al., 1982; Ducklow et al., 1985; Cho and Azam, 1988; Coveney and Wetzel, 1992). The incorporation of [^{3}H]leucine and [^{3}H]thymidine in subarctic waters is greater after additions of 0.5 µm amino acid than when a similar amount of glucose was added (Kirchman, 1990).

The use of the thymidine method assumes (1) the label is specifically for bacteria (Karl, 1982), (2) the principal route of incorporation of thymidine into DNA is via the "salvage pathway" (Moriarty, 1988), and (3) a large and constant fraction of dividing cells takes up thymidine (Fuhrman and Azam, 1982). A large fraction of bacterial cells does not take up thymidine even when heterotrophic activity takes place. Hence, thymidine does not measure a substantial proportion of the active bacteria (Douglas et al., 1987). The bacterial population in nature may not exist in a state of unbalanced growth and heterotrophic activity is not necessarily synonymous with cell division. Unbalanced growth in the marine environment may help explain the low percentages of thymidine, as compared to glutamate, incorporation in bacteria (Douglas et al., 1987). Unfortunately, production rates change over periods shorter than the doubling time, hence the growth state may vary at a given time (Riemann et al., 1984; Moriarty and Pollard, 1982). It is also difficult to relate thymidine incorporation to open ocean environments (Moriarty, 1988). There is also the need to convert thymidine uptake to bacterial biomass production as well as to convert cell volume to cell carbon content. Estimates resulting from this conversion may vary by a factor of 5 (Kirchman et al., 1982; Kemp, 1990).

When the thymidine method is applied to sediments, it appears that bacterial production measured by thymidine incorporation and ^{32}P-phosphate incorporation are in good agreement (Moriarty, 1988; Moriarty et al., 1985). However, the difficulties encountered with this procedure are:

1. The added thymidine may not disperse evenly during the mixing of the sediments (Moriarty, 1988; Montagna and Bauer, 1988) and mixing may bring about the perturbation effect on microbes.

2. Abiotic adsorption of label onto sediment particles results in the label not being available to the microorganisms; hence it may require a very short incubation period (Kemp, 1988).

3. There is greater difficulty in extracting DNA due to its adsorption on the sediment particles (Findlay et al., 1984).

4. It is impossible to saturate thymidine incorporation kinetics so that the data obtained are incorrect (Fallon and Newell, 1986; Kemp, 1988).

5. Some sediment bacterial populations extensively catabolize thymidine (Hollibough, 1988; Carman et al., 1988; Dobbs et al., 1989).

6. Some bacteria in anaerobic sediments do not take up thymidine efficiently (Moriarty, 1988).

It should also be mentioned that the uptake and incorporation of [^{3}H]adenine as a measure of total microbial nucleic acid synthesis used by Karl (1979) is also controversial (Karl and Winn, 1986; Fuhrman et al., 1986).

We are apparently far away from the point where unequivocal interpretation of results obtained by these thymidine method and microbial techniques is always possible (Van Es and Meyer-Reil, 1982). Consult Riemann and Bell's (1990) review of the accuracy of the thymidine method for further information. Ducklow and Carlson (1992) also review the methods to a very limited degree. Robarts and Zohary (1993) deal with thymidine incorporation and provide more details concerning the pros and cons of the thymidine method.

Finally, starved cells also take up thymidine (Märden et al., 1988). The highest rates of incorporation occured during the initial starvation phase and two strains actually increased their thymidine uptake during the first 2 hours of starvation. However, there was a decrease in thymidine uptake with increasing time of starvation.

4.5.8 RNA

In E. coli, both rRNA/cell and the RNA:DNA ratio were highly correlated with specific growth rate over the range of $\mu = 0.3$ to 1.6/hours (DeLong et al., 1989), and this relationship was also noted in E. coli, S. typhimurium, and A. aerogenes as well as in phytoplankton by Dortch et al. (1983). Kemp et al. (1993) examined this relationship in four marine isolates at different growth rates ($\mu = 0.01$ to 0.25/hour; five times lower than studies by DeLong et al., 1989) and found that the RNA contents (RNA/cell, RNA:biovolume ratio, RNA:DNA ratio, RNA:DNA:biovolume) were significantly different among the isolates. When they normalized RNA content to cell volume, it did not reduce these differences, but on average, the correlation between μ and the RNA:DNA ratio accounted for 94% variance when isolates were considered individually. However, in data pooled across isolates (analogous to an average measurement for a community), the ratio of RNA:DNA μm^{-3} (cell volume) accounted for nearly half of the variance in μ ($r = 0.47$). DeLong et al. (1989) concluded that RNA content is likely to be a useful growth rate predictor for slowly growing marine bacteria but in practice may be most useful when applied at the level of individual species and not to a bacterial community. Kemp (1990) found thymidine-based production estimates may be accurate only within a factor of 5, and the RNA method is comparable in accuracy (Kemp et al., 1993). However, these investigators state it is possible that this relationship may not hold for studies where the growth rates are much lower than those seen in these studies and the relationship becomes meaningless when growth is effectively zero as it is for cells in starvation conditions. A transient increase in cellular rRNA content has been observed in laboratory cultures of marine bacteria recovering from nutrient limitation and this pattern probably reflects the dynamics of changing cell physiology over the course of growth in batch culture (Kramer and Singleton, 1992, 1993). Kramer and Singleton (1993) observed variation in rRNA only after heavy enrichment, which may be a function of an enhancement of a subpopulation of bacteria within the original community. Kemp (1994) warns that we may find that rRNA content can be interpreted meaningfully only at the level of individual species and not for an entire community.

4.6 Conclusions

Karl (1993) summarizes our current difficulties in addressing microbial ecology by stating the following:

1. Currently we are methods limited.
2. The limitations of each method must be identified so that we can interpret our field results with caution, self-criticism, and accuracy. Assay "assumptions" too readily become "facts" with extensive and uncritical use of any method, technique, or instrument.
3. Nevertheless, although unreliable methods can only provide questionable data, if we lose sight of the ecological processes we endeavor to understand while we scrutinize and debate the validity of experimental approaches, microbial ecology will stagnate. The investigator must keep in mind that even questionable data may be better than none at all.

With the above in mind, we must, in the future, address some of the most difficult controls to use experimental procedures. These controls need to take care of priming our samples, perturbation effects, effect of energy sources in the atmosphere, etc. This coupled with data provided in Chapter 3 on the low organic matter content of ecosystems makes this author question the accuracy of the microbial data in the literature because it is premature to assume the techniques are accurate. In all probability the rates of activity reported are too fast when extrapolated back to the environment.

Whatever method we use on environmental samples, we must be aware of its limitations. Most methods do not account for the perturbation of the sample, priming of the sample when substrates are added, and organic substrates in the air being absorbed into environmental samples. Starved cells react to added substrates immediately. Because of the foregoing, the methods of analysis must also undergo scrutiny. "No single approach to the study of microbial ecology is universally accepted" (Karl, 1980). The standing stocks of cellular carbon pools and the rate of biomass production and loss are two of the most fundamental measurements necessary to describe the microbial community. Karl (1994) states that these parameters cannot be measured with the precision or accuracy that is often required for modern hypothesis testing.

It is easy to criticize any method employed but the criticism(s) must focus on the positive so that the difficult task of developing new methods can be achieved. With this in mind we are a step closer to reality with each new technique tried.

5

Starved Bacteria in Oligotrophic Environments

> The ecology of a species, as exhibited by its distribution, its survival mechanisms, and the role it plays in the ecosystem, is a product of many factors, including its developmental and morphological characteristics and its biochemical and physiological mechanisms.
>
> —Goodfellow and Dickinson (1985)

5.1 Introduction

Oligotrophic environments dominate the Earth, because of the presence in ecosystems of various physiological types of microbes that utilize the organic matter present (Morita, 1986). If conditions are appropriate, the metabolic rate of microbes can be extremely fast. It is well known that *Escherichia coli* has the ability to divide once every 20 minutes; yet under ideal conditions, *Vibrio natriegens,* a marine bacterium, can reproduce in less than 10 minutes (Eagon, 1962). If growth of either of these organisms occurred for 2 or 3 days, then one can easily see why microbes are mainly responsible for the oligotrophic conditions on earth. If a surplus of energy comes into the environment, then the microbes, except in rare cases, utilize this surplus until the oligotrophic conditions return. Under oligotrophic conditions, the metabolic rate can be very slow due to the lack of proper substrates. Thus, heterotrophic bacteria in oligotrophic environments may be in one of several metabolic states (dormant, inactive, starved, or oligotrophic).

All heterotrophic microorganisms compete for the limited energy source within their own niche. When faced with the stress of nutrient deprivation, the microbes have three alternatives: formation of dormant spores or vegetative cells, the continuation of growth (even at a slow rate), or death (Gray, 1976). The phase called starvation-survival should be added to this list. It is now recognized that the amount of organic matter (energy source for heterotrophic bacteria) in soil is insufficient to support the growth and reproduction of all the soil heterotrophic microorganisms. Microorganisms in the environment are constantly facing downshifts in nutrient levels that they themselves cause. As an example, detachment of cells from a particle or fecal pellet represents a downshift from an environment of plenty to starvation conditions, from a growing to a nongrowing survival phase (Kjelleberg and Hermansson, 1987). Conversely, escape from oligotrophic environments onto surfaces (attachment), such as marine snow, detritus, fecal pellets, flocculated organisms, or solid inorganic surfaces, is a strategy to overcome the scarcity of nutrients. Another method would be to escape into a more hospitable environment such as on or inside a host plant or animal, or the predatory mode of life (e.g., *Bdellovibrio* which attacks its specific prey).

The input of energy (sunlight) on this Earth is limited. The Earth is not a eutrophic system. Microbes are mainly responsible for the utilization of the green plant photosynthetic products, thus leaving most of the Earth grossly oligotrophic. Not all physiological types of bacteria can utilize this material and must rely on syntrophy for their respective energy and/or nutrient sources. If all the organic matter that is recalcitrant to microbial oxidation were suddenly used by microorganisms, there would be a terrific output of carbon dioxide so that the greenhouse effect would be tremendously accelerated. The ocean itself has approximately 2.10×10^{17} g of dissolved organic matter and 2.10×10^{16} g of detritus. Fortunately, most of it resists microbial decomposition.

The use of the organic matter in ecosystems relies on syntrophy. Instead of employing the term *syntrophy,* Russian microbiologists use the label *microbial succession* (see Mishutin, 1975). The term *symbiotic* has also been used to show that one organism is dependent on the excretory products of another organism. In some cases, a one-carbon compound can be metabolized to a more complex carbon substrate. Interspecies dependence occurs in nature. Syntrophy naturally brings about a succession of different species in the environment. As a result, microbial growth in ecosystems is sporadic, at best.

Unfortunately, currently usage lumps together all the various species of bacteria in the various ecosystems as heterotrophs (Kjelleberg et al., 1993; Moriarty and Bell, 1993). Among the heterotrophs are species with different nutrient requirements and some are very fastidious. The concentrations of the various organic substrates vary in time and space and it follows that the population of various species should be different in time and space. Thus, most species in ecosystems undergo growth, starvation, and recovery (Morita, 1982, 1985). The starvation period may vary from a short to a very long period, may be seasonal, diurnal, etc., depending on the environmental factors.

Although many microbial studies have been made on natural environmental samples, these studies generally ignore the oligotrophic nature of the environment. The data in the literature concerning the occurrence, activity, etc., of bacteria in the environment are too voluminous to be covered in a chapter, hence, most of the data reported in this chapter are mainly concerned with the relationship of the energy limitation of the environment to the so-called oligotrophic bacteria.

5.2 Creating Oligotrophic Environments

Microbes create oligotrophic environments: each time a molecule is used by a microbe, the resultant molecule always contains less energy. Radioactive leucine and glucose added at natural concentrations to seawater were transformed to high molecular weight dissolved organic materials that persisted through 6 months incubation (Brophy and Carlson, 1989). At the end of the 6-month incubation period, only 1–17% of the high molecular weight materials were respired when reincubated with the microbial population of seawater, whereas 40–75% of the monomers were respired over the same period of time. In situ transformations of biologically available carbon, according to the investigators, may be an important source of refractory dissolved organic carbon in the ocean. Labile dissolved organic matter has been shown to undergo microbial transformation into humiclike matter in seawater (Tranvik, 1993).

The indigenous microbes within any ecosystem do not utilize all the organic matter present. Humus and humuslike material are essential to soil. Signature compounds exist in geological material, which also indicates that microbes do not utilize all the organic matter within their habitat.

5.3 Bacterial Size in Oligotrophic Environments

Pirie (1973) wrote a review entitled "On Being the Right Size," but unfortunately he did not discuss the size of microorganisms in relation to the environment. Small bacteria have been observed in the laboratory as well as in the natural environment. Small bacteria have been observed and mentioned in the early literature (Henrici, 1928; Rahn, 1932). Bissett (1932) stated that nearly all bacteria, although he observed them only in *Pseudomonas, Bacterium (Escherichia), Salmonella, Proteus,* and *Rhizobium,* produced microcysts (a resting cell state) in response to nutrient limitations. Small coccoid cells are characteristic of the indigenous soil microflora (Conn, 1948). Cell size is dependent upon the medium and temperature (Schaechter, 1968). Filterable forms of *Cellvibrio gilvus* were reported in 1959 (Tuckett and Moore, 1959) employing Morton ultrafine bacteriological fritted glass filters. Some of the filterable forms were capable of passing through a 0.35-μm pore radius filter but not through a 0.30-μm filter. These were observed mainly by electron micrographs, but with the advent of the acridine orange direct count (AODC) method, the dominance of small cells in the environment was stressed further. The particle counters (especially the Elzone Monitor Particle Counter), image analysis, and flow cytometer have greatly added to the knowledge that bacteria in various environmental systems are mainly below the 1-μm range. It should be noted that the size of bacteria will differ according to the technique used for their determination (Fry and Davis, 1985). However, the knowledge that small cells can result from starvation was definitively established by Novitsky and Morita (1976).

The shape, form, and size of bacteria are known to change with environmental conditions. The ability of microbes to become larger when nutrients and other conditions are optimal, as well as becoming smaller when faced with the lack of nutrients, illustrates the species' variability and plasticity. The cells of most species that undergo starvation generally decrease in size. As a result, in most oligotrophic environments the dominant size of the cells is smaller than cells observed in the laboratory. This is to be expected because laboratory cultures are grown under optimum physical and chemical conditions; one might even term them "giant cells" compared to their counterparts in nature. Thus, the small cells are the dominant or normal size cells found in the environment.

The resulting small cells have been termed dwarf-, mini-, pico-, nano-, and ultramicrobacteria. The term "ultramicrocells" has been used to describe small cells 0.3 μm in diameter or less (Torrella and Morita, 1981). However, upon starvation, the species may not reduce their size to 0.3 μm but may be in the 0.4- to 0.8-μm range. I have not used the term "dwarf cells" mainly because some ultramicrocells, upon prolonged starvation, will become even smaller (dwarf). Although other investigators have used the term "dwarf" bacteria, this usage may cause the cells to be confused with cells from a dwarf colony. In addition, in the starvation-survival studies, we have the dwarfing process with individual cells that are actually getting smaller with time. Minicells or minibacteria can be confused with the minicells produced by *E. coli* that are anucleoid (Torrella and Morita, 1981). Minicells (*E. coli*) are deficient in DNA and do not reproduce (Adler et al., 1967), but their oxygen uptake is linear with time, the same as a dividing population. The terms "nano-" and "pico-" plankton (bacteria are also included), proposed by Sieburth et al. (1978), are employed in biological oceanography erroneously, mainly because they do not represent 10^{-9} and 10^{-12} of a plankton. Sieburth et al. (1978) justified the terms on the basis of volume instead of length. Sieburth and Estep (1985) suggest that the nanoplankton refer to planktonic protist cells in the 2- to 20-μm diameter range while the picoplankton cells are in the 0.2- to 2.0-μm size range and do not represent 10^{-9} and 10^{-12} m, respectively.

Another benefit to the bacterium for being small is to increase its surface/volume ratio so that there can be an increased uptake of substrate. In order to increase the surface/volume ratio, cell size does decrease in many bacteria examined but this factor can also be achieved by prostheca formation in some bacteria. One important aspect of being small is that small organisms are less limited by diffusion kinetics than larger organisms; this may provide an important insight into cell morphology and nutrient uptake (Jackson, 1987a). The part diffusion plays in the biology of bacteria is discussed by Koch and Coffman (1970) and Koch (1990) and being small definitely helps the diffusion process. Jackson (1987b) further discusses diffusion rates of organic matter in the marine environment as well as chemotaxis. Because starvation may reduce the size of the degree of reduction of the cell size depends on the species and the length of the starvation process. Size reduction may serve as a mechanism of predator avoidance, as protozoans consume larger bacteria (Pomeroy, 1984; Morita, 1985; Andersson et al., 1986; Gonzalez et al., 1990). Wright (1984) suggested that grazing would control the upper end and that survival mechanisms during periods of nongrowth would set the lower limit of bacteria numbers in natural waters.

According to Grossman and Ron (1989), *E. coli* has a certain minimum size below which the cell cannot divide; there is no minimum cell size for DNA replication. The starved cells were at least 30% smaller than unstarved newborn cells. These studies were done in two successive steps with amino acid starved cells. In a chemostat, Ant-300 cells have the ability to reproduce when the cell volume is as low as 0.478 ± 0.060 μm^3 (Moyer and Morita, 1989a). In all probability, each species has its own minimum size before division of the cell ceases.

Starved cells retain their morphology to a large degree (Amy and Morita, 1983a). Torrella and Morita (1981), employing the slide culture method, were able to observe cells that always remained small (ultramicrocells) in freshly collected Yaquina Bay water, which had very slow growth rates on nutrient media. Furthermore, these ultramicrocells formed microcolonies and their size did not increase significantly.

5.3.1 Soil

In determining the size of the natural microflora by thin-sectioned and freeze-etch preparations, Balkwill et al. (1975) determined that gram-positive and gram-negative cells ranged from about 0.15 μm to just over 1.0 μm, with the average diameter for all cells ranging from 0.47 to 0.51 μm; whereas dwarf cells (Bae et al., 1972) ≤ 0.3 μm in diameter accounted for 27% to 32% of all the cells observed. When *Agromyces ramosus* was grown on different nutritionally limiting nutrients, the cell size ranged from 0.18 to 0.36 μm in diameter (Casida, 1977). Faegri et al. (1977) found that the average bacterial diameter of soil bacteria in soil is ca. 0.5 μm (0.01 pg dry weight/cell). In the latter study, soils were stored to make certain that dormant cells would comprise a major portion of the microflora. Upon incubation of these stored unamended soils at 50% moisture holding capacity, the size distribution of these cells changed as the incubation time increased (Fig. 5.1). Nevertheless, the ≤ 0.20-size cells decreased and then increased after 48 hours. The same could be said for the ≤ 0.30-μm cells; whereas the 0.31- to 0.50- and ≥ 0.51-μm cells increased in number and then decreased. The total number of cells recovered from the soil samples remained constant. However, if nutrients were added to soil there was an apparent more rapid increase in cell size of the ≤ 0.20- and ≤ 0.30-μm cells as indicated by the percent decrease in numbers

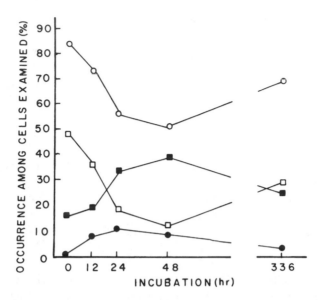

Figure 5.1   Effect of cell size of 30 C incubation of unamended soil A at 50% moisture holding capacity. A total of 450 cell sections was examined. Cell diameters: □, ≤0.20 μm; ○, ≤0.30 μm; ■ 0.31 to 0.50 μm, ●, ≥0.51 μm. The ≤0.30 μm curve includes those cells ≤0.20 μm in diameter. Reprinted by permission from *Journal of Bacteriology* 113:1462–1473, H. C. Bae and L. E. Casida, Jr., 1973; copyright 1973, American Society for Microbiology.

(Table 5.1). Nutrient conditions, as well as moisture, definitely influence the size of bacteria in soil.

The ability to scavenge nutrients from an oligotrophic environment may be related to the surface to volume ratio of the organism. There appears to be a selective advantage for cells having a relatively high surface/volume ratio in systems having very low concentrations of limiting nutrients (Kuenen et al., 1977). The smaller the organism, the higher the surface to volume ratio. However, if the surface/volume ratio is greater than 20 to 30%, a genetic adaptation seems to be required (van Gemerden and Kuenen, 1984). This increase in surface to volume ratio is mainly the result of low substrate concentration. Likewise the formation of prosthecae of *Ancalomicrobium,* which also increases the surface/volume ratio, is due to the low substrate concentration but the formation of the prosthecae in other organisms is not easily explained (Dow and Whittenbury, 1980).

When carbon input to soil is eliminated for 27 years, the number of small cocci is reduced only 22% and the number of larger cells is reduced 60 to 100% (Schnürer et al., 1985). Changes in cell size in a forest soil and a peat was observed

Table 5.1 Effect of nutrient amendments to soils A and B on percentage occurrence of small bacterial cells.

Soil	Nutrient Additions	Cells $\leq$ 0.20 µm Diam (%)[a]			Cells $\leq$ 0.30 µm Diam (%)[a]		
		0 h	12 h	24 h	0 h	12 h	24 h
A	Water[b]	55	36	18	83	73	56
	Glucose	55	18	10	83	48	47
	NH$_4$Cl	55	18	16	83	55	36
	Glucose + NH$_4$Cl	55	23	7	83	57	26
B	Water	39	20	11	74	40	26
	Glucose	39	11	13	74	26	24
	NH$_4$Cl	39	10	11	74	27	25
	Glucose − NH$_4$Cl	39	16	9	74	26	25

Source: Reprinted by permission from *Journal of Bacteriology* 113;1462–1473, H. C. Bae and L. F. Casida, Jr., 1973; copyright 1973, American Society for Microbiology.
[a]Percent occurrence among 90 cell-section photographs examined for soil A and 45 for soil B.
[b]Soil moisture adjusted before incubation to 50% moisture-holding capacity with distilled water or 1% aqueous solutions of each nutrient.

after a rainfall, which may indicate that only 15–30% of the bacteria are active, but during the dry seasons small coccoid-bacteria dominated (>90%) (Clarholm and Rosswall, 1980). Most of the biomass increase was due to the presence of large rods found mainly after rainfall; these disappeared in a couple of days (Clarholm and Rosswall, 1980). According to Conn (1948), it is doubtful whether true cocci exist in soil, since isolated coccoid cells have been shown to turn into rods in richer laboratory media, thus cocci in soil may be starved cells. If nutrients are depleted in *Arthrobacter,* the rods continue to divide, each turning into a group of coccoid cells within 48 hours (Veldkamp et al., 1963; Schaechter, 1968). As a result, the relative proportion of cocci to rods could be an indication of the proportions of active and nonactive bacteria in soil.

Bakken (1985) found that most of the bacteria separated from clay loam (CL) by centrifuging, employing a Ludox gradient, were in the <0.5-µm range, regardless of the supernatant fraction (S) (Table 5.2). The 0.5- to 0.8-µm size fraction was second and very few were in the >0.8-µm size fraction. The biovolume of dwarf cells represents only 10–20% of the total bacterial biovolume (Bakken and Olsen, 1986). Bakken and Olsen (1987a) noted that the number of bacterial cells in soil that form colonies on nutrient agar comes from the larger size cells. Only 0.2% of the cells smaller than 0.5 µm formed colonies, 10% of the 0.4- to

Table 5.2   Release of cells from clay loam by repeated blending centrifugation steps: yield and size distribution (microscopic count) of cells in the supernatants.

Supernatant and Residue	Total no. of Cells in each Supernatant ($\times 10^9$ per g [dry wt] of soil)[a]	% of Total in Soil Sample	Size Distribution (%) among Cocci of Diam[b]:		
			<0.5 μm	0.5–0.8 μm	>0.8 μm
S1	1.6	16	68	28	4
S2	1.0	10	71	27	2
S3	1.1	11	71	26	3
S4	1.1	11	72	26	1
S5	0.7	7	61	32	6
S6	0.6	6	53	36	10
S7	0.6	6	60	35	6
S8	0.4	4	58	41	1
RS8	2.3	23	59	35	6

Source: Reprinted by permission from *Applied and Environmental Microbiology* 49:1482–1487, L. R. Bakken, 1985; copyright 1985, American Society for Microbiology.
[a]Expressed as numbers per gram (dry weight) of soil applied.
[b]Rods were included by volume.

0.6-μm size cells, and 30 to 40% of cells larger than 0.6 μm. The size change does not explain the high number of dwarf cells seen in the soil, and therefore the authors suggest that dwarf cells are not the result of small forms resulting from larger forms. Approximately 80–90% of the colony dwarf cells retain a small diameter during growth, thus Bakken and Olsen (1986) state that there is little support that the dwarf cells are derived from "normal" sized cells in soil. Because their data do not fit in with the data obtained from the aquatic environments, Bakken and Olsen (1986) believe different populations exist in soil vs. aquatic environments. The soil environment contains mainly gram-positive (ca. 97%) while the aquatic environment contains gram-negative bacteria (ca. 97%). Furthermore, the nutritional requirements or incubation conditions may not fulfill requirements of these dwarf cells in soil. Many compounds may serve as triggering agents for germination. These include L-alanine, tyrosine, inosine, and adenosine. This may be why Bakken and Olsen (1987b) could not get many of their ultramicrocells in soil to grow. Incidently, thiosulfate appears to interfere with the normal dwarfing response in *Pseudomonas testosteroni* (Kieft et al., 1990).

   Bååth (1994) studied the incorporation of thymidine and leucine into macromolecules of soil bacteria of different sizes. The bacteria were then extracted by homogenization-centrifugation and passed through different size filters. The specific thymidine incorporation was highest in unfiltered and 1.0-μm filtered suspension (approximately 10×10^{-21} mol thymidine/bacteria/hour) but decreased to 1.39×10^{-21} mol thymidine/bacteria/hour for bacteria passing through the 0.4 μm filter. The portion of culturable bacteria (percent colony-forming units [CFU]/AODC) also decreased with bacterial cell size from 5.0% for unfiltered bacterial suspensions to 0.8% in the 0.4-μm filtrate, giving a strong linear correlation ($r^2 = 0.995$) between thymidine incorporation rate and the proportion of culturable bacteria. Leucine incorporation gave similar results as the thymidine incorporation. These experiments help verify the results of Christensen (1993) that the larger bacteria were more responsible for thymidine incorporation after drying and rewetting of root-free soil than the smaller bacteria. This also substantiates the results of Berland and Bakken (1991) who found that the larger bacteria respond more rapidly to altered nutrient conditions of the soil.

5.3.2 Freshwater Environments

Between 40% and 95% of the bacterial cells in Lake Constance, Germany, fall into the ultramicrobacteria size range (Simon, 1987). Martin (1963) reported filterable *Vibrio* from freshwater. In lake water, the difference between the number of bacteria that could be counted on 0.2- vs. 0.45-μm filters is quite large, especially in the deeper regions (Fig. 5.2). Of the total bacteria in freshwater or seawater 4–11% have the capacity to pass through a 0.2-μm pore size filter (Velimirov et al., 1992).

Using microcomputer assisted biomass determination of bacteria on scanning electron micrographs, Krambeck et al. (1981) examined the biomass of bacteria in ten different lakes and found that the mean cell volumes ranged from 0.015 to 0.022 μm³, most being in the 0.015-μm³ range. During two diurnal cycles, there is a change in the biovolume, probably due to uptake of extracellular primary products. Solonen (1977) found in Finnish lakes that the biovolume ranged from 0.04 to 0.24 μm³ and Bjørnsen (1986a) found the biovolume to be 0.188 to 3.38 μm³ in two Danish lakes. Fry and Zia (1982a) found the mean volume in freshwater to be considerably larger (0.32 to 1.0 μm³) but did find bacteria as small as 0.02 μm³. A negative correlation was found between the percentage of ultramicrocells and the concentration of soluble carbohydrates in freshwater (Fry and Zia, 1982b).

Transmission electron microscopy revealed that the bacteria in a pristine shallow subsurface aquifer in Oklahoma averaged 0.6 μm in diameter and 85% were coccoid or short rod-shaped gram-positive cells (Bone and Balkwill, 1988). These subsurface microbes appear to be greatly different than those from the surface. Colony diversity decreased with depth.

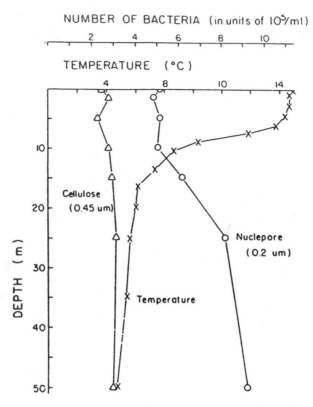

Figure 5.2.   Temperature and numbers of bacteria at various depths in Loon Lake, British Columbia, in September 1975. The bacteria were counted on Nuclepore filters (0.2-μm pore size) or Sartorius (cellulose) filters (0.45 μm pore size). Reprinted by permission from *Applied and Environmental Microbiology* 33:1225–1228, J. E. Hobbie, R. J. Daley, and S. Jasper, 1977; copyright 1977, American Society for Microbiology.

5.3.3   Marine Environment

Before Millipore membrane filters were on the market, Oppenheimer (1952) published a paper dealing with microbes capable of passing through a 0.4-μm pore size filter. These filters were the forerunner of the current Millipore membrane filters. Although the number of culturable bacteria capable of passing through the 0.4-μm membrane filter was only a small fraction compared to the bacteria retained, it definitely illustrates the small size of some bacteria. Oppenheimer (1952) reported 12 bacteria/ml of seawater capable of passing through a pore size of 0.4 μm. Because the water employed in this study was presumably taken off the Scripps Institution of Oceanography pier, the water is relatively high in organic carbon compared to that of the open ocean. It should also be noted that the current flows southward and the sewage disposal systems (the Los Angeles area) north of Scripps' pier enter the system. Anderson and Huffernan (1965), a little over a decade later, demonstrated that many species of marine bacteria can pass through a Millipore membrane filter with the pore size of 0.45 μm. *Spirillum* sp. appeared in the greatest number. Other genera that were identified were *Leucothrix, Flavobacterium, Cytophaga,* and *Vibrio.* There were also other unidentifiable organisms that were capable of passing through the 0.45-μm Millipore membrane filter. The number of filterable bacteria per 40 ml of seawater varied from 123 to 1000 (Anderson and Huffernan, 1965).

With the advent of the use of fluorescent dyes and the epifluorescent microscope, the small size of the bacteria in environmental samples was noted (Daley and Hobbie, 1975; Watson et al., 1977; Zimmermann and Meyer-Reil, 1974). In the marine environment, there were many investigators (e.g., Ferguson and Rublee, 1976; Watson and Hobbie, 1979; Azam and Hodson, 1977a,b; Wiebe and Pomeroy, 1972) who noted the small size and that larger cells are found in more organically rich environments, attached to particles or at physical interfaces. In cultural slide experiments using nonnutrient agar, Wiebe and Pomeroy (1972) demonstrated that these small cells enlarged and began to divide like media cultured cells. The biggest support for the idea that starvation-survival was real (Novitsky and Morita, 1976) was the demonstration of the fact that small ultramicrocells could be the result of starvation.

Samples of seawater were concentrated on a 0.2-μm-polycarbonate membrane filter and transferred to a small block of nutrient agar mounted on glass slides, incubated and observed microscopically with time (Torrella and Morita, 1981). By this microslide technique, three distinct patterns of growth were observed: (1) small cells increased in size and grew relatively quickly, forming microbial colonies; (2) small cells grew slowly without increasing in size and developed into microcolonies, which usually stopped growing after a few cell divisions, and (3) small cells, which made up the majority of those visible under the microscope as evidenced by the photomicrographs, did not grow. Those bacteria from pattern 1 are probably the typical ones isolated from the marine environment and studied in the laboratory. The second pattern was interpreted as indicating the presence of bacteria adapted to the very low nutrient concentrations encountered in the marine environment, whereas pattern 3 indicates that the conditions for growth (physical and chemical) were not adequate for the growth of most bacteria or that the cells had lost their ability to divide. Patterns 2 and 3 may result because of the high organic concentration of the microslide technique.

In a study of a Gulf Coast estuary, MacDonell and Hood (1982) were able to recover many bacteria, which appear to be in the genera *Vibrio, Aeromonas, Pseudomonas,* and *Alcaligenes,* from seawater passed through a polycarbonate membrane filtrate (0.2 μm). Recovery of these organisms required incubation in a dilute nutrient broth for about 2 days, a form of resuscitation for the organism. Numerous cocci and coccobacilli, approximately 0.2 μm, were caught on the filter and could be observed. Of the 27 isolates they obtained, 89% were incapable of growth in any of several full-strength nutrient-rich broths, including Trypticase soy broth plus marine salts, marine 2216 broth, thioglycolate broth plus marine salts, or any full-strength carbohydrate test medium. The growth response of UM106 is shown in Figure 1.2, which indicates that newly isolated UM106 does not grow well in a higher concentration of Trypticase broth concentration but can adapt.

The abundance of ultramicrocells that could pass through a 0.2-μm filter in a subtropical Alabama estuary was determined during a 1-year period by Hood and MacDonell (1987). Many starved species probably will not form cells sufficiently small to pass through a 0.2-μm filter, so it still remains to be seen how many of them are in the starved state. The population of ultramicrocells was dominated by the *Vibrio* species, but *Listonella* and *Pseudomonas* were also abundant. Laboratory studies demonstrated that strains of *Vibrio* could be induced to pass through 0.2- and/or 0.4-μm filters. Low nutrient exposed cells became very small and some grew on both oligotrophic (5.5 mg C/l) and eutrophic (5.5 g C/l) media, while a few cells grew only on oligotrophic media. The data suggest that the ultramicrobacteria in estuaries may be nutrient-starved or low-nutrient-induced forms of certain heterotrophic, eutrophic, autochthonous, estuarine bacteria.

By using image analysis, Maeda and Taga (1983) compared the size of bacteria from four different locations and found that most of the bacteria were between 0.4 to 0.8 μm (Table 5.3), but nevertheless, there was a difference between the cell sizes in the four locations. Cells measuring <4 μm were also found, accounting for roughly 1%. The Pacific Ocean and Tokyo Bay sites, in general, had the small bacteria where the DOC was the lowest and highest respectively. However, it should be pointed out that Tokyo Bay received quite a bit of terrestrial runoff and therefore many of the ultramicrocells may be due to starvation of terrestrial microorganisms or microorganisms from humans. Unfortunately, no determination was made concerning the culturability of these organisms.

In natural seawater, Kogure et al. (1979) observed that most cells were coccoid forms about 0.3 μm. Plate counts were about 0.1% of the direct viable count (use of nalidixic acid and yeast extract) and the direct viable counts were about 5% to 10% of the total direct counts. Watson et al. (1977) examined samples from waters near Woods Hole, the Sargasso Sea, and the southwest African coast and demonstrated the abundance of very small cells (>0.3 μm dia.), which were actually found to be rods and vibrios by transmission electron microscopy (TEM). Rods under TEM were ≤0.2 μm in diameter and the length could reach 0.5 μm.

The cell volume of bacteria in Otsuchi Bay, Japan, surface waters as determined by the Elzone Particle Counter, ranged from 0.2 to 0.4 μm^3 (Kogure and Koike, 1987), which is several times larger than the average size of the bacteria found in the open sea by Watson et al. (1977). Kogure and Koike (1987) also demon-

Table 5.3   Size distribution of bacteria at four stations in different sea areas.

Station (Area)	Length (µm)					
	<0.4	0.4–0.8	0.8–1.2	1.2–1.6	1.6–2.0	>2.0
T-2 (Tokyo Bay)	1.4%	90.9%	6.9%	0.6%	0%	0%
A-1 (Sagami Bay)	1.2	58.9	29.4	9.1	1.6	0
O-6 (Ohtsuchi Bay)	1.0	48.3	37.4	9.4	2.4	1.4
P-8 (Pacific Ocean)	2.2	73.7	22.6	1.6	0	0

Source: Reprinted by permission from *La Mer* 21:207–210, M. Maeda and N. Taga, 1983; copyright 1983, La Mer.
Note: Counted bacterial numbers were 272 at Station T-2, 265 at Station A-1, 211 at Station O-6, and 206 at Station P-8.

strated that the volume of cells does increase when the seawater sample is incubated, again showing that cells have the ability to grow. Koike et al. (1990) examined oceanic particles between 0.38 and 1.0 µm, which would be classified as "dissolved" by the marine chemists. In this fraction, 95% of the particles were nonliving. They found that they could remove 99.5% of the particles by filtration through a 0.02-µm filter, and the filtrate, when inoculated with a small amount of seawater filtered through a 0.6-µm glass-fiber filter, produced more bacteria, but no submicrometer particles. Even the filtrate from a 0.6-µm filter included bacteria and nonliving particles. The latter probably provided sufficient nutrients for the bacteria to grow but the question that begs to be answered is whether the filtration process released nutrients that were complexed to other substances or adsorbed loosely on particles. If the inoculum was the filtrate from a 5.0-µm filter, then small animals began eating the bacteria and presumably produced submicron particles. These submicrometer particles are produced by egestion or by lysing bacterial membrane.

Fuhrman (1981) compared the apparent size of the bacterioplankton when viewed by epifluorescent and scanning electron microscopy (SEM). SEM estimates of the average size of bacteria were smaller than those made with epifluorescent microscopy. This comparison can be found in Figure 5.3 for the waters off Scripps' pier and the California Bight. Unfortunately, no determination was made with open ocean waters or waters from the deep ocean. Nevertheless, it can be seen that the size of most of the bacteria are less than 0.5 µm and that the most frequent size of the cells of bacteria taken off the Scripps' pier is larger than cells from the California Bight, which may be due to the greater dissolved organic

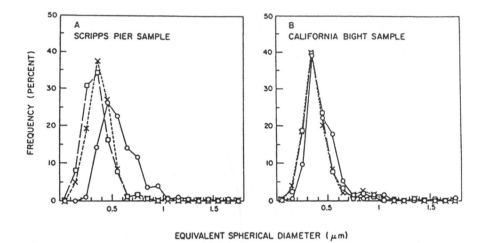

EQUIVALENT SPHERICAL DIAMETER (μm)

Figure 5.3   Apparent size frequency distributions of bacterioplankton, plotted as percent of the total number of cells in a particular size class versus the mean equivalent spherical diameter of that size class. ○ epifluorescence microscopy; □ epifluorescence microscopy with each individual shrunken proportionally to obtain same mean volume as SEM estimates × SEM. Reprinted by permission from *Marine Ecology Progress Series* 5:103–106, J. A. Fuhrman, 1981; copyright 1981, Inter-Research.

matter (DOM) next to the Scripps' pier than in waters from the California Bight. The rare large cells account for a large portion of the total volume, more than one-fifth of the total volume of 208 cells measured in the California Bight (Azam and Hodson, 1977b). When the bacterial activity was measured against the cell size, the activity more closely followed the bacterial numbers rather than the bacterial volume.

Autotrophic picoplankton, capable of passing through a Nuclepore (1 μm) filter, are active rather than dormant (Li et al., 1983). In waters obtained from the eastern Canadian arctic and from the Sargasso Sea, picoplankton accounted for the utilization of ≥65% of the organic substrates (amino acids, glucose, bicarbonate, and phosphate) used (Li and Dickie, 1985a). From the Caribbean and Sargasso Seas, bacteria that could pass through a 0.2-μm Nuclepore filter were demonstrated to accumulate amino acids and to grow at a rate of 0.18/hour (Li and Dickie, 1985b). Zimmermann (1977) estimates that a 0.1-μm pore size filter misses about 10% of the total bacteria.

Observations on the size of deep-sea and sediment bacteria obtained and fixed with glutaraldehyde in situ were determined by Carlucci et al. (1976) and found to be mainly coccoid or rodlike, less than 0.4 μm in diameter or width. Of the bacteria taken from the Puerto Rico Trench, 80% were capable of passing through a 0.45-μm filter (Tabor et al., 1981). Among the ultramicrobacteria were species of *Pseudomonas, Vibrio, Alcaligenes,* etc. They also noted small cells in the deep ocean that had the ability to grow into large cells when cultivated on laboratory media. They incubated two isolates in nutrient-free artificial seawater for 9 weeks and found that 10% of the initial population was capable of passing through a 0.40-μm filter. Because of the length of time required for cells to sink to depths greater than 6000 m it was estimated to be less than 1 week to hundreds of years (McCave, 1975; Smayda, 1969). Tabor et al. (1981) concluded that bacterial populations in the deep sea can be considered to have existed under low nutrient conditions (starvation) for long periods of time. In their study, Tabor et al. (1981) noted that bacteria from surface waters were, on average, larger than those from deep water.

Button and Robertson (1989) reported that cells, filterable through a 0.4-μm filter, dominated the bacterial population. They employed flow cytometry to characterize the aquatic bacteria according to population, cell size, and apparent DNA content. A cytogram of bacteria from a Resurrection Bay, Alaska, sample was compared with a pure culture of *Pseudomonas* T2 (previously starved to reduce the size). The natural bacterial population definitely is smaller than the starved culture of *Pseudomonas* T2. The size distribution of the cells was evaluated by isolating the forward scatter vs. frequency to give the histogram and the mean size of the indigenous bacteria was found to be 0.051 μm³. The DNA of the indigenous sample and the starved *Pseudomonas* were compared and the DNA content of the indigenous bacteria was less then one-third that found in the starved *Pseudomonas*. The starved *Pseudomonas* had a viability of only 50%, according to plate counts. According to the authors, the flow cytometric determinations of bacterial populations are in agreement with those of epifluorescent microscopy.

Employing the dilution to extinction method, Schut et al. (1993) found the bacterial population varied from 0.11×10^9 to 1.07×10^9 cells/l; the mean cell volume ranged from 0.042 to 0.074 μm³, and the apparent DNA content of the cells ranged from 2.5 to 4.7 fg/cell. These determinations were made by high-resolution flow cytometry.

Of the bacteria in the ocean that take up glucose, 70% are 1 μm or less in size (Williams, 1970). Ninety percent of the uptake of [3H]-thymidine is due to the oceanic bacteria that can pass through a 1-μm Nuclepore membrane (Azam and Hodson, 1977a). By use of radioautography and membrane filters, Hoppe (1976) found that most of the active bacteria capable of taking up labeled amino acids were found to be in the 0.2–0.6-μm size fraction but they could not be cultured. Those that had the ability to form colonies in the appropriate media were in the 0.6- to 1.0-μm fraction.

Based on epifluorescence microscopy and [3H]thymidine incorporation in water samples taken from the North Pacific gyre, Cho and Azam (1988) concluded that 98% of the bacterial population that passed through a 1-μm Nuclepore filter were less than 0.1 μm^3 in volume. They also estimated that 95% of the bacteria were free living and that nearly the entire carbon flux passed through these free-living bacteria.

Hoppe (1976), Bowden (1977), Wilson and Stevenson (1980), and Ferguson and Rublee (1976) found in marine and estuarine waters that the average volumes of bacteria range from about 0.05 to 0.19 μm^3. Hoppe (1976) suggested that the bacteria are unculturable, because they do not form colonies. This finding was supported by Fry and Zia (1982a). However, the microculture technique does not permit the bacteria to undergo resuscitation. For the bacteria isolated by the dilution technique by Button et al. (1993), the size ranged from 0.002 to 0.1 μm^3 with apparent DNA content ranging from 1 to 8 fg/cell.

All microbial ecologists recognize the fact that it is only possible to culture a small percentage of the microbes in any given ecosystem. This, coupled with the fact that most cells are ultramicrocells, makes one wonder if most of the bacteria present are not truly indigenous but are those that intruded due to dispersal by natural forces from their original environment. Having no energy and an unfavorable chemical and/or physical environment, they are starved and become ultramicrocells. If not that, they could be dead cells or cells that are metabolizing but not capable of reproduction.

The current view is that in the pelagic ecosystem, most of the activity (production and metabolism) is carried on by very small cells (Pomeroy, 1974; Williams, 1981; Platt et al., 1984; Li et al., 1983). Knowing the size of the bacteria in natural environments is important in order to determine the bacterial biomass as well as the bacterial C.

The percentage of the ultramicrobacteria in the various ecosystems as the result of nutrient deprivation is not known. However, I believe it is safe to say that the vast majority of the ultramicrobacteria are the result of nutrient deprivation. This is based on the fact that most of the main genera found in the ocean have the ability to form ultramicrocells when starved. It is also possible that the larger cells are efficiently predated (Pomeroy, 1984; Wiebe, 1984).

The bacteria inhabiting aggregates were up to 25 times larger than free-living bacteria (Alldredge et al., 1986), suggesting that the bacteria associated with aggregates have access to a greater concentration of nutrients. There are diel changes in metabolism as well as cell size where substantial activity occurs in the morning and evening (Riemann et al., 1982).

The size of bacteria is also influenced by temperature. The cell volumes of marine bacteria were greater at 0°C than at 10°C (Wiebe et al., 1992). These investigators also found the cell volumes of bacteria were greater at high nutrient concentrations and in complex media than at low concentrations or in a medium with a single source of carbon and nitrogen (proline), and their data suggest that the small size of bacteria in situ could be an effect of organic composition as well as concentration. In upwelled water, the bacteria were small (Painting et al., 1989).

According to Moriarty and Bell (1993), the size of the bacterial cell is an indicator of the trophic status of bacteria because starved bacteria are smaller than well-fed bacteria. Bacteria in oligotrophic environments tend to be smaller than in eutrophic environments, but the differences are not large (Fuhrman et al., 1989; van Duyl et al., 1990; van Duyl and Kop, 1988). Many increase in size when incubated with added nutrients, which suggests that they are living at suboptimal concentrations of organic matter (Morita, 1982; Cho and Azam, 1988; Kirchman et al., 1991). Natural planktonic bacteria often do not contain sufficient rRNA to generate a measurable level of fluorescence due to their small size and slow growth (Lee et al., 1993). As a result, multiple 16S rRNA-targeted fluorescent probes were used to increase the signal strength (Lee et al., 1993).

5.4 Isolation of Oligotrophic Bacteria in Various Ecosystems

Winogradsky (1924) noted the difference between zymogenic and autochthonous flora in that the former was mainly in a resting phase with short bursts of activity in the presence of suitable substrates and the latter was more or less continually active. The majority of bacteria at a given instant are in a latent state and only a small number are in an active state (Winogradsky, 1949). Only a minority of microbes can grow on usual media. Furthermore, this minority appears to be inactive (latent) in normal soil. Thus, it is possible to use the following terms synonymously: copiotrophic and zymogenic; and oligotrophic and autochthonous (Williams, 1985).

Williams (1985) cautions that concepts are not always fully supported by experimental evidence and that the categories are not necessarily mutually exclusive. He further warns that the failure to isolate a bacterium known to be present in the soil does not provide unequivocal proof that it is an oligotroph. Although some investigators have isolated the so-called oligotrophic bacteria from various environments, the question that must be asked is how active they are. Since most of the organic matter in nature is recalcitrant, do oligotrophic bacteria make the necessary enzymes to utilize the organic matter present? How well do they compete with other organisms in their immediate niche for the energy that is present? Furthermore, the method of isolation must be carefully taken into consideration as well as the problem of resuscitation when attempting to cultivate oligotrophic bacteria.

5.4.1 Soil

The major environmental stresses that occur in soil include the unavailability and low rate of supply of nutrients, unavailability of water or oxygen, development of acidity, and toxic compounds. Soil is considered a nutrient-poor environment, even though there is a paucity of evidence for the occurrence of strict oligotrophs in soil (Williams, 1985). Various oligotrophic bacteria have been isolated from soil, including *Caulobacter* (Krasil'nikov and Belyaev, 1967; Poindexter, 1979, 1981a,b), and *Hyphomicrobium* (Poindexter, 1979). Isolation of bacteria from the rhizosphere will not be considered because root exudates do furnish energy to the rhizosphere.

A number of bacterial cultures from soil had the ability to grow on different concentrations of nutrient broth (NB), NB/10, NB/100, NB/1000, and NB/10,000 (Suwa and Hattori, 1984). Type I had the ability to grow only on media having a dilution range of NB to NB/1000. Type II had a wider range (from NB to NB/10,000). Type III had a range of NB/10 to NB/100; whereas Type IV grew only on NB/10,000. Types II and IV cultures on NB/1000 were recultured in order to confirm their ability to grow on the same medium. A portion of each of the first cultures on NB/1000 medium were diluted sufficiently in sterile distilled water to obtain several hundred cells per ml, transferred into fresh NB/10,000 medium, and incubated at 27°C for one week (Table 5.4). (Isolates labeled H were isolated from upland soil and those designated S from paddy soil.) It can be seen that some isolates did not show any appreciable growth although a few did resume their growth. The authors also point out that the number of viable cells of some isolates may be due to fragmentation and not to growth. The authors considered Types II, IV, and most of the the DNB to be oligotrophs. The DNB isolates did not possess any cellulolytic activity and few could hydrolyze starch or casein. Because cellulose is one of the major components of plant debris, especially after the more labile components are quickly decomposed, these DNB organisms would have to depend on the cellulose digestors for their energy. The percentages of DNB microbes among those isolated were 58% from Japan Alps soil, 30–40% with river and field samples in Japan, and 5–10% from soil surrounding street trees and roadside sand in Japan. Half of the DNB microbes were gram negative, 10% were gram positive, and the rest were irregularly gram-stained organisms. All the DNB microbes were non-spore-formers. The inability of certain microorganisms to grow on high-nutrient media is due to the suppression of growth caused by the rapid overgrowth of other organisms (Hattori, 1976; Mishustin, 1975).

Some of the DNB bacteria were capable of growth in distilled water indicating that their growth was supported by material in the air (Hattori, 1984). Later, Ohta and Hattori (1983) isolated DNB organisms living on organic debris as well as rice roots, yet these areas contained the most utilizable organic matter. These DNB bacteria were predominant in the bacteria communities, but a transient decrease was noted immediately after the application of manure. These DNB microbes fell into five groups: (1) regular rods, (2) filament-forming bacteria, (3) irregular rods, (4) prosthecate organisms, and (5) large oval rods. Regular rods were 42% of the total DNB organisms, and irregular rods were the most abundant

Table 5.4 Growth of isolates of types II and IV on NB/10,000 medium and in distilled water.

	Growth on					
	NB/10,000		Distilled water		Double-distilled water	
Isolates[a]	1[b]	2[c]	1	2	1	2
H532, H101, H111, S1516, S1014, S1016, S1026, S1030	+[d]	−[d]	−			
H504, H515, H518, H519, H520, H521, H522, H528, H529, H534, H537, H538, H539, H544, H546, H547, H554, H107, H115, H117, S1003, S1511, S1512, S1513, S1010, S1020, S1021, S1022, S1032	+	+	−			
H502, H514, S1012	+	+	|	−		
H509, H525, H543, S1004	+	+	+	+	−	
H516	+	+	+	+	+	+

Source. Reprinted by permission from *Soil Science and Plant Nutrition* 30:397–403, Y. Suwa and T. Hattori, 1984; copyright 1984, Japanese Society of Soil Science and Plant Nutrition.

[a]Isolates from S1511 to S1516, from H502 to H554, and S1004 belong to type II and others to type IV.

[b]First cultivation.

[c]Second cultivation.

[d]+, growth was observed; −, growth was not observed. In these experiments growth was confirmed by plate count.

(46%). Only a few oligotrophic bacteria can hydrolyze macromolecules like starch and none can decomposed cellulose (Whang and Hattori, 1988).

When soil bacteria are grouped by the analysis of colony formation, Hashimoto and Hattori (1989) found that most copiotrophic bacteria are fast growers and most oligotrophic bacteria are slow growers. However, the question remains as to their growth rate in nature because the amount of bioavailable energy for the copiotrophs and/or the oligotrophs may be lacking in their immediate environment.

Viability of soil microorganisms has been described by the terms "microhabitat" and "molecular environment." In general, "microhabitat" refers to small volumes of soil that are favorable for growth, the size of the volume depending on the sensitivity of the microbes to environmental change, the size of the organism, and the steepness of the gradient of environmental conditions. By implication, the remaining volume of the soil is unfavorable for growth. As yet there are no methods to measure these volumes. Hence, a microhabitat can be a molecular environment.

5.4.2 Subsurface Environments

The only bacterial families found in the deeper formation from deep sediment samples of the Savannah River site were the Pseudomonaceae and Acinetobacteriaceae (Jiménez et al., 1990). These authors suggest that a long period of adaptation to the environmental conditions of the deep subsurface does exist. However, is it a matter of adaptation or a matter of survival? The composition of the microflora in the sediments in any given geological formation differed from one drilling site to another (Balkwill, 1990). His study indicated that the subsurface microorganisms differed markedly from those in the overlying topsoil and distinct strains of subsurface microorganisms were isolated on concentrated and dilute plating media. Balkwill (1990) obtained 10^5 to 10^8 cells/g dry weight in nontransmissive aquifer-bearing sediments and this number did not decrease with depth; whereas, in nontransmissive confining layers (between aquifers), less than 10^3 cells/g dry weight were detected. In reviewing the literature, it becomes evident that each subsurface sample is different.

The composition of the microbial community in deep sediments should contain a remnant population with fewer species than that typically found near the surface. This concept is suggested by the research of Bone and Balkwill (1988). The population of the subsurface environment as a whole may be capable of utilizing a wide range of organic compounds, which is suggested by the work of Fredrickson et al. (1991). Nevertheless, groundwater is considered to be oligotrophic and only 1 to 10% of the groundwater bacteria are metabolically active (Federle et al., 1986; Ghiorse and Balkwill, 1983; Marxsen, 1988). There is also a reduction in microbial diversity in subsurface soil and aquifers (Bone and Balkwill, 1988), probably due to the decrease in the variety and quantity of organics in these environments compared to surface soil or waters.

Hazen et al. (1991) compared the bacteria from deep subsurface sediment and the adjacent groundwater and found that bacterial densities in sediment were higher, by both direct and viable count, than in groundwater samples. Sediment bacterial densities ranged from less than 1.00×10^6 up to 5.01×10^8 bacteria/g dry weight for direct counts, whereas viable counts were less than 1.00×10^3 CFU to 4.07×10^7 CFU/g dry weight. Bacterial densities in groundwater were 1.00×10^3 to 4.07×10^4 CFU/ml and 5.75 to 4.57×10^2 CFU/ml for direct and viable counts, respectively. Isolates from sediment were also found to assimilate a wider variety of carbon compounds than groundwater bacteria.

In subsurface waters, which are the result of seepage of water from the surface, the amount of organic matter should be extremely low because the microbes on the soil particles in the various strata act as a fixed bed reactor with a multitude of enzyme reactions. The more readily utilizable material will be acted upon and degraded. Alexander et al. (1991) selected 19 bacterial strains that differed in their ability to be transported through soil. Measurements were made of sorption partition coefficient, hydrophobicity, net surface electrostatic charge, zeta potential, cell size, encapsulation, and flagellation of the cells. Only sorption and cell length were correlated with transport of bacteria through soil. Their results also suggest that bacterial movement through aquifer sand was enhanced by reducing the ionic strength of the inflowing solution. The theoretical consideration concerning dispersal (movement away from the habitat of origin) of bacteria in groundwater habitat is presented by Lindquist and Bengtsson (1991). This dispersal of bacteria in saturated, porous soils can be characterized by the partitioning of cells between the aqueous and solid phases, as a result of the physical and chemical nature of the soil and water, and cell surface modification.

Ultramicrobacteria formation by bacteria isolated from water within rocks appears to be a universal process (Lappin-Scott et al., 1988a,b; Lappin-Scott and Costerton, 1990). This should be expected since penetration into rocks and dispersion of microbes into subsoil environments are aided by starvation, which results in cell size reduction, shape, and less exopolysaccharide production.

The data on subsurface microbiology are so voluminous that important aspects cannot be covered in this book. As a result, Fliermans and Hazen (1990), Kaiser and Bollag (1990), Ghiorse and Balkwill (1983), Ghiorse and Wilson (1988), and Chapelle (1993) should be consulted. Subsurface environments contain many different microhabitats (Hirsch and Rades-Rohkohl, 1983).

It is interesting to note that investigators working in subsurface microbiology seldom refer to their isolates as oligotrophs.

5.4.3 Oligotrophic Lakes

In nonoligotrophic lakes, the amount of sunlight (energy) for photosynthesis provides sufficient organic matter so that the heterotrophic bacteria are not starving. However, in oligotrophic lakes, phosphate and or nitrogen is generally the limiting factor for phytoplankton production. Carbon fluxes in oligotrophic lakes may be larger than in eutrophic lakes and studies suggest that pelagic bacteria are a large component of the carbon budget (Cole and Caraco, 1993). As a result, the hypothesis that has been developed is that bacteria consume a significant fraction of bacterial production, either by direct predation or through consumption of bacterial exudates or dissolved compounds during lysis. Cole and Caraco (1993) list three prominent questions concerning bacteria in oligotrophic lakes:

1. Are organic inputs sufficient to support the high rates of bacteria secondary production that we measure? (Scavia, 1988; Strayer, 1988)
2. Why do measurements of bacterial production frequently exceed measures of mortality due to grazing? (Pace, 1988)
3. How can bacteria metabolize a relatively large fraction of primary production and at the same time not be an important link to high trophic levels? (Ducklow et al., 1986).

Other thoughts concerning the foregoing are: (1) What percentage of the bacteria found in the oligotrophic lakes are in the starvation-survival stage? They may have originated in the lake itself, be brought into the system by land drainage, or from the atmosphere, and (2) In the measurement of bacterial productivity, how much of the productivity can be attributed to perturbation of the water sample, organic matter in the air, priming of the environmental sample, etc.?

The TFI medium for Ishida and Kadota's (1979) procedure for isolating oligotrophic bacteria contains 5 g trypticase (BBL), 0.5 g yeast extract (Difco) in 1000 ml of artificial lakewater (ALW) (1.9 mg C/l). Good growth rates are noted in diluted medium for five obligate oligotrophs. At low concentrations of medium good growth occurred but when the concentrations of the medium were higher, growth was reduced. However, these data must be confirmed by other investigators. Whether these oligotrophs can be adapted to a higher concentration of organic matter must be investigated, in studies like those conducted by MacDonell and Hood (1982; see Fig. 1.2). Ishida and Kadota (1981a,b) claim that good growth of the obligate oligotrophs also occurred in artificial lake water containing only magnesium sulfate, calcium chloride, and potassium chloride. Is this "growth" due to fragmentation resulting from starvation or due to the use of nutrients absorbed from the laboratory air? Unfortunately, Ishida and associates did not take the precaution of eliminating the possibility of organics in the air being absorbed into their medium.

Ishida et al. (1980) later provided an improved version of the method (Fig. 5.4), which makes use of LT 10^{-1} broth (0.5 g trypticase [BBL]) and 0.05 g yeast extract/l of aged lakewater. They also demonstrated that the number of obligate and facultative oligotrophs varies with the season as well as depth from which the sample was obtained. [*Note*: Oligotrophs are defined as those organisms that grow in medium containing 1 mg of organic-C/l. In these papers, the organic C is in the form of trypticase and yeast extract, whereas, in the environment, most of the organic C may not be available.] Later, Ishida and Kadota (1981a) determined obligate oligotrophs show growth in 1 mg organic-C/l and no growth in rich medium of 5 g organic-C/l. They also observed the growth dependence of obligate oligotrophs on various concentrations of nutrients. Growth rates appear to be slow (days and not hours) to reach the maximum cell numbers and, at higher organic C, growth ceases. Oligotrophic bacteria were found to be dominant in Lake Mergozzo, Italy (Ishida and Kadota, 1977).

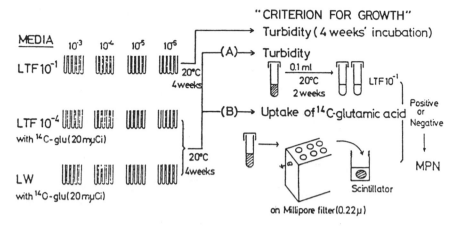

Figure 5.4   Schema simply modified for enumeration and detection of oligotrophic bacteria. Reprinted by permission from *Bulletin of the Japan Society for Scientific Fisheries* 46:1151–1158, Y. Ishida, K. Shibahara, H. Uchida, and H.Kadota, 1980; copyright, 1980, Japanese Society of Scientific Fisheries.

5.4.4   Marine Environment

One of the most important factors governing the distribution of bacteria in the sea is the availability of nutrients (ZoBell, 1946, 1968). In nearshore environments, such as bays and estuaries, the organic matter may be high due to terrestrial runoff, pollution, etc. As a result, the bacteria counts reach 10^9 cells/ml and anoxia may develop due to the microbial utilization of the organic matter. Nevertheless, there will be some physiological types of bacteria in such rich environments that may be in the starvation state. For further information concerning bacterial production in nearshore environment, White et al. (1991) and Shiah (1993) should be consulted.

Yanagita et al. (1978) investigated the nearshore environment at several locations. The picture of the occurrence of eutrophs and oligotrophs were dependent upon depth, temperature, salinity, river flow, and dissolved oxygen in relation to the months samples were taken in Toyoma Bay, Japan. The number of oligotrophs and eutrophs for the same area should be examined in light of the various chemical and physical factors present, as well as the season. This bay is also influenced by the Jinzu-gawa, Jaganji-gawa, Kamiichi-gawa, and the Hayatsuki-gawa rivers, which, in turn, influence the various forms of nitrogen available. The number of oligotrophs at station 1 exceeded or was comparable to that of eutrophs with some exceptions through the seasonal survey. At stations 2 to 4, a reverse situation was noted, which can be attributed to the influx of water from the Jinzu-gawa river.

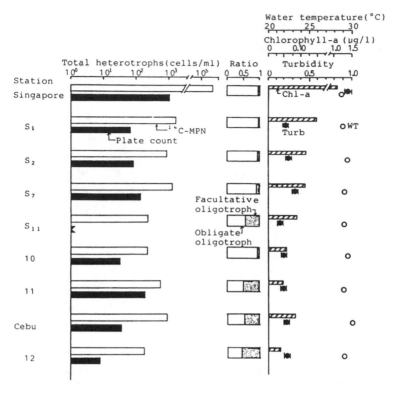

Figure 5.5   Horizontal distribution of total heterotrophs and ratios of obligate and facultative oligotrophs to total heterotrophs. Reprinted by permission from *Marine Ecology Progress Series* 30:197–203, Y. Ishida, M. Eguchi, and H. Kadota, 1986; copyright 1986, Inter-Research.

The mean O.I. index (total number of oligotrophs to total number of bacteria) for Toyama Bay was 0.34, indicating that more eutrophs were present in this near-shore environment than oligotrophs. They caution that the O.I. index cannot be taken as an indicator for water eutrophication.

By use of the ^{14}C-MPN method with medium containing 0.2 mg organic C per liter, the total number of heterotrophic bacteria in the South China Sea and the western Pacific Ocean ranged from 1.3×10^2 to 2.5×10^3 cells/ml of which 45% to 98% were obligate oligotrophs (Ishida et al., 1986). The horizontal distribution of obligate and facultative oligotrophic bacteria is presented in Figure 5.5, where the ^{14}C-MPN were positively correlated with the chlorophyll *a* content and the turbidity of the seawater. As in freshwater, the uptake of amino acids by the oligotrophs is much more than the uptake of glucose and acetate. These authors present a rough estimation that a bacterial cell (10^3/ml) in the ocean with the substrate concentration of glutamic acid (10 nM; 1.5 µg/l) has a chance to meet eight molecules of glutamate. Taking into consideration the Brownian movement of glutamate and neglecting the bacteria movement, the number of collisions of the glutamate molecules with the bacterial cell is more than 1×10^6 per second. This collision value would naturally be increased if the flagella movement and chemotactic behavior of the bacteria is taken into consideration. According to Purcell (see Anonymous, 1979), "small particles diffuse so rapidly in the solution surrounding a bacterium that with 0.1 seconds all of the molecules are replaced."

Eguchi and Ishida (1990) found 9.2×10^3 to 5.4×10^4 cells/ml, with an average of 2.2×10^4 cells/ml in pelagic water; whereas in coastal waters the number of heterotrophs was an order of magnitude higher (3.5×10^5 cells/ml) as determined by the ^{14}C-MPN method. Of the bacteria in pelagic waters, 85% were obligate oligotrophs and in coastal waters eutrophic and facultative oligotrophic bacteria were predominant. Obligate oligotrophic bacteria in pelagic waters had a high specificity for the utilization of amino acids, especially glycine, but not for acetate. This was also noted for oligotrophic bacteria in Lake Biwa (Ishida and Kadota, 1981b) and differs from the characteristics of nutrient uptake for "model oligotrophs" proposed by Hirsch et al. (1979). They determined that the total amino acid concentration in pelagic water (Station D) was 191.1 nM, of which 57.7 and 34.5 nM were glycine and serine, respectively. However, they claim that the obligate oligotrophs do not need a broad range of substrates. The predominance of glycine and serine in seawater depends less on the utilization by heterotrophs, mainly because of the low energy yield per mol of these amino acids. Nevertheless, their results show glycine to be a useful nutrient for heterotrophs in pelagic waters. These authors suggest that the majority of eutrophs and and small number of facultative oligotrophs are in a starvation-survival mode (nongrowing) but not in an inactive mode as suggested by Kjelleberg et al. (1987), while the rest of the facultative oligotrophs and obligate oligotrophs are probably growing actively in oligotrophic waters. One should note again that the level of organic matter is very low, 0.3 mg C/l for their ^{14}C-MPN, because many investigators consider 10 mg C/l as the threshold level of organic C needed for growth of microorganisms.

Oligotrophic bacteria are dominant in Antarctic waters (Eguchi and Ishida, 1986). These oligotrophic bacteria were determined using five radioactive carbon compounds (glutamate, glucose, acetate, glycolate, and glycine). The numbers ranged from 2.0×10^1 to 2.4×10^4 cells/ml; most were obligate oligotrophs.

In comparing the number of heterotrophic bacteria in pelagic waters and coastal waters, Eguchi and Ishida (1990) found 9.2×10^3 to 5.4×10^4 cells/ml and 3.5×10^5 (av.) cells/ml, as determined by the ^{14}C-MPN method. Of the bacteria in pelagic waters, 85% were obligate oligotrophs.

Baxter and Sieburth (1984) isolated a *Vibrio* sp. (strain G1) and an *Acineto-*

bacter sp. (strain GO1) by the extinction dilution technique from seawater collected off the School of Oceanography, University of Rhode Island's pier in Narragansett Bay. The seawater was prefiltered though a Nytex (20 µm) screen and finally through a sterile 0.5-µm Nuclepore filter before use in the isolation procedure. The dilution series contained seawater supplemented with 0.01 and 0.1 mg C/l in the form of glucose. However, only facultative oligotrophs, called "eurytrophic" by Baxter and Sieburth, were isolated by this procedure.

The difficulty in isolating membrane-filterable low-nutrient bacteria or oligotrophs may have been caused by use of inappropriate medium (Carlucci, 1974), but may be possible by use of very dilute medium containing a low concentration of peptone and yeast extract (0.175%) (Torrella and Morita, 1981).

The lack of psychrophiles found below the thermocline is probably due to the lack of bioavailable energy in that area. If a psychrophile and a mesophile are grown together in media at 4°C, the psychrophile will outgrow the mesophile (Morita and Burton, 1970). This is a good indication that there is a lack of bioavailable energy below the thermocline. The same results were obtained by Harder and Veldkamp (1971) in chemostat studies, where the psychrophilic *Pseudomonas* sp. (optimum growth temperature at 15°C) outgrew the psychrotrophic *Spirillum* sp. at all dilution rates employed at $-2°C$, but at temperatures above 15°C the *Spirillum* sp. outgrew the *Pseudomonas sp.*

5.5 Metabolic Activity in Oligotrophic Evironments

Microorganisms seldom live under conditions in which the nutrients are evenly distributed (Wimpenny, 1981; van Es and Meyer-Reil, 1982). Growth and generation time are a function of the food supply (Penfold and Norris, 1912). If the bacterial productivity and specific growth rates are low, it is likely that many or all of the heterotrophic bacteria are starved (Kjelleberg et al., 1993; Moriarty and Bell, 1993). Because bacterial productivity is more directly related to the turnover of organic carbon than either the specific growth rate or changes in numbers, it is a better indicator of the trophic status than abundance (Kjelleberg et al., 1993; Moriarty and Bell, 1993). For the bacterial growth rate in different environments, consult Kjelleberg et al. (1993) and Moriarty and Bell (1993).

The question as to whether small cells are active or dormant still has to be answered. Because it is believed that there is no continuous active species in soil, then there is probably no continuous species in the marine environment; but no distinction is made between the various species in any environment—mainly because to correlate species identification and activity in an environmental sample is too difficult. Many species of bacteria are capable of utilizing the same substrates for energy. For example, the number of species that can utilize glucose is enormous. Why then in nature are there so many species that can use the same substrates for energy? Each species reacts differently in terms of competitive interactions (competition, amensalism), cooperative interactions (symbiotic associations, mutualistic associations), cyclic reactions (e.g., in the sulfur cycle between aerobic sulfur oxidizers and anaerobic sulfate reducers; between aerobic nitrifying and anaerobic denitrifying bacteria) and chains (syntrophy), habitat, and niches (for more details, see Wimpenny [1981]). Therefore, when substrate is added to an environmental sample, not all cells respond equally. The ultramicrocells are thought to have equal or even greater activity per unit cellular volume because of the larger surface-volume areas (Hodson et al., 1981a; Azam and Hodson, 1977a). Williams (1970) and Derenbach and Williams (1974) demonstrated that most of the heterotrophic activity resided in the fraction of filtered seawater that passed through a 3.0-μm filter; whereas Azam and Hodson (1977a) reported highest activity with water that had passed 0.6- and 0.10-μm filters.

The importance of utilizing low concentrations of nutrients for the study of the microbial ecology of marine ecosystems has been pointed out by various investigators (Carlucci and Shimp, 1974; Jannasch, 1967; Jannasch and Jones, 1959; Poindexter 1979, 1981b). Williams (1985) states that "by general definition an oligotrophic bacterium predominates in a nutrient-poor environment, is isolated using a low nutrient medium and shows relatively high growth rates in culture at low concentrations of energy-yielding substrates." Yet the greater number of viable bacteria isolated from oligotrophic environments appear to be more mesotrophic forms. Some isolates of oligotrophic bacteria appear to be intolerant of elevated nutrient concentrations (Poindexter, 1979). The properties needed by an oligotroph are the presence of efficient uptake mechanisms of high affinity and low specificity, constant uptake capability, and the ability to husband a wide variety of nutrients (Shilo, 1980). Of the bacterial enzymes, surveyed by Matin (1979), 47% increased their activity with decreasing dilution rates in chemostat studies.

Poindexter (1981b) discusses the nutrient uptake of oligotrophs and states that two physical means of increasing uptake capacity would be (1) high surface:volume ratio, and (2) high density of transport sites. Many heterotrophic bacteria, when starved, reduce their size so that they become ultramicrocells. In terms of the high density of transport sites, the loss of the phospholipids in the membrane (Guckert et al., 1986) would indicate a membrane change that probably would reflect a differential permeability. Furthermore, starved cells have displayed higher affinity systems.

Although we use readily available substrates in laboratory culture, in nature the organisms may not have the opportunity to utilize such substrates due to the fierce competition for substrates among the indigenous population. Extrapolation of laboratory utilization of substrates by pure culture does not necessarily mean it occurs in the field (Richard, 1987). High concentrations of easily metabolizable compounds have been shown to cause repression of mineralization of less readily metabolizable compounds (Clarke and Ornston, 1975; Magasanik, 1976). Poindexter (1987) suggested that those organisms that adapt to nutrient limitation through mutation are themselves descendants of populations in which the phenotype changes accomplished adaptation to fasting conditions. It should also be noted that starvation can result from the wrong composition of nutrients, and not necessarily a result of low concentrations of organic matter (Enger, 1992a).

Are all the various species of bacteria in any ecosystem actively metabolizing? The obvious answer is "no." According to Gray and Williams (1971) ". . . there is a growing realization that many of the organisms isolated from soil by conventional methods were present in a dormant condition and considerable attention has been paid to developing a method for the study of the so-called active organisms." To reaffirm this notion, McLaren (1973a) concluded that it has not yet been shown that there is any continually active biomass in soil. The main reason for this situation is the lack of nutrients, especially energy. However, other stress factors must also be taken into consideration. Even under favorable environmental conditions, only 15–30% of the bacterial population is active (Clarholm and Rosswall, 1980). Microbial populations in soil were largely dormant, with only small maintenance energy rates comparable to those in laboratory culture (Paul and Voroney, 1980). This dormant population has two unique features: an enormous richness of species and an ability to survive (Jenkinson and Ladd, 1981). This statement also applies to the aquatic environment.

Estimates of in situ bacterial productivity and growth vary depending on the method used and the assumptions made regarding the growth state of bacteria (Christian et al., 1982). Growth in the natural environment is sporadic at best. One of the reasons for this sporadic growth is that most organisms in nature rely on syntrophy, especially the oligotrophic bacteria, because very few oligotrophs produce hydrolases (Semenov, 1991). The reason why so many physiological types of bacteria coexist is also due to syntrophy and as Konings and Veldkamp (1980) stated, the heterogeneity of the environment yields rapid growth of different organisms for short periods. Nevertheless, microbiologists employ different methods to measure the growth and biomass of bacteria in natural populations. All microbiologists recognize that the direct count of bacteria in natural samples is much greater than the viable count. Any method that adds energy to the natural sample becomes suspect as well as any method that relies on exponential growth on the bacteria in a natural sample. In the former case, the addition of energy may permit the cell to break its starvation-survival mode. This can also be true in systems where no energy is added but the sample is perturbed. Perturbation of the sample releases utilizable energy from the system.

The energy budget of soils provides a solid basis for the concept that fungi and other microorganisms in soil exist in an environment characterized by insufficient resources for sustained growth (Lockwood, 1981). Classical thermodynamics is, unfortunately, limited in that it does not allow predictions of reaction rates except under certain specialized conditions (McCarty, 1971).

It is well known that the enumeration of the microbial population in any environmental sample is difficult because of the numerous different requirements of all the physiological types of bacteria. As a result, we can only enumerate an infinitely small fraction of the microbes present in environmental samples by use of various media. As for the heterotrophic bacteria, it appears that those enumerated by use of media are geared to an opportunistic life in that they survive on very, very low levels of nutrient when they go into a starvation mode; they are capable of going into a zymogenic mode of growth if they are exposed to high concentrations of media. The common practice of using media having high concentrations of organic matter can be easily faulted. Methods using high concentrations of organic matter in media were developed by medical microbiologists who wished to obtain a large mass of cells and their use has been perpetuated ever since. Microbial ecologists also recognize that media containing low concentrations of organic matter will permit the elucidation of more colonies on an agar medium than the use of a high concentration of organic matter in the agar medium. However, this may be a method of resuscitation and is discussed in chapter 7.

Soil and sediments as matrices for microbial growth are well discussed by Nedwell and Gray (1987). Ducklow and Carlson's (1992) comprehensive review of bacterial production in the ocean should be consulted for the marine environment.

Because oligotrophic environments exist mainly due to the action of the indigenous microbes (Morita, 1986), energy and carbon sources are very limited in soil (Hattori, 1984; Williams, 1985), in many freshwater environments (Kuznetsov et al., 1979; Fry, 1990a) and in the marine environment (Williams, 1971; Morita, 1988, 1990). Poindexter (1987) suggested that those organisms that adapt to nutrient limitation through mutation are themselves descendants of populations in which the phenotypic (size and shape) changes represent an adaptation to fasting conditions.

There are many investigations indicating that populations of bacteria placed into the natural environment decline with time. Most of these studies employ cultured bacteria that are grown under optimal conditions. Although there may be many factors (desiccation, predation, pH, temperature, etc.) responsible for their decline in numbers, one of the most important factors is the bioavailabilty of labile energy. It may be the most important factor for the decline of bacteria placed into various ecosystems.

Most of the natural environments are grossly oligotrophic and the organic matter present is polymeric. Ectoenzymes are needed to hydrolyze the polymeric material into utilizable substances. However, it does appear that ectoenzymes are located in the periplasmic space of starved bacteria (Amthor, 1989). Although free enzymes do occur in nature, it is not known whether they are complexed to other substances or adsorbed to clay particles and not free to carry out their enzymatic activities. Does perturbation release these enzymes so that they can react? Are these ectoenzymes released from lysed cells? If the microbes produce ectoenzymes in an oligotrophic environment, then the question of the energy source becomes important. Enzymes are proteins; hence, many acylations of amino acids must take place and each acylation takes energy. Without substrate (energy), the metabolic state of microbes comes to a standstill. Turley (1994) provides a short review on the subject.

5.5.1 Soil

As early as 1923, Waksman and Starkey concluded that nutrient availability was the chief factor affecting microbial numbers and survival in soil. Bacon (1968) reviewed the chemical environment in *The Ecology of Soil Bacteria* (Gray and Parkinson, 1968) and came to the conclusion that it is the accessibility of the organic matter present, rather than its lack of abundance or diversity, that limits the life of bacteria in soils. Soil is a complicated three-phase system presenting many chances for spatial and temporal variations in nutrient concentrations and environmental conditions and all microorganisms present in soil are not equal in their biomass, ATP content, metabolic activities, etc. (Williams, 1985).

Even if it is assumed that measures of substrate input and biomass are accurate, the reported growth rate of bacteria in any natural environment must take into account cryptic growth, yield, and energy required for maintenance, because these factors have a profound effect on the calculation of generation times (Chapman and Gray, 1986). Yet, the supply of readily utilizable carbon is apparently severely limited (Clark, 1965; Lockwood, 1977), and, as a result, there is intense competition among all soil organisms for utilizable carbon. Competition for substrate should be viewed both in terms of a given substrate and the specific conditions under which that substrate is presented (Clark, 1965).

According to Richard (1987) there can be little doubt that the primary limiting factor in the control and distribution of chemoheterotrophic microorganisms is the availability of substrate. He further stated that:

> When the soil ecosystem approaches a steady state, the gross pattern of distribution of substrates does not change appreciably with the passage of time. On a microscopic scale however, there are still vast changes, and these changes occur from place to place at any one time, and from time to time in any one place. Of paramount importance is the fact that substrates for chemoheterotrophs in soil are ephemeral. Consequently the distribution of such microbes or their propagules is determined not only by the present, but also by the past, disposition of their substrates.

In soil, the scarcity of food, or lack of a suitable and available energy source is the principal factor limiting bacterial growth (Clark, 1967). As a result, any addition of fresh energy-yielding material to soil will almost invariably elicit an increase in bacterial activity. Growth of soil biomass is theoretically controlled by the C supply above that required for maintenance energy (Smith and Paul, 1988). Plant litter is undoubtedly a major source of energy input to soil, but, in some ecosystems, the litter layer is distinct from the soil system. The amount of litter in relation to its accumulation varies considerably in different climatic zones (Jensen, 1974). Much of the litter is decomposed in the litter layer, hence the amount of readily utilizable energy that seeps into the soil is probably very little, but is mainly humic material. The rhizosphere will not be considered in this section, because it is known that living roots release into the soil a variety of low molecular weight substances, including simple sugars, oligosaccharides, and amino acids (Sparling et al., 1982). All of these compounds will have a great impact on the growth of organisms in the rhizosphere (Hunt et al., 1985), but the exudates cannot be demonstrated unless the system is bacteria-free. This latter situation only demonstrates the fierce competition for substrate by the bacteria living in the rhizosphere.

Soil studies indicate that one-fifth of the total substrate available annually was required for each generation of new cells produced (Clark and Paul, 1970). Maintenance energy consumed one-third or more of the available substrate, reducing even further that available for growth and reproduction. In studies by Clark and Paul (1970), a comparison was made with stationary laboratory cultures and field studies in terms of the amount of carbon evolved. On an equivalent biomass, the evolution of CO_2 from field studies was only 1/35 to 1/70 of the laboratory culture. They concluded that the growth and activity of microorganisms must be severely restricted. Laboratory grown cells, even in the stationary phase, have quite a bit of endogenous metabolism, especially if they have built up energy reserves. Laboratory grown cells probably do not resemble cells in nature in terms of their size, their energy reserves, etc.

The Q_{O_2} values, calculated on the basis of microscopic counts give a much better representation of the amount of metabolically active bacteria in the soil than plate counts (Faegri et al., 1977). However, Gray and Williams (1971) concluded that any microbial growth rates calculated from respiration measurements are overestimates. Some energy-rich materials, including lignin and other recalcitrant substances, are not easily digested in soil without addition of soluble organic substances. These substances can allow new enzyme synthesis (McLaren, 1973b).

According to Norton and Firestone (1991) in their studies on active and inactive bacteria, as demonstrated by the [2-(*p*-indophenyl)-3-(*p*-nitrophenyl)-5-phenyl tetrazolium chloride] (INT) method, "the maintenance C requirement based on the total biomass C exceeds the C input value for soil further than 5 mm from any root; however, when based on the active biomass, the requirement is the same magnitude as the input value, indicating that sufficient C inputs exist to support the observed active biomass."

The concept of sequential development of microbial populations after the addition of carbonaceous materials was first suggested in 1924 by Winogradsky (cf. Stotzky and Norman, 1961a) in his concept of "autochthonous" and "zymogenous" populations and later affirmed by Garrett (1956; see Stotzky and Norman, 1961a). According to Stotzky and Norman (1961a), there is ample evidence supporting the concept of sequences of organisms in the decomposition of heterogenous substrates, such as plant residues. Today, this process is termed *syntrophy*. Food (substrate) specialization makes possible the existence of a large number of ecological niches within a given habitat (Clark, 1965).

When a substrate (e.g., glucose) is added to soil, there is no good correlation between CO_2 evolution, substrate disappearance, formation of intermediates, and changes in pH with the numbers of bacteria and fungi as estimated by the dilution plating technique (Stotzky and Norman, 1961a). These results indicate, according to the authors, that the decomposition of a simple substrate is complex, and the kinetics of glucose decomposition is composed of a series of first-order reactions.

Clark (1965) compared the output of carbon dioxide to the bacterial population in soil. He concluded that a good part of the soil microflora must be in the dormant condition and this dormant condition was attributed to the lack of suitable available energy material. The reason Clark (1965) gave for the large number of bacteria in soil was that food specialization makes possible the existence of a large number of ecological niches within a given habitat but they are in a state of chronic starvation. Furthermore, he states that competition in soil can be viewed both in terms of a given substrate and the specific conditions under which that substrate is present. The balance between various members of the soil microbial population at any given time depends on the availability of compounds essential for their growth (West and Lochhead, 1950). Common humic substances are degraded rapidly by bacteria from pristine sites (Ghiorse and Wilson, 1988). The lack of a suitable electron acceptor and the surface effects might explain why not all organic carbon is metabolized by bacteria (Ghiorse and Wilson, 1988).

When ^{14}C-labeled substrates were incubated in soil, glucose was most readily decomposed, followed by hemicellulose, cellulose, maize straw, and barley straw, respectively (Sørensen, 1963). Of the glucose-C left the soil as CO_2 in the first 10 days of incubation, there was 60% as compared to only 23% of the barley-C. The humified material, resulting from the addition of the various materials remaining in the soils after 3 months decayed at almost the same rate, regardless of the origin of the organic matter (glucose, hemicellulose, cellulose, or straw) and the half-life of the labeled C ranged from 5 to 7 years, depending on the type of soil. The half-life of the native soil C was estimated from CO_2 evolution and ranged from 13 to 29 years. When glucose was used, the amino acid–C tended to be the largest and the amino acid–C increased with the clay content of the soils. Sørensen (1963) found, after 20 years, 9.3% still remained in the soil. Between an 8- and 20-year period, the labeled organic matter had a decay rate corresponding to a half-life of 15 years. An average of 15% of the labeled C in amino acids remained almost constant throughout the study. The native C in the organic matter (1.9%) decayed at a rate corresponding to a half-life of 91 years, but the native C in amino acids increased from approximately 14% to 16%. During the 8- to 20-year period the microbial biomass, which was determined by the chloroform fumigation technique, remained constant. In the laboratory, the carbon became stabilized when 30% or less of the original carbon remained (McGill et al., 1973).

The maintenance energy demands of soil microbes surpasses values derived from pure culture studies (Anderson and Domsch, 1985a,b). Anderson and Domsch (1986) studied different microbial populations in three different types of soil in terms of various kinetic parameters such as substrate affinity, growth rate, growth yield and turnover time, and the metabolic quotient of basal respiration. They used glucose as the carbon source. Specific growth rate (μ) was found to vary between 0.0037 and 0.015/hour, depending on soil type and glucose concentration and was far below the potential μ_{max}. Other kinetic parameters fell into the range observed by others.

The ratio of microbial biomass carbon (C_{mic}) to total organic carbon (C_{org}) was measured in 134 plots in 26 experimental sites and no universal equilibrium constant was found. The wide spectrum C_{mic} to C_{org} found was attributed to differences in soils, vegetational cover, and soil management as well as variations in sampling time and analytical methods.

Three fractions of soil organic matter were recognized by Clark and Paul (1970): the bioavailable fraction that turned over at least once every 4 years, the relatively stable products of microbial metabolism with a half-life of 5–25 years, and the very resistant humic material with a half-life of 250–2500 years. Decomposition rate constants per day for unprotected and physically protected decomposable organic matter were calculated by Paul and Voroney (1980) and found to be 8×10^{-2} and 8×10^{-4}, respectively. The equivalent rates for recalcitrant organic matter were 8×10^{-6} and 8×10^{-8}.

"Estimates of the overall amounts of energy available to bacteria in soil mass are useful, but by their nature are based on the assumption that there is a steady supply of nutrients, which is evenly distributed throughout the soil for use by randomly situated microbes" (Williams, 1985). The same can be said for all environmental systems. One must then consider spatial and temporal discontinuities in the energy supply in all ecosystems and the reaction of bacteria to the spatial and temporal discontinuities. Thus, the heterotrophic bacteria must have experienced periods in which organic matter was in very short supply or nonexistent.

Table 5.5 is a compilation of data by Nedwell and Gray (1987) indicating the percent of active bacteria and fungi found in various environments. The percentage of active bacteria is much lower than the inactive ones, much of which is due to the lack of organic matter. Generally, the level of active fungal biomass is only about 10% that of bacteria. Because carbon compounds represent the main energy source for heterotrophs, the carbon cycle plays a major role in controlling the transformation of N and P in ecosystems, and understanding these transformations requires a complete understanding of the C fractions and their differing stabilities. Because all cycles of matter are interlinked, energy sources become extremely important in the dynamics of these cycles. For instance, in the process of denitrification, the bioavailability of organic carbon affects denitrification activity in the soil, because most denitrifying bacteria are heterotrophs (Beauchamp et al., 1989). The genera *Alcaligenes* and *Pseudomonas* are ubiquitous in soil (Gamble et al., 1977); the specific C requirements for these species have not been documented and it may be expected that wide ecological diversity exists not only between species but also between strains within species (Beauchamp et al., 1989). When *Arthrobacter globiformis* and a *Pseudomonas* soil isolate were inoculated into autoclaved soil and examined by plate counts and ultrastructural analysis, by Labeda et al. (1976), the *Arthrobacter* appeared to be in a nongrowing coccoid-rod resting state while the *Pseudomonas* possibly suffered a nutritive deficiency. Total microbial biomass in soil does not necessarily reflect microbial activity (Clark and Paul, 1970). This is to be expected when so many bacteria are in the starvation-survival state. It takes energy for many of the functions of microbes in the different cycles of matter.

Although many investigators have discussed the concept of a microbial microenvironment (niche), the microbial activity predominates on, or in close proximity to, the surfaces of soil particles and this phenomenon can be demonstrated by adsorption experiments with microorganisms, substrates, and soil components (Burns, 1983). Even growth in the rhizosphere is slow. For instance, in plant solution culture, the growth rate of bacteria in the rhizosphere of barley may be only 0.029 per hour for the first 4 days and 0.007 per hour at 7–16 days (Barber and Lynch, 1977).

The measured ATP values of the soil biomass strongly suggest that the concentration in largely resting soil populations is little different from that in the actively growing organisms (Jenkinson and Ladd, 1981). The data obtained by Brookes et al. (1983) indicated that the dormant soil microbial biomass maintains both adenylate energy charge (AEC) (0.85) and ATP at levels characteristic of exponentially growing organisms in vivo, even during prolonged incubation without fresh substrate and that roots of plants make a negligible contribution to total ATP extracted from fresh-sieved soil. Subsequently, Martens (1985) modified the extraction of ATP by using $NaHCO_3$ reagent to make it more appropriate to soil samples so that he could determine whether the bacteria in soil were in a starving or resting state because they live in a substrate-depleted system. For six unplanted

Table 5.5 Percentage of the viable biomass thought to be active in various environments.

Organisms	Habitat	Technique	% active	Reference
Bacteria	Forest soil			
	A1 horizon	FDA	34	Lundgren and
	A2 horizon		54	Söderström, 1983
	B horizon		52	
	Agricultural			
	Barley field	ETS	15	Macdonald, 1980
	Manured soil		25	
	Turfed soil		11	
	Compost		23	
	Vegetable soil		23	
	Field soil		31	
	Aquatic			
	Water over	Tritiated	2.3–56.2	Meyer-Reil, 1978
	sand	glucose	average 31.3	
Fungi	Pine forest			
	A1 horizon	FDA	2.4	Söderström, 1979
	A2 horizon		4.3	
	B horizon		2.6	

Source: Reprinted by permission from the *Symposium of the Society for General Microbiology* 41:21–54, N. B. Nedwell and T. R. G. Gray, 1987; copyright 1987, Society for General Microbiology.
Note: FDA = hydrolysis of fluorescein diacetate. ETS = electron transport system activity.

soils, he obtained AEC values from 0.3 to 0.4, thus indicating either a low metabolic activity of the soil populations or a starving population. Martens (1985) also added low glucose supplements (up to 500 µg C/g soil) to low biomass and high biomass soil and obtained AEC charges from 0.34 to 0.50 and 0.32 to 0.37, respectively. The data of Martens (1985) appear to be in agreement with the other findings that the soil microbial population is mainly a resting state. Others (Atkinson, 1977; Atkinson and Walton, 1967; Chapman et al., 1971; Karl, 1980), employing the $NaHCO_3$ extraction method, demonstrated AEC values typical of a moribund population.

According to Teteno (1985), only organisms that possessed a high AEC (ca. 0.9) were considered to be able to survive in soil and if the AEC fell to some critical level, the cells would immediately be decomposed by other soil organisms. The AEC values of 0.8 to 0.9 obtained in soils give an incorrect view of the overall soil microbial activity, hence in situ soil microbes behave differently than those grown in a chemostat or in the laboratory (Brookes, 1986). Do in situ aquatic bacteria behave like the in situ soil bacteria? After reviewing all the literature, Brookes (1986) concluded that (1) acid reagents, like TCA should be used to extract ATP from soil because neutral or alkaline reagents do not prevent dephosphorylation of ATP immediately, (2) the soil biomass maintains a high ATP concentration and AEC level characteristic of exponentially growing cells, although it is mainly a resting population, and (3) although the biomass of soil increases or deceases in size in response to inputs of fresh organic material, it can maintain itself for prolonged periods without fresh substrate, probably by using soil organic matter as an energy source. However, it should be pointed out that starved cells do have a high ATP level (see Chapter 7).

Shields et al. (1973) suggested that the mean generation time for microorganisms in prairie soils is also slow and they calculated a maintenance coefficient for soil bacteria growing in situ to be about 0.002/hour. Gray et al. (1974) proposed a mean generation time of 4.07 days, on the assumption that there was no fungal and animal activity; in the likely occurrence of such activity, the minimum generation time for bacteria would be greater (10.2 days). Generation times for bacteria in soil are as follows: 12 days in deciduous woodland (Hissett and Gray, 1976), 26 days in grassland soil (Hunt, 1977), 316 days in Australian clay soil (Jenkinson and Ladd, 1981), 0.03 to 632 days in a classical Rothamsted Broadbalk plot (Jenkinson and Ladd, 1981), 39 hours in tundra peat (Parinkina, 1974), 66 and 55 hours in organic and mineral layers of podsol profile, respectively (Parinkina, 1973), 14–17 hours for actively dividing bacteria in forest soil (Clarholm and Rosswall, 1980), and 3.3 days in Meathop Wood in Cumbria, England (Chapman and Gray, 1986). The microbial turnover rate of different types of soil ranges from 0.07 to 0.32 per year. In a grassland prairie soil, the energy input was sufficient to allow bacteria to divide only a few times a year (Babiuk and Paul, 1970).

Various methods have been used to estimate the microbial biomass in soil. Only the direct count, viable count, and carbon dioxide evolution provide an indication of the bacterial biomass. An assessment of the various methods employed to determine microbial mass is discussed by Williams (1985). One of the main methods used is the evolution of the carbon dioxide (Jenkinson and Powelson, 1976) based on the following equation:

$$B = F/k$$

Values for k have been calculated to be 0.5 when tested at 25°C (Jenkinson, 1976) and 0.41 at 22°C (Anderson and Domsch, 1978). The major problem is the value assigned to k.

Smith and Paul (1988) raised the question "What do our microbial biomass values represent and how can we use them to study nutrient cycling?" In soil, biomass measurements are based on the chloroform fumigation incubation method, the respiratory response method, and the ATP measurement. All these methods do have inherent errors. In addition, these measurements ignore the interaction of the many different individual microbial components that make up the entire system (Jenkinson and Ladd, 1981). These biomass values have been used mainly for comparative purposes (Smith and Paul, 1990). Biomass studies have been used successfully in studying disturbed and nondisturbed ecosystems, biodegradation of toxic chemical, forest clearcutting, etc. The accumulation of organic matter generally increases the microbial biomass (Jenkinson and Ladd, 1981). The general rule is that the 2–5% of the organic carbon in soil is microbial biomass. Jenkinson and Ladd (1981) noted that, in soil, the size of the living biomass relative to annual input of substrate is such that only a small part of the biomass can be in active growth at any one time.

A change of 1 μg biomass C/g soil is equivalent to about 10^5–10^6 bacterial cells (Sparling et al., 1981). The main contribution of organic matter in soil to soil fertility is its capacity to supply nutrients for plant growth (Duxbury et al., 1989). The accumulation and distribution of organic matter in soils lies in its capacity to *protect* (italics by author) the organic matter from microbial decomposition. The stability of organic matter in soil results from protective processes occurring within soil rather than from the creation of recalcitrant chemical structures (Duxbury et al., 1989). Biomass C may also be a source of readily available C, especially when some of the biomass is dead (Soulides and Allison, 1961; Jenkinson and Powlson, 1976; Draycott and Last, 1971; Shields et al., 1973; Jenkinson, 1976; Marumoto et al., 1977, 1982a,b; Brookes et al., 1984). The average dry weight of soil bacteria has been estimated by various investigators and ranges from 2 pg (Gray et al., 1974) to 0.06 pg/g soil (Nikitin, 1973).

Any environmental change in an ecosystem will kill some of the microbes within a microbial community. These dead cells are a source of nutrients for the surviving microbes. This has been shown in many studies where soil has been fumigated with $CHCl_3$ as well as for bacteria in sediment (Novitsky, 1983b).

Bacillus and *Clostridium* in soil are mainly in the spore state. Because spore formation is the result of the lack of energy in these organisms, it is another indication that soil is "grossly oligotrophic," a term employed by Williams (1985). This point is further discussed by Gray and Williams (1971). Because of the lack of bioavailable organic matter, most bacterial cells in the soil are inactive and most are in the starvation mode.

In summary, whatever the cause may be, the concentration of organic bioavailable matter in natural environments is extremely low, especially when compared to the media used by microbiologists. However, there are rare exceptions where eutrophic environments occur in nature. One also has to wonder how much of the organic matter analyzed by the chemists results from bacterial cells in the sample.

5.5.2   Subsurface Environments

Data on the physiology of deep subsurface bacteria, with a few exceptions, support the hypothesis that they are in a state of starvation or are relatively inactive (Ghiorse and Wilson, 1988). Thorn and Ventullo (1988) studied the bacterial growth rates in near-surface (<20 m) subsurface sediments and concluded that the microbial communities were "nutritionally stressed." Smith et al. (1986) also present evidence that deep subsurface microorganisms are under nutrient stress, as indicated by fatty acid analysis. Webster et al. (1985) estimated that 1 to 10%, by both the ATP and INT methods, of the subsurface bacteria in Lulu, Oklahoma, and Conroe, Texas, sites is metabolically active.

Microbial studies at the Savannah River Project are in deep contrast with the above studies. High rates of bacterial recovery, rapid growth on agar plates, and some high rates of acetate uptake suggest that the microbes in deep sediments are in high numbers and very active metabolically (Balkwill, 1989). These data suggest that either: (1) there was a unique set of circumstances at this site causing unusually high bacterial activity, (2) there was contamination by the drill process, or (3) the indigenous microbial population became more active as the result of sample storage and handling. The following investigations suggest that the bacteria in the deep sediments at the Savannah River site were initially in a starvation-survival mode and that their activity was greatly enhanced during sediment storage, handling, and subsequent preparation for analysis.

1. Balkwill (1989) reported that, although some microcolonies were found in the cores, most bacteria occurred as small free cocci or coccoid rods, typical of the gram-negative cells under starvation.
2. Phelps et al. (1989) found that anaerobic incorporation of acetate into cellular lipids increased threefold during the first 24 hours after collection. This suggests that there was an increase in microbial activity within this short period caused by the priming effect of adding a utilizable substrate, as shown by Williams (1985).
3. Analyses of freshwater dissolved organic carbon (DOC) indicated that by far the largest components of DOC are recalcitrant fulvic and humic acids. This was also reported in shallow well water by Wallis and Ladd (1982).

Further studies on the Savannah River subsurface sediment employing the heterotrophic potential method by Hicks and Fredrickson (1989) indicate that over 70% of the added acetate and almost 60% of the added phenol were respired over 21 days. One significant problem with long incubation periods is the possibility of adaptation to the added substrate.

Marxsen (1988) studied groundwater from sandy and gravelly deposits (near

Fulda, Federal Republic of Germany) which were unpolluted to slightly polluted. In 11 samples, the total number of bacteria ranged from 2.0 to 30 $\times$ 10^6 bacteria/ cm^3, while the number of respiring bacteria by the INT method ranged from 0.055 to 0.49 $\times$ 10^6 bacteria/cm^3. Thus, the proportion of respiring bacteria to the total number ranged from 0.66 to 7.4%. The respiring bacteria were observed to be the larger cells with hardly any active cells in the small cocci. He could not find a correlation between the dissolved organic substances and the quantity of respiring bacteria. No correlation was observed between the quantity of the respiring bacteria and the heterotrophic bacterial activity. The latter was measured employing glucose.

Wilson et al. (1983) reported that microbes in the deep subsurface can degrade some, but not all, of the organic pollutants commonly encountered in groundwater. The degradation of trichloroethylene and tetrachloroethylene was noted but Wilson et al. (1983) suggest that it is likely degraded by abiotic processes. Previously, Dilling et al. (1975) also reported the degradation of these organics to be abiotic.

Chapelle and Lovely (1990) studied the microbial metabolism rates in Atlantic coastal plain aquifers of South Carolina employing radiolabeled acetate and glucose. The turnover of acetate and glucose, as well as the estimated rates of acetate oxidation are shown in Table 5.6. Although $^{14}CO_2$ from [U-^{14}C]glucose gave detectable linear rates, the production of CO_2 from [2-^{14}C]acetate was only detectable within 77 days of incubation in some of the sediments. Chapelle and Lovely (1990) could find no clear relationship between sediment organic matter concentrations and the turnover rates of acetate and glucose pools (Table 5.6). They also showed that all the clayey sediments of the confining beds had low rates of $^{14}CO_2$ production from radiolabeled glucose or acetate compared to those observed in the active sandy sediments from the Black Creek and Middendorf aquifers. If the laboratory rates were correct, all the organic carbon should be consumed in a matter of years. A geochemical model was formulated to determine rate estimates. In those instances in which the production of $^{14}CO_2$ was sufficiently rapid to permit an estimate of the rate of [2-^{14}C] acetate turnover, the estimated rate of CO_2 production was two to four orders of magnitude faster than rates estimated by geochemical modeling. It should be noted that the sediments of this hydrologic system were deposited between 70 and 80 million years ago (Gohn, 1988). Chapelle and Lovely (1990) strongly "suggest that laboratory studies are not reliable indicators of the potential for in situ biodegradation in the deep subsurface since laboratory incubations of deep subsurface sediments may greatly overestimate in situ rates of microbial metabolism." Furthermore, the rates obtained in the laboratory are, at best, potential when substrates are added to the system and what is measured is substrate-induced respiration. In addition, one does not know what perturbation of samples does to the microbes. Nevertheless, Chapelle and Lovely (1990) state that measurements of CO_2 accumulation in bottle incubations greatly overestimate the rates of in situ metabolism.

Table 5.6 Acetate and glucose turnover and estimated rates of acetate oxidation to carbon dioxide.

Site	Depth (m)	Hydrologic Unit	Sediment Texture	Turnover Rate Constant (yr^{-1})		Acetate Concn in Groundwater (μm)	Estimated CO_2 Production from Acetate Turnover[a] ($mmol\ liter^{-1}\ yr^{-1}$)	Sediment Organic Carbon (%)
				Acetate	Glucose			
Florence	38	Black Creek (aquifer)	Sandy	118	174	1.0 ± 0.25	2.4×10^{-1}	0.46
Florence	92	Black Creek (confining bed)	Clayey	<0.025	NA[b]	NA	NA	1.70
Florence	113	Middendorf (aquifer)	Sandy	6.8	NA	0.8 ± 0.07	1.1×10^{-2}	0.040
Florence	141	Cape Fear (aquifer)	Sandy	<0.025	NA	0.5 ± 0.03	$<2.5 \times 10^{-5}$	0.025
Myrtle Beach	123	Black Creek (aquifer)	Sandy	94	73	1.8 ± 0.9	3.4×10^{-1}	NA
Myrtle Beach	165	Black Creek (aquifer)	Sandy	2.6	74	1.8 ± 0.9	9.4×10^{-3}	0.33
Myrtle Beach	240	Black Creek (confining bed)	Clayey	<0.025	0.05	NA	NA	NA
Myrtle Beach	295	Middendorf[c] (aquifer)	Sandy	<0.025	15	NA	NA	0.06
Myrtle Beach	305	Middendorf (confining bed)	Clayey	<0.025	7.7	NA	NA	0.09
Myrtle Beach	388	Cape Fear (aquifer)	Sandy	<0.025	0.2	1.4 ± 0.8	$<7 \times 10^{-5}$	0.21

Source: Reprinted by permission from Applied and Environmental Microbiology 56:1865–1874, F. H. Chapelle and D. R. Lovely, 1990; copyright 1990, American Society for Microbiology.

[a] Calculated from equation 1 as described in Materials and Methods.

[b] NA, Not analyzed.

[c] Sample contaminated with drilling fluid.

The metabolic rates, as measured by CO_2 production, obtained by geochemical modeling for an aquifer were compared with metabolic rates of other environments (Fig. 5.6) (Chapelle and Lovely, 1990). They found that their rates were comparable to other aquifers in the same region (Chapelle et al., 1987). Of interest is that the measurements of the metabolism of [^{14}C] organics in deep ocean waters (Williams and Carlucci, 1976) are greater than those obtained through geochemical modeling (Chapelle and Lovely, 1990). From their geochemical modeling, the rate of organic carbon metabolism in deep subsurface aquifers are on the order of 10^{-4} to 10^{-6} mmol of C/l/year, but this assumes that all the organic C is bioavailable. According to these authors,

> the microorganisms in these aquifers could be remnants from the microbial population that were present at the time of deposition (70–80 million years ago) or may be transported in the groundwater, which has been underground for 10,000 to 50,000 years. In either case, these results suggest that the subsurface microbial population can remain metabolically active for long periods of time while metabolizing organic matter at very slow rates.

Why is it that the microbial population is in a starved state, not metabolizing, but when readily bioavailable substrates, such as glucose or acetate, are added metabolism occurs? Growth is probably extremely sporadic and metabolism could also be very sporadic—both occurring when sufficient energy is present. The microbes would then utilize the energy immediately and pass again into a stage of metabolic arrest.

White et al. (1983) analyzed the fatty acids from a shallow aquifer at Fort Polk, Louisiana, and found the groundwater sediments exhibited different fatty-acid profiles than the surface sediments, with no indication of long-chain polyenoic fatty acids characteristic of microeukaryotes such as algae, protozoa, or fungi. High levels of poly-β-hydroxyalkanoic endogenous polymers were observed indicating nutritional stress. Their analyses also indicated a more sparse and qualitatively different microbiota from the surface sediments. A ratio of 64 pmol PHA/ nmol muramic acid for this aquifer was obtained by White et al. (1983), indicating unbalanced growth. Thus, many studies by White and his associates indicate that the amount of PHA/cell is much greater in aquifers than shallower microbiota. White et al. (1983) state that "the nutritional status of the microbiota as reflected in the accumulation of poly-β-hydroxyalkanoates per cell and uronic acid containing extracellular glycocalyx per cell indicates increasing stress with increasing depth." Extracellular polysaccharide glycocalyx accumulates with nutrient deprivation in prokaryotes (Costerton et al., 1981; Sutherland, 1977; Uhlinger and White, 1983). In a subsurface sediment of a pristine shallow subsurface aquifer in Oklahoma, Bone and Balkwill's (1988) finding indicates that the organisms are nutrient-stressed in situ, being specifically adapted for growth under near-

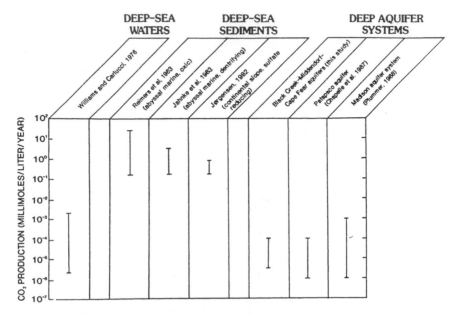

Figure 5.6 Comparison of carbon dioxide production rates from deep-sea waters, deep-sea sediments, and deep aquifer systems. Estimates for sulfate-reducing deep-sea sediments were calculated from sulfate reduction rates and the assumption that 2 mol of carbon dioxide are produced per mol of sulfate reduced. Reprinted by permission from *Applied and Environmental Microbiology* 56:1865–1874, F. H. Chapelle and D. R. Lovely, 1990; copyright 1990, American Society for Microbiology.

starvation conditions that probably prevail in their oligotrophic environment. Furthermore, no evidence for long-chain polyenoic fatty acids characteristic of microeukaryotes such as algae, protozoa, or fungi were noted in the subsurface groundwater from Bucatunna aquifer (410 m, near Pensacola, Florida) (White et al., 1983). Bacteria containing high cellular levels of poly-β-hydroxyalkanoic or uronic acid containing exopolysaccharides polymers are indicators of unbalanced growth and stressed nutritional status (White et al., 1983). Phospholipid analysis of bacteria from subsurface Middendorf samples indicates highly stressed cells (Wilson et al., 1986). Nutritional stress is indicated in uncontaminated subsurface sediments as evidenced by high levels of poly-β-hydroxybutyrate and extracellular polysaccharides (Smith et al., 1986).

Thus, most of the data on bacteria in subsurface environments indicate that they are under nutrient stress. In cases where nutrient stress is not indicated, bacterial activity may be due to contamination (e.g., drilling) or from storage and handling.

5.5.3 Marine Environments

Watson et al. (1977) suggested that most bacteria observed in 186 samples obtained during a cruise were in the dormant state. By using an in situ method, Hobbie et al. (1977) found that most cells in seawater were inactive. In sediments, Novitsky (1987) found that over 90% of the sediment-water interface microbial community was not actively growing. Meyer-Reil (1978) found that only 2.3–56.2% of the microbes in the water over sand were active; the average being 31.2%. Tabor and Neihof (1982), by use of an improved INT method, determined that 60% were active in Chesapeake Bay. Novitsky (1983b) reported that, in marine sediments, most of the bacteria were inactive. On the average, 20% of the bacteria are being active at any given time (Williams, 1985), but the inactive bacteria are not dormant with respect to substrate utilization (Kjelleberg and Hermansson, 1984). Studies on natural samples by microautoradiographs definitely show that there is considerable variation in the proportion of metabolically active bacteria in different habitats (Meyer-Reil, 1978; Douglas et al., 1987). Deming and Yager (1992) question whether a very active but small portion, or a less active but large portion of the bacterial biomass accounts for the overall bulk rate measurements. Nevertheless, they point out that the correlation between bacterial DOC utilization rates and parallel biomass measurements over the abyssal range of 3700–5840 m in the North Atlantic Ocean is significant and strongly positive. The doubling time of mixed microbial populations in some deep-sea sediments ranges from 5.7 to 693 days (DeLong, 1992).

When *V. cholerae* serotypes 01 and non-01 isolates, considered to be indigenous residents of estuarine environments, were placed in sterile sediment, sterile waters, and nonsterile waters, all obtained from Apalachicola Bay, Florida, it was noted that growth and extended periods of survival occurred (Hood and Ness, 1982). When placed in nonsterile estuarine water, the organism declined in numbers — an indication that there were insufficient nutrients to support growth. On the other hand, in sterile (autoclaved) estuarine water, growth did occur in some instances. Autoclaving could have made some of the recalcitrant organic matter bioavailable to the organism. Autoclaving sediments releases bound nutrients from them (Gerba and MacLeod, 1976). The increase of cells in sterile sediments may also be partially due to the absence of competing organisms (Hood and Ness, 1982).

Wright and Coffin (1984b) predict that, in the absence of predators, bacterial growth is a function of substrate concentration. As a result of this growth in the absence of predators, production of bacterial cells will be low and most of the substrate consumed by the bacteria will be used for maintenance energy (Wright, 1988b). In the absence of grazing by predators, the growth rate should approach zero (Coffin and Sharp, 1987).

The generation time of 1 month (range 14 to 60 days) was found by Newell and Christian (1981) for bacteria in Drake Passage. By use of the thymidine method, Moriarty and Pollard (1981, 1982) determined the generation time of 3.8 to 125 hours in seawater and 144 to 1008 hours in Zoostera beds, reflecting the organic matter concentrations in these two different sites. At three different marine sites, Craven and Karl (1984) determined the generation times to be 4 to 166 hours by the adenine method and 212 to 350 by the ATP method. Pomeroy et al. (1991) found the generation time of 30 to 86 days during three spring blooms (1986, 1988, and 1990) in Conception Bay, Newfoundland, and the percentage of dividing cells was 3.19 and 2.22. Thus, when one compares the generation times between the environment and laboratory studies (e.g., *E. coli* in nutrient broth), it can be seen that the generation times in the environment are much lower—mainly because of the lack of organic matter.

The oligobacteria isolated by the dilution technique by Button et al. (1993) often showed doubling times of a month or more, and cultures sometimes took as long as 2 months to attain populations of 10^5, but doubling time was generally 1 day to 1 week in vivo. According to these authors, the concept of viability acquires new dimensions—mainly because viability is an operational term that means the ability of a single cell to attain a population discernible by the observer. For these oligobacteria, there is a slow estimate of viability because they reach the stationary phase before attaining visible turbidity. These oligobacteria sometimes reach their sensitivity limits of 10^4/ml. They point out that sterile seawater is capable of supporting 10^6 to 10^7 cells/ml. The authors took extraordinary precautions to exclude organic matter contamination but their medium was seawater that was filtered and then autoclaved. The latter could bring about hydrolysis of recalcitrant polymeric organic compounds. When nutrients (0.01 to 1 mg C/l) were added to Resurrection Bay water, no change in viability was noted, but when seawater from the North Sea was used, 5 mg/l or more amino acids actually inhibited growth.

Employing dialysis bags in situ at 4 and 60 m in the vicinity of a shallow sea tidal mixing in the Irish Sea, Turley and Lochte (1986) found an increase in the mean cell volume preceded an increase in specific growth rates, with the resultant decrease in cell volume. The doubling time of the 4 m microbial population was once per day while at 60 m, it was once per 2 days. What role the bottle effect played, if any, was not discussed. Their data suggest that many bacteria are only active for part of the day in the relatively shallow waters of their study.

Chocair and Albright (1981) noted that about 5% of the total number of near-shore bacteria were active. Novitsky (1983b) observed that about <5% of the sediment-water interface microbial community was active in glutamate uptake, as determined by autoradiography. At the sediment-water interface, marine sediments sometimes contain more than 10^9 bacteria/g (Dale, 1974; Novitsky, 1983b); the amount of nutrients available for such a large population was not present. Even if it had been available, such a large population would have immediately reduced the nutrient availability. Novitsky (1987) set about to determine the number of nongrowing cells in marine sediment; whereas other studies (Meyer-Reil, 1978; Zimmermann et al. 1978; Tabor and Neihof, 1982; Novitsky, 1983a) have determined the number of metabolically active cells. Employing biomass estimations by ATP and DNA synthesis ([2-^{3}H] adenine) and the specific growth rates from the kinetics of [^{3}H]ATP pool labeling, his data showed that a large and active microbial community exists in the interface, and the greatest amount of microbial activity occurs within the first few millimeters of the sediment. However, the ATP and DNA synthesis methods established that over 90% of the sediment-water interface community was not actively growing. The above data plus the fact that killed microbial cells in marine sediments are very labile support the hypothesis that the nongrowing portion of the microbial community is mainly dormant (Novitsky, 1986).

By enriching seawater with 10 μM amino acids to determine nitrogen regeneration, Hollibaugh (1976, 1978, 1979) observed a response similar to that reported by Williams and Gray (1970). He also hypothesized that the observed phenomena was due to activation of a dormant population (starved cells?) or to an enzyme induction rather than to a biomass increase of the population. Both Hollibaugh (1979) and Williams and Gray (1970) remarked on the lack of any short-term bottle effect and it was noted that, in bottles, there were no increases in the bacterial population in the first day or two of the experiments.

Wright (1984) also noted a time lag in bacterial growth (as measured by AODC) when estuarine and coastal seawater (prefiltered through a 3-μm Nuclepore filter) was enriched with glutamic acid at 1, 10, and 100 μg/l and 1 and 10 mg/l. His data indicate that all the glutamate was used within 8 hours when the initial concentration was 100 μg/l or lower and within 24 hours glutamate was exhausted in all samples. The AODC reached a peak in 16 hours with the exception of the samples with 10 mg/l glutamate. The AODC then declined until the 36-hour sampling followed by a slow increase and final decline. Of great interest in estuarine waters was the fact the initial rapid increase and decline in samples with the lower concentrations of glutamate was due to a proliferation and then disappearance of the smaller bacteria (>0.6 μm) and that the increase in the highest concentration of glutamate (10 mg/l) was due to large cells. Wright (1984) concluded that (1) the bacteria in a given sample ranged from dormant and inactive to highly active and rapidly growing and (2) they had been removed from most sources of substrate. In the latter case, there is a variable amount of readily available substrate and additional amounts of substrate are made slowly available through cell death and from natural enzymatic breakdown of polymers. Thus, Wright (1984) suggested that the estuarine population was very active but closely tied to substrate supply. When the supply was cut off by removal from the natural environment the cells divided once (within 16 hours) and then a large proportion of those new small cells apparently died and lysed. Their death and other unknown processes supplied more substrate and allowed the second increase to occur during the second and third days of the experiment.

By use of microradiography with ^{3}H-glucose, glutamic acid, and thymidine, Novitsky (1983b) showed that there was a decrease in the percentage of active cells with depth from 35% to <1%, whereas the number of active cells in the sediment-water interface and sediment averaged <10% of the total bacterial population in Halifax Harbor, Canada. However, it should be noted that in Figure 5.7, the percentage of active cells is different with respect to the substrate employed. Thus, it can be seen in Figure 5.7 that thymidine does not give the highest number of active cells. The samples in Novitsky's study were taken by hand by scuba divers to provide accurate depth location and to provide the least disturbed samples. This latter situation is very important to ensure that samples are not perturbed to any great degree.

In the nearshore euphotic zone, Azam and Cho (1987) gave the following approximate figures for each milliliter of seawater: 1000 bacteria, 1–10 cyanobacteria, 1 detritus particle, 1 heterotrophic microflagellate, with ciliates, macrozooflagellates present for a short period of time, and 1 ng DOC of which 0.05 ng is utilizable (or 50 fg C). Without any further input of the DOM, bacteria can reproduce once or twice (Ammerman and Azam, 1981). Since each bacterium contains roughly 20 fg C (Bratbak, 1985; Bjørnsen, 1986b), this would allow the bacteria to reproduce once or twice, as noted by Ammerman and Azam (1981). Although the above is given as an approximation, I believe we need to examine the foregoing closely, especially if we are to take into consideration the possible energetics involved. If I assume the DOM is glucose, and 1 mol glucose produces 868×10^3 cal, then each bacterium in the foregoing milliliter of water would have 190×10^{-12} cal. This does not take into consideration either how efficient microbes are in using the energy (not 100% efficient because we know that bacteria can give off heat during metabolism) or the energy of maintenance, if any. What it does point out is that we do not know how much energy it takes for a bacterium to multiply, especially in nature where cells are much smaller than those grown in nutrient media in the test tube. In an earlier paper, Ammerman et al. (1984) state: "The production of 2×10^6 bacteria represented the consumption of less than 10% of the dissolved organic carbon (DOC) in seawater (20 fg C cell^{-1}, 50% assimilation efficiency; 1.5 µg DOC ml^{-1} seawater, or 5.3% DOC utilized.)" In these studies, the authors rule out (1) organic matter from the air because the bottles were tightly capped and not shaken, and (2) the bottle effects, but not the possibility of cell breakage during filtration leaking out nutrients or cryptic growth. However, it should be mentioned that these studies were done with Scripps Pier water.

On the other hand, Pomeroy et al. (1991), working in the cold, more pristine environment of Conception Bay, Newfoundland, during three spring blooms

% ACTIVE CELLS

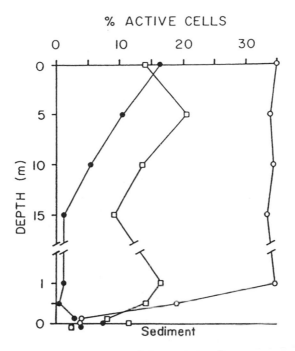

Figure 5.7 Active cells as a percentage of the total population for samples taken at the Western Passage sampling site. Water depth at this site was 22 m. The three deepest water column points represent samples taken 1 m, 0.5 m, and 1 cm above the sediment surface. Symbols: ○, glutamate; ●, glucose; □, thymidine. Reprinted by permission from *Applied and Environmental Microbiology* 45:1753–1760, J. A. Novitsky, 1983; copyright 1983, American Society for Microbiology.

(1986, 1988, and 1990) using three different methods, seldom found more than 10^3 bacteria per ml. The percent of cells dividing was low (3.19 and 2.22% were reported). The generation time of the bacteria was reported to be 30 to 86 days in Conception Bay. These investigators, examining the frequency of dividing cells data of Hanson and Lower (1983) and using a long-transformed regression (Newell and Christian, 1981), estimated a doubling time of 1 month, with a range of 14–60 days for bacteria in Drake Passage. In terms of bacterial productivity, as measured by the ^{3}H-leucine method, figures were never over 1 mg C/m^3/day, but most were in the range of a few hundredths of a mg C/m^3/m/day. Thus, their results generally suggested a low growth rate or that only a small part of the bacterial population was actively growing. They also postulated that a limit of bacterial growth at low temperature at the cellular level, such as a physicochemical limit to membrane permeability, limited temperature ranges of enzyme activity, or a combination of both. The Q_{10} rule also applies in this situation because many of the bacteria were capable of good growth at a higher temperature with substrates added. However, it appears that the membrane permeability decrease at low temperature is probably the major factor (Morita and Buck, 1974; Goodrich and Morita, 1977). In the Bedford Basin, Nova Scotia (Li and Dickie, 1987), the Bransfield Strait, and Antarctica (Bird and Karl, 1988), no direct effect of temperature on the bacteria was reported; while Gillespie et al. (1976) reported an effect of temperature on bacteria in the Antarctic waters where the heterotrophic potential was lower at environmental temperature but above 5°C above the environmental temperature was detrimental. Wiebe et al. (1993) subjected many isolates from the subtropical southeastern continental shelf to a matrix of temperatures and substrate concentrations equal to and exceeding natural ones. They found that, at the annual minimum temperatures, the heterotrophic bacterial isolates required higher concentrations of dissolved substrates for active growth than are usually found in seawater. Thus, bacterial and protozoan utilization of phytoplankton production during winter and early spring is low, permitting a greater energy flow to zooplankton and benthic energy flux, while in late spring, summer, and fall, the microbial loop dominates energy flux and organic carbon utilization. Communities of microorganisms had previously been found to grow actively when concentrations of labile substrates were micromolar but not when they were nanomolar (Pomeroy and Deibel, 1986; Pomeroy et al., 1990, 1991). At nanomolar concentrations, increased respiration and growth rates occur at increased temperature. Although psychrophiles have the ability to grow and respire at low temperature, they are not the dominant thermal group found in marine samples. Approximately one-third of the bacteria isolated from deep Norwegian Sea sediments were psychrophilic when samples were incubated at 5°C (Norkrans and Stehn, 1978). According to Wiebe et al. (1993), the basis for the seasonal cycles of bacterial activity is the temperature-substrate interaction. Could we then be dealing with completely different microbial populations in the various samples examined? Prolonged incubation (6 months) was necessary in order to obtain the maximum colony-forming units (CFU), especially at 5°C compared to 20°C (Norkrans and Stehn, 1978).

The bacteria are in microspheres or microzones, and Bell and Mitchell (1972), Azam and Hodson (1981), and Azam and Ammerman (1984a) proposed that the bacterial activity occurs in clusters around phytoplankton and zooplankton. Microspheres that have utilizable DOC do exist (Azam and Ammerman, 1984b). Sloppy feeding (Azam and Cho, 1987) and fecal pellets can also add to the nutrient content of microspheres. There can be many microspheres in a given environment, and the microbial growth then occurs around these phytoplankton and zooplankton as well as other detritus particles. Thus, some species of bacteria are active while in the microsphere, while other species are in a virtual state of starvation outside the microsphere. Nevertheless, the bacteria in the microsphere that have energy will consume the more readily utilizable forms of the organic matter, leaving behind the more recalcitrant molecules. For a discussion of microenvironments (microzones, microspheres), consult Azam and Ammerman (1984a). The reason for the low DOM or DOC values of seawater is probably a reflection of the fact that the labile DOM or DOC is taken up so fast that by the time analysis takes place, it has disappeared from the water sample or that the more energy-rich microsphere when combined with the low energy microspheres, give the low organic matter values. A tight coupling between the production and utilization of dissolved DOM is suggested by the data of Vaccaro et al. (1968), Azam and Hodson (1977a,b), and Azam and Ammerman (1984a,b). Mitchell et al. (1985) calculate that there are utilizable DOC concentrations between 10^{-9} to 10^{-7} M in microspheres having the size of 10^{-1} cm. The microbial populations in non-energy-rich microspheres are virtually in a starvation state. If microspheres or microzones do occur that have utilizable nutrient, then possible chemotaxis of bacteria must be considered.

On the other hand, working in freshwater environments, Chróst and Rai (1993) postulated that the rates of bacterial production are tightly coupled with enzymatic degradation of polymeric compounds of the DOC pool only when bacteria are limited by availability and/or by a low concentration of excreted organic carbon from phytoplankton. These enzymes are termed "ectoenzymes" by Chróst (1990), because they are periplasmically located or associated with the cell surface and do not diffuse into the immediate environment like exoenzymes or extracellular enzymes. Accordingly, most of the ectoenzymes studied in aquatic ecosystems are inducible catabolic hydrolyases under repression/derepression control of synthesis (Chróst, 1989, 1990, 1991; Chróst and Overbeck, 1987; Hoppe et al., 1988; and Meyer-Reil, 1981). The rates of these ectoenzymes have been positively correlated to the depletion of readily utilizable substrates for aquatic bacteria (Chróst, 1991) and/or the influx of polymeric DOC into aquatic ecosystems (Chróst, 1989, 1991, 1992; Cunningham and Wetzel, 1989). As a result, low concentrations of readily utilizable substrates depressed synthesis of the ectoenzymes and the high content of polymers in water induced synthesis. The formation of end products of the ectoenzyme hydrolysis of polymers is tightly coupled with uptake of the end products (Azam and Cho, 1987; Chróst, 1988, 1990, 1991; Hoppe et al., 1993). Enhanced production of β-glucosidase and aminopeptidase by aquatic bacteria when their growth is limited by readily utilizable substrates excreted by phytoplankton during photosynthesis was postulated by Chróst (1991). Algal populations are known to excrete a variety of low molecular weight photosynthetic products (Fogg, 1983) that can support bacteria growth and metabolism (Chróst and Faust, 1983; Cole et al., 1982). The foregoing results were obtained in mesotrophic lakes. Does this process occur in oligotrophic environments, where the cells are already starved and no energy is left for the synthesis of the ectoenzymes or in the deep sea, where no algal growth occurs due to the lack of light?

Contrary to the foregoing, 80 to 90% of bacterial growth is due to the free-living bacteria (Williams, 1970; Azam and Hodson, 1977b; Fuhrman and Azam, 1980, 1982) and rapid bacterial growth occurs in particle-free (0.22 μm-filtered) enriched seawater with a doubling time of 8–12 hours (Fuhrman and Azam, 1980; Ammerman and Azam, 1992). Fuhrman and Azam (1982) showed that at least 50% of bacteria were growing by use of thymidine microautoradiography. However, it should be pointed out again, these studies pertain to nearshore environments and to the euphotic zone. According to Chin-Leo and Kirchman (1990), the time scale for bacterial growth varies, depending on whether the source of organic matter is exudate material from normal phytoplankton during photosynthesis, organic matter from decaying or stressed phytoplankton, by excretion from zooplankton, or lysis of cells during feeding by zooplankton.

Bacteria probably maintain efficient control over their metabolism in response to rapid fluctuations in nutrient availability (van Gemerder and Kuenen, 1984; Williams, 1984; Kirchman and Hodson, 1986). Thus they have the ability to start or stop growth, depending on the availability of nutrients. Bacterial growth in the sea may be limited not only to the supply of DOM but also its quality. The stimulation of bacterial growth is greater when dissolved free amino acids (DFAA) are added than when glucose and ammonium are added (Kirchman, 1990). From an energetics viewpoint, this appears logical because energy is needed to synthesize amino acids when glucose and ammonium are added. DFAA is the main nitrogen source for bacterioplankton (Keil and Kirchman, 1991b), and there is agreement between chemical analysis (HPLC) and microbial measurements. Very rapid turnover rates have been observed when tracer amounts of DFAA are added to seawater samples (Ferguson and Sunda, 1984; Billen, 1984; Fuhrman and Ferguson, 1986). The analysis of DFAA by use of HPLC may also cause a higher DFAA value because the amino acids are derivatized. Furthermore, the use of pressure in the analysis may also cause release of adsorbed amino acids. Tentative calculations by Suttle et al. (1991) indicate that the DFAA pools are small and rapidly recycled and that they could be a significant source of C and N for bacterial growth in the Sargasso Sea.

The efficiency at which bacterioplankton convert substrate into biomass, known as the conversion or gross growth efficiency, is well discussed by Ducklow and Carlson (1992). This growth efficiency is central to the concept of the Microbial Loop. Ducklow and Carlson (1992) point out that one of the shortcomings of the models is that they do not take into account the metabolic costs of particle breakdown and polymer hydrolysis. When the conversion efficiencies are determined by POC and/or DOC changes, the efficiencies are low. Thus, Ducklow and Carlson (1992) pose the following questions: (1) Is bacterial production a high fraction of primary production only when most of the DOM is fresh, easily utilized material (Hagström et al., 1984) or (2) Do bacteria switch to less readily degraded substrates (Kirchman et al., 1991) to maintain market share?

Assimilation efficiencies of different bacteria were examined by Ho and Payne (1979) and two different groups emerged. They found the assimilation efficiency of the first group to be 80% (assimilation efficiencies increased as the substrate concentration decreased). The second group did not change its assimilation efficiency over the range of substrate concentration employed and this group also displayed metabolic diversity as evidenced by the ability of each to catabolize hydrocarbons. Generally, conversion efficiencies range between 50% and 80%, but these data were obtained with pure and mixed cultures of bacteria growing on simple and complex, but no refractory, substrates (Williams, 1984). On the other hand, Newell et al. (1981) and Stuart et al. (1981) obtained low efficiency values around 10% when natural organic detritus from kelp and phytoplankton detritus was employed. Iturriaga and Hoppe's (1977) data, obtained using algal organic exudates, support the low conversion rate. Thus, conversion efficiencies will vary with the substrates (Williams, 1984). The highest conversion efficiencies were obtained when labile substrates such as amino acids were employed. For *Pseudomonas denitrificans*, Koike and Hattori (1974) found that, under anaerobic conditions, the assimilation of substrate was 60% of that found under aerobic conditions. Growth efficiencies greater than 60% were found by Williams (1970) with natural populations using labile substrates; whereas Cole et al. (1989) found it to range from <19% to >60% and a 50% growth efficiency has frequently been used in calculations of carbon budgets (Cole et al., 1988; Lignell, 1990). In oligotrophic clear-water and humic lakes and in a eutrophic estuary, 20% to 26% growth efficiencies have been reported (Tranvik, 1988; Bjørnsen, 1986a), while ranges from 2% to 40% have been reported for the open oceans (Bjørnsen and Kuparinen, 1991; Griffith et al., 1990).

Much effort has been directed at studying the growth rate of bacteria in the natural environment (Bell et al., 1983; Christian et al., 1982; Fuhrman and Azam, 1982; Hagström, et al. 1979; Moriarty and Pollard, 1982; Newell and Fallon, 1982; Pedros-Alio and Brock, 1982). Their methods include the Frequency of Dividing Cells (FDC), ^{3}H-thymidine uptake, isotope dilution, etc. Each of these methods suffers from the fact that samples are perturbated, or that energy is added to the system. There are methodological errors in the determination of bacterial growth efficiency. For instance, Bjørnsen (1986b) calculated growth efficiency by measuring CO_2 and particulate organic carbon (POC) production. This involved stripping of CO_2 from the system by acidification and readjusting the pH. On the other hand, in Coveney and Wetzel (1992) and Kirchman et al. (1991), efficiency was based on uptake of radiotracers or assumed conversion factors for POC determinations. The biases in these latter methods were due either to the isotopic dilution factor or choice of the biovolume-to-carbon conversion factor (Bjørnsen, 1986b; Linley and Newell, 1984). If estimates are based on uptake of a single substrate, that single substrate may represent only a small fraction of the total natural DOC (Meyer et al., 1987).

In order to get away from methodological biases, Kroer (1993) examined bacterial growth efficiency on DOC by a direct method independent of radiotracers and biovolume-to-carbon conversion factors. On the average, he found 6% of the DOC was consumed and when the DOC was converted to bacterial biomass, efficiencies ranged from 26% to 61%. These efficiencies did not correlate with either the concentration of DOC or the temperature. Growth efficiencies did increase with added nitrogen (ammonium) suggesting that the differences in growth efficiencies may be attributed to differences in the concentration of the usable N in the substrate. The average incorporations of C from DCAA (dissolved combined amino acids), DFAA (dissolved free amino acids) and D-DNA (dissolved DNA) from a shallow estuarine environment were found to be 0.13, 0.77, and 0.10 and the corresponding incorporation of N was 0.10, 0.81, and 0.09 (Jørgensen et al., 1993). Again enrichment of the cultures with NH_4^+, glucose, methylamines, and high molecular-weight dissolved organic matter increased bacterial production.

Growth yields of bacteria is summarized by Wiebe (1984), but the data have been taken from pure culture studies. The questions that have been posed (Williams, 1973; Payne, 1970; and Forest, 1969) are: Are the values taken from pure culture studies close to the correct one for the microbes in the environment, and are natural populations in the environment growing at low substrate concentration more efficient than those of microbes reported in cultures?

5.6 Conclusions

Until we know what the energy sources, their quantity, and their bioavailability are for the various species of bacteria in any ecosystem, it will be difficult to fully understand the role of bacteria in microbial ecology. The success of heterotrophic bacteria depends on the bioavailability of fixed carbon needed to generate ATP and provide for the carbon skeletons. If all the organic matter that is recalcitrant to microbial decomposition were suddenly used by the microorganisms, there would be a terrific output of CO_2 tremendously accelerating the greenhouse effect. The ocean itself has approximately 2.10×10^{17} grams of dissolved organic matter, and 2.10×10^{16} grams of detritus. Fortunately, most of it resists microbial decomposition.

In many publications dealing with microbes in oligotrophic environments, the term "active biomass" or "active bacteria" are used to describe the indigenous population. The term "active" is a relative term. Just how active can the bacteria be in systems where the organic matter is scarce and most of it is refractory? Recognizing the recalcitrant nature and paucity of the organic matter, coupled with the low conversion efficiency, the growth of bacteria in oligotrophic environments must be slow. In a review, van Es and Meyer-Reil (1982) gathered published data on the active proportion of the bacterial population; the active portion fell in the range between 10 and 50% of the total bacterial community in the sea. Conversely, it means that 50% to 90% of the total bacterial community is inactive and this inactivity is probably due to the lack of the proper energy source to the metabolism of this inactive bacterial community.

Most bacteria in all oligotrophic environments probably fluctuate between being active and inactive, due mainly to the bioavailability of organic matter for energy. The time period between active and inactive may be very short to very long. Microbial multiplication in natural habitats must be assumed to alternate with dormancy (Jannasch, 1979). As mentioned previously, microorganisms in oligotrophic environments rely on syntrophy, and the microflora of higher forms in the starvation mode can also bring about the starvation state in their intestinal microflora (Conway et al., 1986). This inactive state is starvation-survival since there are no substrates for the microorganisms. The starvation-survival state is analagous to the "torpor" state mentioned by Smith and Baldwin (1982).

It is the opinion of this author that microbial activities as reported by past investigators are too great, except in those reports where microbes are interacting with the phytoplankton or other organisms where they encounter greater amounts of organic matter as energy. Consequently, the fluxes of carbon, nitrogen, etc., reported in the literature must be scaled downward. These measurements have not ruled out perturbation, surface (bottle or volume) effects, priming by substrates, or errors in the method of measurements. Recognizing these pitfalls, we must strive for more realistic measurements.

6

Starvation of Various Bacterial Groups

> It is a fact of microbial life that the majority of the bacteria can survive for considerable periods, in the absence of nutrients. The survival characteristics of starved bacteria depend upon the organisms and such factors as the growth phase from which they were taken, growth rate nutritional status, population density, biological history, and the nature of the environment.
> —Dawes, 1976

6.1 Introduction

Bacteria may have different mechanisms to resist starvation and their activity is related to the environment in which the cells are suspended. Therefore, research should focus on survival of the heterotrophic bacteria in oligotrophic environments, rather than their growth. The reason for this attention on the survival of microorganisms is that more ecosystems are oligotrophic and the microbes themselves create these oligotrophic environments, by growing and reproducing when sufficient energy is available. Unfortunately, the survival portion of starvation-survival has not been addressed by most researchers. This chapter deals mainly with starvation in various groups of bacteria because starvation is the first step in the starvation-survival process.

Most microorganisms on earth are not spore-forming bacteria. Spores are usually formed as nutrients are being depleted, following a period of vegetative growth in nutrient-rich conditions (Sussman and Halvorson, 1966). According to Gray and Williams (1971), spore formation is probably not due to antibiotics, but to the lack of energy created by microbial competition. The spore lies dormant until the substrates become available, and during this period, it can resist other unfavorable environmental conditions such as desiccation and toxic substances. Non-spore-forming bacteria can also resist desiccation. Bollen (1977) showed that sulfur oxidation can take place with dried soil that had been stored for 54 years. Miller and Simons (1962) dried cultures in a vacuum over calcium chloride at room temperature which were then stored in the refrigerator at 10°C and were rehydrated after 21 years. Only 13 of the 202 cultures representing 67 different species failed to grow. None of the heterotrophs examined by Miller and Simons (1962) were spore-forming bacteria and there were many pathogens among the organisms used. Spore formation evolved in some organisms and not in others. In nature, there must be a physiological state analogous to the spore for the non-spore-forming bacteria; the starvation state is the physiological state analogous to the spore.

In reviewing the literature concerning starvation effects on the organisms that follow, bear in mind that the great majority of these studies were conducted on the various organisms for a very short starvation period (hours mainly) and concentrated on the metabolism of endogenous reserves of the cell. A further point that must be made is that all the organisms were grown on rich laboratory media in order to have cells for experiments. As stated elsewhere, the natural environment is predominately oligotrophic and endogenous reserves will probably not be formed. These studies also concentrated on viability of the entire clone, but not on a few cells capable of surviving the starvation process. Starvation in microbes has not been a subject addressed too frequently in the literature and, thus, the data are incomplete.

6.2   Plant Pathogens

Distilled water has been found to be suitable for the storage of important plant pathogens such as *Corynebacterium insidiosum, Agrobacterium tumefaciens,* and a few species of pathogenic *Pseudomonas* (Crist et al., 1984; DeVay and Schnathorst, 1963; Kelman, 1954; Wakimoto et al., 1982). Seventy-five isolates of fluorescent *Pseudomonas* spp. and 30 isolates of *A. tumefaciens* were stored in distilled water at 10°C and tested after 24 and 20 years respectively (Iacobellis and DeVay, 1986). After this period of storage they found that 92% of the *Pseudomonas* isolates and 90% of the *A. tumefaciens* isolates were still alive. The method of preparation for storage was simple; a loop containing the bacteria from an agar surface was placed in a screw cap test tube containing 5 ml of distilled water, the cap secured and placed at 10°C. Table 6.1 shows the results dealing with the viability of *P. syringae* subsp. *syringae* after incubation for 24 years. The authors noted that, initially, there was a rapid decline in viability followed by a very slow decline in viability. After 24 years of incubation, the cells appeared smaller and had an electron dense cytoplasm. All the *Pseudomonas syringae* subsp. *syringae* retained their pathogenicity and their ability to produce syringomycin, indicating genetic stability of these traits during starvation.

Bacterial wilt of solanaceous plants caused by *Pseudomonas solanacearum* readily loses its virulence when subcultured on enriched media, less in poor media, but could survive in sterile tap water, distilled water, or phosphate buffer (pH 7.0) for more than 220 days (Kelman, 1954). *P. solanacearum,* when placed in

Table 6.1   Viability of cells of *P. syringae* subsp. *syringae* after storage in distilled sterile water at approximately 10°C.

Isolate	CFU/ml after:			
	Zero time[a]	6 mo[a]	20 mo[a]	24 yr
B3	1.0×10^8	1.0×10^5	8.1×10^5	6.6×10^2
B15	5.5×10^7	1.3×10^5	9.5×10^5	0
B59	1.9×10^8	5.4×10^6	1.3×10^6	1.4×10^6
B60	2.7×10^8	3.0×10^6	8.8×10^6	6.4×10^5
B61	1.1×10^8	1.4×10^6	1.3×10^6	3.9×10^6
B68	3.5×10^8	2.9×10^6	1.9×10^6	1.0×10^5
B70	3.6×10^8	8.4×10^6	8.6×10^7	2.9×10^5
B98	7.4×10^8	5.8×10^5	9.3×10^7	5.0×10^5
B99	1.0×10^9	1.2×10^6	4.1×10^8	8.0×10^5
B100	9.6×10^8	2.4×10^6	4.2×10^7	8.9×10^5
B101	6.7×10^8	1.3×10^6	3.4×10^7	5.3×10^5
B104	1.2×10^9	2.8×10^7	5.0×10^8	2.6×10^6
B105	8.7×10^8	2.3×10^7	7.6×10^7	3.5×10^6
B124	7.7×10^7	2.0×10^4	2.4×10^5	1.0×10^5

Source: Reprinted by permission from *Applied and Environmental Microbiology* 52:388–389, N. S. Iacobellis and J. E. DeVay, 1986; copyright 1986, American Society for Microbiology.
[a]Data from DeVay and Schnathorst (1963)

water, retains its viability and infectivity toward its host plants over an extended period of time (Kelman and Jensen, 1951). One hundred isolates of *P. syringae* were stored in water for at least 20 months with little or no decrease in viable cells (DeVay and Schnathorst, 1963). The same results were obtained with *A. tumefaciens* and *C. insidiosum*. DeVay and Schnathorst (1963) showed that no marked decrease in cell viability was evident from temperature fluctuations during storage within the limits for these experiments (22–32°C). *P. syringae*, both ice- and non-ice-nucleating strains, subjected to starvation by Buttner (1989), were found to be capable of surviving under starvation conditions. These organisms showed an increase in viable cell numbers when starved and after 5 weeks a decrease in viable cell counts were noted. Cell size decreased, changing from rods to coccobacilli.

Wakimoto et al. (1982) were also able to demonstrate that this organism did not lose its virulence when placed in pure water. These cells (different strains of *P. solanacearum*), when placed in a sterile distilled water (starvation menstruum) at 10^3 cells/ml, increased rapidly to a level of 10^6 cells/ml but when a higher concentration of cells was initially added to distilled water, the number of cells diminished. This was previously reported by Novitsky and Morita (1978b). Thus, there is an inverse correlation between the number of cells employed as the inoculum and the increase in cell numbers when cells are placed in a starvation menstruum. A redilution of the original cells placed in the starvation menstruum, five consecutive times after incubation, also produced an increase in the number of cells, indicating that the cells have propagated 10^{15} times (Wakimoto et al., 1982). This experiment can be seen in Figure 6.1. No decrease in cell size could be observed between bacterial cells that had multiplied 10^{15} times in pure water and those before multiplication. However, in all probability the density of the cells was probably different.

Wakimoto et al. (1982) proposed the following as possibilities for the increase in number of bacteria in pure water:

1. False multiplication due to gradual dispersal of the aggregated bacterial cells.
2. Repeated binary fission by using stock energy source such as poly-ß-hydroxybutyrate with getting nutrients from outside.
3. Use of a trace amount of nutrients accompanied with the inoculum or originating from dead cells which may be continuously supplied during the incubation period.
4. Utilization of a trace amount of available nutrients which may be contained even in pure water.
5. Fixation of nutrient from the air.

They then proceeded to eliminate the foregoing for the following reasons:

The results obtained by serial subculture (redilution) experiment in which the bacteria could multiply ca. 10^{15} times during subculture repeated 5 times in pure water suggest that the items 1, 2, and 3 could be discarded. The similarity in size of the bacterial cells as shown by electron microscopy before and after multiplication is also suggestive of the noninvolvement of item 2. Furthermore, the fact that the bacteria could multiply in pure water even if they were incubated under anaerobic conditions may be suggestive of the nonparticipation of item 5.

No further speculation is made by Wakimoto et al. (1982). Thus, only item 4 is left. However, the amount of energy in pure water would not account for the tremendous increase in the number of cells. Raindrops on plant pathogens may also be a way of increasing their numbers so that there is a better chance of one organism surviving.

P. solanacearum was found to grow readily in steam-sterilized deionized water (pH 7.0), river water (pH 7.23), and in sterilized Hatano soil (volcanic ash, pH 6.10) at 20°C or above and survives over 360 days (Tanaka and Noda, 1982). This organism was also found to survive at different pH values in sterilized McIlvain buffer solutions at 30°C as follows: pH 4.0, 1 day; pH 5, 6–100 days; pH 7, 350

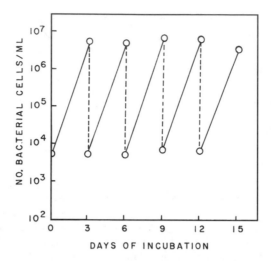

Figure 6.1   Multiplication of *P. solanacearum* isolate KU 7501-1 during serial subcultures in pure water. ○——○, multiplication; ○---○; 1000-fold dilution. Reprinted by permission from *Annual of the Phytopathological Society of Japan* 46:620–627, S. Wakimoto, I. Utatsu, N. Matsuo, and N. Hayashi, 1982; copyright 1982, Phytopathological Society of Japan.

days or more; pH 8, 110 days (Tanaka and Noda, 1982). The foregoing makes laboratory air the possible greatest source of outside energy.

A. tumefaciens can survive in many soils for a long time, providing that they remain at equable temperatures, do not dry out, and remain close to neutrality. In unsterilized soil at pH 7.2, cells declined from 1.3×10^7/g soil to 1.93×10^5 per gram of soil after 90 days and 2.55×10^4/g soil in 146 days (Dickey, 1961). Under laboratory conditions, *Xanthomonas citri* survived less than 9 days in unsterilized soil, but for more than 14 days in sterile soil (Lee, 1920). Survival for more than 3 days is rare under field conditions. However, in air-dried soil they can survive for 166 days, which suggests that soil microflora play an active role in removing these bacteria from soil (Fulton, 1920). Survival of *C. insidiosum* is much like *X. citri* (Nelson and Semeniuk, 1963). It is possible that they remain viable for much longer periods but in reduced numbers (Gray, 1976). However, many plant pathogens survive longer in plant debris on the soil surface but decline to low levels in 40–107 days when the debris is buried in the soil (Brinkkerhoff and Fink, 1964).

The longevity of plant pathogenic bacteria in soil is reviewed by Crosse (1968). Most plant pathogens are short-lived in natural soil, whereas in sterile soil they remain viable indefinitely. These data were obtained with organisms that were probably grown in laboratory media before they were placed in nonsterile soil, and, therefore, do not represent the indigenous plant pathogens.

6.3 Rumen Bacteria

Wachenheim and Hespell (1985) examined *Ruminococcus flavefaciens,* grown in batch or continuous cultures (cellobiose limited), starved under strictly anaerobic conditions. The average half-life at 39°C of this batch-grown organism was 10.9 hours under starved conditions, whereas continuously grown cells declined faster when the initial cell suspensions were grown at faster dilution rates. A lower pH (5.5) resulted in a poorer ability to survive. When starved, cells from batch- or continuous-grown cultures lost protein, carbohydrate, RNA, and DNA. The poor ability of other rumen bacteria (*Megasphaera elsdenii* [Mink and Hespell, 1981b], *Selenomonas ruminantium* [Mink and Hespell, 1981a]), and mixed rumen bacteria [Wachenheim and Hespell, 1985]) to survive starvation has also been noted. This poor starvation-survival characteristic has been attributed to the fact that the organisms do not need to survive extended periods of starvation because the rumen feeds once or more a day (Wachenheim and Hespell, 1985).

6.4 Fish Pathogens

There are many pathogens of fish but the survival of these pathogens in nature has not been fully investigated. Survival of the pathogens must occur if they continue to infect wild fish. The data pertaining to survival of fish pathogens, as they relate to starvation-survival, are presented below. The little data dealing with the ability of the various pathogens to survive in nature are given in *Bacterial Fish Pathogens: Diseases in Farmed and Wild Fish* by Austin and Austin (1987).

The survival potential of a specific bacterium in seawater may be defined as the time it takes before the number of colony-forming units (CFU) decreases to undetectable levels in bottle experiments where the bacterium in question has been introduced into autoclaved seawater (Austin and Austin, 1987; Rose et al., 1990b; McCarty, 1971; Enger, 1992a). According to Enger et al. (1992), such a definition is clearly situational, because the starved time for a given species will vary with the initial number of cells introduced into the starvation regime. Thus, Enger (1992a) proposed that a more consistent method to report and compare survival potentials obtained by such experiments was to report the data in terms of a reduction time when only 10% of the CFUs remained viable. In this manner, Enger (1992a) compares the reduction time of different fish pathogens (Table 6.2) from the marine and freshwater environments in bottle experiments. The data

Table 6.2 Comparison of calculated decimal reduction times of different fish pathogenic bacteria in bottle experiments with sterilized seawater or freshwater.

Species	Seawater		Freshwater		Reference
	Decimal Reduction Time (days)	Salinity	Decimal Reduction Time (days)	Salinity	
Renibacterium salmoninarum	7.6[a]	32‰	15	Fresh water	Evelyn 1987
Aeromonas salmonicida	1.4[b]	Sea water	3.4[b]	Fresh water	McCarthy 1977
Aeromonas salmonicida	3.4[b]	Brackish water	—	—	McCarthy 1977
Aeromonas salmonicida	1.4–2.2[b]	Sea water	—	—	Rose et al. 1990b
Yersinia ruckeri	4.2–7.2	35‰	93.6–106.4	0‰	Thorsen et al. 1991
Vibrio salmonicida	> 420[b]	31‰	0.3[b]	10‰	Hoff 1989
Vibrio anguillarum	> 420[b]	31‰	0.5[b]	0‰	Hoff 1989

Source: Reprinted by permission, O. Enger, 1992, Ph.D. thesis, University of Bergen.
[a]Value is calculated on the basis of number of cfu.
[b]Value is calculated on the basis of data in published graphs.

reveal two types of groups based on their adaptation to marine conditions. *Vibrio salmonicida* and *V. anguillarum* demonstrate a type of adaptation capable of survival for many years in bottles with autoclaved seawater without any external energy source (see Enger et al., 1987). This group shows a very low tolerance to low salinities. The second group includes *Renibacterium salmoninarum* and *Yersinia ruckeri* that die off quickly in seawater but exhibit long decimal reduction time in freshwater. *Aeromonas salmonicida* does not fit into either group and displays a short reduction time in seawater or freshwater. Enger et al. (1990) hypothesized that *V. salmonicida* and *V. anguillarum* are autochthonous in the marine environment, whereas *Yersina ruckeri* and *R. salmoninarum* are possibly allochthonous in the marine environment. The latter two are opportunistic fish pathogens in the marine environment (Hoff, 1989), whereas the former two could hypothetically be regarded as opportunistically fish-pathogenic freshwater bacteria.

The early studies, dealing with the survival of *A. salmonicida* surveyed by Austin and Austin (1987), show that the organisms survive from 12 hours to 63 days, depending on the experimental conditions. This organism, causing furunculosis, when placed in filtered sterilized river water (0.22 µm), after being washed twice, declined in platable numbers within 7 to 19 days (Allen-Austin et al., 1984). When 0.01% tryptone soya broth (w/v) was added on day 24, platable cells again appeared and the recovery was temperature dependent. At 22°C good recovery was noted in 6 hours with confluent growth being achieved at 24 hours. At lower temperatures, the response was slower. Thus it appears that *A. salmonicida* had entered an unculturable dormant state. This was later disputed by Rose et al. (1990a) who demonstrated that, upon addition of nutrient broth, growth was due to a small number of viable cells too few to be detected by the sampling protocol employed. Later, Rose et al. (1990b) noted that the spread of furunculosis was probably more dependent on the rate of shedding from infected fish than on the ability of the organism to survive. Nevertheless, they did find that survival was extended in sterile seawater to which tryptone soya broth was added. This was confirmed by Morgan et al. (1992), who also found that this organism survived as well in naturally eutrophic water. Flow cytometry, fluorescence light microscopy, scanning electron microscopy, and transmission electron microscopy indicated that nonviable cells were present in sterile lake water and that the morphological integrity found in viable cells was also maintained in nonculturable cells (Morgan et al., 1992). In sterile lakewater, both autoclaved and filtered, Morgan et al. (1993) found that *A. salmonicida* did enter a viable but nonculturable stage. Viability was determined by flow cytometry with the dye rhodamine 123, which is taken up and maintained within cells with a membrane potential.

Alteromonas denitrificans and two fish pathogens, *Vibrio salmonicida* and *Vibrio anguillarum*, exhibit considerable survival capacity in laboratory experiments where predators are removed (Enger et al., 1990). Predation pressure will decimate the number of a given nongrowing bacterium within three days. A subpopulation growing at a rate of 0.5 generations/day will be reduced by 60% within 72 hours. Studies with *Salmonella typhimurium* and *Klebsiella pneumoniae* show that these cells are effectively removed from a natural system if the total bacterial counts exceed 10^6 to 10^7 cells per ml (Mallory et al., 1983). Presumably, this rate of removal is an effect of nonselective predation.

Autoagglutinating cells of *A. salmonicida* possess a negative net charge while the nonautoagglutinating cells were positively charged (Sakai, 1986). In microcosms, the autoagglutinating cells were capable of attaching to positive charged particles added (humic acid and sand) when a small amount of tryptone was supplied, whereas attachment did not occur in nonautoagglutinating cells. The presence of sand was crucial to long-term survival of the organism. Sakai (1986) postulated that the autoagglutinating cells attached to positively charged particles and utilized the negatively charged humic acids and amino acids adsorbed on particles. In artificial freshwater, *A. salmonicida* was found to be able to survive longer than 300 days but a viable but nonculutrable state was not detected (Eberhart et al., 1992).

In seawater microcosms, *V. salmonicida* (cold water vibriosis) survived for more than 14 months, but the lethal salinity for cells harvested in the late-exponential growth phase was probably in the vicinity of 10‰ (Hoff, 1989). In a starvation menstruum, the organism was found to remain quite stable for 60 weeks (end point not determined) and the CFU counts exhibited a series of fluctuations with a periodicity of around 8 weeks (Enger et al., 1990). The CFU fluctuations in starving *V. salmonicida,* as well as in *A. denitrificans,* during the first weeks of starvation appear to be too rapid to be the result of periodic death and growth (cryptic growth).

V. anguillarum, which causes vibriosis, was found to have the capacity to survive for more than 50 months in a seawater microcosm (Hoff, 1989). A salinity of 5‰ was lethal when the organisms were harvested from the late-exponential growth phase, while a salinity of 9‰ proved to be lethal after the organisms had been starved at a salinity of 30‰ for 67 days. After 11 days, *A. anguillarum* lost its exoproteolytic activity.

Under anaerobic conditions, the fermentative strain of *Listonella* (= *Vibrio*) *anguillarium* responded to nutrient depletion with rapid reduction of their cell size and a decline of viable cell counts by three orders of magnitude (Heise and Reichardt, 1991). Nevertheless, after nearly 1400 hours of starvation, a little over 10^4 cells remained viable from an original cell population of over 10^7.

Three strains of *Y. ruckeri,* the etiological agent of enteric red mouth disease of salmon, survived starvation in unsupplemented water for at least 4 months (Thorsen et al., 1992). No detectable changes in CFU during the first 3 days of starvation at salinities of 0 to 20‰ were noted and only a small decrease in CFU during the following 4 months. However, at a salinity of 35‰ the survival potential was greatly reduced. Flow cytometric methods, combined with direct viable counts, showed that genomic replication was initiated before the onset of starvation was completed during the initial phase of starvation. It was also noted that starved cells could contain up to six genomes per cell. A viable but nonculturable state of *Y. ruckeri* was not found.

The causative agent for bacterial kidney disease in salmonid fish, *R. salmoninarum,* is excreted in feces of infected fish. After the last fish had died, no evidence of viable *R. salmoninarum* could be found in the sediment/feces of the fish tank for up to 21 days (Austin and Rayment, 1985). In normal, nonsterile river water the renibacterial cells declined to zero, whereas in filter-sterilized river water, no appreciable reduction in cells was noted but vanished on day 35. *R. salmoninarum* survived longer than 180 days in artificial freshwater, and best survival took place at salinity of 5 or 10‰ (Eberhart et al., 1992). Survival decreased with increased salinity and the organisms remained pathogenic following survival but the viable but nonculturable state was not detected.

6.5 Chemolithotrophs

Gibbs (1919) and Winogradsky (1949; research published originally in 1931) were able to obtain nitrifying bacteria from soil maintained in the laboratory for 7 years. Nitrifiers are still present in air-dried soil after 2 years (Wilson, 1928). When ammonium sulfate was added to dry soil samples held in bottles for more than 15 years, nitrification was demonstrated (Fraps and Sterges, 1932). Soil samples kept at 20° to 30°C for 2 years conserved their nitrifying power (Greaves and Jones, 1944). No decrease in the number of nitrifying bacteria could be detected in air-dried soil maintained at a normal temperature for 1 year (Hale and Halversen, 1940). According to Ziemiecka (1957), nitrifying bacteria will not survive desiccation when originally grown in liquid culture, but will survive for 2 years if water is added from time to time. Nevertheless, nitrifying bacteria can survive for long periods of time in anoxic environments even through they cannot grow under reducing conditions (Painter, 1970).

Garbosky and Giambiagi (1962) took 25 different soils from a depth of 0–40 cm from Argentine Patagonia and placed them in 5-l sterilized bottles. These were maintained for 5 years in the dark. Survival was shown in nearly all samples. Soil temperature and soil type were shown to be important factors in the survival of the nitrifiers as well as the joint action of lime and potassium in quantities not less than 200 mg% CaO and 30 mg% K_2O respectively. The content of organic matter, organic N, and pH of the soils employed did not influence the survival.

Johnstone and Jones (1988a), when studying *Nitrosomonas cryotolerans,* found that the levels of protein, RNA, and DNA remained essentially unchanged when the organism was starved for 10 weeks. An active electron transport system remained throughout the study. The energy charge was initially very low (0.68) at the onset of starvation and decreased with starvation until 5 weeks when it stabilized at 0.50 and remained constant throughout the remainder of the study. A nearly constant ATP level held for 2 weeks then decreased until 4 weeks when it stabilized to 0.85 fmole/ml at 10^6 cells/ml. Johnstone and Jones (1988b) studied the recovery of the organisms when its energy source was supplied. During recovery after 5 weeks of starvation, the organism immediately responded to the addition of ammonium-producing nitrite at a constant rate. During the recovery period, the electron transport system remained steady. They concluded that the organism was well adapted to the oligotrophic nature of the ocean where substrates are often limiting. *N. cryotolerans* adapts readily to nutrient deprivation by lowering its endogenous respiration and anabolic processes to undetectable levels. These starved cells are less resistant than freshly starved cells to light inhibition but both freshly starved and long-term-starved cells were unaffected by CO (Johnstone and Jones, 1988c). The nitrifying bacteria do not take up tritiated thymidine (Johnstone and Jones, 1989).

6.6   The Prosthecate Bacteria

The prosthecate bacteria have long been recognized to be capable of growth and survival in oligotrophic environments (Henrici and Johnson, 1935; Houwink, 1955). The prosthecate length reflects the nutrient status of the water mass (Boltjes, 1934) and this variation in length is most reflected in *Caulobacter* (Schmidt and Stanier, 1966). This correlation has been observed for *Hyphomicrobium* (Harder and Atwood, 1978), *Rhodomicrobium vannielii* (Whittenbury and Dow, 1977), and in an isolate of *Ancalomicrobium* (Whittenbury and Dow, 1977). Kuenen et al. (1977) noted that organisms showing a selective advantage at very low concentrations of limiting substrate appear to have a relatively high surface area to volume ratio. Dow and Lawrence (1980) considered the prosthecate bacteria as oligocarbophilic because they have been shown to utilize trace amounts of C_1 and C_2 hydrocarbons in the atmosphere. These trace amounts of hydrocarbons in the atmosphere are sufficient to repress prostheca expression in closed systems.

The prosthecate bacteria *Caulobacter* and *Hyphomicrobium* occur in nature where the nutrient concentration and total bacterial counts are low, whereas the other genera in this group, *Asticcacaulis, Prosthecobacter,* and *Ancalomicrobium,* are not so commonly encountered (Poindexter, 1979; Semenov and Staley, 1992). *Caulobacter* can also be found in all oligotrophic fresh- and seawater but also in stored distilled water. In natural environments, up to 45% of the organisms attached to surfaces can be *Caulobacter,* having appendages typically up to tens of micrometers in length. She states that the prothecal development constitutes a morphological adaptation to environments that provide only transient, low concentrations of nutrients. In excess of 1% organic matter, there is a reduction or inhibition of growth, typically accompanied by changes in morphology. The cells become involuted (swollen, elongated, branched, and irregular) and the appendages become reduced in length and/or number. Nevertheless, the growth at adequate nutrient concentration is not significantly dependent on the prosthecate development. Thus, a change in the surface/volume ratio is made by the formation of the prosthecate in order to more efficiently capture nutrients from a low-nutrient environment.

Yet, on the other hand, less than 0.1 mM of phosphate greatly enhances the stalk development in *Caulobacter crescentus* and *Asticcacaulis excentricus* (Schmidt and Stanier, 1966) and the cells appear similar to those seen in natural environments. They also possess a high quantity of poly-ß-hydroxybutyrate and a lower content of ribonucleic acid than cells grown in adequate phosphate medium. Subsequent studies (Schmidt and Samuelson, 1972) reported more than a 10-fold decrease in the size of the intracellular pool of ATP in phosphate-limited cells. Thus, starving conditions and/or limited phosphate brings about the formation of the prosthecate so that the surface/volume ratio favors the capture of limited substrates in the environment.

The concept of the "shut down" or "growth precursor" cell was proposed by Dow and Whittenbury (1980). The "shut down" concept is a survival mechanism especially suited for organisms whose habitat is aquatic environments where nutrients and growth conditions may be unfavorable for a long period of time. This "shut down" cell is in a different physiological state to that of the parent or sibling cell, the latter formed at division. These "shut down" cells have been observed in the swarmer cells of *R. vannielii, Hyphomicrobium,* and *Caulobacter.* The "shut down" or "growth precursor" cell is analogous to cells in metabolic arrest for heterotrophic bacteria, the latter term employed by the author.

The generalized cell cycle for the prosthecate bacteria is shown in Figure 6.2. The growth precursor cell is formed from the swarmer cell and is represented by A* in Figure 6.2. The protein synthesis, RNA synthesis, DNA synthesis, and DNA-dependent RNA polymerase of the swarmer cells of *R. vannielii, C. crescentus,* and *Hyphomicrobium* sp. are reviewed in detail by Dow et al. (1983). One of the main reasons why these organisms are easier to study in relation to their energy requirements is that the swarmer cell has a distinct morphological characteristic compared to the rest of the cell cycle. Hence, it can be isolated from the rest of the cells in the cell cycle.

6.7 Methanotrophic Bacteria

Methylosinus trichosporium OB3b and *Methylobacter albus* BG8 were subjected to starvation conditions under oxic and anoxic condition by Roslev and King

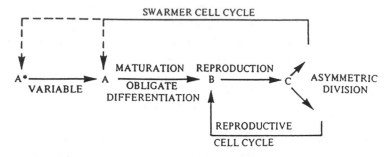

Figure 6.2 Generalized cell cycle. *A**, growth precursor cell—dispersal phase: *A*, swarmer cell committed to differentiation; *B*, prosthecate cell—reproductive unit; *C*, dividing cell. Reprinted by permission from *Symposium of the Society of General Microbiology* 34:187–247, C. S. Dow, R. Whittenbury, and N. G. Carr, 1983; copyright 1983, Society for General Microbiology.

(1994). They found that survival and recovery of these organisms were generally highest for cultures starved under anoxic conditions, determined by post-starvation measurements of methane oxidation, tetrazolium salt reduction, plate counts, and protein synthesis. Under anoxic conditions and carbon deprivation, *M. trichosporium* OB3b survived up to 6 weeks while maintaining a physiological state that permitted relatively rapid methane oxidation after substrate addition. Only a small fraction of cells (4 to 10%) survived more than 10 weeks, but required several days for recovery on plates and in liquid culture. Strain WP 12, a non-spore-forming methanotroph, displayed 36 to 118% of its initial methane oxidation capacity after 5 days of carbon deprivation. Very few changes were noted in biomass and morphology for methanotrophic bacteria under anoxic conditions, but under oxic conditions, morphological changes and loss of cellular protein occur. According to the authors, the data suggest that methanotrophs can survive starvation under anoxic conditions by using maintenance energy derived solely from an anaerobic endogenous metabolism, and this capacity partly explains a significant potential for methane oxidation in environments not continuously supporting aerobic methanotrophic growth.

6.8   Enterobacteriaceae

6.8.1   *Escherichia coli*

E. coli added to soil naturally or artificially have been observed to die out at a relatively rapid rate (Glathe et al., 1963; Rieves, 1958; Skinner and Murray, 1926;

Unger and Wagner, 1965; Waksman and Woodruff, 1940). When the organisms were added to a loam soil, cell numbers were maintained or slightly increased before die-off occurred; whereas in sandy soil, die-out occurred immediately (Glathe et al., 1963). The difference between delayed and immediate die-out of the *E. coli* in the two soils was attributed to the levels of available nutrients. If glucose was added to the soil, protection against die-out occurred, but it also caused the multiplication of the indigenous microflora (Klein and Casida, 1968). This added glucose also provides the indigenous microflora to produce toxic compounds against *E. coli* but the formation of toxic compounds was not the major controlling factor in the die-out of *E. coli*. The authors concluded that a major factor in this organism's die-out from natural soil is its inability to step down its metabolic rate to meet the low availability of usable organic carbon in the soil. Cell size reduction at low growth rate has been reported in *E. coli* (Donache et al., 1976).

There is a difference between *E. coli* and *Aeromonas* (isolated from river waters) in their ability to utilize glucose after being starved. The oxygen uptake of *E. coli* does not return to the level before starvation, whereas in *Aeromonas* it does (Wu and Klein, 1976). In lake water, *E. coli* is capable of growing when supplemented with synthetic sewage (Lim and Flint, 1989) at 15 or 37°C even in the presence of other bacteria. If competitors are present, there is a rapid decline of viable *E. coli*.

E. coli, when placed in sterile and nonsterile Cayuga Lake (New York) water, declined rapidly and reached undetectable levels in nonsterile water at 24 and 72 hours when counted on eosine–methylene blue and half-strength trypticase soy agar (Henis et al., 1989). It was found that, upon the addition of 10 µg amino acids/ml or 0.1 M phosphate, *E. coli* multiplied exponentially for more than 24 hours. Their data suggest that *E. coli* grown on rich media suffers a shock when introduced into lake water because of hypotonicity.

The survival of *E. coli* in nutrient-free seawater is dependent on the age of the cells and on some physicochemical conditions during their prior growth (Gauthier et al., 1989). For instance, cells from the late stationary phase had a better rate than those cells harvested from other phases in the log growth phase of the culture. Other factors were pH, temperature, and oxygen concentration. Previous studies by Gauthier et al. (1987) and Munro et al. (1987) showed better survival if *E. coli* had been grown previously in seawater medium.

García-Lara et al. (1993) subjected *E. coli* adapted and nonadapted to seawater to starvation in seawater. Based on plate counts, the nonadapted cells declined significantly while the population of adapted cells remained constant. In the latter situation, the drop in viable counts was not mirrored by the epifluorescent counts and [^{3}H]thymidine-labeled cells. Neither population lost its exoproteolytic activity or electron transport system. However, the cell-free exoproteolytic activity decreased in both populations. On the other hand, adapted cells showed higher electron transport activity and thymidine incorporation. They concluded that

E. coli subjected to seawater stress survives to a greater extent than the CFU counts demonstrate and these CFU counts overestimate the real death or lysis rate.

Certain members of the family Enterobacteriaceae were found to remain viable when placed in distilled water or lake water (Porter et al., 1991). These strains used were isolated from the lake water as well as from the stock culture collection. On the basis of their ability to survive in distilled water and lake water for 24 hours, these investigators divided the organisms into two groups. Those that persisted at a level of 40% or better were called survivors and belong to the genera *Klebsiella, Enterobacter,* and *Serratia.* The other group (*Acinetobacter, Aeromonas, Alcaligenes, Erwinia, Escherichia, Flavobacterium,* and *Pseudomonas*) were found to persist at a level of 10% or less after 24 hours. These nonsurvivors had very little ribonuclease. *K. pneumoniae* and *Enterobacter cloacae* were capable of surviving for more than 5 days. The first mechanism that comes into play during this stress involves the conversion of cellular macromolecules such as ribonucleic acid into essential cellular components by means of an active ribonuclease (Porter et al., 1991). Therefore, cells surviving longer periods probably maintained viability during RNA degradation which supplied energy for cell maintanence.

6.8.2 Salmonella

Cryptic growth was observed when a cell density equivalent to 1 to 10 mg (dry weight)/ml was used but was not observed when they were starved on the surface of membrane filters or in suspensions equivalent to 20 µg (dry weight)/ml (10^5 cells/ml) (Druilhet and Sobek, 1976). The survival time varied according to the buffer used as the starvation menstruum, giving half-life survival times of 77 to 140 hours. A greater number of cells survived when higher cell concentrations were used due to cryptic growth. Approximately 30% of the cells survived 8 days when starved on membrane filters.

Starving cells of *Salmonella typhimurium* were found to preferentially degrade supercoiled plasmid DNA while they retained copies of the plasmid in relaxed open circular form (Robert, 1986; in Kjelleberg and Hermansson, 1987). Some of the proteins synthesized in *E. coli* and *S. typhimurium* during carbon starvation are also observed in heat-shocked and other stress responses (Gottesman, 1984; Spector et al., 1986; Groat and Matin, 1986). In *S. typhimurium,* Spector et al. (1986) found only a few proteins produced during heat shock or anaerobiosis that could also be identified as starvation inducible. The effects of gentamicin and tetracycline (protein synthesis inhibitors) and rifampin (RNA inhibitor) were more pronounced with extended time of nutrient deprivation, while inhibitors of DNA and cell wall synthesis had no inhibitory effect on 24 hour starved cells (Stenström et al., 1989).

6.8.3 Enterobacter

According to Postgate and Hunter (1962), the death of 50 cells of *Aerobacter aerogenes* (and subsequent use of the cellular material) will only allow for the doubling of one survivor cell. Furthermore, cryptic growth only occurs at cell densities of 10^5 bacteria/ml or higher (Druilhet and Sobek, 1976). Nevertheless, the amount of energy available by dead cells for cryptic growth is not available in any of the experiments.

There is much literature concerning the end point of survival on enteric microorganisms, mainly because some are used as indicators of fecal pollution. However, this literature is not discussed because they deal with the end point of survival and not with the mechanism(s) of starvation process. Yet, when enterics are placed in natural waters, they face a starvation situation due to the lack of organic matter in natural waters.

6.9 Other Heterotrophic Bacteria

6.9.1 Acinetobacter

An *Acinetobacter* sp. isolated from 324 m depth was found to retain viability under starvation conditions in sterilized aquifer material, even when subjected to desiccation (Kieft et al., 1990). These authors state that "starvation-survival appears to be a requisite trait for existence in subsurface systems."

6.9.2 Alteromonas

A. denitrificans survived up to 7 years in unsupplemented seawater (Nissen, 1987). Incorporation of thymidine into DNA decreased rapidly upon starvation, indicating a cessation of DNA synthesis. Variations in the protein, carbon, and nitrogen content, in CFU, and in the uptake of arginine indicate that *A. denitrificans* does not rapidly adapt to starvation but undergoes a series of cellular alterations. One of the unusual observations was the presence of a "slimy" layer on the bacteria, which was present in all stages of starvation. This "slimy" layer may aid in binding to surfaces, which may be a mechanism to aid survival in some cases.

6.9.3 Arthrobacter

Arthrobacter sp. has a growth cycle characterized by sphere-rod-sphere morphology (Ensign and Wolfe, 1964) and can survive at least 4 weeks (Boylen and Ensign, 1970a). When *Arthrobacter globiformis, A. simplex, Arthrobacter crystallopoietes,* and *A. atrocyaneus* were grown in a chemostat, their morphology was related to growth rate (Luscombe and Gray, 1974). Rods grown in a chemostat, when placed in nutrient-free solutions, divided and produced cocci that survived for about 56 days, whereas cocci grown in the chemostat survived for about 70 days. The respiration rate of freshly harvested rods was higher than that of cocci but within 2 days both fell to low levels. During the initial stages of starvation, various *Arthrobacter* sp. increased their cellular DNA content (Scherer and Boylen, 1977). The exceptionally long period of survival by starving *A. crystallopoites* was attributed, in part, to the exceptional low basal rate of endogenous metabolism (0.03% cellular carbon per day) (Boylen and Ensign, 1970a,b). Low endogenous metabolism of starving *Nocardia carallina* appears to be the reason for the long half-life (Robertson and Batt, 1973).

Other details of this species, when starved, are found in Chapter 7.

6.9.4 Azotobacter

The introduction of *Azotobacter,* as a soil inoculant, has not been very successful and its survivability is enhanced by liming, the addition of carbohydrates, or both (Katznelson, 1940). The need to add carbohydrates to soil is an indication that the system is poor in energy for the *Azotobacter.*

Azotobacter agilis does not use ribonucleic acid, deoxyribonucleic acid, cellular carbohydrates, and cold trichloroacetic acid fractions of the cell as endogenous reserves but poly-ß-hydroxybutyric acid is used (Sobek et al., 1966). Viability in these experiments was dependent upon the initial levels of poly-ß-hydroxybutyrate and the growth substrate. Cellular protein served as a secondary endogenous reserve substrate for this organism.

Filterable forms of *Azotobacter* (~0.3 μm dia.) have been observed and the small size of these forms may represent an adaption to a low-nutrient environment or may function in the microbe's survival (Lopez and Vela, 1981). Filterable forms of Azotobacteriaceae include *Azotobacter vinelandii, Azotobacter chroococcum, Azomonas macrocytolgenes,* and *Beijerinckia indica.*

Vela (1974) showed that azotobacters retained viability in dry soils for more than 12 years while cysts survived in dried agar cultures for 10 years. Later, *Azotobacter chroococcum* and other members of the family Azotobacteriaceae were isolated from dry soils stored in glass containers in the laboratory and protected from contamination for periods of 22 to 24 years (Moreno et al., 1986). No azotobacter cysts were found in various soil samples (Vela and Wyss, 1964). Although the soil was dry, available substrates need to be in a water phase before they are bioavailable to the *Azotobacter* cell. Thus, the cells are deprived of substrates in dry soil. On the other hand, *Azotobacter* was detected in soil samples stored in the laboratory for more than 10 years (Vela, 1974). Bollen (1977) observed that *A. chroococcum* retained viability for 8–10 years in shrunken dried agar slants that had become so hard that removal from the tube carried adherent flakes of glass.

6.9.5 Bdellovibrio

Rittenberg (1979) considered *Bdellovibrio* to be in its free-living phase in oligotrophic environments and in its parasitic phase within the energy-rich host. The high rate of endogenous respiration correlated with the rapid death of starving cells of *Bdellovibrio bacteriovorus* and degradation of cellular RNA and protein (Hespell et al., 1974). Generally other bacteria usually degrade 70% to 90% of their cellular RNA before degrading cell protein, but starving bdellovibrio degrade protein and RNA simultaneously. Dunn et al. (1974), Engelking and Seidler (1974, 1975), and Hespell and Mertens (1978) demonstrated that starving *B. bacteriovorus* cells in phosphate buffer could utilize exogenously supplied ribonucleoside monophosphate and to a lesser extent, ribonucleosides as energy sources. These exogenous supplied substrates prolonged viability of starving cells, reduced degradation of cellular RNA and protein, reduced overall loss of cellular carbon, and formed $^{14}CO_2$ from UL-^{14}C-substrates. The survival of this genus appears to be in the bdelloplast (Sánchez-Amat and Torrella, 1990) because low nutrient concentrations favor an increased yield of stable bdelloplasts. In addition, these latter investigators found that infecting attack-phase bdellovibrio lost viability rapidly in basal salt solution; whereas, the bdelloplast viability remained higher, even after 180 days.

6.9.6 Brevibacterium

Brevibacterium linens cells isolated from Camembert cheese by Boyaval et al. (1985) were subjected to starvation. Viability dropped 50% in 24 hours, followed by a considerable decrease in mortality. Intracellular RNA was rapidly consumed during the first few days of starvation, although magnesium levels remained high. The quantity of DNA initially increased by 17% within 24 hours and then remained stable during the 30 days of the starvation experiments. After 30 days the protein dropped to 60% of the original value and the amino acid pool dropped to 60% of the original value after 30 days of starvation. During the starvation process, RNA, protein, and amino acids are degraded.

6.9.7 Bordetella

Porter et al. (1991) showed the ability of *Bordetella bronchiseptica* to increase in cell numbers in phosphate-buffered saline as well as in various natural freshwaters and seawater. This ability is shared with *Bordetella avium* but the human pathogenic species *Bordetella pertussis* and *Bordetella parapertussis* die out. When cells were placed in phosphate-buffered saline, in lake or pond water, within 48 to 72 hours at 37°C, stationary-phase cells developed. These organisms remained viable for at least 3 weeks. A 1000-fold increase was recorded in phosphate-buffered saline, lake, or pond waters. A five- and eightfold increase in cell number was observed in reagent-grade water and in seawater, respectively, from the same sizes of inocula. The authors attribute the difference in cell number increases in the various menstruums to the nutrient concentrations in the respective fluids. Yet no energy source is added to either reagent-grade water or to the phosphate-buffered saline.

6.9.8 Campylobacter

Six strains of *Campylobacter jejuni* and six strains of *Campylobacter coli,* isolated from cows and pigs, were tested for their survival in lake water (Korkhonen and Martikainen, 1991a). *C. jejuni* survived longer in culturable form than *C. coli* in untreated and membrane-filtered lake water at 4° and 20°C. It was postulated that the difference in survival time may be the reason why *C. jejuni* is generally isolated from surface water more frequently than *C. coli.* Both species survive better in filtered than unfiltered water, suggesting that predation and competition for nutrients affect the survival of both organisms in the aquatic environment. *C. coli* survived better than *C. jejuni* under all conditions tested (Korkhonen and Martikainen, 1991b).

6.9.9 Chromatium

The specific growth rate of *Chromatium vinosum* dropped as a function of the incident light intensity (its energy source), and at ca. 15–40 lux, although no growth could be detected (van Gemerden, 1980), the culture was nearly fully viable (> 90%). With decreasing light, the bacteriochlorophyll content increased significantly, representing a logical strategy to ensure light energy capture. Thus, photosynthetic bacteria also can face starvation from the lack of its energy source.

6.9.10 Cytophaga

When cell suspensions of *Cytophaga johnsonae* were washed and reincubated in a mineral salts solution containing no organics, the cells continued to divide (Rei-chardt and Morita, 1982). After 5 days in the mineral salts solution, a considerable increase in the CFUs were noted. Motility loss occurred with time, which occurred concomitantly with the loss of slime and convolutions from the outer membrane of the cells, especially at lower temperatures (12 to 15°C) after 5 days. Three different morphological types were noted, which were the result of temperature-dependent starvation. Short, viable rods (1–2 × 0.3–0.6 µm) were observed microscopically when starved at suboptimal temperatures, viable coccoids (0.8–1.2 μm) at optimal temperatures, and elongated thin moribund rods (0.1–0.2 × 10–14 μm) at temperatures above 28°C. Cell divisions or transformation to coccoids did not occur when the cells were starved at near maximal temperatures. The DNA increased as much as 65 and 32% at 15 and 25°C in exponentially grown rods starved in mineral salts. After 6 days at 25°C, an advanced state of coccoid formation resulted and both cellular DNA and protein dropped rapidly. As starvation continued, so did the formation of coccoid forms resulting in cellular DNA to protein ratios of 1.4 to 2.0 times higher than in rods harvested during the exponential growth at 15°C, indicating conservation of DNA. The maximal DNA to protein ratios were found in suspensions of starvation-induced moribund rods held at 29°C.

6.9.11   Cellulomonas

Trehalose and glycogen are stored products when *Cellulomonas* sp. is grown under conditions of energy and carbon excess (Schimz and Overhoff, 1987). When carbon starved, the half-lives of glycogen and trehalose are 1.6 hours and 34 hours, respectively. The half-life of the adenylate energy charge is 52 hours. These investigators attributed short-term survival to glycogen and long-term survival to trehalose because they provide energy to stabilize the adenylate energy charge. During the starvation process, the relative protein concentration increases from 0.405 to 0.465 mg/mg cell (dry weight). There is an initial increase in ATP, which then decreases. The initial increase may be due to the utilization of reserved energy sources.

6.9.12   Desulfovibrio

Over a period of 800 hours, *Desulfovibrio vulgaris* showed no decline in cell numbers when starved concomitant with a 5% decrease in cell volume (Heise and Richardt, 1991). This is accompanied by a sharp decrease in cellular proteins per cell volume. When one considers the rather specific substrate requirements of sulfate-reducing bacteria, these obligate anaerobes are likely to incur frequent periods of nutrient depletion in their environment (Heise and Richardt, 1991).

6.9.13   Legionella

The decline in culturability of *Legionella pneumophila* does not necessarily correspond to cell death (Hussong et al., 1987). This organism, as measured by plate count (CFU/ml), declined to a greater extent than cell lysis, assessed by thymidine, DFA, and AODC counts, suggesting that the organism survives in aquatic habitats to a greater extent than revealed through culturable counts (Paszko-Kolva et al., 1991). If grazing does not occur, the organism is capable of surviving long periods in organic matter deficient drinking water, river water, and seawater.

6.9.14 Leptospira

Under starvation conditions, *Leptospira biflexa* serovar *patoc* I, did not undergo a major reduction in size (Kefford et al., 1986). However, greater adhesiveness did result from starvation and this adhesion may provide a strategy for the survival of leptospires in oligotrophic habitats, mainly because these bacteria can scavenge fatty acids adsorbed at surfaces. When provided with an energy substrate, the adhesive ability of the organism decreases and cell size and motility increases.

6.9.15 Micrococcus

Fifty percent of the cells of the population of *Micrococcus luteus* in a 75-day culture were in a dormant state, because they could not initially form colonies (Kaprelyants and Kell, 1993). These cells were able to convert to CFUs after resuscitation. For resuscitation, 8 to 10 ml of culture was centrifuged and washed twice with sterile growth medium lacking lactate. The cells were then inoculated into fresh medium contain 0.5% (wt/vol) L-lactate and 15 µg penicillin G (Sigma)/ ml. After incubation for 10 hours, the cells were washed three times with fresh medium (no lactate) and then inoculated into fresh medium, These investigators also found that in an extended stationary phase following batch culture, the decrease in the number of CFUs was not logarithmic and the percentage of CFUs reached an apparent steady value (0.05%) after 60 to 80 days of storage. After 20 days of storage, the cell population became much more homogeneous in size and DNA content than the population at the beginning of the stationary phase, as revealed by flow cytometry. Also, after 20 days of storage, no metabolic activity was detected. One noteworthy observation was that approximately 10 hours of storage of stationary phase cultures of *M. luteus* was critical for cell stability. Storage of cells for 2 hours followed by washing in phosphate buffer led to massive lysis during the next 15 days, while cells that had been stored for 20 hours or more were stable under the same conditions for at least 2 months. They surmised that dormancy may be far more common than death in starving cultures.

6.9.16 Peptococcus

Excessive loss of RNA was noted in the starvation of *Peptococcus prévotti* (Montague and Dawes, 1974). This loss produced a large flux of adenine nucleotides.

The adenylate energy charge remained essentially constant during the period of rapid RNA decline but fell when the residual RNA was being degraded more slowly. When the adenylate energy charge fell below 0.4 to 0.5 from starvation, the loss of viability became more rapid.

6.9.17 Pseudomonas (other than Plant Pathogens)

When *Pseudomonas aeruginosa* is introduced into sterile nutrient-free seawater, the culturable cells decreased progressively over a period of time and developed cells capable of passing through a 0.45 μm pore membrane (Bakhrouf et al., 1989). These filterable cells, produced as a result of starvation, were capable of producing normal size cells when grown in regular media. Cell size reduction has also been noted at low growth rates of the organism (Robinson et al., 1984). Cell size reduction is a good indicator of the starvation process.

Fan and Rodwell (1975) labeled the protein in log-phase *Pseudomonas putida* cells with L-[4,5-^{3}H] leucine. During log growth of these cells at 30°C, intracellular protein was degraded at a rate of 1 to 2% per hour. However, when the cells were starved, the degradation rate increased to 7 to 9% per hour. A lower rate of protein degradation was not observed with starved cells in the presence of rifampin, chloramphenicol, and tosyllysine chloromethyketone. If the cells were incubated in the presence of a nutrient that cannot be used immediately for growth, the starvation-induced intracellular protein degradation was lowered. Cell-free extracts of growing and starving cells degraded protein at the same rate. In starved cells, the transport and dehydrogenase activities remained stable, whereas there was a loss of 20 to 30% in the oxidase and decarboxylase activities, and a loss of 5 to 8% of the hydrolase activity. The longest period of these studies was 240 hours. They stress that they cannot attribute loss of enzymatic activity in vivo to protein degradation.

Various strains of *Pseudomonas fluorescens,* when starved at an initial cellular concentration of 10^5 cells/ml, showed an initial increase in numbers; but when starved at a concentration of 10^8/ml, the cell numbers decreased gradually, so that at the end of a 12-week incubation period, only about 10^5 cells/ml had survived (Jørgensen and Tiedje, 1993). These organisms also had the capacity for long-term survival without O_2 or NO_3^-. When these cells were starved in phosphate buffer, more of the C of [^{14}C] glucose added was partitioned to fatty acid production than to $^{14}CO_2$ production; but when the cells were suspended and starved in pore water, the $^{14}CO_2$ production and ^{14}C-fatty acid formation were similar. They appear to be capable of low levels of fermentation under extreme electron acceptor starvation, which provides for their maintenance.

Pseudomonas sp. no. 200, which has the ability to corrode mild steel, is inhibited when the available energy (substrates) are low (Obuekwe et al., 1987).

6.9.18 Rhizobium

In the old literature (see Chapter 1), *Rhizobium* is known to remain viable in soil. Various investigators have shown that some plant pathogens and *Rhizobium* sp. can survive in water, buffers, and buffers with added ions for many years. Nevertheless, these systems are devoid of any organic matter as a source of energy for the various microorganisms.

Three strains of *Rhizobium japonicum* and two slow-growing cowpea-type of *Rhizobium* remained viable and able to produce nodules on their respective host after being stored in purified water at ambient temperature for a period of 1 year or longer (Crist et al., 1984). The cells increase in numbers when placed in water (Fig. 6.3). The capacity to nodulate, after 1 year of starvation-survival, remains high in three *R. japonicum* strains. Unlike the vibrios, *Rhizobium* does not decrease in size during starvation in water as evidenced by electron micrograph on both water-stored and freshly cultivated organisms. Like Wakimoto et al. (1982), Crist et al. (1984) were able to show an increase in starved cells when rediluted in distilled water (Fig. 6.4).

R. meliloti starved in water or buffer suspension (pH 6.5) with and without added salts, exhibited constant or slightly increased cell numbers during storage over a 2-month period (Bioardi et al., 1988). Phosphate buffer (pH 6.5) with added $CaCl_2$, $MgSO_4$, and $FeCl_3$ gave the best survival, cell numbers remained rather constant through the 2-months period (Fig. 6.5). The presence of Ca^{2+}, Mg^{2+}, and Fe^{3+} enhanced cell viability (Postgate and Hunter, 1962) at pH 5.5 better than at pH 6.5. Thus, we see that the ionic environment has much to do with the starvation process.

6.9.19 Rhodospirillum

Illumination prolonged viability and suppressed the net degradation of cellular material of phototrophically grown *Rhodospirillum rubrum* cells but had no effect on chemotrophically grown cells (Breznak et al., 1978). This is to be expected because chemotrophically grown cells had no bacteriochlorophyll. The half-life survival times in the light were 17 and 14.5 days for the carbohydrate-rich anaerobically or aerobically grown cells respectively. These values were 3 days (aerobically) and 0.5 days (anaerobically) in the dark. On the other hand, the chemoorganotrophically grown cells had half-life survivals of 3 and 4 days during starvation aerobically in the light and dark respectively, and 0.8 days during starvation anaerobically in the light or dark. During starvation, cellular carbohydrate was extensively degraded, but slowest in phototrophically grown cells. Phototrophically grown cells containing poly-ß-hydroxybutyrate as a carbon reserve

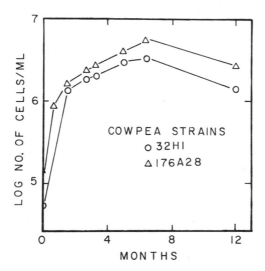

Figure 6.3   Increase in cell number and retention of viability during water storage of cowpea-type *Rhizobium* sp. strains 32H1 and 176A28. The water was purified by reverse osmosis and deionization. The samples were stored at ambient temperature (23 to 25°C) in the dark. At times indicated, samples were removed from the stored samples, and the viable cell counts were determined. Reprinted by permission from *Applied and Environmental Microbiology* 47:895–909, D. K. Crist, R. E. Wyza, K. K. Mills, W. D. Bauer, and W. R. Evans, 1984; copyright 1984, American Society for Microbiology.

were less able to survive starvation anaerobically in the light than carbohydrate-rich cells under similar conditions, but the authors could not attribute the cause of death to degradation of any specific cell component.

6.9.20   Sarcina

A washed suspension of aerobically starved *S. lutea* in phosphate buffer at 37°C decreases its endogenous oxygen consumption to a very low level after 10 hours, even though many of the cells survive for 40 hours (Burleigh and Dawes, 1967). If starvation is prolonged, approximately 1.5% of the initial viable population dies per hour. Oxidation of the intracellular free amino acids accounts for most of the observed endogenous oxygen uptake, but RNA is also utilized. Survival of *S. lutea* is correlated with the ability of aerobically starved cells to oxidize exogenous L-glutamate and glucose. Those cells rich in polysaccharides survive less well than those cells deficient in the polymer, indicating that those cells rich in polysaccharides are less fit physiologically to withstand starvation.

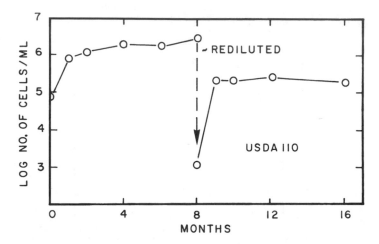

Figure 6.4 Effect of redilution on growth and viability of *R. japonicum* USDA 110, previously stored for 8 months in water. A 4-day-old culture of USDA 110 was first diluted to a density of 10^5 bacteria/ml in purified water and stored at an ambient temperature for 8 months. At that time a sample of the stored bacteria was diluted into fresh water to a density of 10^3 bacteria/ml and stored again for an additional 8 months. Storage conditions and viable determinations are described in Figure 6.3. Reprinted by permission from *Applied and Environmental Microbiology* 47:895–909, D. K. Crist, R. E. Wyza, K. K. Mills, W. D. Bauer, and W. R. Evans, 1984; copyright 1984, American Society for Microbiology.

6.9.21 Streptococcus

Streptococcus lactis, when starved in phosphate buffer, rapidly releases its intracellular amino pool into the external menstruum but the lactic dehydrogenase and DNA do not appear in the menstruum until the organisms begin to lose viability (Thomas et al., 1969). Thus, we are seeing permeability changes in starved cells. Upon starvation, in the absence of Mg^{2+}, electron micrographs show that ribosome particles rapidly disappear and the death rate of starved *S. lactis* in phosphate buffer depends on the presence of Mg^{2+}. Agitation and aeration tend to decrease survival (Thomas and Batt, 1968). When ^{14}C-valine is added to a suspension of *S. lactis,* protein synthesis is barely detectable but synthesis of RNA depends on exogenous glucose (Thomas and Batt, 1969b). In the same organism, soluble protein and UV-absorbing material from RNA degradation are released into the suspending buffer when the cells are starved (Thomas and Batt, 1969b).

The presence of intracellular glycogen favors survival of *Streptococcus mitis* (van Houte and Jansen, 1970).

The ATP pool of *Streptococcus cremoris* is rapidly depleted when grown in a chemostat when lactose becomes limiting (Otto et al., 1985). The adenulate en-

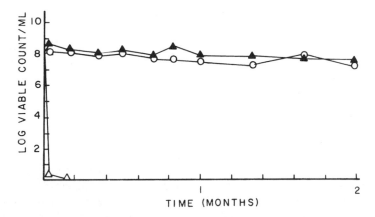

Figure 6.5 Survival of *Rhizobium meliloti* B323 in different buffers and stored on a rotatory shaker at 200 rpm. ○, phosphate buffer at pH 6.5; △, phosphate buffer at pH 5.5; ▲, phosphate buffer pH 5.5 + CaCl₂, MgSO₄ and FeCl₃. Reprinted by permission from *Journal of Applied Bacteriology* 65:189–193, J. L. Boiardi, N. Moreni, M. L. Galar, 1988; copyright 1988, Blackwell Scientific.

ergy charge (AEC) of growing cells is 0.8 but drops to 0.2 when the cells enter starvation. In spite of low phosphoenolpyruvate in 24-hour starved cells, they retain the ability to take up lactose, to synthesize ATP, and quickly generate an electrochemical proton gradient.

6.9.22 Sphaerotilus

Sphaerotilus discophorus, with large intracellular stores of poly-ß-hydroxybutyrate, survived longer in phosphate buffer than cells with little or no poly-ß-hydroxybutyrate (Stokes and Parson, 1968). The addition of Mg^{2+} or increased O_2 supply, both of which stimulated rapid oxidation of the poly-ß-hydroxybutyrate, led to a more rapid death of the cells. These authors also found that exponential phase cells were more resistant to death than stationary phase cells. Their explanation was that poly-ß-hydroxybutyrate provided energy and carbon for cell maintenance during starvation.

6.9.23 Staphylococcus

When *Staphylococcus aureus* is starved under anaerobic conditions, it loses nearly all the amino acids in the intracellular pool in 7.75 hours, leaving only glutamate, aspartate, and lysine. In 24 hours, all the amino acid in the amino acid pool is lost (Horan et al., 1981). Viability of starved *Staphylococcus epidermidis*, under anaerobic conditions, declined to $> 10\%$ within 12 hours and during this period, RNA was slightly degraded (Horan et al., 1981). The main substrates for endogenous metabolism were protein and the intracellular amino acid pool. The adenylate energy charge and the ability to transport serine declined markedly within the first 6 hours of starvation. The membrane potential also declined during this period but no correlation between the magnitude of the membrane potential and viability was noted.

6.9.24 Vibrio

Growth and extended periods of survival of *V. cholerae* were observed in sterile sediments, sterile water, and nonsterile water but not in nonsterile sediments (Hood and Ness, 1982). *V. cholerae,* once in sediment, has been shown to survive for an extended period of time. Different affinities are demonstrated by nutrient-deprived and nutrient-enriched *V. cholerae* cells toward $CaCO_3$, chitin, and silicate surfaces (MacDonell et al., 1984). Initially, the cells have a greater ability to associate with $CaCO_3$, chitin, and silicate surfaces, but after 72 hours in nutrient-free seawater this ability decreases sharply.

The survival time of *V. cholerae* was shown to be enhanced in lake water by the presence of both the soluble forms of chitin, glycol chitin, and the insoluble particulate form of chitin (Amako et al., 1987). Some amino acids showed the same protective effect. Although Amako et al. (1987) stressed the survival in terms of low temperature ($0°C$), the studies in actuality are starvation-survival studies, mainly because the cells were placed in physiological saline. All strains of *V. cholerae* gave the same results.

6.9.25 Zymomonas

Neither *Zymomonas anaerobia* nor *Zymomonas mobilis* degraded endogenous carbohydrate, DNA, and protein upon prolonged starvation (Dawes and Large, 1970). However, both organisms had a high RNA (22% w/w), which was degraded upon starvation. The RNA decreased linearly to 5% of the dry weight in 125 hours in *Z. anaerobia;* whereas in *Z. mobilis,* half the RNA was degraded in the first 24 hours, after which it declined much more slowly. After a prolonged starvation period of 7 days, the viable population had fallen to 3%. This 3% possessed the ability to produce ATP from glucose but, with longer starvation periods, this ability was impaired. Thus, starvation may lead to death of the cells due to its inability to produce ATP.

6.9.26 Marine Spirochaete

When a marine spirochaete, which did not require amino acid for growth, was starved in the presence of L-valine, L-isoleucine, and L-leucine, survival was prolonged (Harwood and Canale-Parola, 1981). Amino acid fermentation took place, which permitted the spirochaete to survive longer because the energy released was used (55-hour study). The dissimilation of the branched-chain amino acids can provide this bacterium with maintenance energy for cell functions not related to growth. The authors suggested that in its natural environment, spirochaete, designated at MA-2, may catabolize branched-chain amino acids as a strategy for survival when growth substrates are not available.

6.10   Conclusions

Many bacterial species can survive in an energy-deprived environment. However many of the studies cited address only the starvation aspect and do not deal with the survival question. As a result most of the studies cover periods of hours. I am quite certain that if others were investigated, many more species would undergo starvation-survival. The question that remains to be investigated is how long can any of the above organisms survive without the presence of energy. Each species differs in the physiological mechanism(s) in response to the stress of the lack of energy. The data on starvation and starvation-survival on the various species are fragmentary and no definitive conclusions can be made on most species. The fragmentary nature of this chapter only indicates that much research needs to be undertaken.

7

Physiology of Starvation-Survival

Organisms only have a limited capacity to control their environment and they almost invariably respond to environmental change by changing themselves.

—Harder and Dijkhuizen, 1983

7.1 Introduction

According to Sykes (1963), survival is akin to resistance. He stated that resistance implies an active state of opposition to an attack on the cell by an outside agent, whereas survival carries the idea of a more passive state of resignation induced by the natural turn of events, and the foregoing differentiation is not at all precise. He further states that survival can only take place if the cell either retains its metabolic function intact or is able to bypass or bridge in some way any gaps created in this cycle and no time factor is involved. Death by senescence simply means that metabolic function(s) or the bypass system has been destroyed by an oxidation or other process (Sykes, 1963). Within any species of organism, there are differences between individuals in the ability to withstand chemical or biological conditions. Also, within a given population, not all cells must survive. As long as one cell of any species survives in any ecosystem, survival of the species is met.

In the investigation of starvation-survival of microorganisms, we have been studying the system in terms of the response of bacteria to oligotrophic or starvation conditions. Since most of Earth is in an oligotrophic state and most bacteria are in a starvation-survival mode, we should be studying the response of the natural population as well as starved pure cultures to eutrophic conditions. From an ecological point of view, the response of the indigenous microbial population to eutrophic conditions is most important for the perpetuation of the species. Thus, in the life cycle of any species, there will be a period of growth and nongrowth, starvation and recovery. Most likely, bacteria undergo a relatively short-term life of growth, starvation, and recovery (Morita, 1982, 1985) but the longevity of starved bacteria is also intriguing. This cycle allows for the ability of the species to compete, persist, colonize other habitats, and proliferate (Kjelleberg et al., 1993). For a microorganism to survive the stress of total starvation, it must maintain all its critical cellular processes and endogenous metabolism to ensure the maintenance of all essential processes (Dawes, 1984). However, I maintain that endogenous metabolism is not necessary to maintain the essential processes of the cell, mainly because all processes are "shut down" in the metabolic arrest stages, a situation analogous to a bacterial spore.

7.2 Degrees of Starvation-Survival

There are degrees of starvation-survival, which range from the first moment the cell encounters an energy deprivation to long periods of energy deprivation leading to cell death of most of the bacteria. Thus, the period of energy deprivation may be very short or very long, with a few cells of the species having the ability to survive long periods of time until the environment becomes favorable for their growth and multiplication. It only takes one cell per environment for the survival of the species (Morita, 1980c). In long term studies of Ant-300, there remains a certain minimum percentage (0.3%) of cells that remain viable (Moyer and Morita, 1989a). Because most research has been done with cells grown in the laboratory and then subjected to nutrient deprivation, most studies have not dealt with those cells in ecosystems that are growing slowly, due to the insufficient supply of energy in the ecosystem, and then subjected to starvation conditions. It is also possible the organisms grow in spurts when energy becomes available.

Unfortunately, the vast majority of investigations dealing with starvation-survival address the immediate response of bacteria to starvation conditions. In these situations, the starvation periods addressed are in terms of hours, days, or a few weeks, not in terms of months and years. It should also be pointed out that carbon (not nitrogen or phosphorus) starvation leads to the development of starvation- and stress-resistant cells (Kjelleberg, 1993).

7.3 Early Studies on Starvation-Survival

Aerobacter aerogenes, taken from the log phase and washed twice in phosphate buffer and placed in phosphate buffer that had no nutrients, were observed for survival characteristics at the same temperature at which they were grown (Harrison, 1960). The population with the higher cell density (e.g., 10^9 cells/ml) showed the best survival over a period of 6 days, whereas the population with a cell density of 10^5 cell/ml died within 2 hours. This is termed the *population effect.* In cell densities of 10^7 cells/ml, the death curve showed fluctuations. Maximum stationary phase cells survive better than log phase cells. At high cell densities, some of the cells expire and produce cellular effluent. This cellular effluent provides the remaining cells a source of energy, whereas in low cell densities, the amount of cellular effluent is not sufficient to provide the remaining cells a nutrient source. In a nonnutrient environment, death by starvation is characterized by a population effect that is not clearly understood (Harrison, 1960). Postgate (1976) and Strange (1967) both point out that an apparently simple stress may well be complex and that removal of the population from a stress and its transfer to a recovery environment may well create new stresses. However, in nature, it is rare that cell densities of any one species of 10^9 cells/ml are encountered, let alone a cell density of any one species of 10^5 cells/ml, especially cells in the log phase of growth. Postgate and Hunter (1962) confirmed the fact that denser populations survived longer than sparse populations. This population effect was noted in the storage of *Serratia marcescens* in water at room temperature by Bateman and White (1963). Harrison (1960) also reported that a sparse population actually multiplied in a buffer in which a dense population died. Death in dense population may be due to anoxia (Harrison, 1960), which was later confirmed by Postgate and Hunter (1962).

Postgate and Calcott (1985) stress that an important distinction must be made between clonal survival, survival of a population as a whole, and the survival of the individual within that population. The survival characteristics of a population of bacteria depend on the growth phase, growth rate, nutritional status and biological history of the population, composition of medium for previous growth, pH, temperature, starvation fluid, and population density (Harrison and Lawrence, 1963; Postgate and Hunter, 1962, 1963; Ribbons and Dawes, 1963; Strange et al., 1961; Thomas and Batt, 1968; Gauthier et al., 1987, 1989; Munro et al., 1987).

Further investigation is needed to determine whether cells inherit longevity. In *A. aerogenes,* Postgate and Hunter (1962) reported that longevity was not inherited. On the other hand, Harrison and Lawrence (1963) demonstrated that starvation-resistant mutants could be obtained from batch cultures and that they were different from the wild type. Harrison and Lawrence (1963) isolated a starvation-resistant mutant of *A. aerogenes* indicating that the ability to withstand starvation may be genetically determined along with its growth rates under more favorable conditions. The wild type and the "starvation-resistant" mutant were isolated and again cultivated, harvested during the log growth phase, and starved. A slow decline in the viability of the "starvation-resistant" mutant was observed; whereas, the wild type demonstrated a rapid decrease in viability. According to Dawes (1976), there are vast differences in resistance to starvation exhibited by different bacteria.

In most of the studies reported in this chapter, one should realize that cells are first grown in rich media under ideal conditions before the starvation process is initiated. Also one should recognize that the starvation period is short compared to what might happen in nature. Hence, many of the studies reported in this book deal with short time periods (usually hours or days). What would the data points look like if the cells were starved for a year, 10 years, etc.? Furthermore, cells are generally taken from the log phase of growth and high cellular densities are used. Neither of these reflect the cells' condition in the environment.

The half-life (ST_{50}) or 50% survival time under nutrient deprivation for most bacteria studied ranges from a few hours to a few days (Table 7.1). Boylen and Ensign (1970a) demonstrated that the half-life survival time for *Arthrobacter crystallopoietes* approaches 100 days, but the other bacteria mentioned in their study had a half-life under 5 days. The half-life survival time has little value in considering the survival of the species, because the gene pool of a species can be conserved by the survival of only one organism (Novitsky and Morita, 1977). The half-life of the survival time due to nutrient deprivation appears to be species-specific.

Sinclair and Alexander (1984) suggested that "starvation-susceptible bacteria will not persist in environments that are nutrient poor or in which they fail to compete for organic nutrients and that starvation resistance is a necessary but not sufficient condition for persistence in environments that are nutrient poor or that support intense interspecific competition."

7.4 Adaptation to Starvation

There are a variety of responses exhibited by microorganisms when challenged with nutrient deprivation (Morita, 1985). This has been shown by various species within the same genera or among the same species, depending upon how starvation was induced (Kramer and Singleton, 1992). More than one generation may be required for complete adaptation to low-nutrient environments (Law and Button, 1977; Höfle, 1982, 1983). Many microbial habitats and complex communities exist and, consequently, microorganisms inhabiting such ecosystems are under severe selective pressure to evolve means of coping with the changing nutrient and physiological conditions (Dow and Lawrence, 1980). As a result, bacteria have come to terms with nutrient variability and the severity of an oligotrophic environment by the following means: (1) movement by flagella and gas vacuole production (chemotactic and phototactic responses), (2) adhesion to surfaces, (3) colonization of favorable ecological niches, (4) specialized dispersal phases, and (5) formation of prosthecae (Dow and Lawrence, 1980).

Faster growing *Streptococcus cremoris* have been reported to die off much faster than slower growing ones (Poolman et al., 1987). Another adaptation, seen in an organism under chronic starvation, is the ability to take up nutrients quickly when they are locally available, even though they may not be immediately catab-

Table 7.1   Comparison of 50% survival times[a] of starved bacteria

Organism	50% Survival Time, h	Reference
Sphaerotilus discophorus	12	Stokes and Parson 1968
Starvation-susceptible soil isolates	12	Chen and Alexander 1972
Bacillus M153[b]	16	This study
Streptococcus mitis	22	van Houte and Jansen 1970
S. lactis	30	Thomas and Batt 1968
Escherichia coli	36	Dawes and Ribbons 1965
Aerobacter acrogenes	45	Strange et al. 1963
Azotobacter agilis	50	Sobeck et al. 1966
Sarcina lutea	65	Burleigh and Dawes 1967
Pseudomonas aeruginosa	96	Mackelvie et al. 1968
Starvation-resistant soil isolate R5	100	Chen and Alexander 1972
Arthrobacter M51[b]	160	This study
Nocardia corallina	480	Robertson and Batt 1973
Pseudomonas M216[b]	1030	This study
Arthrobacter globiformis	1370	Luscombe and Gray 1974
Arthrobacter sp.	1680	Zevenhuizen 1966
Arthrobacter crystallopoietes	2400	Boylen and Ensign 1970a

Source: Reprinted by permission from *Canadian Journal of Microbiology* 24:1460–1467, L. M. Nelson and D. Parkinson, 1978; copyright 1978, Research Council of Canada.

[a]Time required for half of the population to die; the longest 50% survival time at 30–37°C reported in each study is presented.

[b]Fifty percent survival times at 30°C are estimated from 50% survival times obtained at 15°C by means of a correction factor.

olized (Koch, 1979). Cells have the ability to take up enough glucose in 2 or 3 minutes to last for an average of 17 minutes, which may be a significant advantage in an environment where fluctuation in nutrient levels occur minute by minute (Koch, 1979).

Other adaptions to starvation include the formation of ultramicrocells, induction of low endogenous respiration, change in chemotactic abilities, use of endogenous reserves, and the regulation of protein turnover, etc. This definitely signifies that cellular reorganization must occur during starvation. After reading this chapter it will become clear that in the initial stage of starvation, cells are not truly dormant but much metabolic activity is occurring within the cell to prepare it for long-term starvation. Even in the starvation state, organisms are capable of capturing and incorporating organic substrates that occur at very low concentrations in the environment.

Many factors can influence the survival of the species during starvation. Starvation patterns may be influenced by the diluent used to suspend the bacteria (distilled water vs. saline, where saline was the choice) (Postgate and Hunter, 1962). Cultural conditions prior to starvation can influence the starvation process (Postgate and Hunter, 1962; Thomas and Batt, 1968) as well as the starvation conditions. Slow-growing cells are adapted much better for starvation than fast-growing cells (Höfle, 1984). Generally, starvation-survival is best at lower temperatures but not when the organism is exposed to a cold shock (Nelson and Parkinson, 1978). Decreasing the temperature of starvation by a factor of two prolongs survival by a factor of six to eight (Postgate and Hunter, 1962; Thomas and Batt, 1968). Enhanced survival is noted for slower growing cells (Postgate and Hunter, 1962; Dawes and Senior, 1973; Harrison, 1961; Nelson and Parkinson, 1978) while Mg^{2+} or Ca^{2+} (Thomas and Batt, 1968; Postgate and Hunter, 1962) markedly improved survival.

Strange et al. (1961) also studied the survival of *A. aerogenes,* mainly with stationary phase cells. They noted that composition of the medium on which the cells were grown influenced the survival time. In phosphate buffer, survival was longer with cells grown in tryptic meat broth or tryptone-glucose medium. Bacteria grown in tryptic meat broth contained a relatively large amount of protein and less RNA than the bacteria grown in defined medium. The Q_{O_2} (μl O_2/hour/bacterium) was higher in cells grown in defined medium. When the organism was starved, there was a loss of dry weight, glycogen, RNA, and protein from the cells (Strange et al., 1961). It was also noted that the addition of glucose did not favor survival.

The ability to survive starvation is not necessarily correlated with the growth rate during active proliferation (Harrison and Lawrence, 1963). These authors starved *A. aerogenes* in phosphate buffer harvested during the logarithmic and stationary phases of growth. The stationary phase cells remained viable longer and the logarithmic phase cells died rapidly. Nevertheless a fraction of the cells remained viable, which appeared to consist of the original wild type and a "starvation-resistant" mutant. When these "starvation-resistant" mutants were isolated, grown again to the logarithmic phase, and starved, a relatively slow decline in viability was noted, whereas with the wild type, viability decreased rapidly. The growth rates of the mutant and the wild type were 1.9 and 2.5 divisions/hour respectively.

Significant changes can be seen when a bacterial population is subjected to nutrient deprivation (Dawes, 1976; Trinci and Thurston, 1976). The shift between growth and survival metabolism undergone by starving cells is frequently traumatic, especially for slow growing cells characteristically low in endogenous reserves, and gives rise to a higher mortality rate during the initial starvation period, commonly followed by a much lower death rate for cells successful in making the transition from growth to survival metabolism (Postgate and Hunter, 1962;

Dawes, 1976). The possibility exists that metabolic injury may precede death by starvation, although Postgate and Hunter (1962) did not observe it in their experiments.

7.5 Starvation-Survival Process: Laboratory Studies

7.5.1 Patterns of Starvation-Survival

Sixteen freshly isolated marine bacteria were tested for starvation-survival and three patterns of survival were noted (Amy and Morita, 1983a). A fourth pattern of starvation-survival was noted by Jones and Morita (1985). These four patterns of starvation-survival are shown in Figure 7.1. Line C of Figure 7.1 shows an initial increase in cells due to fragmentation (reductive division), followed by a decline. The pattern indicated by line C was initially determined by Novitsky and Morita (1977, 1978a) and verified by Amy and Morita (1983a), Oliver and Stringer (1984), and Moyer and Morita (1989a). Other organisms that display the line C pattern include another vibrio species (Kjelleberg et al., 1982; Dawson et al., 1981; Nystrom et al., 1990a) in a *Pseudomonas* sp. (Kurath and Morita, 1983) as well as other bacteria (Tabor et al., 1981; MacDonell and Hood, 1982; Wakimoto et al., 1982; Porter et al., 1991; Schimz and Overhoff, 1987; Nissen, 1987). It has also been observed in various strains of *Vibrio cholerae* (Baker et al., 1983). Line D of Figure 7.1 depicts a rapid die-off of the freshly isolated bacteria (Amy and Morita, 1983a) and may be representative of most bacteria that grow in nutrient-rich environments, such as *Escherichia coli*. In many situations, *E. coli* is released into the environment along with defecated material, which represents an energy source. Line A has been observed in several isolates (Amy and Morita, 1983a) but, unfortunately, not as yet studied to any great extent. Line B was found, thus far, only in a nitrifying bacterium and, in this case, the bacterial cell did not increase or decrease in size (Jones and Morita, 1985).

The starvation pattern with time was initially thought to occur in two stages (Kurath and Morita, 1983). During the first stage with a *Pseudomonas* sp., there are changes in the direct count, respiring cell count, viable cell count, ATP, glucose and glutamate uptake/ml, and endogenous respiration. Between the 17th to 20th day, the foregoing stablized, which was the beginning of the second stage and remains stablized throughout the second stage. Later, Moyer and Morita (1989a) demonstrated three stages of starvation in Ant-300. During the first stage lasting 14 days, large fluctuations in viable counts occurred. This was followed by the second stage (14 to 70 days) in which the viable population decreased 99.7%. The third stage was marked by a stabilization of viable cell population (3% of the total count numbers). If energy is supplied any time during the three stages, growth and reproduction will again occur. Each species will dictate the

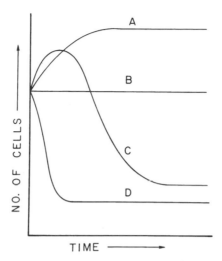

Figure 7.1   Starvation-survival patterns of microorganisms. Reprinted by permission from *Bacteria in the Natural Environment: The Effect of Nutrient Conditions,* M. Fletcher and G. Floodgate (eds.), R. Y. Morita, 1985, "Starvation and miniturisation of heterotrophs, with special reference on the maintenance of the starved viable state," pp. 111–130; copyright 1985, Society for General Microbiology.

time element for the occurrence of each stage as well as the length of each stage. These stages are discussed next section.

7.5.2   Optical Density, Cell Numbers, and Cell Size during Starvation

In his classical book, *Morphological Variation and the Rate of Growth of Bacteria,* Henrici (1928) details changes in bacterial cell morphology and size, depending on the environmental conditions. Bacteria in the oligotrophic environment are mainly microcells. Small cells resulting from nutrient starvation of Ant-300 were noted by Novitsky and Morita (1976). Figure 7.2 illustrates the change in the size of the bacteria at 0 time, 1 week, and 6 weeks after starvation. These size changes are better illustrated in Figure 7.3, where the size of the starved cells was determined by passage through Nuclepore filters of different pore sizes. Size reduction is most rapid during the first two days of starvation and, after 2 weeks of starvation, 100% of the cells were capable of passing through 3.0- and 1.0-μm Nuclepore filters but none could pass through a 0.2-μm filter. After 4 weeks of starvation, the initial number of cells were still viable. Electron micrographs of thin sections of starved cells are illustrated in Figure 7.4. The ultrastructures of cells (Fig. 7.4a) appear typical for gram-negative bacteria and show no unusual

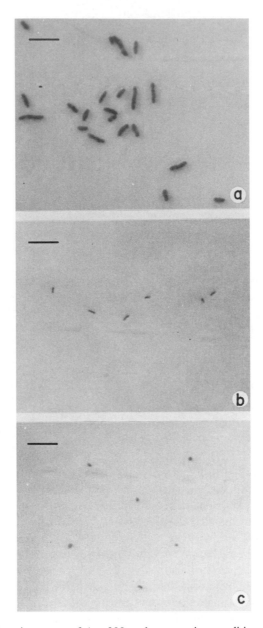

Figure 7.2   Light microscopy of Ant-300 under starvation conditions. Bars represent 5 μm. (*a*) Zero time; (*b*) 1 week after starvation; (*c*) 6 weeks after starvation. Reprinted by permission from *Applied and Environmental Microbiology* 32:619–622, J. A. Novitsky and R. Y. Morita, 1976; copyright 1976, American Society for Microbiology.

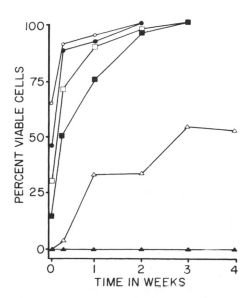

Figure 7.3   Effect of time of starvation on the size distribution of Ant-300 cells. Portions of the starving culture were passed through filters of various pore sizes. Plate counts of the filtrates are expressed as a percentage of the unfiltered counts. Filter pore sizes: ○, 3.0 μm; •, 1.0 μm; □, 0.8 μm; ■, 0.6 μm; △, 0.4 m; and ▲, 0.2 μm. Reprinted by permission from *Applied and Environmental Microbiology* 5:619–622, I A, Novitsky and R. Y. Morita, 1976; copyright 1976, American Society for Microbiology.

structures. The small starved cells (Fig. 7.4b, c) appear small and roughly spherical. The ultrastructures are generally the same as the nonstarved cells except for an enlarged periplasmic space containing material observed in all of the starved cells examined. The fact that these microcells are able to grow indicates that they are complete cells, unlike the minicells of *E. coli* that contain no deoxyribonucleic acid and are unable to divide (Adler et al., 1967).

Novitsky and Morita (1977) noted when Ant-300 was placed in a starvation menstruum, an increase in viable cells was noted during the first week, yet the optical density (OD) of the microbial suspension decreased 62% (Fig. 7.5). The increase in cell number is the result of fragmentation (Novitsky and Morita, 1976). After the second week, the viable cell count decreased but the total direct cell count did not (indicating the cells have remained structurally intact), while the OD declined slightly more, yielding an additional value of 21% after 7 weeks of starvation. After a year of starvation, 10^3 viable cells/ml could still be demonstrated. This increase in cell number over the initial inoculum persisted for 70 weeks. At low cell densities of Ant-300, when suspended in natural seawater or artificial seawater, a 200-fold increase in cell numbers occurred (Novitsky and

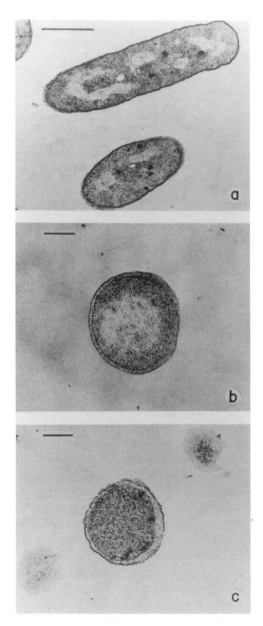

Figure 7.4   Electron micrographs of Ant-300 cells. (*a*) Nonstarved cells; bar represents 1.0 μm. (*b*) and (*c*) Cells starved for 5 weeks; bar represents 0.2 μm. Reprinted by permission from *Applied and Environmental Microbiology* 5:619–622, J. A. Novitsky and R. Y. Morita, 1976; copyright 1976, American Society for Microbiology.

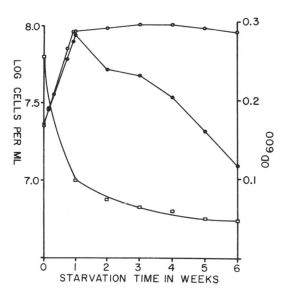

Figure 7.5   Starvation-survival of Ant-300. Cells harvested from exponential growth were starved in a buffered-salt mixture. Total direct counts, ○; plate counts, •; and optical density, ▫. Reprinted by permission from *Applied and Environmental Microbiology* 33:635- 641, J. A. Novitsky and R. Y. Morita, 1977; copyright 1977, American Society for Microbiology.

Morita, 1978b). These data initiated our studies on starvation-survival, especially long-term survival without energy. Similar results were obtained by Amy et al. (1983a). In *Vibrio* S14, 40% of the total decrease in cell size occurs after the fragmentation period (Kjelleberg and Hermansson, 1984). Thus, one of the phenotypic responses to starvation is the decrease in cell size resulting in a surface/volume change.

The increase in cells when placed in menstruum without energy was demonstrated with *Pseudomonas solanacearum* in pure water (Wakimoto et al., 1982) and for *Bordetella bronchiseptica* in phosphate-buffered saline or in different lake waters (Porter et al., 1991). An increase in the number of cells did not occur regularly when *E. coli* was placed in river water (Flint, 1987). It is very difficult to determine where the energy for such repeated increases in cells come from. Some of this energy comes from the degradation of intracellular compounds, but in *E. coli,* Koch (1979) has shown that it can take up enough glucose in 2 or 3 minutes to last for an average of 17 minutes. As a result, this ability to take up excess substrate may be a significant advantage in the minute to minute fluctuations in the nutrient level in the cell's natural environment.

The increase in the number of cells without the presence of energy is difficult to analyze from a bioenergetics viewpoint. Nevertheless, many investigators observed the increase in number of cells (not necessarily biomass) when cells are starved. This may be "Nature's" way to ensure the survival of the species. An analogous situation occurs with other life forms where the number of eggs, seeds, or offspring exceed the number that will survive.

Smigielski et al. (1990) surveyed 21 marine *Vibrio* spp. and all responded to nutrient deprivation by undergoing cell size reduction. Of the *Vibrio* spp., 43% possessed one or more plasmids, thus suggesting that the genes responsible for cell size reduction are located on the chromosome rather than the plasmid. This was supported by the fact that three strains were cured of the plasmids but still retained the ability to undergo fragmentation and subsequent cell size reduction during starvation. As would be expected, these strains lost some of their plasmid capabilities, such as antibiotic resistance and heavy metal resistance. They also demonstrated that, if respiration was blocked when insufficient energy was available for cellular reorganization, reduced viability of the cells did not occur and there was no decrease in cell size. Even when grown in a chemostat with L-lactose limitation, *Spirillum* and *Pseudomonas* decreased in size as the dilution rate decreased, so that the surface to volume ratio increased (Matin and Veldkamp, 1978). Cell size reduction has also been noted in *P. putida* KY2442 (Givskov et al., 1994a). After 24 years of starvation in sterile distilled water, *P. syringae* subsp. *syringae* is smaller and less dense than the same species when grown after it had been starved (Iacobellis and DeVey, 1986).

Not all species of bacteria will reduce in size when starved. Nevertheless, reduction in cell size has been noted in soil bacteria (Bae et al., 1972; Balkwill et al., 1977), estuarine bacteria (MacDonell and Hood, 1982), and bacteria (including deep-sea forms) in ocean (Humphrey et al., 1983; Tabor et al., 1981; Torrella and Morita, 1981) as well as in oil well waters (Lappin-Scott et al., 1988a). Cell size changes with starvation time of several bacteria have been shown using a microcomputerized Elzone-ADC-80XY particle counter (Table 7.2) (Kjelleberg and Hermansson, 1984). From Table 7.2 it can be seen that during the 22 hours of starvation, the major drop in cell volume varied with the species of bacteria employed, but all were 100% viable after 22 hours of starvation. On the other hand, Kjelleberg et al. (1983) noted that if inhibitors of the proton flow, the electron transport chain, and membrane bound ATPase were added to a starving culture, miniaturization of cells is prevented.

As a consequence of forming ultramicrocells during nutrient deprivation, the surface/volume ratio becomes larger and helps sequester nutrients efficiently in low-nutrient environments (Matin and Veldkamp, 1978; Poindexter, 1981b). This is listed as an important character of the model oligotrophs (Hirsch et al., 1979). By growing a *Spirillum* in a chemostat at dilution rates from 0.35/hour to 0.01/hour, Matin and Veldkamp (1978) were able to show that the surface/volume ratio changed from 6.3 to 9.3. By taking an ultramicrobacterium having a diameter of 0.2 μm (surface:volume ratio of 31) and growing it in rich medium, the cell size

Table 7.2   Changes in cell volume during starvation for various times.

Organism	Cell Vol (μm^3) at:			
	0^a	15 min	5 h	22 h
S14, not identified	1.23	1.16	0.64	0.37
Ol14, *Spirillum* sp.	1.10	0.89	0.76	0.49
DW1, *Vibrio* sp.	—^{b}	1.61	0.89	0.64
BMS1, not identified	1.30	1.10	0.53	0.51
EF190, *S. marcescens* (nonpigmented)	1.04	0.94	0.89	0.61
DW2, *Flavobacterium* sp.	1.37	1.16	1.16	1.04
S9, *Pseudomonas* sp.	1.16	0.94	0.46	0.39

Source: Reprinted by permission from *Applied and Environmental Microbiology* 48:497–503, S. Kjelleberg and M. Hermansson, 1984; copyright 1984, American Society for Microbiology.
^{a}0 designates start of starvation.
^{b}—, Not determined.

increased to 0.4 × 1.5 μm, which results in a surface to volume ratio of 11 (MacDonell and Hood, 1982). Ant-300 appears as a large cell (6 μm^3) when grown in batch culture, but when grown in a chemostat at D − 0.015/hour, its size is 0.045 μm^3 (Fig. 7.6) (Moyer and Morita, 1989a). These cells become much smaller when starved (Fig. 7.7). Gottschal (1992) calculated the surface/volume ratio for Ant-300 from the data of Moyer and Morita (1989a) to be 3.8 in batch culture and 14 in the chemostat. It is of interest to note that ultramicrocells grown at D = .015 still retain the ability to grow; thus, in oligotrophic environments, ultramicrocells dominate. The difference in volumes between starved and unstarved cells grown in a chemostat can be noted in Table 7.3.

The relationship of viable cell counts (plate counts), total cell counts (acridine orange direct count) (AODC), and respiring cells counts [2-(p-indophenyl)-3-(p-nitrophenyl)-5-phenyl tetrazolium chloride] (INT) during starvation of a *Pseudomonas* sp. is shown in Figure 7.8 (Kurath and Morita, 1983). The initial cell density was approximately 3 x 10^8 viable cells/ml and increased to 4.2 x 10^8 cells/ml in the first 4 days of starvation. After 21 days, the viability was stable at approximately 5 x 10^5 viable cells/ml and remained at the same level for the remainder of the 40-day experiment. The total direct count at the onset of the experiment was slightly higher than the viable count at the onset of starvation and showed no significant variation during the entire experimental period. At the onset of the experiment, 97% of the cells in the suspension were actively respiring, as indicated by the INT method. The number of respiring cells decreased rapidly to 27% in 6 days of starvation and after 18 days, the number of respiring cells stabilized at 1.0% of the total bacteria present. The difference between viable and

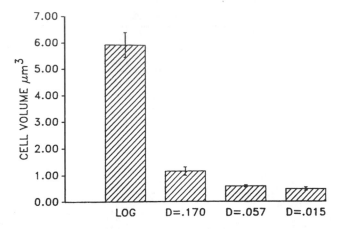

Figure 7.6   Average cell volume of unstarved cells. Cell population samples were as follows: Batch culture (labeled LOG), D = 0.170/hour, D = 0.057/hour, and D = 0.015/hour. The hatched bars represent the means of 20 cell volume determination from each population; the vertical bars represent the standard errors of the means. Reprinted by permission from *Applied and Environmental Microbiology* 55:1122–1127, C. L. Moyer and R. Y. Morita, 1989; copyright 1989, American Society for Microbiology.

INT counts suggests the existence within the starving population of a subpopulation of nonviable but actively respiring cells that is approximately 10-fold more numerous than the viable cells. This phenomenon has been observed in natural populations by autoradiography (Hoppe, 1976, 1978). The difference between the number of respiring cells and the viable counts represents the number of cells not culturable on the medium employed—the first laboratory experiments dealing with nonculturable bacteria. After 6 weeks of starvation, 49% of the Ant-300 cells demonstrated the ability to respire, as indicated by the INT method (Amy et al., 1983a).

V. cholerae, when exposed to nutrient-free artificial seawater and filtered natural seawater, decreased in volume by 85% and developed into small coccoid forms surrounded by remanent cell walls (Baker et al. 1983). Initial cell counts in nutrient-free microcosms (culturable and direct counts) increased 2.5 logs within 3 days and even after 75 days, the number of cells was still 1 to 2 logs higher than the initial inoculum size (Baker et al., 1983). After 48 hours of starvation, the cells began to exhibit small regions of electron transparency at the periphery but there was no significant change in overall size. After 5 days, there was an increase in the amount of electron transparency at the polar regions, which appeared opposite the flagellar pole. After 75 days of starvation, a separation of the cell wall and the cytoplasmic membrane appears. In another paper, Hood et al. (1986) presents electron micrographs of starved *V. cholerae,* which indicate that the cell

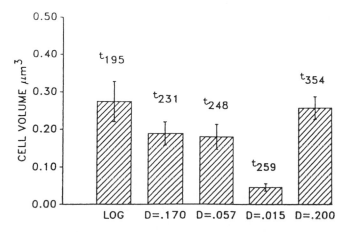

Figure 7.7   Average cell volume of starved cells. Cell populations sampled were as follows: Batch culture (labeled LOG) starved for 195 days, D = 0.170/hour starved for 231
days, D − 0.057/hour starved for 248 days, D = 0.015/hour starved for 259 days, and D
= 0.200/hour starved for 354 days. The hatched bars represent the means of 20 cell volume
determinations from each population; the vertical bars represent the standard errors of the
mean. Reprinted by permission from *Applied and Environmental Microbiology* 55:1122–
1127, C. L. Moyer and R. Y. Morita, 1989; copyright 1989, American Society for Microbiology.

Table 7.3   Average cell volume measurements for starved and unstarved Ant-300 cells.

Dilution rate[a] D (h[−1])	Cell Volume (μm[3]) ± SEM		% Reduction in Volume
	Unstarved	Starved[b]	
Batch culture[c]	5.94 ± 0.465	0.275 ± 0.053	95.4
0.170	1.16 ± 0.156	0.189 ± 0.030	83.7
0.057	0.585 ± 0.038	0.181 ± 0.033	69.1
0.015	0.478 ± 0.060	0.046 ± 0.010	90.4
0.200		0.258 ± 0.030	

Source: Reprinted by permission from *Applied and Environmental Microbiology* 55:2710–2716,
C. L. Moyer and R. Y. Morita, 1989; copyright 1989, American Society for Microbiology.
[a]Cells were grown in continuous culture with SLX medium.
[b]Cells from stage 3 starvation survival.
[c]Cells were grown with Lib-X medium.

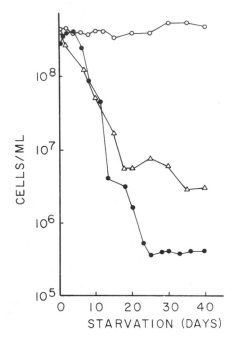

Figure 7.8 Viable, direct, and respiring cell counts after various periods of starvation. All counts were carried out on the same starvation culture. Viability (•) was determined by spread plates, direct counts (∘ were determined by epifluorescence microscopy, and respiring cell counts (△) were determined by the INT method. Reprinted by permission from *Applied and Environmental Microbiology* 45:1206–1211, G. Kurath and R. Y. Morita, 1983; copyright 1983, American Society for Microbiology.

loses more than 90% of its original volume in 30 days. All granules and inclusion bodies were lost and the three-layered integrity of the outer membrane and the cell membrane was lost, but remnants of the structures remained, the nuclear region was compressed into the center of the cell surrounded by a denser cytoplasm and the cell wall formed an extended or convoluted structure that pulled away from the cell membrane. The ribosomal structure appeared to be conserved as it exhibited no change in the starved cell.

Other than a decrease in cell size, the formation of periplasmic spaces are noted when cells are starved, presumably due to the shrinkage of protoplasm. This has been noted in soil bacteria (Bae et al., 1972), marine isolates Ant-300 (Novitsky and Morita, 1976) and *Vibrio* S14 (Mårdén et al., 1985), *Arthrobacter crystallopoites* (Boylen and Pate, 1973), *V. cholerae* (Hood et al., 1986), *E. coli* (Reeve et al., 1984a), and *Klebsiella pneumoniae* (Lappin-Scott et al., 1988a). After 24 days of starvation, *K. pneumoniae* cells, as revealed by transmission electron microscopy (TEM), changed from rods to small rods or cocci between 0.5 by 0.25 μm and 0.87 by 0.55 μm (Lappin-Scott et al., 1988b). Also by TEM, cells of *Pseudomonas* sp. Strain S9 and *Vibrio* S14 formed membrane vesicles (probably formed from blebs) on initial starvation and the loss of poly-β-hydroxybutyrate in *Vibrio* S14 cells (Mårdén et al., 1985). In some cells, cell surface roughness, as revealed by high textured layers in electron micrographs, can be noted when starved (Kjelleberg and Hermansson, 1984) but not in all cells. However, it is interesting to note that Ant-300 ultramicrocells grown at various dilution rates give rise to other ultramicrocells (Moyer and Morita, 1989a). In oligotrophic environments, small cells reproduce other small cells, which is the result of the lack of energy in these environments. Naturally, it takes less energy to produce a smaller cell.

There are similar and dissimilar characteristics between *V. cholerae* and Ant-300 (Novitsky and Morita, 1976; Amy et al., 1983a). Ant-300 did not respond to starvation in the same way as *V. cholerae* in terms of the total size reduction and granule disappearance; the nuclear regions became more compact, but intracellular integrity decreased in both organisms. Ant-300 (Novitsky and Morita, 1976) showed little distortion in the cell wall, and *V. cholerae* (Hood et al., 1986) exhibited considerable convolution of the cell wall. Märden et al. (1985) subjected three marine vibrios to starvation and found the one showed little cell wall distortion, whereas two produced vesicles in 24 hours and released these vesicles with time and that these vesicles were related to the continuous size reduction during starvation. Hood et al. (1986) also noted some vesicle formation early in the starvation period (within 24 hours) with *V. cholerae* but these disappeared quickly. In freshly isolated bacteria from the open ocean, morphological changes (flagellation and shape) during starvation were noted by Amy and Morita (1983a). An increase in numbers for Ant-300 has been noted by Oliver and Stringer (1984) also. In *V. cholerae,* nearly 10 times more viable cells (direct viable counts) resulted after 30 days of starvation, whereas 2 to 2.5 log increases were noted when lower inoculation sizes were used (Hood et al. 1986). Log increases have also been noted in other vibrios, by Dawson et al. (1981) as well as by Kjelleberg and his associates. It also occurs in a species of *Pseudomonas* (Kurath and Morita, 1983).

In *A. crystallopoietes* (both rod-shaped and spherical cells), the size and morphology of the cells do not undergo significant changes during 8 weeks of starvation in phosphate buffer at 30°C, but there is a decline in their glycogen reserves and the number of ribosomes and an increase in the number of vesicular membranes within the cell and the nucleoplasm expands in volume to fill the emptying cytoplasm (Boylen and Pate, 1973). Very little change in cell shape or size was detected in electron micrographs of *Arthrobacter globiformis, Arthrobacter nicotianae, Bacillus linens, Corynebacterium fascians, Mycobacterium rhodochrous,* and *Nocardia roseum* when these organisms were starved in phosphate buffer up to 56 days (Boylen and Mulks, 1978). There appeared to be a gradual disappearance of intracellular material, but not resting structures.

Dry weights during starvation decreased rapidly in the first 20 to 40 hours and slowly therafter in the *Pseudomonas* sp., *Bacillus* sp., and *Arthrobacter* sp. studied by Nelson and Parkinson (1978). Even in *B. subtilis* the cell numbers increase by 50% upon glucose starvation (Maruyama et al., 1977). This increase was observed 20 minutes after the initiation of the starvation process. An environmental actinomycete was found to fragment into small cells when starved and, when resuscitated, increased in size (Herman and Costerton, 1993). These authors also found an intracellular space in the starved actinomycete.

Simon and Azam (1989) examined the size, dry weight, and protein content of pelagic bacteria and found that, when cells become smaller, they become richer in carbon, protein, and dry weight but poorer in water. The cell protein:dry weight and cell protein:carbon were essentially constant at 63% and 54% respectively throughout the entire cell size range. There are also changes in the DNA, RNA, and cell wall with cell size that can be noted in Table 7.4 (Simon and Azam, 1989). Gottschal (1990) suggest that the surface/volume ratio coupled with the high density of ultramicrocells brings about the repression of many metabolic enzymes and substrate uptake systems in very slowly growing or starving cultures, thereby aiding considerably in maintaining nutrient pool sizes in the nmol/l range in an environment in which the nutrient concentrations are on the order of 10^9 to 10^{-12} mol/l.

The starvation mechanism takes place over a period of time, the time depending on the microorganisms in question. Thus, to obtain a complete state of metabolic arrest may take a longer period. In Ant-300, the period appears to be 70 days (Moyer and Morita, 1989a), whereas in a *Pseudomonas* sp., it takes approximately 20 to 25 days (Kurath and Morita, 1983). In Figure 7.9, it can be seen that the first stage of the starvation process takes approximately 14 days; the second stage takes place between the 14th and the 70th day; the third stage takes place after the 70th day. The three stages can also be noted in cells grown at different dilution rates as well as in batch grown cells (Moyer and Morita, 1989a). The last is the stage of metabolic arrest and some cells do die during this period. Metabolic arrest is analogous to the bacterial spore and metabolic processes do not take place due to the lack of substrates for the various enzyme systems. Cells in this stage are termed "shut down" cells by Dow et al. (1983). As will be seen later, there was much endogenous metabolic activity in stage 1 and 2, especially during stage 1. These stages are akin to sporogenesis and the cell does not become truly "dormant" until stage 3 is reached. Thus, to describe cells when initially starved as dormant is erroneous.

Can this increase in cell numbers when organisms are starved happen in natural seawater or in the natural environment? Ant-300 does increase in cell number when starved in natural seawater, or for that matter in seawater supplemented with

Table 7.4 Macromolecular composition, dry weight (dw), carbon, and cell volume: carbon conversion factor of an average marine bacterium in the size range 0.026 to 0.4 μm³.

Volume (μm³)	Protein[a] (fg)	(% vol)	(% dw)	Cell Wall[b] (fg)	(% vol)	(% dw)	Cell Membrane[c] (fg)	DNA[d] (fg)	RNA[e] (fg)	dry wt (fg)	Carbon[f] (fg)	(% dw)	Carbon volume (mg ml⁻¹)
0.026	12.1	46.5	61.4	2.5	9.6	12.7	0.7	2.5	1.9	19.7	10.4	53.0	400
0.036	14.7	40.8	62.5	3.1	8.6	13.2	0.9	2.5	2.3	23.5	12.6	53.6	350
0.050	17.7	35.4	63.0	3.9	7.8	13.9	1.2	2.5	2.8	28.1	15.2	54.2	304
0.070	21.6	30.9	62.8	4.8	6.9	14.0	1.5	3.0	3.5	34.4	18.7	54.3	267
0.100	26.7	26.7	62.8	6.2	6.2	14.6	1.9	3.5	4.2	42.5	23.3	54.7	233
0.200	40.3	20.2	63.5	9.9	5.0	15.6	3.0	4.0	6.3	63.5	35.0	55.1	175
0.400	60.6	15.2	63.3	15.8	4.0	16.5	4.9	5.0	9.5	95.8	53.3	55.7	133

Source: Reprinted by permission from Marine Ecology Progress Series 51:201–213, M. Simon and F. Azam, 1989; copyright 1989, Inter-Research.

[a] Calculated by the power function with cell volume and corrected for the amount of amino acids not detected by HPLC (18%, Reeck 1983)

[b] Calculated for a coccus assuming a thickness of 11 nm and corrected for the protein proportion (44%, Stanier et al. 1986)

[c] Calculated for a coccus assuming a thickness of 8 nm and corrected for the protein proportion (75%, Stanier et al. 1986)

[d] Hoffman et al. (unpubl.)

[e] 11% of dry weight (Herbert 1976)

[f] Reconstituted from the % carbon of all macromolecules [protein 53%, cell wall and membrane 77% (assuming to be exclusively lipids), nucleic acids 36%; Herbert 1976]

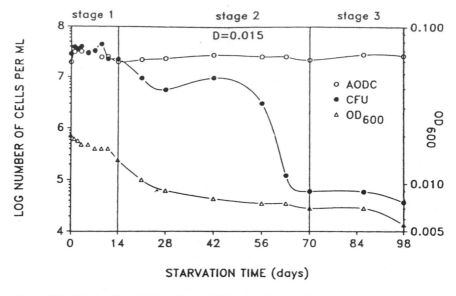

Figure 7.9   Total cells, viable cells, and OD_{600} with starvation time for cells from D = 0.015/hour. Reprinted by permission from *Applied and Environmental Microbiology* 55:1122–1127, C. L. Moyer and R. Y. Morita, 1989; copyright 1989, American Society for Microbiology.

10 times the normal amount of amino acids found in seawater (Novitsky and Morita, 1978b). If the initial increase in cell numbers due to starvation was a result, for the most part, of impurities in the menstruum, a larger increase would be expected in seawater and seawater amended with amino acids. When the concentration of amino acids was increased by 10 times, the initial increase in Ant-300 cells was less than either seawater or the starvation menstruum supplemented with single strength amino acids, and the survival was slightly less than with no additions. The growth rate and doubling time of Ant-300 at different dilution rates are shown in Table 7.5.

Suwa and Hattori (1987) determined the growth rate of their cultures in NB/10, NB/100, and NB/1000 and noted that there is a rapid increase in the number of cells. In comparing this with the data of Novitsky and Morita (1978b), who showed that identical fragmentation patterns could occur when Ant-300 is subjected to a starvation menstruum or at concentrations of amino acids equal to and 10 times that in seawater, one can wonder if their data show that fragmentation is occurring and not growth. Just how dilute must the energy source be before fragmentation occurs? Previously, Suwa and Hattori (1984) did state that some of the cells underwent fragmentation and not growth. Similarly, the data on the

Table 7.5   Ant-300 dilution rates, with corresponding growth rates and doubling time.

Dilution Rate[a] D (h^{-1})	Growth Rate μ (h^{-1})	Doubling Time t_d (h)
0.015	0.015	46.2
0.057	0.057	12.2
0.170	0.170	4.1
0.200	0.200	3.5
Batch culture[b]	0.144	4.8

Source: Reprinted by permission from *Applied and Environmental Microbiology* 55:2710–2716, C. L. Moyer and R. Y. Morita, 1989; copyright 1989, American Society for Microbiology.
[a]Cells were grown in continuous culture with SLX medium.
[b]Cells were grown with Lib-X medium.

growth of *Pseudomonas acidovorans* in inorganic salts solution with and without phenol may be the result of fragmentation and not growth (Schmidt and Alexander, 1985).

The initial increase of cells due to fragmentation, when cells are starved, is inhibited by nalidixic acid, which suggests that DNA synthesis, rather than an excess of nuclear bodies, allows the fragmentation process to occur (Märdén et al., 1988).

7.5.3   Cryptic Growth and Cell Lysis

Cryptic growth (cannibalism) is where a living proportion of a population utilizes dead bacteria as nutrients for its survival (Ryan, 1959). It was first observed by Winslow and Falk (1923a). Strange et al. (1961) employed the term "regrowth" while Harrison (1960) employed the term "cannibalism." Cryptic growth occurred in *Aerobacter aerogenes* after 60 to 70 hours of starvation (Postgate and Hunter, 1962). On the other hand, DNA or cell lysis products were not detected in the starvation menstruum (Boyaval et al., 1985; Boylen and Mulks, 1978; Boylen and Ensign, 1970b), strongly suggesting that cryptic growth did not occur. Thomas and Batt (1968) concluded that cryptic growth probably did not occur in *S. lactis* ML3, because ML3 had specific nutrition requirements not likely to be met by cryptic growth.

According to Postgate and Hunter (1962), survival studies at ordinary temperatures in nonnutrient aqueous solutions are complicated by three factors: (1) Buffers prepared from highly purified reagents contain impurities that permit limited growth of bacteria; (2) as soon as a portion of the population dies, nutrients may be released into the medium enabling the survivors to multiply; (3) reagents, laboratory glassware, and distilled water may contain materials actively toxic to bacteria. To emphasize item 1, they cite the data of Garvie (1955) and Strange et al. (1961) that showed the apparent "reactivation" of "dead" bacteria. However, the ability of cells placed in a starvation menstruum to increase in cell numbers should not be discounted. Although Postgate and Hunter (1962) demonstrated cryptic growth in *A. aerogenes,* Nioh and Furusaka (1968) found that the ability to grow cryptically was not universal. However, cryptic growth did not take place when cells of Ant-300 or a *Pseudomonas* sp. were starved (Novitsky, 1977; Kurath and Morita, 1983), indicating that no cell lysis occurs during starvation. This is in agreement with previous reports by Clifton (1967), Dawes (1976), Gronlund and Campbell (1963), and Harrison and Lawrence (1963). Cryptic growth, according to Postgate and Hunter (1962), presumably accounts for the survival of populations in culture long after growth has ceased. Thus, cryptic growth may be species dependent. Item 3 should include the possibility of airborne organics entering into the nonnutrient menstruum. Postgate and Hunter's (1962) review discusses other factors (temperature, ion concentration, pH, etc.) that influence cryptic growth.

Novitsky (1977) labeled Ant-300 cells with ^{14}C-glucose (uniformly labeled). These cells were suspended in starvation menstruum in the usual manner and permitted to starve. Filtrates were prepared during various periods of starvation by passage of the starvation menstruum through a 0.2-μm Nuclepore filter and the filtrate inoculated with starved bacteria. No production of $^{14}CO_2$ was noted, hence the possibility of cryptic uptake and respiration on the released material was ruled out. Because several cells must die to permit the doubling of each remaining cell (Postgate and Hunter, 1962; Druilhet and Sobek, 1976), cryptic growth was ruled out as a possible explanation for the increase of cell number during the initial phase of starvation. Cryptic growth could not be demonstrated in a *Pseudomonas* sp. by Kurath (1980), who used the same technique as Novitsky (1977).

During resuscitation of starved cells, the increased numbers of CFUs in the population may be at the expense of other cells (cryptic growth or cannibalism). Postgate and Hunter (1962) calculated that 50 cells need to die to support the division of one. Kaprelyants and Kell (1993) stressed that this possibility is doubtful for the following reasons:

1. Given the constancy of the total counts, it is implausible that lysis could be so exactly balanced by growth.

2. During resuscitation the percentage of small cells in the population decreased and there was an exactly equivalent increase in the percentage of larger cells, indicating that small, dormant cells were converted into larger, active cells.

3. The transient character of the accumulation of larger cells during resus-

citation suggests that the increase in the level of viable cells in this period was not due to typical growth but to an unbalanced increase in cell size during this period.

4. Even if all the cells at 30 hours (5×10^3 cells/ml) started to grow with a doubling time of 4 hours, the increase in viable counts after an additional 20 hours (end of the resuscitation period) should have resulted in approximately 1.6×10^5 cells/ml.

7.5.4   Utilization of Intracellular Energy

ENDOGENOUS RESPIRATION

Endogenous metabolism relies on intracellular components of the cell as a source of energy and is carried on in the absence of external sources of energy. When a cell is derived from exogenous substrate, the starvation process is initiated. Foster (1947) stated that endogenous metabolism occurs simply because the organism cannot help it, and therefore it bears no relationship, direct or indirect, to the period of survival. "In nutritional terms, endogenous metabolism can be viewed as encompassing all those chemical activities engaged in by the starving cell" (Lamanna, 1963). Endogenous metabolism permits the cell to rearrange its internal cellular structure and physiology, permitting the cell to survive long periods of nutrient deprivation. There are numerous data on endogenous metabolism in the older literature dealing with the Warburg respirometry, where the endogenous respiration (no substrate) was subtracted from the cells undergoing metabolism with a specific substrate in the Warburg vessel.

Dawes and Ribbons (1963) defined endogenous respiration as "the total metabolic reactions which may serve specifically as exogenous substrates." Although substrates of endogenous metabolism have received much attention, the data on the modification of cellular energetics, rates of processes, and respiratory functions in response to starvation are limited. In most cases, these data were determined by manometry, expressed as Q_{O_2} in µl O_2/mg dry weight/hour. After reviewing the literature Dawes (1976) concluded that the "evidence suggests that prolonged viability of starved bacteria is associated with a low rate of endogenous metabolism." Nelson and Parkinson's (1978) studies support this concept. They studied three Antarctic isolates and found a *Pseudomonas* to be most starvation resistant, *Arthrobacter* somewhat less resistant, and a *Bacillus* to be most susceptible to death from starvation. The *Bacillus* did not rapidly establish a low rate of endogenous respiration, whereas the starvation-resistant organisms did. On the other hand, Chen and Alexander (1972) found no correlation between a low rate of endogenous metabolism and the ability to survive in soil microbes.

Strange (1967) states that the endogenous metabolism varies in different organisms, with the chemical composition and physiological state of the same or-

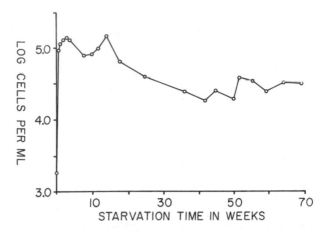

Figure 7.10 Endogenous respiration of starving Ant-300 cells. Respiration was determined at various times during starvation by the amount of $^{14}CO_2$ evolved from a previously labeled starving culture. Values were expressed as a percentage of the total cellular carbon respired/hour. Reprinted by permission from *Applied and Environmental Microbiology* 55:635–641, J. A. Novitsky and R. Y. Morita, 1977; copyright 1977, American Society for Microbiology.

ganisms, and with the physicochemical conditions of the environment. He further suggests that the rapid utilization of cellular constituents during starvation may not have a maintenance role. ATP formation results from endogenous metabolism but ATP content did not correlate with survival properties (Strange et al., 1963). Loss of the ability to synthesize β-galactosidase by starved bacteria was not due to ATP deficiency (Strange et al., 1961). The control and economy of the energy provision mechanisms in starved bacteria are ill defined (Strange, 1967).

Endogenous respiration was reduced by over 90% in *Nocardia corallina* starved for 20 days (Robertson and Batt, 1973) and greatly reduced when *A. globiformis* starved for 56 days (Luscombe and Gray, 1974). When spherical cells of *A. crystallopoietes* underwent starvation, the endogenous metabolism declined 70% during the first 40–50 hours and then remained constant at a base level of 0.03% of the cell carbon converted to CO_2/hour (Boylen and Ensign, 1970a). This was accompanied by a 30% decrease in dry weight of the cells, yet the cells remained 100% viable for 30 days. Boylen and Mulks (1978), Burleigh and Dawes (1967), Dawes and Holmes (1958), and Postgate and Hunter (1962) reported that a decrease in endogenous respiration is not associated with a long survival time, but with rapidly dying populations. However, it is difficult to determine with certainty whether the decrease in respiration is actually a survival mechanism or the result of cell death. Reduction in endogenous respiration has also been noted in marine bacteria (Humphrey et al., 1983) and *Coryneform* spp. (Boylen and Mulks, 1978).

The endogenous respiration of Ant-300, when starved, is shown in Figure 7.10 (Novitsky and Morita, 1977). There is a drop in endogenous respiration of 62% in the first week of starvation, but after 7 weeks the endogenous respiration is reduced to 0.0071%/hour (greater than 99%) and appears to remain constant thereafter. However, when this figure is extended to 2.5 years (culture known to be viable when an electrical outage terminated the experiment) this value of 0.0071%/hour is much too high. If the experiment had been extended longer, it is felt that endogenous respiration would cease and the cell would enter a period of "metabolic arrest" so that no endogenous respiration would occur. Respiration or any enzymatic activity of the cell ceases because the various enzymes no longer have substrates to act on. When cells reach a stage of metabolic arrest, in all probability, endogenous respiration ceases. With no substrates present, the metabolic processes are on "hold" until substrates become available. What percentage of the original population can undergo such a change without losing its ability to grow again remains unknown. Forty-five to 60% of 6-week-starved Ant-300 had the ability to respire without the addition of any substrate, as determined by the reduction of tetrazolium to colored formazan (Amy et al. 1983a).

Log phase cells of *Vibrio* sp. strain DW1 endogenous respired about 150 ng of oxygen/10^9 viable cells but dropped to 58% of this value when starved for 5 hours and to 6% in 5 days (Kjelleberg et al., 1982). The decrease in size of the cells after fragmentation was accompanied by intense microbial activity (Kjelleberg et al., 1983). This intense microbial activity is in the first phase of starvation, as described by Moyer and Morita (1989a). This activity was measured by oxygen uptake, the use of inhibitors of proton flow, and the electron transport chair and membrane-bound ATPase. In *Vibrio* sp. S14, the rate of endogenous metabolism decreased gradually until it became undetectable (Nyström et al., 1990a).

Amy et al. (1983a) demonstrated that both the endogenous and respiratory potentials (with added substrate) of marine vibrios decrease immediately upon starvation, but temporary increases occur in ca. 7 days and cell death continues in spite of the increased endogenous respiration or because of it. Subsequent periods of starvation brought about lower rates of both the endogenous and potential respiration. Nevertheless, starved cells retain the ability to respire when challenged with a substrate. The ability of starved Ant-300 to take up various other substrates after various periods of starvation was documented by Glick (1980).

Employing calorimetric procedures, Gordon et al. (1983) could not detect any heat from starved *V. alginolyticus* at a concentration of 10^5/ml. These cells were grown in liquid medium for 18 hours, washed twice with mineral salts solution (no organic nutrients), and suspended to the desired concentration before use. Hence, the cells were not starved for any length of time.

In his excellent article, Dawes (1976) raised the following questions:

1. Does survival bear any direct or indirect relationship to endogenous metabolism?
2. Is endogenous metabolism wholly catabolic or do available precursors and energy liberated in the process support some degree of macromolecular synthesis? If so, are cellular components whose loss is particularly likely to result in death selectively reformed from dispensable materials?
3. Does the possession of specialized reserved materials exert a sparing action on the degradation of protein and RNA, and does it confer longevity?
4. Can death be attributed to the loss of any specific cellular constituent, or to the energetic state of the cell?

CELLULAR ENERGY RESERVES

The endogenous metabolism data in the literature address cells grown under optimal conditions compared to cells growing in any ecosystem. In such cases, intracellular reserve materials (polyphosphate, poly-β-hydroxybutyrate, carbohydrate polymers, or lipids) are formed, mainly because there is an excess of energy present in the system. In addition, the energy utilized is in relation to the viability of the entire cell population but within any given population some cells are more resistant to starvation than their counterparts. When endogenous reserves are depleted, the starvation process continues but whether the energy of maintenance is involved remains problematical. For most of the literature cited in the following discussion, it should be realized that the work has been done on cells grown in laboratory media where the energy is plentiful. In such cases, amino acid pools, intracellular energy reserves, and extracellular energy reserves are formed. Naturally these are the first energy sources utilized when the cell meets a condition where energy becomes limiting. Endogenous respiration requires various cellular components to undergo degradation (glycogen, macromolecules, protein, etc.), but, as previously mentioned in Chapter 1, most cells in the oligotrophic environment generally do not have these reserve energy sources. According to McCann et al. (1992), the key to starvation survival is the ability to regulate the rate at which macromolecules are degraded, but this statement does not take into consideration survival of organisms in ancient and subsurface environments. As stated previously, in many of these studies, the organisms were deprived of nutrients for a matter of hours, not days, weeks, or years. Wilkinson (1959) hypothesized that the production of storage compounds in bacteria is a result of "hothouse" conditions (growth in laboratory media) and, therefore, questioned the existence of such compounds in nature. Is the level of endogenous metabolism the same for laboratory cultured cells and cells in the environment?

Harrison (1960) stated that "log phase cells in the absence of food die as the result of loss of cellular substance." Intracellular energy reserves, such as polysaccharides, glycogen, and poly-β-hydroxybutyric acid, formed in energy-rich systems, are degraded, thereby postponing the energy deprivation process. Wilkinson (1959) puts forth three points that must be met in order to show the energy storage function of a compound. These are:

1. The compound is accumulated under conditions when the supply of energy from exogenous sources is in excess of that required by the cell for growth and related processes.
2. The compound is utilized when the supply of energy from exogenous sources is insufficient for the optimal maintenance of the cell, either for growth and division or for maintenance of viability and other processes.
3. The compound is broken down to produce energy in a form utilizable by the cell, and it is, in fact utilized for some purpose which gives the cell a biological advantage in the struggle for existence over those cells which do not have a comparable compound.

He further states that it is important not to rely on the first two criteria for the evaluation of a storage function because some substances may be produced as an attempt to detoxify end products of metabolism that may otherwise accumulate too rapidly and prove toxic. A shunt product normally requires no energy for its synthesis whereas a storage compound usually requires energy for synthesis.

Glycogen promotes *E. coli* and *A. aerogenes* to survive longer but not glycogen-rich *Sarcina lutea* (Strange, 1968). There is a rapid utilization of the glycogen during starvation that provides for energy of maintenance and survival for a short duration (Dawes and Ribbon, 1964a,b). During starvation, carbohydrates are known to undergo degradation (Boylen and Ensign, 1970b; Boylen and Pate, 1973; Robertson and Batt, 1973; Strange, 1968; Strange et al., 1961). Lipid degradation also occurs during starvation (Robertson and Batt, 1973; Thomas and Batt, 1969a).

The ability of soil isolates to survive carbon starvation was correlated to poly-β-hydroxybutyric acid content and not to the glycogen content of the cells, but the poly-β-hydroxybutyric acid was used rapidly (Chen and Alexander, 1972). Glycogen content of cells does not promote survival of the organisms. Glycogen-rich cells of *S. lutea* do not survive as well as glycogen-poor cells (Burleigh and Dawes, 1967). Glycogen in *E. coli* cells are oxidized within 1 to 3 hours and its presence does not appear to favor their survival (Dawes and Ribbons, 1964a; Strange, 1968). Opposite results were obtained by van Houte and Jansen (1970) in *S. mitis* where the presence of glycogen did favor survival.

Nelson and Parkinson (1978) noted in their studies on starvation-survival that the hexose carbohydrates declined with starvation time, with nitrogen-limited cells having an initial higher concentration of hexose carbohydrate than carbon-limited cells. The decline was most rapid in the first 20 hours. This decrease in carbohydrates was noted in a *Pseudomonas* sp., a *Bacillus* sp., and *Arthrobacter* sp. The decline in the hexose carbohydrate pattern was similar to that of RNA, protein, and ATP. Lower levels of carbohydrates were noticed with starvation time in marine isolate 95A (Jones and Rhodes-Roberts, 1981), *Cellulomonas* sp. (Schimz and Overhoff, 1987), *V. cholerae* (Hood et al., 1986), *Klebsiella* from oil well waters (MacLeod et al., 1988), and *A. crystallopoites* (Boylen and Pate, 1973).

In *V. cholerae,* after 7 days of starvation, 88.7% of the carbohydrate disappeared followed by a small decrease at 30 days (Hood et al., 1986). Although the total carbohydrates may decline rapidly during nutrient deprivation, there is a relative increase in the six-carbon sugars from a value of 32% to 58% concomitant with a decrease in the three- and five-carbon sugars (glyceraldehyde, ribose, and the amino sugar, *N*-acetylglucosamine). The outer membrane of *V. cholerae,* according to Hood et al. (1986), is composed of the essential lipid A (hydrophobic in nature), an R core (no 2-keto-3-deoxyoctulosonic acid), and side chains of many tetra- and pentasaccharide units (hydrophilic in nature). It is thus possible that the more hydrophilic molecules of the O–side chains, which are composed of the three- and five-carbon sugars in *V. cholerae,* are more readily utilized under starvation conditions, whereas the six-carbon sugars, which probably make up the oligosaccharides of the R core, are relatively conserved. This concept is in keeping with the data of Kjelleberg and Hermansson (1984) who found that the outer membranes of certain vibrios became more hydrophobic under starvation conditions.

Copious amounts of extracellular mucus-exopolymers are secreted by microorganisms (Geesey, 1982). Biofilm formation by glycocalyx containing polysaccharides is considered to be a universal strategy for bacterial survival (Costerton et al., 1987). Lappin-Scott et al. (1988b) noted that, during long-term starvation, *K. pneumoniae* cells reduced their extracellular polysaccharide production. Extracellular exopolysaccharides were also found to be produced during starvation of a *Pseudomonas* sp. strain S9 by Wrangstradh et al. (1986, 1989, 1990). This exopolysaccharide material produced during starvation in S9 was composed of 28% glucose, 35% *N*-acetylglucosamine and 37% *N*-acetylgalactosamine. However, this polysaccharide material was not formed in S9 under agitated conditions (Wrangstadh et al., 1989). Exopolysaccharide production was noted by Vandevivere and Kirchman (1993), especially when the attached cells were compared to unattached cells. This increase in polysaccharide production was not influenced by starvation since 100-μM citrate was used. They suggest that the solid-water interface may be more favorable than the surrounding bulk liquid, especially under oligotrophic conditions and that the production of exopolymer for some cells may prevent desorption of daughter cells. Thus it is suggested that a more favorable environment is secured for the cell's progeny. The question remains as to how much of the exopolysaccharide material found in nature is due to the starvation of the indigenous microorganisms. For further information concerning exopolysaccharides excreted by microorganisms and utilization of this exopolysaccharide material by other organisms, Decho (1990) should be consulted.

Because poly-β-hydroxybutyrate, as an intracellular store of energy and carbon, provided energy and carbon for cell maintenance, cells resisted starvation better. Thus the presence of poly-β-hydroxybutyrate (PHB) in cells of *B. megaterium* (Macrae and Wilkinson, 1958), *Micrococcus halodenitrificans* (Sierra and Gibbons, 1962), *Hydrognomonas* (Hippe, 1965), and *Azotobacter agilis* (Sobek et al., 1966). Mårdén et al. (1985) showed that PHB was utilized during starvation. In *Vibrio* S14, the PHB that accumulated during growth was utilized in 3 hours from the onset of starvation (Malmcrona-Friberg et al., 1986). *M. halodenitrificans,* rich in poly-β-hydrobutyric acid, survived 10 times as long as the cells without the energy reserve (Sierra and Gibbons, 1962) and increased survival times were reported for *Spherotilus discophorus* (Stokes and Parson, 1968), *Azotobacter agilis* (Sobek et al., 1966), and microbial cells containing poly-β-hydroxybutyric acid (Dawes and Senior, 1973). Dawes and Senior (1973) reviewed the role and regulation of energy reserve polymers in microorganisms and stated that organisms accumulating reserve compounds during a period of plentiful nutrition are protected from starvation. Dawes and Senior (1973) further stated that the physiological role of reserve compounds cannot be precisely answered. Reserve compounds are not synthesized by all the various species of microbes and microbes in oligotrophic ecosystems probably do not synthesize these reserve compounds, yet they have the capacity to survive. During starvation, the substrates of the endogenous metabolism of *A. crystallopoietes* are the ribonucleic acids and protein (Boylen and Ensign, 1970b). The loss of protein and ribonucleic acids with starvation time is noted in the next section.

CELLULAR MACROMOLECULAR DEGRADATION

The degradation of cellular macromolecules contributes to the endogenous metabolism that occurs when cells no longer have an external source of energy. Furthermore, it supplies the energy for cells undergoing the starvation-survival process in the first two stages of starvation. Most of the studies dealing with endogenous metabolism or the utilization of cellular components do not take into consideration the possibility of a mechanism of survival against the lack of nutrients on a long-term basis.

RNA and Ribosomes

When cells are grown in the laboratory, they contain excess polymers such as carbohydrates, protein, RNA, etc. Only two-thirds of the cell population of *E. coli* grown at low dilution rates in continuous culture (doubling time of 16 hours) was shown to be actively synthesizing protein at a given time (Koch, 1979; Koch and Coffman, 1970). The protein synthesizing machinery, which is expensive in terms of energy cost, is fully functional in slow and rapid growing cells because it is needed for reproduction of the cell when conditions become right (Koch, 1971). Starved *E. coli* break down and resynthesize their macromolecular components (RNA and protein) at a rate of 5% per hour, whereas growing cells do not appear to exhibit any appreciable turnover of protein (Mandelstam, 1960). In washed cells of *E. coli*, ^{35}S methionine can be incorporated (Melchior et al., 1951) and the protein synthesis can occur. Even nitrogen-starved cells can synthesize inducible enzymes (Mandelstam, 1958a,b; Pollock, 1962).

Upon starvation, the cell makes major adjustments that include a reduction in RNA synthesis and accumulation; increased protein turnover; reduced membrane transport; reduced endogenous synthesis of nucleotides, glycolytic intermediates, lipids, fatty acids, polyamines, and peptidoglycans; increased control over protein translation; and cAMP accumulation. The foregoing results from the accumulation of guanosine 5'-diphosphate 3'-diphosphate (ppGpp) due to an idling reaction of the ribosomes and uncharged RNA by way of a protein known as *stringent factor* (Cozzone, 1981). This signal molecule rapidly accumulates at the onset of starvation, indicating a stringent response was provoked (Nyström et al., 1990b). In *Vibrio* sp. S14, this stringent control appears to be relieved as the ppGpp level decreases accompanied by relative increases in the respiratory activity and rates of macromolecular synthesis. A threshold concentration of ppGpp must be exceeded for the response to be manifest. According to Stouthamer (1984), three phases of the growth rate can be distinguished: first, with sufficient tRNA, followed by a phase in which aminoacyl tRNA becomes limiting and ppGpp formation begins, and the final phase where the stringent response is manifested. In the latter phase, the cellular metabolism is put in check at the expense of some limited energy expenditure from cAMP and ppGpp formation and from the proofreading of proteins (Mason et al., 1986). It is not clear whether the energy expenditure is equivalent to the maintenance energy (Mason et al., 1986).

Ammerman and Azam (1987) found cAMP to be 2 to 7% of the total cellular mass (10^4 to 10^5 cells/ml) in waters from the Southern California coast. cAMP was taken up by a high-affinity transport system. The cAMP uptake increases if energy becomes limiting and causes cells to enter the stationary phase. However, if the stationary phase is caused by nitrogen limitation, then there is no increase in the cAMP uptake. From their study, it appears that cAMP uptake capacity varied greatly under different degrees of energy limitation. According to these authors, marine bacteria living in the dilute ocean environment may transport and accumulate cAMP for regulatory purposes when energy is limiting. Could the detection of cAMP in a natural microbial population be a method to determine the existence of starved cells?

Herbert (1961) discusses the chemical composition of microorganisms as a function of their environment. The nucleic acid content of bacteria can vary widely and in different bacteria the values for RNA content may be as low as 1.5% and as high as 40%, whereas within a single strain, a 16-fold variation in RNA content has been observed under different conditions of growth.

The mean RNA level was found to be considerably influenced by the culture conditions (Sepers, 1984). Chemostats were adjusted to provide a sinusoidal and stepwise variation at different dilution rates. The dilution rates varied from 0.03 to 0.17 and from 0.08 to 0.18 on the sinusoidal curve with 1 and 10 hours for a complete sinus respectively. The dilution rate was varied stepwise from 0.03 to 0.18 and vice versa with periods of 1 and 10 hours respectively. The mean RNA level (mgRNA/mg protein) was the lowest (0.213) at a dilution rate (D = 0.10/hour) compared to 0.253 and 0.232 at D = 0.13/hour and 0.15/hour respectively. The protein variation was considered to be very small in all experiments, especially when the error of protein measurement was taken into consideration. The respiratory activity increased with higher dilution rates, which probably reflects the total enzyme system involved in the uptake and respiration of aspartate used in the system.

When growth ceases, the net synthesis of protein, RNA, cell wall, lipid, and DNA ceases, even though turnover of these molecules continues (Dawes, 1976; Strange, 1967). When cells are starved, they use their own cellular material. RNA appears to be the polymer that is most readily used during starvation. The amino acid pool, adenine nucleotides, and probably mRNA initially decreases but then stabilizes. Nonessential carbohydrates are used first, then RNA and DNA. When approximately one half of the RNA is consumed, the cells begin to die.

Montague and Dawes (1974) reported an 80% loss of cellular RNA in *Peptococcus prevotii* after 15 hours of starvation and an 85% loss of RNA from *A. crystallopoietes* after 30 days of starvation. Degradation of RNA seems nearly universal in starving bacteria (Burleigh and Dawes, 1967; Campbell et al., 1963; Clifton, 1966, 1967; Dawes and Large, 1970; Gronlund and Campbell, 1963; Holden, 1958; Lovett, 1964; MacKelvie et al., 1968; Robertson and Batt, 1973; Strange et al., 1963; Thomas and Batt, 1968; Jones and Rhodes-Roberts, 1981). Reduction in RNA levels has also been noted in *B. linens* (Balkwill et al., 1977), *E. coli* (Rybkin and Ravin, 1987), *P. prevotii* (Montague and Dawes, 1974), *Corynebacterium* spp. (Boylen and Mulks, 1978), *A. aerogenes* (Postgate and Hunter, 1962), *Lactobacillus arabinosus* (Holden, 1958), *Arthrobacter* spp. (Scherer and Boylen, 1977), and *Zymomonas* spp. (Dawes and Large, 1970). In *V. cholerae*, there is little decrease in the RNA level upon starvation—only a 2% decrease in 14 days and 20% in 30 days (Hood et al., 1986). RNA level declined most rapidly in the first 20 hours in *Bacillus, Arthrobacter,* and *Pseudomonas* (Nelson and Parkinson, 1978).

In the water space, the rod-shaped cells of *A. crystallopoietes* degraded approximately 40% of their carbohydrate reserve, 20% of their protein, and 60% of their RNA in the first 70–80 hours of starvation, followed by much lower rates of degradation, but the water space remained nearly constant during the starvation period (Boylen and Pate, 1970b). These cells, on the other hand, degraded 5% of their protein, 10% of their RNA, and 40% of their carbohydrate during the first 80 hours of starvation, whereas the spherical cells degraded significantly less protein and RNA but utilized their carbohydrate reserve at approximately the same rate (Boylen and Ensign, 1970b). Neither cGMP or cAMP were found in starving cells (Leps and Ensign, 1979b).

RNA values of Ant-300, as measured by the orcinol test, initially decreased and then increased linearly during starvation (Amy et al., 1983a). Synthesis of RNA during starvation is not common (Maaløe and Kjeldgaard, 1968; Schaechter, 1961), but one case has been reported (Borek and Ryan, 1958) with a methionine-requiring mutant of *E. coli* which is incapable of the stringent response. It is possible that the orcinol-positive material was not RNA but another molecule containing a pentose, such as aldoheptoses and hexuronic acid (Dische, 1955). The form in which RNA is stored is not known but thin sections of 5.5-week starved cells did not show an increase in ribosomal bodies (Amy et al., 1983a). Cells labeled with [2-^{14}C]uracil during growth showed no respiration of the ring structure during starvation but did show some leakage of nonmacromolecular labeled molecules into the starvation menstruum (Amy et al., 1983a). A similar breakdown of RNA with cellular leakage in starved bacteria occurred with the release of phosphate and base fragments into the menstruum, and the utilization of most of the ribose was reported by Postgate and Hunter (1962). It is possible that RNA does not function during starvation but either is held in reserve to function when the proper environment presents itself (Gronlund and Campbell, 1963) or is the result of regulation in the cell (Gallant, 1979).

When *E. coli* is leucine starved, 40% of the leucyl-t RNA is maintained by protein turnover (Morris and DeMoss, 1965). Respiratory capacity usually increases with increasing growth rate (Tempest and Neijssel, 1978). During the first phase of the starvation of *Vibrio* sp. strain S14, Nyström et al. (1990b) found that there is an accumulation of guanosine 5'-diphosphate 3'-diphosphate (ppGpp) and a decrease in RNA and protein synthesis. In the second phase, there is a temporary increase in the rates of RNA and protein synthesis between 1 and 3 hours of starvation, paralleling a decrease in the ppGpp pool. In the third phase, after 3 hours, a gradual decline in macromolecular synthesis occurs.

Ribosomes are also known to be degraded during starvation (Boylen and Pate, 1973; Gronlund and Campbell, 1965; Maruyama and Okamura, 1972). In the course of starvation in enteric bacteria, the degradation of ribosomes has been noted (Mandelstam, 1960; Mandelstam and Halvorson, 1960). Ribosome dissociation and subsequent rRNA degradation to nucleotides occurred when *E. coli* was starved (Kaplan and Apirion, 1975a,b). In *E. coli,* under phosphate limitation, loss of viability was followed by ribosome degradation (Davis et al., 1986). In

Vibrio spp., the washed cells in a starvation menstruum retained their viability over a much longer period (e.g., 10 days) in spite of the decreases in total RNA (Albertson et al., 1990a; Flärdh et al., 1991). Thus, cells retained a residual ribosome population that was apparently important in the recovery process. The substrate for long-term endogenous metabolism of *A. crystallopoietes* are ribosomes (Boylen and Ensign, 1970b), In *E. coli,* Mandelstam and Halvorson (1960) and Willetts (1967) noted that massive amounts of RNA and proteins are produced during starvation. Thus, Leps and Ensign (1979a) suggest that the key to long-term starvation-survival may be an efficient control of autodegradative enzymes and control of these enzymes may be exerted by reducing the energy level of the cell.

During starvation, the ribosomes of *Vibrio* sp. strain CCUG 15956 were lost slowly (half-life, 79 hours) and existed in large excess over the apparent demand for protein synthesis (Flärdh et al., 1992). After 24 hours of starvation, the total rate of protein synthesis was 2.3% of the rate during growth and after 3 days, this rate was 0.7% of the rate during growth, while the relative amounts of ribosomal particles at these times were 81 and 52%, respectively. These ribosomal particles consisted of 90% 70S monoribosomes. No polyribosomes were detected in starved cells. In starved *E. coli,* there is a rapid breakdown of polyribosomes to single 70S ribosomes having a half-life of 1.5 to 2.5 min at 37°C (Dresden and Hoagland, 1967). According to Flärdh et al. (1992), the absence of polyribosomes in carbon-deprived cells indicates that either the availability of mRNA or the rate of translation initiation limits protein synthesis. During starvation, the 70S monoribosomes were responsible for most of the protein synthesis; however, some protein synthesis was detected in the polyribosome region on sucrose density gradients (Flärdh et al., 1992). The authors concluded that nongrowing carbon-starved *Vibrio* S14 possesses an excess protein synthesis capacity, which results in an immediate synthesis when there is an upshift in nutrients. In spore-forming bacteria, there is a conservation of structural ribosomes and a decrease in total lipids, DNA, and proteins (Setlow, 1983). This also occurs in *V. cholerae* (Hood et al., 1986).

When examining *Vibrio furnissii* and *Vibrio alginolyticus,* Kramer and Singleton (1992) found that they differed in response to starvation. *V. furnissii* responded to starvation in a way that is more or less typical of most *Vibrios,* whereas *V. alginolyticus* decreased in viable cell numbers with time. Both organisms showed cell size reduction when starved. Cells retained 10 to 26% of their original rRNA content after 15 days of starvation. Strange et al. (1963) also found that degradation of rRNA did not increase the RNA concentration in the supernatant fraction. RNA degradation products, mainly deaminated bases, were quickly released into the suspending fluid (Strange et al., 1963). Starvation resistant mutants of *A. aerogenes* are smaller in size, lower in RNA:DNA ratio, have greater light-scattering ability and are more heat resistant than logarithmic cells (Harrison and Lawrence, 1963),

Cells in a carbon-limited state of slow and balanced growth have elevated levels of mRNA, tRNA, amino acids, and RNA polymerase (Koch, 1971); hence, starving and slow-growing cells divert their energy toward synthesis of the macromolecules that are necessary for protein synthesis. The mechanism that leads to the degradation and reutilization of the constituent moieties of ribosomes and possibly of other proteins associated with the protein synthetic machinery of the cells is activated when *E. coli* is starved (Koch, 1979). The ribosome content of a cell should be inversely proportional to the doubling time (Koch, 1976). The extra RNA story is well discussed by Koch (1971, 1976). (Extra RNA is excess RNA in a cell not needed for protein synthesis.) Now it is known that starvation proteins can be made when substrate becomes exhausted. Nevertheless, this excess stable RNA, not involved in protein synthesis, may increase the cell's ability to respond to sporadic inputs of nutrients (Koch, 1971).

The amount of 16S rRNA of a natural microbial community better estimates the physiologically active microorganisms that are present than the absolute genetic diversity and community structure (Moyer et al., 1994). Hence, estimating the diversity at the DNA level, rather than at the RNA level, theoretically provides a more accurate measurement of taxonomic group variability by potentially detecting slowly growing or dormant microorganisms present within a community (Moyer et al., 1994). However, Moller et al. (1995) states that using rRNA content as the sole indicator of the physiological states of cells in complex natural systems may, in the presence of starved or nongrowing cells, give misleading results.

The synthesis of mRNA continues during amino acid starvation of stringent *E. coli* cells (Koch, 1971). In energy- or nutrient-starved *Vibrio* S14, the mean half-life of mRNA is increased from 1.7 to 10.3 minutes, which is due both to an overall mRNA stabilization and the presence of extremely long-lived starvation-specific messages (Albertson et al., 1990c).

DNA

Brdar et al. (1965) demonstrated that in starving *E. coli* cells there was a net increase in DNA synthesis, but no net increase in protein or RNA. These authors point out the possibility that the limited potential for utilization of energy for anabolic processes in starving cells is oriented mainly toward DNA synthesis, because the amino acid pool was not utilized for protein synthesis. Thus, DNA synthesis is favored over the synthesis of other macromolecules.

DNA is not usually degraded during starvation (Boylen and Ensign, 1970b; Burleigh and Dawes, 1967; Dawes and Large, 1970; Dawes and Ribbons, 1964a,b; Holden, 1958; Thomas and Batt, 1969b; Warren et al., 1960) or increases slightly (Boylen and Ensign, 1970b; Brdar et al., 1965; Campbell et al., 1963;

Gronlund and Campbell, 1963; MacKelvie et al., 1968; Strange et al., 1963) but Harrison and Lawrence (1963) reported that *A. aerogenes* degraded 48% of its DNA in 19 hours when only 1% of the population was viable. However, Postgate and Hunter (1962) reported no significant reduction in DNA for the same organism.

Conservation and protection of the genome is essential for the survival and recovery of nongrowing bacterial cells (Lebaron and Joux, 1994). In starving cells of Ant-300, Novitsky and Morita (1977) noted that there was a decrease in the DNA content of starving cells. This decrease was rapid during the first 14 days and thereafter decreased slightly. After 6 weeks of starvation, the cellular DNA decreased 46%, while the DNA in the menstruum could account for only 5% of this decrease. However, in exponentially growing cells of Ant-300, the average number of nuclear bodies per cell, as determined by Feulgen staining, was 1.44 and 4.02 after 63 and 95 hours after inoculation, respectively. There is a linear relationship between the average number of nuclear bodies per cell and the percent increase in cell numbers during the first week of starvation. During the first week of starvation when Ant-300 cells increase in numbers, its DNA is degraded. Nalidixic acid, an inhibitor of DNA synthesis, had no effect on the starvation process (Novitsky and Morita, 1977). After starvation of Ant-300, Amy et al. (1983a) reported only one nuclear body per cell was observed. However, in *Arthrobacter* spp. (Scherer and Boylen, 1977), *Zymomonas* spp. (Dawes and Large, 1970), *A. aerogenes* (Postgate and Hunter, 1962), and *B. linens* (Boyaval et al., 1985), the DNA level was stable or increased when the organisms were starved.

DNA in *V. cholerae* declined at a constant rate from 20 to 80% in 7 to 30 days of starvation (Hood et al., 1986). Hood et al. (1986) assumed that the initial reduction in DNA/cell may be related to the increase in cell number which occurs during fragmentation. Because there is no increase in the number of cells after the first week of starvation but a continuing loss of DNA/cell, they suggest that it represents a reduction in extra DNA copies, configuration changes in the molecule, or some other unknown process. According to Lebaron and Joux (1994), neither DNA synthesis nor degradation seems to occur during starvation.

The relationship between cell size and DNA content of indigenous soil microbes was determined by Bakken and Olsen (1989), where they found that microbes passing through 1.0-, 0.8-, 0.6-, and 0.4-μm polycarbonate filters contained 2.0. 1.8, 1.7, and 1.8 fg DNA/cell. These DNA/cell values can be contrasted with 8.5 fg DNA/cell from an *E. coli* (Bakken and Olsen, 1989). They also noted that small indigenous cells from soil were more resistant to sonication, especially the cells that passed through the 0.4-μm polycarbonate filter. This may be due to the presence of more gram-positive cells, which are presumably more resistant to sonication.

Amounts of DNA, RNA, protein, and carbohydrates per cell increase progressively with growth rates (Maaløe, 1960). "If there are demands upon the cell's genetic content imposed by a need to be physiologically plastic, and the genome size necessary must be kept to a minimum, there cannot be a permanently redundant (spare) DNA" (Tempest et al., 1983). High growth rates may be incompatible with a large genome size (Cavalier-Smith, 1981). Both DNA and RNA do increase during glucose starvation in *B. subtilis* (Maruyama et al., 1977) but the changes in genetic marker frequencies during the course of starvation decreases.

As stated previously, Ant-300 contained up to 6 nuclear bodies per cell when grown in batch culture (Novitsky and Morita, 1976), whereas *S. typhimurium* contained one or two genomes (as evidenced by flow cytometric analysis) (Lebaron and Joux, 1994). In the latter organism the genomes are fully replicated at the onset of starvation. Sixty percent of the cells contained one genome and 40% contained two genomes. These two genomic populations persisted for up to 35 days of starvation. *Alteromonas haloplanktis,* like Ant-300, contains up to 6 nuclear bodies per cell and most contained at least two genomes (Labaron and Joux, 1994). Each replication cycle was completed in the early stage of starvation by stopping cells in the partition step of the cell cycle prior to division (Lebaron and Joux, 1994). The fragmentation process, during the initial stages of starvation observed in Ant-300, results in cells containing one genome (Novitsky and Morita, 1976). The occurrence of multigenome content per cell is attributed to nonsynchronous initiation of replication in log phase cells (Bove and Lobner-Olesen, 1991; Neidhardt et al., 1990). *A. denitricans* also has multicopies of the genome (Thorsen et al., 1992) and probably has the same traits as *A. haloplanktis* (Lebaron and Joux, 1994).

Tempest et al. (1983) point out that free-living microbes must be physiologically highly flexible, but this flexibility would require that the microbe carry all the relevant extra genetic information. High growth rates may not be compatible with a large genome size (Cavalier-Smith, 1981). Genome size must then be kept to a minimum and the organisms cannot carry much redundant (spare or accessory) DNA (Tempest et al., 1983). Ecophysiological consideration requires that the genome size be kept to a minimum, yet functionally replete, but this would seem to require the presence of mutable material (Clarke, 1974; Hartley, 1974). Yet, in starvation, the genome size does decrease (Moyer and Morita, 1989b). It may be possible that starvation results in the cell getting rid of the redundant DNA that one sees in glutonous bacteria. Approximately 80% of the genetic content of *E. coli* is committed to maintaining physiological flexibility (Koch, 1976).

Under balanced exponential growth conditions, only 20% of the DNA was read (Kennell, 1968) and Bishop (1969) found 28%. Thus, Koch (1976) postulated that 80% of the genetic content of *E. coli* is committed to maintaining physiological flexibility or that the DNA was not read every generation and may be simply nonfunctional DNA that has not yet been eliminated by selection. Koch (1976) argues that this DNA contains the information that is needed to handle important problems previously handled by *E. coli* but not needed under cultivation conditions.

Proteins

During starvation, the free amino acid pool of organisms disappears (Horan et al., 1981). Removal of the beginning amino acids of newly formed or forming proteins during starvation also results (Koch, 1979) and this process may involve as much as 5% of the peptide bonds formed by the bacterial cell (Pine, 1973). The energy resulting from the foregoing could provide for 0.05% new protein (Koch, 1979). The degradation of "ill-made" and/or "self-destruct" proteins also occurs and could provide for the synthesis of 1 to 5% new protein (Koch, 1979). The third group of proteins that undergoes degradation during starvation includes lipoproteins, flagella, pili, and small amounts of periplasmic proteins (Koch, 1979). These latter compounds may undergo partial digestion, yielding amino acids that may be reutilized by the cell. Very few proteins not associated with starvation are subject to turnover to a significant degree during starvation and any new protein synthesis is distributed among various kinds of proteins in the same proportions during carbon-limited chemostat growth as during starvation.

RNA and protein degradation are regulated in a coordinated manner. According to Goldberg et al. (1973), for cells in a nutrient-poor environment, protein catabolism appears to increase in order that the cells can synthesize new enzymes appropriate to the new conditions. Thus, it is also possible that starving cells can distinguish between the proteins they need to survive in the starvation process. The accumulation of cyclic AMP in starved cells may be the stimulus for the increased proteolysis. However, the addition of high concentrations of cyclic AMP in growing and nongrowing cells does not augment protein degradation. Intracellular proteolysis is not stimulated by ppGpp in cells permeabilized by cold shock and toluenization (Pine, 1973).

Boylen and Ensign (1970b) found that 40% of the cellular protein of *A. crystallopoietes* was degraded after 30 days of starvation. Protein degradation during starvation has also been reported by others (Campbell, et al., 1963; Fan and Rodwell, 1975; Gronlund and Campbell, 1963; MacKelvie et al., 1968; Strange, 1967; Strange et al., 1961). Robertson and Batt (1973) found that *N. carallina* degraded cellular protein but only after complete degradation of its polysaccharide. Free amino acids are also know to be degraded during starvation (Burleigh and Dawes, 1967).

Contrary to expectation, starvation of various auxotrophs of *E. coli* for any one of several amino acids does not lead to complete deacylation of the affined tRNA molecules. During leucine starvation, some of the leucine tRNA remains acylated with leucine and some of the leucine carries a protector against periodate oxidation (Yegian and Stent, 1969). These investigators also found the same situation for histidine and isoleucine (when the organisms were starved for exogenous threonine). Similar observations have been reported for methionine (Martin et al., 1963), for leucine (Morris and DeMoss, 1965), and for isoleucine (Ezekiel, 1964).

Proteins of *V. cholerae* declined at a constant rate in 30 days of starvation, losing 70% of their original values in the first 20 hours (Hood et al., 1986). Proteolysis during the first hour of starvation was found to be 16 times greater than during the exponential phase of growth in a marine *Vibrio* sp. (Nyström et al., 1988). Protein degradation has been noted in a starving marine isolate 41 (Jones and Rhodes-Roberts, 1981), in starving *E coli* and *E. coli* K12 (Reeve et al., 1984a,b; Rybkin and Ravin, 1987) and starving *S. typhimurium* (Reeve et al., 1984a,b), starving *Corynebacterium* spp. (Boylen and Mulks, 1978), starving *A. aerogenes* (Postgate and Hunter, 1962), starving *B. linens* (Boyaval et al., 1985), and starving *V. cholerae* (Hood et al., 1986). Others (Goldberg and St. John, 1976; Miller and Green, 1983; Nath and Koch, 1970, 1971a,b; Yen and Miller, 1980) also noted protein degradation to occur during starvation. During slow growth and starvation, Koch (1971, 1979) noted that a reduced but rather complete set of proteins remains. Only one-quarter of all the cellular protein in *E. coli* is subject to degradation (Pine, 1973). This is in contrast to animal cells where nearly all of the protein is subject to degradation. Upon long-term starvation, this large inaccessible protein in *E. coli* cannot be recycled because of its structure or intracellular localization.

Although it is recognized that protein degradation occurs during starvation, it appears that enzyme proteins are not so readily degraded. The level of NADH oxidase of spherical and rod-shaped cells of *A. crystallopoietes* dropped to 20 and 30%, respectively, in the first 1 to 2 days of starvation, then remained constant for 9 days (Meganathan and Ensign, 1976). Meganathan and Ensign (1976) also found that catalase activity decreased continuously and reached a low level in 9 days; enzymes involved in the the tricarboxylic acid cycle and glucose metabolism were stable for 1 week. Succinic dehydrogenase, fumarase, and aconitase were stable during the 21 days of starvation. Novitsky and Morita (1978a) were able to demonstrate that starved cells of Ant-300 were capable of utilizing glucose immediately when starved for 7 weeks. *Vibrio* sp. strain S14 (5 hours) and *Pseudomonas* sp. strain S9 (4 or 24 hours) actually produced greater proteases in response to starvation than at the initiation of starvation, then decreased (Albertson et al., 1990a). Nevertheless, significant levels of exoprotease activity remained after 120 hours. De novo protein synthesis was responsible for this increase in exoprotease enzymes. Fatty acid esterase and extracellular alkaline phosphatase was maintained during starvation of Ant-300 (Amthor, 1989).

During the initial phase of starvation in Ant-300 protein, RNA and DNA showed a similar pattern of a rapid decrease with time followed by a slight increase in the protein and DNA (Amy et al., 1983a). These levels stabilized at 2.9×10^{-8} and 2.5×10^{-7} µg per cell, respectively. According to Koch

(1971, 1979), *E. coli* is not capable of degrading certain classes of protein; hence the protein concentration in the culture suspension remained high. Protein is known to be metabolized during starvation in *P. aeruginosa* (Gronlund and Campbell, 1963) and in marine bacteria (Jones and Rhodes-Robert, 1981). When [^{35}S]methionine-labeled Ant-300 cells were starved, there was a linear increase in the level of radiolabel in the cell filtrates for approximately 2 weeks and a concomitant loss of radioactivity in the cells (Amy et al., 1983a). Net protein degradation appears to stop after 2 to 3 weeks. The increase of radioactivity in the cell filtrates suggests that either viable, dying, and/or dead cells degraded or leaked labeled protein through normal metabolism or autolytic processes, or that extensive turnover of proteins took place. Most of these processes were probably not caused by dead or dying cells, because the majority of the original population still remained viable after 3 weeks of starvation.

Reeve et al. (1984a,b) demonstrated that protein degradation was important in the survival of starving *E. coli* to provide the amino acids for protein synthesis, because in its immediate environment, no exogenous amino acids are present. The disappearance of some proteins and the production of new proteins have been noted by various investigators (Amy and Morita, 1983b; Mandelstam, 1958b, 1960; Reeve et al., 1984b; Jouper-Jaan et al., 1986). Thus, the proteins degraded and synthesized during the first few hours of starvation were critical for survival.

During starvation of Ant 300 grown at different dilution rates and batch culture, the concentrations of macromolecules (DNA, RNA, and protein) were monitored on a per total cell basis and on a per viable cell basis (Moyer and Morita, 1989b). On a total cell basis, there is a wide fluctuation in all the macromolecules in stage 1—showing at least 2 to 3 characteristic peaks in concentration level. During stage 2, the DNA levels steadily dropped on a per total cell basis and then stabilized at approximately 42 to 56 days, whereas the RNA concentration remained higher than the DNA concentration with the exception of the D $=$ 0.015/hour cells. Fluctuations in the protein concentration followed the RNA concentrations closely. These fluctuations, especially in stage 1, suggest a redistribution of cellular constituents, which prepares the cells for long-term starvation-survival. (The three stages are depicted in Fig. 7.10.) An initial energy-dependent reorganization involving an increase in endogenous respiration occurs in the marine *Vibrio* sp. strain S14, which is followed by a period of metabolic shutdown (Mårdén et al., 1985). During the first few hours of starvation in *Vibrio* sp. strains S14 and DW1, both DNA and protein synthesis were observed (Mårdén et al., 1988; Nyström et al., 1986). During the adjustment period, all internal energy reserves are utilized. The depletion of poly-β-hydroxybutrate, with *V. cholerae* (Hood et al. 1986) and *Vibrio* sp. strain S14 (Malmcrona-Friberg et al., 1986), and the disappearance of the fatty acids in the membrane in *V. cholerae* (Guckert et al., 1986) are evidences for the utilization of internal energy sources. Evidence for the redistribution of cellular constituents during starvation is the production of high-affinity uptake systems (Geesey and Morita, 1979) and increased chemotaxis in Ant-300 (Torrella and Morita, 1982).

Stage 2 in the starvation of Ant-300 was a period in which the fluctuations of intercellular constituents were buffered from any drastic changes due to the decrease in metabolic activity. As previously stated, Ant-300's endogenous respiration was shown to decrease to 0.0071%/hour after 7 days and remained constant for the rest of the experiment (28 days) (Novitsky and Morita, 1977). During stage 2, the metabolism slows to a point that leaves the cell unable, or with limited ability, to catabolize macromolecules such as DNA, RNA, and proteins, presumably due to the exhaustion of utilizable carbon during the first stage. Cell viability diminishes to only 0.3% of the total cell numbers and remains there for the duration of the starvation-survival (Moyer and Morita, 1989a).

Stage 3 is the final adjustment period and represents the conservation of essential cellular molecules to permit long-term survival. This stage also retains the mechanisms necessary for energy utilization. Thus the cell becomes physiologically adapted to withstand long periods of energy deprivation. This starvation-survival mode is a state of metabolic arrest (Morita, 1988) and the cell remains in this mode, if and until energy becomes available. In all probability most microorganisms undergo the three stages mentioned. However, the time frame for each stage is species dependent.

Even when the DNA, RNA, and protein are normalized to viable cells (Moyer and Morita, 1989b), two to three characteristic peaks in the macromolecules occur in stage 1 for each population. During stage 2, the increases in macromolecule levels reflect a bias due to the magnitude of losses in viable cells. Unfortunately, the inability to separate viable cells from dead or dormant cells in the population results in this bias. Stage 3 showed a stabilization and conservation of the macromolecule concentrations with respect to the viable cell counts. This stabilization and conservation of essential cellular molecules permits the survival for long periods of time without nutrients, as well as to retain the mechanism by which the cell can use energy again when it becomes available. Thus, the cell becomes physiologically fit to withstand the loss of energy in the ecosystem and adapts to the starvation-survival mode. The total cell numbers remaining close to the original levels indicate that the cells do not lyse with the onset of starvation-survival but remain in a state of metabolic arrest, which allows them to survive long periods of time without nutrient input (Morita, 1988). Intact cells, as indicated by an AODC method, retain most of their macromolecules, including DNA and RNA (Amy et al., 1983a; Hood et al., 1986). Since acridine orange is an intercalator of nucleic acids, normalization of nucleic acids to the viable cell counts would produce a greater error than on a total cell count basis.

The DNA from Ant-300 cells was extracted and isolated for electron microscopy by using a modified alkaline lysis (Maniatis et al., 1982) in order to isolate intact chromosomal DNA from the cells while eliminating as much cellular debris as possible. DNA, in unstarved and starved (98 days) logarithmically grown batch cells, is depicted in Figure 7.11a and b, respectively (Moyer and Morita, 1989b). The DNA in the latter is much smaller, indicating that DNA loss occurred during starvation but stabilized in stage 3 between 1.0 and 1.5 fg DNA/cell. The large amount of DNA in Figure 7.11a may represent accessory DNA, because accessory DNA is known to occur (Campbell, 1981). To further emphasize this point, the DNA per total cell was calculated for unstarved and starved populations of Ant-300 (Table 7.6). These figures were then related to the corresponding cell volumes for each of the unstarved and starved cell populations (Moyer and Morita, 1989a). If the assumption is that the DNA packaging efficiency in Ant-300 is equal to that of *E. coli* (0.07 μm^3/5.0 fg), the nucleoid volume for each of the Ant-300 cell populations can be estimated (Table 7.6). The nucleoid volume for starved cells was estimated to be approximately one order of magnitude less than the estimated nucleoid volume for unstarved cell populations. By use of this estimate, the nucleoid volume/cell volume ratio was used in order to achieve a relative volume percentage of the DNA content. Hence, the DNA in the unstarved Ant-300 cell population would be between 30 and 48% for continuous-culture cells, whereas it would be only 6% for batch culture cells. Upon stage 3 starvation, the DNA in the continuous culture cell population of $D = 0.057$/hour and $D = 0.170$/hour (11% and 8%, respectively) decreased to approximately the same level as that of the batch culture (6%). The Ant-300 $D = 0.015$/hour cell population was able to maintain the highest nucleoid/cell volume ratio at 40%, which is nearly the original level of 45%. Thus, we are seeing close to the minimal nucleoid/cell volume needed by the cell in order to maintain all its genetic capabilities. Does this mean that the higher nucleoid/cell volumes noted in fast-growing cells (e.g., cells in the log growth phase) have more (accessory) DNA than needed to carry on all the necessary genetic mechanisms of the cell?

ATP and Adenylate Energy Charge (In Situ Experiments)

In order to gain a proper understanding of the energy status and eventually the growth properties of bacteria, knowledge about the composition and magnitude of the proton-motive force, the magnitude of the adenosine nucleotide pools, and the phosphate potential of bacterial cells under different physiological conditions is required. A proton-motive force is required to maintain the intracellular pools of metabolites; in the starvation phase these pools will gradually disappear (Konings and Veldkamp, 1983). In *S. epidermitis,* Horan et al. (1981) demonstrated that the organism loses all the intracellular amino acids in 24 hours of starvation. Similar observations by Thomas and Batt (1969c) have been made for *S. lactis.* Loss of energy reservoirs and metabolites and ion pools will lead to a loss of viability of the cells, but as long as cells retain their essential cell structure (cell membrane, DNA, RNA, enzymes), they will remain viable (Konings and Veld-

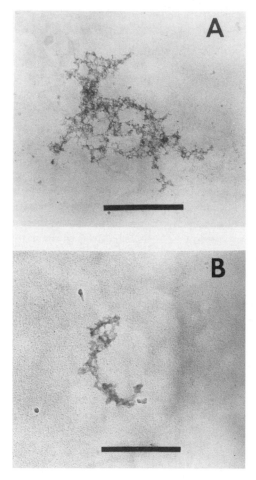

Figure 7.11   Electron micrographs of the DNA molecule from a logarithmic-growth un-starved Ant-300 cell (A) and stage 3 Ant-300 starved cell (B). Bars = 1 μm. Reprinted by permission from *Applied and Environmental Microbiology* 55:2710–2716, C. L. Moyer and R. Y. Morita, 1989; copyright 1989, American Society for Microbiology.

Table 7.6 Effect of growth rate on DNA content in relation to average cell volume for unstarved and starved Ant-300 cells.

Culture	DNA/cell (fg) ± SEM	Cell Vol[a] (μm³)	Estimated Nucleoid Vol (μm³)	Nucleoid Vol/Cell Vol
Unstarved Batch	23.66 ± 0.01	5.94	0.33	0.06
$D = 0.170$ h⁻¹	24.69 ± 0.01	1.16	0.35	0.30
$D = 0.057$ h⁻¹	20.12 ± 0.06	0.59	0.28	0.48
$D = 0.015$ h⁻¹	15.33 ± 0.27	0.48	0.21	0.45
Starved[b] Batch	1.23 ± 0.17	0.28	0.017	0.06
$D = 0.170$ h⁻¹	1.03 ± 0.35	0.19	0.014	0.08
$D = 0.057$ h⁻¹	1.45 ± 0.02	0.18	0.020	0.11
$D = 0.015$ h⁻¹	1.28 ± 0.02	0.05	0.018	0.40

Source: Reprinted by permission from *Applied and Environmental Microbiology* 55:2710–2716, C. L. Moyer and R. Y. Morita, 1989; copyright 1989, American Society for Microbiology.
[a]Data from Moyer and Morita (25).
[b]All cell samples were taken from stage 3 of starvation-survival.

kamp, 1983). The proton-motive force can be built up when energy is supplied. Gradual decrease in the proton-motive force levels during starvation might indicate a gradual increase in the fraction of nonviable cells in the starvation menstruum. Proton motive force and ATPase complex are postulated for survival of *E. coli* by Sjogren and Gibson (1981), which could be coupled with the ability to utilize endogenous reserves. But the time of survival probably rules this out.

There is a quick decrease in the proton motive force in *Thiobacillus acidophilus* (Zychlinsky and Matin, 1983), *Streptococcus cremoris* (Otto et al., 1985), and *Staphylococcus epidermidis* (Horan et al., 1981) when starved in buffer, as determined by the uptake of weak acids and lipophilic cations. On the other hand, Smigielski et al. (1989) maintained a constant apparent proton-motive force for 14 days of starvation.

Starving cells of *T. acidophilus* maintained a ΔpH of 2 to 3 units throughout the starvation period of 200 hours (Zychlinsky and Matin, 1983). After 200 hours, cellular PHB, ATP, proton-motive force, and culture viability were low or not detectable. Thus, a large ΔpH was maintained in the absence of ATP, ATPase activity, respiration, significant levels of proton motive force, and cell viability, which indicates that survival due to starvation was not dependent on chemosmotic ionic pumping. Sjogren and Gibson (1981) postulated that certain enteric bacteria are capable of utilizing acidic conditions (pH 5.5) as an electrochemical gradient to generate necessary high-energy intermediates for prolongation of survival beyond that possible in the environment.

The ATP pool in *Streptomyces faecalis* was shown to decrease during growth and became stabilized as growth ceased (Forrest, 1965), whereas the opposite results were found with *Azotobacter vinelandii* (Knowles and Smith, 1970) and *Nitrobacter winogradskyi* (Eigener, 1976). In the latter case, the ATP level rose during growth and then declined as the cells stopped growing. In other bacteria (Dawes and Large, 1970; Hamilton and Holm-Hanson, 1967; Holm-Hansen and Paerl, 1972; Lee et al., 1971), a constant level of ATP was maintained. The ATP of starving cells declines most rapidly in 20 hours in a *Bacillus, Pseudomonas,* and *Arthrobacter* isolated from Devon Island, Canada (Nelson and Parkinson, 1978). There is a decrease in ATP in *E. coli* K12 also (Tkachenko and Chudinov, 1987).

Kurath and Morita (1983) studied the starvation-survival effect on a *Pseudomonas* sp. and found that log phase cells contained 6.5×10^{-10} µg ATP/cell (Fig. 7.12). In 5 days of starvation, the ATP content decreased by 80%. These starved *Pseudomonas* cells dropped from 248 ng of ATP/ml in 4 days and stabilized at 2–3 ng ATP/ml after 20 days. These data were converted to values of ATP/cell. The respiring cell count (INT) appears to be a better denominator for expressing the ATP/cell than the viable count (plate count) or total direct count (AODC). There is a decreasing amount of ATP/respiring cell for the first 8 days of starvation, followed by a gradual increase back to the original level of 6.5×10^{-10} µg by 40 days of starvation. Kurath and Morita (1983) raised the question of how certain cells are able to maintain ATP pools without exogenous substrates, but it would be most advantageous for starving cells to do so. Amy et al. (1983a) found that ATP levels per viable cell fell during the first 4 days of starvation, rose and then fell, followed by a steady rise again. However, Oliver and Stringer (1984) found that the greatest decline of ATP occurred during the first week of starvation in the same organism. The ATP levels found in growing *Vibrio fluvialis* are at the lower end of the range of ATP concentrations reported for some bacteria by Karl (1980). In active growing cells, 15 to 20% of the ATP is used for active transport of substrate (Stouthamer and Bettenhaussen, 1973). The retention of ATP would ensure that active transport mechanisms are functional if and when nutrient substrate(s) requiring active transport becomes available to cells for energy.

The ATP levels of starving surface-bound and free-living marine *Vibrio* sp. DW1, a rod-shaped motile marine isolate BMS1, and *S. marcescens* EF190 were measured by Kjelleberg and Dahlbäck (1984). They found that free-living DW1 from the midlog phase decreased its ATP/biovolume but increased its ATP level after 4 hours, but when the cells were surface bound, the ATP levels increased after only 2 hours. Thus, cultural condition definitely plays a role in the amount of ATP/biovolume. They also suggest that the increase in ATP/biovolume may be due to cryptic growth or leakage of nutrients from the cell (Kjelleberg and Dahlbäck, 1984).

Webster et al. (1985) starved subsurface isolates and determined the nanogram ATP/CFU. The results they obtained were as follows: *Arthrobacter* (9-week

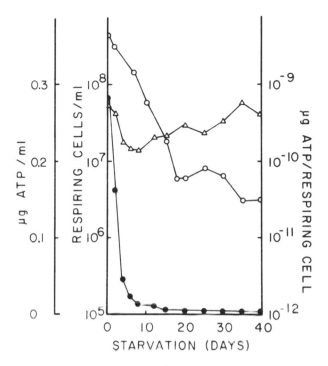

Figure 7.12   ATP content after various periods of starvation. △, ATP/ml of culture; •, ATP/respiring cell; ○, respiring cell count. Reprinted by permission from *Applied and Environmental Microbiology* 45:1206–1211, G. Kurath and R. Y. Morita, 1983; copyright 1983, American Society for Microbiology.

starved), 2.8 ±0.4 × 10^{-7} ng/CFU; *Pseudomonas* (4-week starved), 0.7 ng ±0.2 × 10^{-7} ATP/CFU; *Enterobacter* (8-week starved), 1.3 ±0.1 × 10^{-7}; and *Enterobacter* (13-week starved), 2.9 ±0.5 × 10^{-7} ng/CFU. These values are in the same order of magnitude as a 9-day-starved culture of *A. crystallopoietes* reported by Leps and Ensign (1979b), but 1/10th of that reported by Kurath and Morita (1983). Thus, ATP is retained by the starved cells in order for the cell to remain functional, especially in relation to active transport of substrates.

V. *fluvialis*, studied by Smigielski et al. (1989), formed ultramicrocells of dwarf bacteria upon starvation. After starving for 2–3 days, the viability of the starved cells began to decrease steadily. After 2 weeks, 1.5% of the initial population remained viable. During this period, the respiratory potential of the bacteria decreased by four- or fivefold, most probably as a result of a decrease in the specific activity of the NADH and succinic dehydrogenases in the membranes of starved cells. The data of Smigielsky et al. (1989) suggest that the respiratory activity during starvation is controlled at least partly through a reduction in the specific activity of NADH and succinic dehydrogenases. No reduction in the specific activity of either D-lactate or α-glycerophosphate oxidase was observed in membranes from starved cells. However, the specific activity of ATP synthase was found to be identical in membranes from starved and growing cells, thus, when nutrients become available the starved cell responds immediately by producing ATP rapidly. During short-term starvation, the level of four nucleoside triphosphates fell rapidly and the largest reduction was in the GTP level (Smigielski et al., 1989). The ADP level was higher in starved cells than growing cells and the opposite was true for the GDP. The intracellular water volume per milligram dry cell mass was 0.02 for growing cells, 0.66 for cells starved for 12 hours, and 0.52 for cells starved for 24 hours. The starved cells (1–2 days) maintained a proton motive force that was slightly larger than that of growing bacteria. Thus, the proton motive force (PMF) is maintained in 24-hour-starved cells but it appears reasonable that the membranes of starved cells maintain their energized state for long periods of starvation and that the starved bacteria respond rapidly to the adding of nutrient (Kjelleberg et al., 1982; Amy et al., 1983b). Ellwood et al. (1982) suggest that surface-associated bacteria, especially irreversibly bound cells, are more active due to the high localized proton concentration that results from localized proton concentrations in the chemiosmotic process. In this latter case the protons are passed back through the membrane generating ATP. When the external pH was 7.0, the cells from a growing culture or starved (24 hours) cells had PMF values of -221 mV and -243 mV respectively. Smigielski et al. (1989) noted that, in starved cells, the internal pH was up to half a pH unit higher than in growing cells. Although they state that the observation merits further investigation, an internal pH may be an important feature of starvation-survival response. pH values were found to be identical in starved and nonstarved cells.

There is no apparent correlation between the magnitude of the membrane potential and viability (Horan et al., 1981).

The formula proposed for the energy charge (E.C.) by Atkinson (1971) is: E.C. $= [ATP] + 1/2[ADP]/[ATP] + [ADP] + [AMP]$. Whether prolonged viability of bacteria under starvation conditions is attributed to the E.C. can be debated (Montague and Dawes, 1974; Jones and Rhodes-Roberts, 1981).

Maruyama et al. (1977) determined that the adenylate pool contained mainly ATP (68% at the end of the exponential growth phase), which decreased rapidly to the lowest level at one generation time after the initiation of starvation and then increased again. Their results (Maruyama et al., 1977) suggest that pyruvate, a metabolite of glucose, might be used for the maintenance of the ATP pool. Cellular DNA and protein increases were noted when filaments of *Flexibacter* fragmented (Poos et al., 1972). In *S. epidermidis,* the adenylate E.C. declined markedly within the first 6 hours of starvation (Horan et al., 1981)

Both spherical and rod-shaped cells of *A. crystallopoietes* excreted AMP but not ATP or ADP during starvation but the intracellular E.C. of rod-shaped cells declined during the first 10 hours of starvation and then remained constant for the next 80 hours, then decreased to a value of 0.78 (nearly 1.0 in growing cells) after a total of 168 hours starvation (Leps and Ensign, 1979b). The starvation resistance of this organism could involve the ability to control autodegradation by intracellular nucleases and proteases. In rod-shaped cells, the E.C. decreased in the first 24 hours, then remained constant for the next 80 hours, and then decreased to a value of 0.73 (Leps and Ensign, 1979b). According to Leps and Ensign (1979a,b), the ability of A. *crystallopoietes* (both the spherical and rod-shaped cells) to resist starvation is by their control of autodegradation of intracellular nucleases and proteases.

As long as the E.C. is above 0.50 in stationary phase cells of *E. coli*, the cells remained viable (Chapman et al., 1971). Only when the E.C. fell to 0.10 to 0.30 did starving cells of *S. cerevisiae* lose their viability (Ball and Atkinson, 1975). The viability of *P. prevotii* decreased when the E.C. was below 5.0 (Montague and Dawes, 1974). *P. prevotii* does not form reserve polymers; hence there is a rapid drop in adenine nucleotide content immediately after exhaustion of growth substrates, followed by an increase resulting from RNA degradation to replenish the pool, and then a further reduction in nucleotide content (Montague and Dawes, 1974). The E.C. of *Rickettsia typhi* varies greatly so that when it is starved, a value of 0.2 at zero time is obtained, but with added glucose, it rises to nearly 0.7 almost immediately. When starvation is continued, the values are near 0.1 (Williams and Weiss, 1978). The reason for the maintenance of a high E.C. in starving cells of A. *crystallopoietes* during starvation was the excretion of significant amounts of AMP, and the decrease in ATP was not counterbalanced by enough of an increase in either ADP or AMP (Leps and Ensign, 1979b). Other bacteria have also been reported to excrete AMP into their growth medium (Chapman and Atkinson, 1977). When the E.C. is decreasing in A. *vinelandii* cells, AMP nucleotidase may be involved (Schramm and Leung, 1973; Schramm and Lazorik, 1975). AMP nucleotidase is regulated by an E.C. and becomes more active at low E.C. levels. In most bacteria, starvation results in a decrease in the total adenine nucleotide or ATP content and a lowering of the E.C., but in *Beneckea natriegens* (a marine bacterium), the total adenine nucleotide pool and the E.C. remain unaffected (Niven et al., 1977). In *P. prevoti*, there is a decrease in the E.C. when starved (Montague and Dawes, 1974). In *E. cloacae*, the E.C. in starved cells dropped from 0.80 to 0.40 after 378 days of starvation (Lynch, 1990). Although the E.C. appears to vary among the different species of starving bacteria, it appears to be high enough to take care of the cell's needs when it comes out of the starvation mode when substrates are present.

Wiebe and Bancroft (1975) measured the E.C. on natural microbial populations from the western North Atlantic Ocean water and sediments of a coastal salt marsh and came to the conclusion that E.C. measurements do not show the metabolic state of communities as well as species populations. It is interesting to note that centrifugation brings about a drastic change in the E.C. values, hence it complicates the interpretation of data on microbes in the environment. Previously, Chapman et al. (1977) noted that a variety of physical changes (centrifugation, filtration, deoxygenation) affect the E.C. ratio.

According to Dawes (1976), stabilization of E.C. at the expense of the adenine nucleotide pool size, by the action of AMP nucleosidase or adenylate deaminase, can be only a transient phenomenon in order to minimize the effects of fluctuations in substrate supply. There must be other mechanisms available in the microbial cell to limit ATP utilization when there is a long-term deprivation of energy sources. Any scarce energy sources will be conserved for essential maintenance reactions, which are presumably less sensitive to decreases in E.C. (Knowles, 1977). There are many reports that some bacteria maintain their adenine nucleotide pools at about the same levels as well as exhibit high E.C. values when starved while many bacteria do not retain their adenine nucleotide pools (Knowles, 1977). The foregoing may be a function of the differences between bacterial species in their polymer reserves, functioning of AMP nucleotidase, requirement for energy of maintenance, or differences in experimental techniques (artifacts of cell preparation), etc. (Knowles, 1977). The question must be raised concerning the starvation time used in the above studies. One week certainly is not a long period when one considers the life of microbes in the natural environment. What happens to the E.C. when the cells are starved for a longer period of time? Thus, it appears that E.C. is not a good indicator of the starvation state of the organisms in nature.

Fatty Acids Changes

Harder et al. (1984) suggested that the ability of cells to survive in low-nutrient environments depends on their ability to maintain a functional gradient across the membrane, but this ability to regulate their electrochemical gradients has yet to be investigated (Morgan and Dow, 1986). Certain bacteria have the ability to block their transport systems below critical proton motive force to prevent efflux of accumulated nutrient as has been noted by Konings and Veldkamp (1983). Nickels et al. (1979) and Findlay and White (1983) suggest that poly-β-hydroxybutyrate and poly-β-hydroxyalkanoate concentrations could provide information concerning the nutritional status of microorganisms in nature. In pure culture studies, poly-β-hydroxyalkanoate accumulates during sporulation of *Bacillus* or cyst formation of *Azotobacter* and under conditions of unbalanced growth induced by inorganic nutrient limitation with the supply of carbon and energy from exogenous sources in excess of that required for growth. Unfortunately, PHA and PHB accumulations are limited to environments in which inorganic nutrients, especially N, are limiting and utilizable reduced carbon is in excess (Karl, 1986). Phospholipid breakdown during starvation was noted as early as 1969 by Thomas and Batt (1969a).

Lipids in the membrane were commonly thought not to be energy reserves in the cell (Dawes and Ribbons, 1964a). Although phospholipid breakdown as an energy source during starvation was discussed by Dawes and Ribbons (1962), Dawes and Senior (1973), and Dawes (1976), poly-β-hydroxybutyrate was considered to be the main substrate during starvation. Amy et al. (1983b) suggested that certain bacteria may metabolize membrane lipids during starvation. During long-term starvation of Ant-300, Oliver and Stringer (1984) demonstrated a decrease in the phospholipid content and changes in the fatty acid profiles of the membranes during the induced morphogenesis. More specifically, they found the amount of C16:1 increased during starvation. Malmcrona-Friberg et al. (1986) found that during the first hour of starvation, the total amount of fatty acids decreased. The ratio of monounsaturated to saturated fatty acids decreased and the proportion of short-chain fatty acids increased. As a result, changes in the fluidity and uptake of substrates as well as the use of fatty acids and poly-β-hydroxybutyrate as energy sources took place during starvation.

The vertical profiles of total and particulate lipopolysaccharide were parallel in the seawater column (Maeda and Taga, 1979). Particulate lipopolysaccharide was not shown to be related to the biomass of total bacteria in some cases, but correlated with the numbers of large bacterial cells whose size was greater than 1 μm in length. Since starved bacteria appear to be deficient in lipids, this observation suggests that many of the bacteria in the water column are in a starved state. (Approximately 97% of the bacteria in the sea are gram negative and lipopolysaccharide is a naturally occurring substance in the cell wall of gram-negative bacteria.)

V. cholerae, starved for 7 days, loses 99.8% of its total lipids, which is followed by a small decrease after 30 days (Hood et al., 1986). Phospholipids make up 99.88% of the total lipids while neutral lipids make up only 0.12% of the total lipids in nonstarved cells. Because of the large loss of phospholipids in starved *V. cholerae,* there must be a drastic change in the membrane. Ant-300 cells, when starved, are extremely resistant to disruption compared to nonstarved cells and this is probably due to the loss of phospholipids (M. J. Slominsky, pers. communic.).

The poly-β-hydroxybutyrate disappeared rapidly when *V. cholerae* was starved and there were no detectable levels in 7 days (Hood et al., 1986). The patterns of fatty acid composition changed with starvation—saturated fatty acids increased, while unsaturated and hydroxy fatty acids decreased with starvation. Statistically significant trends of the phospholipid ester-linked fatty acids of 0-day-, 7-day-and 30-day-starved cultures of *V. cholerae* were noted by Guckert et al. (1986). The concentration of the *cis*-monoenoic fatty acids declined with starvation time, while the *trans*-monoenoic fatty acids and the saturated fatty acids (derivatives of the *cis*-monoenoic fatty acids) increased. This, in turn, increased the *cis:trans* ratio. According to the authors, this may be due to the reported high turnover rates of the *cis*-monoenoic fatty acids of the membrane phospholipids and the availability of enzymes for the metabolism of these isomers. Hence, during starvation-induced phospholipid loss, the *cis*-monoenoic fatty acid would be preferentially utilized. Either the ability to synthesize *trans*-monoenoic acid, which are not easily metabolized by bacteria, or modify the more volatile *cis*-monoenoic acids to their cyclopropyl derivatives may be a survival mechanism that helps maintain a functional structurally altered membrane during starvation-induced lipid utilization. The *trans:cis* fatty acid ratio may be used as a starvation or stress lipid index. This ratio would help determine the nutritional status of ultramicrobacteria of cells in the environment. This *trans:cis* ratio has been used by geochemists as an indication of the diagenesis of the *cis*-monoenoic acids, thought to be of direct microbial input (Boon et al., 1978; Perry et al., 1979; Van Vleet and Quinn, 1976; Volkman and Johns, 1977; Volkman et al., 1980). Van Vleet and Quinn (1976) have shown an increase in the *trans:cis* ratio for the total lipid fatty acids with sediment depth and attributed this finding to preferential degradation of the *cis* isomer or clay-catalyzed isomerization to the *trans* isomer. The ratio of *cis:trans* monoenoic compounds from the bacteria in the backflow from a waterflood injection well indicates that the bacteria furthest from the injection site were in a starved state (McKinley et al., 1988). Guckert et al. (1986) suggest that the *trans:cis* data of Van Vleet and Quinn (1976) may be interpreted as a relative measure of membrane stress, such as caused by starvation. Guckert et al. (1986) hypothesize that the viable filterable ultramicrobacteria (thought to represent the in situ marine and estuarine starving bacteria) would have *trans:cis* ratios significantly higher than those reported for bacteria cultures (from 0.01 to 0.08 [Perry et al., 1979; Volkman et al., 1980]) or marine, mangrove, and estuarine sediment surfaces (from 0.01 to 0.09 [Gillan and Hogg, 1984; Guckert et al., 1985; Perry et al., 1979; Volkman and Johns, 1977; Volkman et al., 1980]).

From the study of Guckert et al. (1986), it was found that the amount of phospholipids decreases drastically in starved cells of *Vibrio cholerae*. With such a decrease, the membrane's ability to transport substrates must be drastically altered. It is known that exposure of bacteria to commercial detergents results in a greater metabolic rate due to a greater availability of nutrients, resulting directly or indirectly from a greater permeability of the cellular membrane (Gloxhuber, 1974). Humic products also show considerable membrane surface activity and should act like detergents (Visser, 1982). The ability of certain bacteria to proliferate in the presence of humic substances on substrates that they cannot normally utilize may very well be the result of an action of humic matter on certain enzyme precursors or the initiation of a de novo synthesis of normally unavailable enzymes (Visser, 1985). Zimmerman (1981) also showed that humic compounds help catalyze the transport of thermodynamically normally unavailable electron acceptors or to enhance the rate of already feasible reactions.

White and his associates (Guckert et al., 1985, 1986; Nickols et al., 1979; Short and White, 1971; White, 1993, 1994), in a series of papers, have shown that, with nutrient limitations the PHA (poly-β-alkanoate)/PLFA (phospholipid ester-linked fatty acids) increases and with starvation there is an increase in the (1) ratio cyclo 17:16:1ω7c in the PLFA, (2) ratio cyclo 19:0/18:1ω7c in PLFA, and (3) in cardiolipin/phosphatidyl glycerol. Thus, the fatty acid profile of organisms from any environment can be used to determine if the microbes are in a starved state.

The microbes in high-yield methanogenic digesters increased the rate of ^{14}C-acetate incorporation into total lipid when starved, but this incorporation decreased significantly when the bacteria were overfed (Hedrick et al., 1991a). This starved condition of the microbes in the methanogenic reactors revealed a significant increase in the proportion of the eubacterial stress biomarker *trans/cis* 18:1ω7 (Hedrick et al., 1991b).

The ester-linked phospholipid fatty acid profiles of a *Pseudomonas aureofaciens* strain and an *Arthrobacter protophormiae* strain, isolated from a subsurface sediment, were quantified in a starvation experiment in a silica sand porous medium under both moist and dry conditions by Kieft et al. (1994). These cells were washed before the experiment was initiated; hence they undergo a starvation process even before the desiccation process. Desiccation was by slow drying over a period of 16 days to final water potentials of approximately −7.5 MPa for *P. aureofaciens* and −15 MPa for *A. protophormiae*. Under moist and dry nutrient-deprived conditions, the numbers of culturable cells of both strains fell below the detection level in 16 days but both strains maintained culturability in nutrient-amended microcosms. The dried *P. aureofaciens* cells showed typical ester-linked fatty acid profiles typically associated with stressed gram-negative cells, i.e., increased ratios of saturated to unsaturated fatty acids, increased ratios of *trans-* to *cis*-monoenoic fatty acids, and increased ratios of cyclopropyl fatty acids to their monoenoic precursors. However, *P. aureofaciens* under starved, moist conditions for 16 days showed few changes in the ester-linked phospholipid fatty acid profile, whereas cells incubated in the presence of nutrients showed decreases in the ratios of both saturated fatty acids to unsaturated fatty acids and cyclopropyl fatty acids to their monoenoic precursors. On the other hand, *A. protophormiae* changed very little in response to either nutrient deprivation or desiccation. Although diglyceride fatty acids have been proposed to be indicators of dead or lysed cells, their numbers remained relatively constant throughout the experiments. Only *A. protophormiae,* desiccated for 16 days, showed an increase in the ratio of diglyceride fatty acids to the ester-linked phospholipid fatty acids. Negligible changes in the ester-linked phospholipid fatty acids have been reported in starved *Arthrobacter crystallopoietes* (Kostiw et al., 1972).

Massa et al. (1988) subjected oleic acid- and linolenic acid-grown *E. coli* to starvation in 154 mM NaCl at 37°C and followed the CFU decline by plating on agar medium. The decline in CFU was faster for linolenic acid-grown than for oleic acid-grown cells. However, this decline was not indicative of cell death because culturable CFU was recovered after respirable substrate was added to the starved cell suspension. Their results suggest that cell envelope microviscosity is an important factor in determining the sensitivity of E. coli to nutrient deprivation.

7.6 Starvation Proteins

Ant-300 cells, labeled with [^{35}S]methionine during growth, were starved for various periods up to 30 days and lysed. Fingerprint protein patterns were produced by two-dimensional polyacrylamide electrophoresis on the lysed and starved cells. The autoradiograms of starved cells revealed the disappearence of some protein spots, new protein spots, and the retention of other spots compared to nonstarved cells (Amy and Morita, 1983b). Those proteins that disappeared were degraded and used as a possible source of energy in the starvation process, but there is a degradation of specific proteins with starvation time. Cells starved and permitted to recover in the presence of [^{35}S]methionine for 8 hours showed similar protein patterns as unstarved cells. No starvation proteins appeared to be synthesized prior to 7 days of starvation and the production increased substantially over a period of 30 days, with the greatest increase after 30 days (Amy and Morita, 1983b). Protein synthesis was shown to occur throughout a starvation period of 120 hours (Nyström et al., 1986). It appears that time necessary for the production of the proteins during starvation depends on the species in question. Different patterns of protein degradation and synthesis during starvation of various species of bacteria have been noted (Kjelleberg and Hermansson, 1987). The synthesis of some of the starvation-specific proteins among species is probably the common denominator among the various species. These starvation-induced proteins were later termed "starvation proteins" by Reeve et al. (1984a).

Recognizing that protein synthesis occurs during the starvation process, the ability to incorporate exogenous leucine and methionine during the starvation period was examined. The incorporation rate of leucine and methionine was related to the time of starvation and decreased subsequent to an initial increase during the first few hours of starvation (Nyström et al., 1986). They also found that labeled leucine and methionine were preferentially incorporated into proteins. Because there is an increased activity of the protein-synthesizing system during starvation, it indicates that there is an active cellular response to the starvation process (Nyström et al., 1986).

Protein patterns, as examined by two-dimensional electrophoresis, of whole cell homogenates of growing and nongrowing cells of Vibrio DW1 and S14, were determined by Jouper-Jaan et al. (1986). These two isolates exhibited two different pathways of starvation-survival as evidenced by the protein patterns. *Vibrio* DW1 showed a relatively large decrease in cell volume and general total protein content decrease, including large amounts of 30 kDa protein, and not the loss of just a few proteins. On the other hand, a highly specific protein pattern was noted in 24-hour-starved S14 in which a total of 17–20 proteins disappeared and two novel proteins were synthesized. Later, Nyström et al. (1990b) identified 66 starvation-induced proteins in S14, when starved for glucose, amino acids, ammonium, and phosphate simultaneously, and also found an early temporal class of starvation proteins that were essential for long-term starvation. Nyström et al. (1990) also noted that 13 of the multiple starved proteins were unique in that they were not observed when S14 was starved for any of the individual nutrients.

In order to demonstrate that proteins that occur in higher quantities in starving cells are starvation-specific proteins, Albertson et al. (1987) produced antibodies against starved cells and pulse-labeled the cells with tritiated amino acids at various times during starvation. Starvation antigens appeared in S14 cells in less than 1 hour of starvation and reached a plateau after 14 hours, but no antigens were detected in either the early or late stationary phase. On the other hand, the titer for DW1 cells decreased with starvation time and reached a minimum after 4 hours. Although no starvation antigens could be detected for DW1 cells, several could be detected for S14 cells. It was found that the larger proteins in the whole cell digest were located in the outer membrane and that the larger proteins cor responded to periplasmic proteins.

Later, Albertson et al. (1990d) determined that the starvation-specific proteins of S14 cells are not the product of degradation; this was determined by pulse labeling with ^{14}C-leucine at various times of starvation up to a week. The radioactive proteins were measured in slices from SDS gels made from the various subfractions of the cell, which permitted the determination of a detailed resolution of the time of starvation-dependent modulation of the protein patterns. Thus, five outer membrane proteins were found on the SDS gel in 3 hours but not at 0 time of starvation. Three of the five outer membrane proteins are definitely the result of de novo synthesis. The SDS (sodium dodecyl sulfate) gels demonstrated that there were two periplasmic proteins synthesized during starvation. One of the periplasmic proteins was synthesized after 3 hours of starvation and constituted the dominant protein after 1 week, whereas the other was detected by pulse labeling after 24 hours of starvation. The latter protein also gave rise to a starvation-specific antibody.

Vibrio sp. S14 cells at various times (up to 120 hours) during starvation were examined by one-dimensional gel electrophoresis and revealed the synthesis of 9 membrane, 6 periplasmic, and 4 cytoplasmic proteins (Nyström et al., 1988). Eight of these synthesized proteins were also synthesized by heat- and/or ethanol-shocked cells. Pulse labeling during starvation indicated that the starvation proteins were not products of degradation and the most pronounced changes occurred

...ng the initial hours of starvation. Most of the starvation proteins were found at 3 hours of starvation but new starvation-induced proteins were also found at 5, 10, 24, and 120 hours of starvation, suggesting a sequential induction of genes.

In *V. vulnificus,* Morton and Oliver (1994) found 34 proteins that were induced over a 26-hour period of starvation but were also accompanied by a drastic drop in the total protein synthesis within the first hour of starvation. Twenty-three of these proteins were induced within the first 20 minutes of starvation. The addition of chloramphenicol at sequential times throughout the starvation period revealed that proteins required for starvation-survival are synthesized within the first 4 hours of starvation.

Twenty to 40 specific starvation-induced proteins are found in *E. coli* (Groat et al., 1986; Matin, 1990). Some of these starvation-induced proteins are outer membrane proteins (Albertson et al., 1987; Nyström et al., 1988, 1992). When *Pseudomonas fluorescens* DF57 is starved, a 63-kDa outer membrane protein is formed, while in *Pseudomonas putida* DF14, a 28- and a 29-kDa outer membrane protein is formed (Kragelund and Nybroe, 1994). Also identified were two or three new proteins in the membranes of starved cells of *Vibrio fluvialis* (Smigielski et al., 1989) that were not present in the growing cells. Several peptides were found to be lower in membranes from starved cells compared to those in growing cells. According to these investigators, the observation that cells starved in the presence of 25 µg/ml chloramphenicol did not undergo size reduction suggested that synthesis of proteins not normally found in growing cells occurs during starvation. Protein synthesis is, thus, a requirement for ultramicrocell formation.

Holmquist et al. (1993) examined the induction of stress proteins in three different starving marine *Vibrio* and concluded that induction of stress proteins was strain specific. In other words, the induction of stress proteins like DnaK and GroEL in these organisms might not be a uniform response in other microbes.

7.7 Nutrient Transport and Binding Proteins

In any ecological environment, there are myriad different types of organic matter (most have not been identified) as well as a myriad of different physiological types of bacteria. Each organism seeks its own energy and nutrient source in an ecosystem where the organic matter is scarce; much of it is recalcitrant. Morgan and Dow (1986) make an excellent point concerning substrates in nature, in that there are few data available concerning mixed substrate utilization by bacteria and even fewer regarding the metabolic interactions involved. To illustrate the importance of mixed substrate utilization, the authors cite the data of Harder and Dijkhuizen (1983) where *Pseudomonas oxaliticus* is grown on oxalate and acetate in batch culture and the organism exhibits diauxic growth utilizing acetate first. In carbon-limited continuous culture (below $D = 0.15$ hour) both acids are utilized equally, but acetate is used preferentially and the excess oxalate remaining is unutilized at a level proportional to the dilution rate. Thus, in nature, the situation is likely to be analogous but much more complex. It might also be mentioned that *Neurospora crassa,* under conditions of carbon limitation, is known to produce a powerful system for glucose transport (Slayman and Slayman, 1974).

Starved cells have the ability to utilize a greater variety of substrates than nonstarved cells. When starved cells are given small amounts of energy, resuscitation takes place. This is accompanied by the inability to use some of the substrates that the starved cells could previously use, hence they become impermeable to a greater variety of substrates. This may explain the data obtained by Ishida and Kadota (1981a), where they found less growth of oligotrophic bacteria at high concentrations of nutrients.

The ability of *E. coli* to utilize small amounts of glucose (1.25 mg/ml) was reduced when the cells were starved less than 24 hours, but in a slightly longer starvation period (24 to 48 hours), the cells retained the ability to take up added glucose (Koch, 1979). Starved cells take up glucose at a faster rate than they can use it for growth. Cells subjected to starvation for 2 days are capable of a rapid shift-up in rich medium with very little lag and are capable of full metabolic response and regulation. Koch (1979) also concludes that after *E. coli*'s reserves are gone, there is a period before death when it is nearly metabolically inert and yet capable of responding efficiently and quickly when conditions become more optimal.

Law and Button (1977) found that the ability of the population to scavenge glucose was increased as glucose carbon was replaced by amino acid carbon. When organisms are in low concentrations of glucose, the ability to scavenge glucose is increased greatly.

A low endogenous metabolism channelled principally to support substrate uptake is essential in an oligotroph. To be catabolically versatile (e.g., ability to use many catabolic pathways and ability to use many different substrates) is essential, in addition to having inducible catabolic pathways and constitutive uptake systems with a high and low V_{max} (Shilo, 1980). Furthermore, there should be a minimal level of catabolic repression, permitting simultaneous utilization of several different substrates. The ability to take up substrate also lies in the surface:volume ratio of the cell. Koch (1971) realized that bacteria must have an efficient transport mechanism to take up carbon and nitrogen sources, growth factors, or trace metals when they are confronted with an environment that has these compounds in very low concentrations. Microbes themselves make this situation worse by using any utilizable source of nutrients and with time will deplete the system so that there are no longer sufficient nutrients for the cell to carry on this metabolic process (Morita, 1982). Yet they must be able to utilize energy-yield substrates when they become available. What mechanisms permit bacteria in oligotrophic environments to obtain energy? High-affinity uptake systems, substrate capture (including chemotaxis), and attachment permit the starving microbe to obtain energy from the oligotrophic environment. "The ability to capture nutrients available at very low concentrations at the highest possible rate is of crucial importance to microbes in their struggle for existence" (Gottschal, 1985).

Transport was found to require maintenance of a PMF on a gradient across the membrane, but ATP was not involved (Faquin and Oliver, 1984). In an energy-limited environment, the PMF of bacteria is usually very small. According to Konings and Veldkamp (1983), during a short period of energy supply, the rate of PFM generation on dissimilation of the energy source will largely determine the uptake of essential solutes and consequently the rate of biomass formation. Hence, the rate of PFM generation determines, to a large degree, the reactivity of bacteria in the natural environment.

7.7.1 Chemotaxis

Chemotaxis in bacteria is primarily a mechanism for finding food (energy) (Hazelbauer and Parkinson, 1977). Flagellation sometimes is induced by poor nutrient conditions (Beveridge, 1989). Chemotaxis helps bring the organism to the most favorable position for growth, genetic recombination, or dispersal, but this chemotactic property has been neglected by microbial ecologists (Carlile, 1980). Cell motility and chemotaxis can be important to population dynamics as cell growth kinetics (Lauffenburger, 1991). Further advantages, in addition to the foregoing, are to avoid unfavorable conditions like the presence of toxic agents or to bring together organisms under unfavorable conditions to form resistant aggregates (Rowbury et al., 1983). Chemotaxis also plays a role in plant infection (Wood and Hayasaka, 1981), in *Rhizobium*'s attraction to exudates from the roots of legumes (Schmidt, 1979), in attraction of the fish pathogen (*Aeromonas*) to fish mucous extracts (Hazen et al., 1979) and mammalian pathogens (*V. cholerae, E. coli,* and *Salmonella typhimurium*) to intestinal mucosa (Allweiss et al., 1977; Freter, 1981), in *Bdellovibrio*'s ability to locate a high concentration of prey and also, in a negative sense, to avoid open waters (La Marre et al., 1977). Smith and Doetsch (1969) demonstrated that a motile, chemotactic strain of *P. fluorescens* outgrew a nonmotile strain in stationary broth culture; and Pilgram and Williams (1976) demonstrated that a motile, chemotactic *Proteus mirabilus* outgrew a motile, nonchemotactic mutant in a semisolid, amino acid medium, although the two strains grew equally well in broth. These two studies indicate the competitive importance of chemotaxis in low-nutrient environments. One can also view the concept of chemotaxis from an evolutionary viewpoint. Why would bacteria produce a flagellar structure if it were not needed for the survival of the species, especially before green plant photosynthesis evolved on Earth?

Chemotaxis toward sugars and amino acids depends on a chemoreceptor that recognizes the attractant. Even if the attractant is too dilute to give a chemotactic response, the chemoreceptor is still present in the cells and can then aid in substrate capture of the attractant (Griffiths et al., 1974). Eisenbach (1990) demonstrated that *E. coli* had a secondary machinery for driving flagellar motor movement in the absence of an external energy supply. Thus, this system serves as a backup when the proton-motive force drops below the threshold level. Geesey and Morita (1979) were the first to demonstrate that Ant-300 may not display a chemotactic response for a substrate unless it has been starved for 48 hours. This organism displays the characteristic chemotactic response in relation to the concentration of arginine. In the log growth phase, Ant-300 displays a very poor translational movement, if at all, and electron micrographs reveal nonflagellated cells or cells with very fragile, apparently nonfunctional flagella (Torrella and Morita, 1981). Ant-300 demonstrated chemotactic ability for a variety of amino acids, sugars, and other attractants only after the organism was starved. Negative chemotaxis was also found for some compounds. The threshold for chemotactic response for some marine bacteria appear to be lower (10^{-10} M) than for enteric bacteria (10^{-6} M) and will, in addition, respond to both L and D amino acids (Chet and Mitchell, 1976; Torrella and Morita, 1982).

The chemotactic ability of *Vibrio* S14 was found to be correlated with the utilization of poly-β-hydroxybutyrate during starvation (Mårdén et al., 1985). The chemotactic response by starved *Vibrio* S14 was different, depending on whether the cells were starved or not, and the rate of chemotaxis depended on the different concentrations of the attractants (Malmcrona-Friberg et al., 1990b). They also found that cells starved for 3 hours displayed a stronger response to glucose than those starved for a shorter or longer period, but cells starved for 5 and 10 hours responded more strongly to a lower concentration of glucose than cells starved for 0 and 3 hours. Motility was shown to decrease by over 95% in cell population after 24 hours of starvation, which naturally resulted in a low sensitivity in the chemotaxis assay. Ten to fifty percent of a 22-hours-starved population regained its motility when exposed for 4 hours to nutrients. However, the recovery period may influence the total number of motile cells; the longer the recovery period in nutrients, the more cells become motile. Almost total loss of motility is due to loss of flagella (Malmcrona-Friberg et al., 1990b). When cells are growing in high nutrient niches, chemotaxis is superfluous and wasteful (Morgan and Dow, 1986) even though a rapidly respiring bacterium with a single flagellum expends only approximately 1% of its available energy on flagellar rotation (Rowbury et al., 1983). The so-called model oligotroph, *Caulobacter,* has also been shown to display chemotaxis and has been partially characterized at the molecular level (Shaw et al., 1983; Gomes and Shapiro, 1984).

Torrella and Morita (1982) proposed the concept of "energized chemotaxis" where the cell is energized by the presence of a utilizable substrate so that it can better respond (chemotaxis) to the presence of another attractant. This "energized chemotaxis" aids the organism to locate an essential nutrient when it has a supply of energy. For instance, a situation may exist where there is sufficient amount of carbohydrate(s) (near a decaying kelp bed) but insufficient amounts of a specific amino acid. Chemotaxis would then aid the organisms to obtain the essential amino acid. Random movement coupled with chemotaxis is suggested to be an efficient means for substrate scavenging by marine bacteria (Jackson, 1989). The speed of various organisms by flagellar movement has been compiled by Oguiti (1936). The average speed of *V. cholerae* was cited as 55.5 µm/sec. Since Ant-300 is a marine vibrio, we can assume this rate. If the flagellar movement was unidirectional, then a distance of 4.79 m can be traversed in a day. Thus, if they are able to find a new environment with sufficient substrate, they will have the ability to remain there (Bell and Mitchell, 1972). For Ant-300, the cell has 5 or 6 days to locate a suitable substrate before it goes into starvation mechanism (Novitsky and Morita, 1978a,b). These "energized chemotaxis" studies also indicate that loss of chemotaxis during long-term starvation is, in part, due to the exhaustion of internal cellular energy. Koshland (1977) found that chemotaxis is ATP dependent, whereas motility can function not only with ATP but also if the energized state of the membrane is maintained by a functional respiratory chain. However, it must be remembered that there are threshold concentrations to which an organism will respond chemotactically. In Ant-300, the threshold levels for arginine for 3- or 4-day starvation periods are 10^{-7} to 10^{-8} and 10^{-5} to 10^{-6} M, respectively (Torrella and Morita, 1982). Each attractant has its own threshold level. Chemotaxis has been studied by growing organisms in energy-rich medium and then doing chemotactic response studies. From an ecological viewpoint, we should probably study the chemotactic response of starved organisms more.

Spirochaeta aurantia M1 cells grown under limited energy and carbon sources in a chemostat gave better chemotactic response than when the organism was grown in batch culture not under nutrient limitations (Terracciano and Canale-Parola, 1984). The chemostat-grown cells appeared to sense the attractant (glucose) at 1000 times lower than batch-grown cells. This also occurred with xylose. The organism appears to regulate its chemosensory system to nutrient limitations. When the availability of nutrients becomes severely limited in its environment, this organism has a competitive advantage over the other bacteria within its microniche.

Phytoplankton and zooplankton apparently release dissolved amino acids and/ or proteins into seawater (Hellebust, 1965; Webb and Johannes, 1967). Carbohydrates are excreted by phytoplankton (Hellebust, 1965; Sieburth, 1969). Biotin, thiamine, and vitamin B_{12} are released by algae (Carlucci and Bowes, 1970). Other compounds are also released by the macroorganisms into the seawater (Gagosian and Lee, 1981). Even breakdown products of a *Flavobacterium* sp. are known to produce a chemotactic response by *Vibrio* S14 (Malmcrona-Friberg et al., 1990b). It can be seen that a chemoattractant can be any compound generated by macro- and microorganisms. Thus, a gradient of these compounds may serve as the chemoattractant for microbes. The final result of this chemotactic process may be the attachment of the microbes to the surface of the macroorganisms. Azam and Ammerman (1984a) proposed that the chemotactic abilities of marine bacteria allow them to stay in the high concentration areas around algal cells. The chemotactic response of microbes in seawater toward a dying diatom can be seen in Figure 7.13. The importance of microbes to operate at a micrometer scale has been emphasized by McCarty and Goldman (1979) and Azam and Ammerman (1984b). Alldredge and Cohen (1987) have shown the existence of chemical micropatches surrounding particles that might attract or repel motile bacteria, whereas Jackson (1989) found that temporary clustering around nutrient sources was possible for marine bacteria but there were chemosensory limits to the bacteria response. Particles are also known to have microbes present and the dissolved substrates complexed with the particles may be a chemoattractant for microorganisms also.

Mitchell (1991) examined the influence of Brownian motion on marine bacteria and found that, due to their small size in the environment, they can rotate up to 1400 degrees in one second. As a result, this rapid Brownian motion makes directional chemotaxis difficult or impossible, because a bacterium may point in a particular direction for only a few tens of milliseconds on average. He found that directional movement is possible if the chemotactic speed was 100 μm/sec. Because most in situ marine bacteria have a radius less than about 0.75 μm, he tested to see if ultrasmall bacteria could exceed the speed of 100 μm/sec. His results showed that a marine bacterium would spend in excess of 10% of its total energy budget on movement, which is 100 times greater than the value for enteric bacteria. Thus, he concluded that marine bacteria are likely to be immotile below the critical size-specific nutrient conditions. It should be mentioned that *Vibrio* and Ant-300, when starved, lose their chemotactic response after 1 day (Malmcrona-Friberg et al., 1990b) and 4–5 days (Geesey and Morita, 1979; Torrella and Morita, 1982) respectively. Because of the foregoing situation, the energetic costs to marine bacteria must be weighed against the benefits (Mitchell, 1991). Although not mentioned, it is possible nutrients are obtained due to Browning movement.

The uptake of a substrate can be limited by the rate of transfer by molecular diffusion from solution to the organism's surface, which results in a region near the organisms with low substance concentration and a small concentration gradient (Munk and Riley, 1952). Smaller organisms are less limited by diffusion kinetics than larger ones (Malone, 1980; Sournia, 1982).

Uptake systems of the indigenous bacteria must be maintained in oligotrophic environments in order for the organisms to function when energy sources become available. The maintenance of uptake systems under conditions of nutrient deprivation has been shown for mannitol in a marine *Pseudmonas* sp. and a *Vibrio* sp. (Davis and Robb, 1985). In aquatic systems, 97% of the bacteria are gram negative. The ecological advantage for gram-negative bacteria lies in the associ-

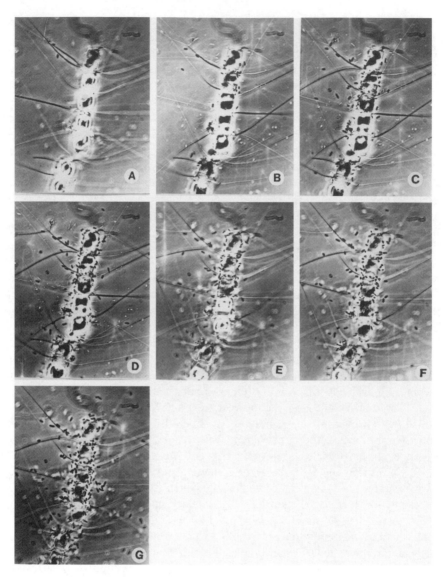

Figure 7.13   Dying diatom colonization by chemotatic bacteria. A = 0 time; B = 20 minutes; C = 30 minutes; D = 40 minutes; E = 1 hour; F = 2 hours; and G = 3 hours. Courtesy of Francisco Torrella.

ation of their extracellular enzymes with the periplasmic space (Amthor, 1989). Periplasmic hydrolysis degrades polymeric substrates to monomeric compounds that are then taken up via specific uptake systems in the plasma membrane (Costerton et al., 1974). During starvation, hydrolases as well as an uptake system would have to be maintained to ensure efficient scavenging capabilities. Hoppe et al. (1988) and Amthor (1989) found that fatty acid esterases and alkaline phosphatases are maintained in Ant-300 after starvation for at least 57 days.

7.7.2 High-Affinity Uptake Systems

In cases of extremely low nutrient concentration, Gorini (1960) was able to demonstrate the concept of catabolic repression. The discovery of the high-affinity mechanism for assimilating ammonia into glutamate via glutamine synthetase and glutamate synthase was reported by Tempest et al. (1970, 1983). Neijssel et al. (1975) reported on the dual mechanisms for the assimilation of glycerol via the glycerol kinase route in carbon-limiting conditions and the glycerol dehydrogenase route in a glycerol-sufficient environment.

In order for microbes to survive in low-nutrient environments, they can either synthesize more of the existing uptake system or acquire the ability to synthesize a different high-affinity system (Harder and Dijkhuizen, 1983). High- and low-affinity enzyme systems are not restricted to the assimilation of NH_4 because such systems have been reported for organic nutrients such as glycerol and glucose (Neijssel et al., 1975; Whitting et al., 1976). The concept of specific affinity (the net specific influx of substrate per unit concentration of that substrate) was proposed by Button's group (Law and Button, 1977; Button, 1978, 1983). However, Gottschal (1985) states that Button's approach may not always be valid. A high- and low-affinity system for arginine was demonstrated in Ant-300 cells when exposed to nanomolar or micromolar concentrations of arginine (Geesey and Morita, 1979). The Lineweaver-Burk plot revealed a bimodal relationship for arginine uptake from 0.034 to 0.59 μM. Assuming two simultaneous reactions, K_t values of 1.7×10^{-8} and 4.5×10^{-6} and V_{max} values of 12 and 51 pmol/minutes/5 $\times 10^7$ bacteria respectively, were obtained. This high-affinity system in Ant-300 was maintained during starvation (Faquin and Oliver, 1984). Similar uptake characteristics have been demonstrated for starved bacteria when faced with low concentrations of mannitol, glucose, and glutamate (Davis and Robb, 1985). This high-affinity system can operate at the amino acid concentrations reported in the open ocean (Lee and Bada, 1975). Two independent arginine transport systems are present in Ant-300 and these have been reported in other bacteria (Reid et al., 1970; Rosen, 1973). Natural marine bacterial assemblages have been shown to have multiphasic uptake kinetics (Fuhrman and Ferguson, 1986). High-affinity uptake systems at low nutrient concentrations have been discussed by Jannasch (1979) and Matin (1979)

Two different uptake systems for leucine were identified in a gram-negative marine bacterium, designated as S14 (Mårdén et al., 1987). These leucine uptake systems developed during starvation. The K_m values for the two affinities for leucine were 0.76 µM and 20 µM. The time of exposure had a marked effect on both uptake systems by altering the velocity of uptake (V_{max}), but not changing the affinity for leucine. For the high-affinity system, a marked increase in the uptake capacity was noted; whereas the uptake velocity decreased for the low-affinity system. Leucine is taken up immediately by starved S14 cells. Osmotic shocked cells lost nearly all of their substrate-binding activity. The osmotic shock supernatant revealed two proteins, 37 and 44 kDa in size that had leucine-binding activity. Ant-300 cells that were osmotically shocked by exposure to 0.15 M NaCl lost 65% of the leucine-binding activity (Glick, 1980). The leucine high-affinity system of S14 was either present or switched on when exposed to energy or amino acid deprivation (Mårdén et al., 1987; Davis, 1985); whereas in Ant-300 the high-affinity arginine uptake system was induced after 50 hours of starvation (Glick, 1980). The data indicates that the high-affinity leucine system is either regulated by induction or repression.

Mårdén et al. (1987) demonstrated a threefold increase in the high-affinity system and the low-affinity system showed a gradual decrease. In *Cytophaga johnsonae,* the low-affinity glucose uptake system becomes inactivated quite rapidly (Höfle, 1984). Carlucci et al. (1986) stated that the growth of oligotrophic bacteria on unsupplemented seawater may demand an affinity for organic substrates at concentrations less than 10 µg/l. These substrate levels correspond to the K_m values of the high-affinity uptake system for a number of bacteria, and thus are in the same order of magnitude as the concentrations considered to be the upper limit for oligotrophic growth (Kjelleberg and Hermansson, 1987).

7.7.3 Substrate Capture

According to Gottschal (1990), there are two possible strategies for capturing substrate when substrates become limiting. Thus to ensure substrate capture at a sufficient rate to sustain growth requires either "an increase in V_{max} (either increased amount of the same enzyme or the synthesis of another enzyme with a higher V_{max} value), or decreased affinity constant (K_m) of the uptake system or of the enzyme(s) that represent the limiting step in the metabolism of the limiting nutrient." There are examples of both (Tempest and Neijssel, 1978; Matin, 1981).

In any ecosystem, the microorganism must first "see" the substrate, then affix it to its surface before it can be transported into the cell. Substrate capture was first proposed by Morita (1984b) for organisms in the marine environment. Substrate capture, then, is the ability of the organism to affix the substrate to its surface and does not include the transport of the substrate into the cell; substrate capture is a prerequisite for transport (uptake). In oligotrophic environments, the cell must do both in order to obtain substrate for an energy source in order to be physiologically active. If no substrates can be captured, the cell will then go into the starvation-survival mode.

Kalckar (1971) suggested that binding proteins located in the bacterial cell envelope may be the means by which microorganisms can scavenge substrates from the environment. In many cases, the chemoreceptor is a binding protein. Several binding proteins participate in transport as well as in chemotaxis. Adler (1976) showed that the functioning of chemoreceptors does not require metabolism or general transport of the compound but only requires recognition of the attractant by the specific binding protein. Table 7.7 is a partial list of binding proteins described by various investigators; many of these proteins are also involved in transport. However, all the organisms from which the binding proteins have been isolated are enteric bacteria.

The binding proteins are loosely bound to the cell in the periplasmic space and can be released by osmotic shock (Hobot et al., 1984). Binding proteins participate in the formation of the substrate pools (reservoirs) in the periplasmic space. Extensive studies have shown that the binding proteins do not exhibit enzymatic activity, nor do the substrates undergo chemical change during the complex formation (Wilson and Holden, 1969). Many of the binding proteins operate at very low substrate concentrations. For example, arginine specific transport in *E. coli* has a K_m of 2.6×10^{-8} M (Rosen, 1973), histidine-specific transport in *S. typhimurium* has a K_m of 2.0×10^{-8} M (Ames and Lever, 1970), and galactose transport has a K_m of 5.0×10^{-7} M (Rotman and Radojkovic, 1964). However, not all substrates are transported by permeases and osmotic shock releasable proteins elicit a chemotactic response (Mesibov and Adler, 1972). Also, some chemoreceptors bind several amino acids. Griffiths et al. (1974) demonstrated that glutamate was loosely bound to the cell surface of Ant-300 and a reduced salinity could affect the retention of glutamate.

7.8 Attachment

Below the threshold level of nutrient for microbes, the presence of surfaces increased the metabolic activity (Heukelakian and Heller, 1940; Jannasch, 1958). This increased metabolic activity was due to the physical nature of a solid-liquid interface that results in the concentration of nutrients and ions (Marshall, 1979, 1980). Subsequent microbial growth leads to a complex situation in which interspecific cross-feeding, cometabolism, proton transfer, and competition become important (Wardell et al., 1983).

The staining of surfaces with ruthenium red or alcian blue has shown the presence of polysaccharides and glycoprotein (Corpe, 1970a, 1974; Fletcher and Floodgate, 1973). These adhesive polymers have been implicated in the attach-

Table 7.7   Partial list of some of the binding proteins elucidated.

Binding Protein	Reference
Sulfate-binding protein[a]	Pardee and Prestidge 1966
Phosphate-binding protein[a]	Mcdveczky and Rosenberg 1969
Galactose-binding protein[a]	Anraku 1967
L-arabinose-binding protein[a]	Hogg and Englesberg 1969; Schleif 1969
Ribose and maltose-binding proteins[a]	Hazelbauer and Alder 1971
Thiamine-binding protein[a]	Iwashima et al. 1971
Cyanocobalamine-binding protein	Taylor et al. 1972
Leucine-binding protein[a]	Piperno and Oxender 1966
Glutamine-binding protein[a]	Weiner and Heppel 1971
Glutamic acid-binding protein[a]	Barash and Halpern 1971
Phenylalanine-binding protein[a]	Kuzuwa et al. 1971
Histidine-binding protein[a]	Rosen and Vasington 1970; Ames and Lever 1970

Source: Reprinted by permission from *Heterotrophic Activity in the Sea,* J. E. Hobbie and P. J. LeB. Williams (eds.), R. Y. Morita, pp. 83–100, 1984; copyright 1984, Plenum Publishing Corporation.
[a]Involved in transport mechanisms of the cells.

ment of bacteria to surfaces (Costerton et al., 1987; Corpe, 1964, 1974; Gibbons, 1977; Harris and Mitchell, 1973; Humphrey et al., 1979). In an organically depleted environment, it is advantageous for microbes to seek out these polymers as sources of energy. The number of attached bacteria ranged from a few percent to 94% of the total microbial population in different aquatic regions, depending on the particle abundance and nutrient conditions of the water phase (Hoppe, 1984). Attached bacteria appear to be provided with special extracellular enzymatic facilities. Attachment is the initial phase for biofilm formation. Attachment is the initial phase for biofilm formation. According to Dawson et al. (1981), the adhesion of small starved bacteria is a survival tactic.

Carbon limitation appears to promote adhesion to surfaces (Brown et al., 1977). Better growth appears to take place by attached bacteria than those in the liquid phase (Jannasch and Pritchard, 1972). Starvation increases the adhesion rates to surfaces (Dawson et al., 1981; Kefford et al., 1986); this was related to the appearance of bridging polymer at the cell surface. Early log, late log, and stationary phase cells show little tendency to attach (Dawson et al., 1981).

The initial attraction of bacteria to surfaces is physical (Wardell et al., 1983), followed by hydrophobicity, electrostatic nature of the cell envelope, and the formation of adhesive polymers. The marine vibrios display increased hydrophobicity and/or increased prevalence for surface binding during the starvation-survival process, especially during the miniaturization phase (Kjelleberg and Hermansson, 1984). Cell surface hydrophobicity was about eight times higher in groundwater than in eutrophic lake water bacteria (Lindqvist and Bengtsson, 1991), indicating the prime importance of hydrophobicity for attachment in low-nutrient environments.

Kjelleberg et al. (1982) showed that bacteria have survival advantages in the presence of surfaces. After attachment of small starved bacteria to surfaces, the addition of a very small amount of nutrients will permit the motile cells to enlarge, divide, and move off the surface within 7 minutes. Marine bacteria adhere to solid surfaces when they are under starvation conditions (Ellwood et al., 1982; Fletcher and Marshall, 1982). Humphrey et al. (1983) noted that 12 rod-shaped, hydrophilic bacteria decreased in size more rapidly at the solid surface than in liquid culture but 3 rod-shaped, hydrophobic bacteria diminished in size more rapidly in the liquid phase than at the solid-water interface. The rapid size decrease (dwarfing process) occurred more rapidly when the cells were in an early stage of logarithmic growth and starvation was initiated. This dwarfing process was found to be inhibited by low temperature and low pH but not by chloramphenicol. Three coccoidal bacteria showed little tendency to become smaller upon starvation in the liquid phase or at a surface (Humphrey et al., 1983).

Kjelleberg et al. (1982) demonstrated that cells from the interface were smaller than those in the bulk liquid and exhibited a marked increase in oxygen consumption (Kjelleberg et al., 1983). Many bacteria, when attached to surfaces, show a decreased rate of oxygen consumption with various substrates compared to cells that are free (Fletcher, 1977; Hattori and Hattori, 1987). *E. coli* and *A. agile* have also shown the same response (Hattori and Furusaka, 1960, 1961). Although Hattori and Hattori (1987) list five possibilities for the decreased oxygen consumption, the most probable is that the cell surface is decreased due to attachment and, hence, the attached area of the cell does not take part in substrate uptake. The induction of smaller cells at interfaces appears to be a constant feature in marine bacteria with hydrophobic surfaces but not with those with hydrophilic surfaces (Humphrey et al., 1983). Humphrey and Marshall (1984) showed that there was a burst of activity (oxygen uptake and heat production by microcalorimetry) when the cells attached to surfaces; during this time the cell reorganizes its morphology to an optimal survival configuration by consuming endogenous substrate. This is especially true when surfactants are incorporated into the dialysis membrane. The heat production was sustained for several hours. The same organism was previously shown to be 100% viable at 20 hours on a membrane surface (Kjelleberg et al., 1983). The oxygen consumption by starved cells at interfaces had previously been shown by Kjelleberg et al. (1982, 1983), Humphrey et al. (1983), and Kjelleberg and Dahlbäck (1984). On the other hand, Gordon et al. (1983) could not demonstrate any stimulation of bacteria on surfaces. This latter situation may be due to the lack of surfactant stimulants such as Tween 85, Tween 80, and Span 20 (Humphrey and Marshall, 1984).

Kjelleberg et al. (1983) examined the activity of the hydrophilic *Vibrio* sp. strain DW1 and the hydrophobic *Pseudomonas* sp. strain S9 undergoing starvation at both nutrient-enriched and nutrient-deficient interfaces. They noted that the dwarfing response was a sequence of two processes: fragmentation and continuous size reduction of the fragmented cells. Hydrophilic organisms become even smaller at nutrient-deficient surfaces than in liquid phase; whereas, in the hydrophobic organism, size reduction did not occur at the nutrient-deficient surfaces. They also found that bacterial scavenging of the surface-localized nutrients is related to the degree of irreversible binding of dwarf and starved bacteria, which in turn may be related to the degree of cell surface hydrophobicity. Cell surface hydrophobicity appears to be well established as an important factor for bacterial adhesion (Dahlbäck et al., 1981; Kjelleberg and Hermansson, 1984). Kjelleberg and Hermansson (1984) showed that several marine bacteria increased their degree of hydrophobicity and binding to glass surfaces during starvation. The role of the overall surface charge and hydrophobicity is uncertain (Kjelleberg et al. 1985; Kjelleberg and Hermansson, 1984). On the other hand, pelagic bacteria have been show to detach from suspended material in response to depletion of nutrients (Delaquis et al., 1989). It has been suggested by van Loosdrecht et al. (1987) that the adsorptive tendencies in an open ocean environment and detachment in soil and sediments were both beneficial strategies for survival, allowing dispersal to more favorable habitats. Kjelleberg et al. (1985) noted that a relatively high number of adhered oligotrophic types were found during the initial phase of adhesion compared to the numbers in bulk water. Most of the adherent bacteria were irreversibly bound. Successive reinoculations of a total of 160 isolates failed to show the existence of an obligatory nature for being either oligotrophic or copiotrophic.

Ishida et al. (1980), Ishida and Kadota (1981a), and Kjelleberg et al. (1982) postulated that true oligotrophic bacteria are those bacteria that are not able to attach to surfaces. Thus, they must take advantage of local increases in nutrients. However, Morgan and Dow (1986) argue that from an ecological point of view, it would be inconceivable that an organism adapted to growth in oligotrophic environments should fail to exploit a local increase in nutrients at a surface. Morgan and Dow (1986) further point out that most bacteria, considered as oligotrophs, exhibit surface attachment and even those exhibiting inhibition at high substrate concentrations are unlikely to be inhibited by the actual concentration of nutrients at surfaces.

Oligotrophic bacteria, isolated by enrichment from marine and estuarine (Seki et al., 1984) and freshwater (Naganuma and Seki, 1985), grown in a chemostat have been shown to be both free living and attached (epibacteria). Nearly all aquatic bacteria can be regarded as having both planktonic and epiphytic phases (Seki, 1982; Aida et al., 1988).

In *P. aeruginosa*, adhesion was shown to be related to the specific growth rate

(Nelson et al., 1985) but increased bacterial attachment under low-nutrient conditions is not a universal phenomenon, as was demonstrated in various pseudomonads when detachment was noted under low-nutrient conditions (Delaquis et al., 1989; Lawrence and Korber, 1993). Thus, surface colonization is a species-specific trait of sessile bacteria (Lawrence and Caldwell, 1987; Lawrence et al., 1992; Lawrence and Korber, 1993).

Attachment (adhesion) of bacteria to surfaces is a strategy for obtaining nutrients from the oligotrophic environment, especially when the nutrients are in very low supply. Attachment is the first phase in biofilm formation. For further discussion of the surface effect, consult Marshall (1976), Ellwood et al. (1980, 1982), and Bitton and Marshall (1980).

7.9   Recovery from Starvation-Survival

The first detectable event, without a lag in the recovery of *E. coli* from starvation, is the production of RNA and this RNA appears to be 70% ribosomal and 30% 4S subunits (Jacobson and Gillespie, 1968). Ribosomes appear immediately during recovery and reach a near-maximal level in 30 to 40 minutes. After a lag of 5 minutes, protein synthesis begins. DNA synthesis in recovering cells appears to be synchronous for the first 20 minutes. Thus it appears that the synthesis of DNA, RNA, and protein are not as interdependent during recovery from starvation as during balanced growth.

When glucose and succinate, but no nitrogen source, are added back to rod-shaped and spherical starved cells (starved 1 week) of *A. crystallopoietes,* the E.C. and ATP rises quite rapidly (Leps and Ensign, 1979b). Chapman and Atkinson (1977) also observed the same results with many other starving bacteria. The ATP concentration increase cannot be accounted for by the phosphorylation of ADP only, because the ADP concentration in the cells is not sufficient to permit this. Hence, it appears that some ATP is synthesized de novo or that release of AMP from RNA degradation occurs (Leps and Ensign, 1979b). Oxidation of RNA may also profile the cell with ribose, which is then another source of energy for ATP generation.

The ability of starved Ant-300 cells to respond quickly to the addition of a low concentration of nutrient was shown by Novitsky and Morita (1978b). There is an increase in cell numbers (200-fold), but it is followed by a decline in numbers to the preaddition levels. This increase is different from the initial increase as shown by an increase in turbidity, cell size increase, and a morphological change back to the shape normally observed in laboratory grown cells. Cells of *Vibrio* sp. strain DW1 had the ability to utilize glucose or casamino acid (or both) after being starved for 5 days without any detectable lag (Kjelleberg et al., 1982). *Vibrio* sp. S14 immediately responded to nutrient addition displaying an 11–12-fold

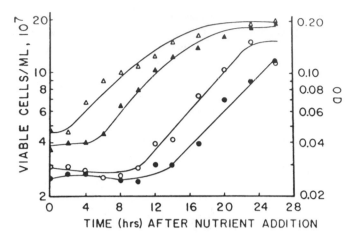

Figure 7.14   Optical density (OD) (△, ▲) and cell viability (○, •) during recovery from starvation for cells starved for 1 (△, ○) and 2 (▲, •) weeks. Viable counts are the averages of duplicate plating. Reprinted by permission from *Applied and Environmental Microbiology* 45:1685–1690, P. S. Amy, C. Pauling, and R. Y. Morita, 1983; copyright 1983, American Society for Microbiology.

increase in respiratory activity, as well as recovering from the starvation-induced ultramicrocells to typical vibrios (Holmquist et al., 1993). A more detailed discussion of the recovery of *Vibrio* sp. is presented by Östling et al. (1993). Usually the addition of nutrients results in substrate-accelerated death (Postgate and Hunter, 1964; Strange and Dark, 1965; Thomas and Batt, 1969c; Calcott and Postgate, 1971). If only glucose is added to the starved cells, then substrate-accelerated death occurs (Novitsky, 1977). Generally substrate-accelerated death is demonstrated with a large dose of a single compound; hence Novitsky and Morita (1978b) questioned its validity in relation to the natural environment where new substrates are introduced in low concentrations, except for accidental introduction. Substrate-accelerated death may be an artifact in the laboratory, which may be due to the unbalanced utilization of certain compounds, whereas nutrients added to the natural microbial population will be mineralized and utilized for the production of bacterial biomass. No detectable death results when the nutrients are added in low concentration.

Amy and Morita (1983a) reported that 16 freshly isolated open-ocean bacteria, when starved (up to 9 months), retain the ability to metabolize ^{14}C-glutamic acid added at a low concentration. By adding one volume of Lib-X medium to nine volumes of the starvation menstruum containing Ant-300 cells starved for various periods, Amy et al. (1983b) studied their recovery. Viable cell counts, optical density (OD), cell size (as measured by electron micrographs), and the cellular concentrations of protein, DNA, and RNA were monitored during recovery time. The OD and cell viability data on recovery are shown in Figure 7.14. During recovery, the cells behaved like a culture in balanced growth, but cells starved for 1 week entered the logarithmic growth phase 4 hours before the cells starved for 2 weeks. This difference was noted in all aspects of the study by Amy et al. (1983b). The relationship between the starvation period and the delay before cellular biomass increase was a linear function for up to 6 weeks, but when the cells were starved as long as 8.5 weeks, the number of viable cells remaining in the starvation menstruum played a role in the recovery of the cells. *E. coli* displays a similar effect when starved longer than 48 hours and this effect is attributed to the viability or to cellular damage (Koch, 1979). Protein, DNA, and RNA were synthesized after the input of nutrients with 0 to 14 hours delay and continued to be synthesized at an exponential rate (Amy et al., 1983b). This delay in synthesis was the result of the length of starvation.

Recovery of *Vibrio* and *Aeromonas* species after long-term starvation was noted by Hood and MacDonell (1987). Most of the cells were starved for 2.5 years, whereas *Vibrio cholerae* CA401 was starved for 3 years, *V. cholerae* E507 and *V. parahaemolyticus* were starved for 1 year. Each organism was evaluated for its ability to grow on standard plate counts (SPC) but the direct counts were generally higher. Many of the *Vibrio* species had the ability to pass through a 0.2-μm filter. They also noted that these organisms had the ability to grow on their oligotrophic media (5.5 mg C/l) and eutrophic media (5.5 g C/l). After 75 days of starvation, *V. cholerae* resumed its bacillary shape within 2 hours and began to divide within 5 hours after nutrient supplementation (Baker et al., 1983).

The starved *Pseudomonas* cells' ability to take up both exogenous ^{14}C-glucose and ^{14}C-glutamic acid after a different period of starvation can be seen in Table 7.8 (Kurath and Morita, 1983). Although there is a 60% difference between the uptake of these substrates, this difference was noted in natural mixed marine populations (Gillespie et al., 1976). Cells under starvation-survival stress would be expected to shift their metabolic balance away from biosynthesis and repro-duction toward the acquisition of energy for existing biological functions. This concept was demonstrated by the increase in the percent respiration of both glu-cose and glutamic acid in response to starvation. Log phase cells showed that 23.4% of the ^{14}C-glucose and 67.3% of the ^{14}C-glutamic acid that was taken up was subsequently respired and given off as ^{14}CO$_2$. In 4 days of starvation, the respiration of ^{14}C-glutamic acid increased to 92% and then gradually declined and finally stabilized at 70% at 28 days—close to the log phase value. The respiration of ^{14}C-glucose increased to a maximum of 50% at 13 days of starvation and subsequently declined to 22% after 36 days.

When nutrients are added to starved cells, growth and cell division resume, but only after a lag phase (Albertson et al., 1990d). Starved bacteria are able to utilize ^{14}C-glutamic acid when it is added at low concentration. The longest starved

Table 7.8   Uptake rates for exogenous glucose and glutamate after various periods of starvation.

Starvation Time (days)	[^{14}C]Glucose Uptake[a]		[^{14}C]Glutamate Uptake[b]	
	per ml	per 10^7 respiring cells	per ml	per 10^7 respiring cells
0	558.2	13.1	323.3	7.6
4	191.6	4.9	103.4	3.2
8	101.8	13.3	35.5	4.7
13	39.5	11.1	15.9	4.4
17	15.0	26.3	6.4	11.2
25	18.5	24.0	3.9	5.1
32	21.3	40.6	1.6	7.6
40	12.1	36.4	3.3	10.00

Source: Reprinted by permission from *Applied and Environmental Microbiology* 45:1206–1211, G. Kurath and R. Y. Morita, 1983; copyright 1983, American Society for Microbiology.
[a]Rate of D-[U-^{14}C]glucose uptake in nanomoles ($\times$ 10^{-6}) per hour.
[b]Rate of D-[U-^{14}C]glutamate uptake in nanomoles ($\times$ 10^{-4}) per hour.

period in this study was 9 months (Amy and Morita, 1983a). Upon recovery after starvation, an orderly pattern of protein synthesis was observed during the lag phase in *Vibrio* S14 (Nyström et al., 1990a). When nutrients (leucine) are added to starved *Vibrio* sp. S14, an instantaneous severalfold increase in the rates of protein and RNA synthesis occurs (Flärdh et al., 1992). If glucose minimal medium is added to *Vibrio* S14 after 24, 120, and 200 hours of starvation, there is an immediate sevenfold increase in the rate of protein synthesis (Albertson et al., 1990d). Also noted were changes in the rate of respiration, the kinetics of RNA and DNA, and changes in the median cell volume occurring at different times after the addition of glucose. Since the increase in protein synthesis was not dependent on a parallel increase in RNA synthesis, Albertson et al. (1990d) indicated that starved cells may have an excess of protein-synthesizing machinery, including stable RNA and functional ribosomes. At the onset of the nutritional upshift, most of the starvation-induced proteins were immediately repressed, while 11 of the 24 recovery-induced proteins were expressed exclusively during the maturation phase and subsequently repressed at the onset of regrowth. The half-life of mRNA in *Vibrio* sp. S14 decreased rapidly with the addition of glucose minimal medium during recovery from starvation (Albertson et al., 1990c).

Kramer and Singleton (1992) found that in *V. furnissii* and *V. alginolyticus,* recovery pattern at the level of rRNA accumulation depended upon the duration of nutrient deprivation and the manner in which it was imposed. Stationary cells (those cells starved by consumption of limiting nutrients resulting from growth into the stationary phase) starved for 2 days had only slight relative increases in rRNA levels after excess nutrients were added. As the duration of starvation was extended to 8 and 15 days, increasingly greater amounts of rRNA (30 and 70 times preenrichment values, respectively) were transcribed after nutrient enrichment. Starved cells were recovered from 2 and 8 days of starvation without extensive rRNA production; but after 15 days, nutrient enrichment caused 16S RNA levels to increase 30-fold. Recognizing that stationary phase starved cells and shift down starvation cells differ, the question that arises is which pattern occurs in nature? Both could occur depending on the environmental conditions.

Recovery of *M. luteus* cells starved for 3 to 6 months could be resuscitated by incubation in the appropriate medium (Kaprelyants and Kell, 1993). Resuscitation of starved cells in liquid culture has been used by Allen-Austin et al. (1984), Kaprelyants et al. (1993), Kaprelyants and Kell (1993), and MacDonell and Hood (1982). Votyakova et al. (1994) carried these studies further using 3-month-starved cultures. When 9 cells (± 5 cells) were placed in a 30-ml flask in liquid medium, 10 to 40% of the cells could be resuscitated. The lag period before the appearance of a population of cells showed a significant accumulation of the fluorescent dye rhodamine 123, indicating the restoration of the membrane permeability barrier of the starved cells during resuscitation. Membrane energization is needed for the uptake of the dye. Resuscitation was favored by the use of a 1:1 mixture of fresh lactate medium and supernatant from late-logarithmic-phase *M. luteus* cultures. The authors believe that a population effect occurs in the resuscitation process where some excreted factor(s) promotes the transition of the cells from a state in which they are incapable of growth and division to one in which they are capable of colony formation. Thus they concluded that the presence of a small fraction of viable cells at the onset of resuscitation facilitates the recovery of the majority of remaining starved cells.

Employing 48-hours-starved cells of *Vibrio* sp. strain S14, Flärdt and Kjelleberg (1994) noted an immediate response in the first minute of glucose addition, resulting in a rapid increase in the ATP and GTP pools with the [ATP]/ ratio reaching a typical level for growing cells within 4 minutes. During the first 4 to 5 minutes, the total rates of RNA and protein synthesis increased after the addition of glucose. They suggest that carbon-starved cells are deficient in amino acid biosynthesis and that ppGpp and the stringent response are involved in overcoming this deficiency, probably by depressing the synthesis of amino acid biosynthesis enzymes. They further suggest that starved cells are primarily starved for energy.

Recovery of *S. typhimurium* (28-days-starved) occurs when sufficient organic matter is supplied, but DNA synthesis was undetectable throughout the recovery experiment (Lebaron and Joux, 1994). Lebaron and Joux (1994), from their flow cytometric analysis, proposed that the ability of *S. typhimurium* to recover from their capacity to grow in culture medium could be linked to changes in the top-l state of their DNA. However, 28-day-starved *A. haloplanktis* recovered rapidly when 5 mg of peptone per liter was added (Lebaron and Joux, 1994). The DNA histograms show that the increase in DNA resulted from the division of the multigenomic cells in agreement with the lack of thymidine incorporation in DNA at the same time. Thymidine was detected by 130 hours of recovery, suggesting a renewal of DNA synthesis and higher DNA heterogeneity. The positioning of the multigenomic starved cells ensured the cells the ability to undergo rapid division when growth conditions became available.

Starved cells of *Klebsiella pneumoniae,* isolated from oil well waters, were able to grow rapidly when stimulated by nutrients (Lappin-Scott et al., 1988b).

7.10 Viable but Nonculturable

Postgate (1976) stated that bacteria can lose the ability to multiply but remain biologically completely functional as individuals. For instance, 20% of the population of *Klebsiella aerogenes,* when starved for 24 hours, retained its viability and yet 80% of this population had an intact osmotic barrier, with a complete amino acid pool, and was able to respond to a mild osmotic shock (Postgate and Hunter, 1962). Postgate (1976) asked the question whether such organisms are any more dead than, say, a woman past menopause. He also raises the question concerning the nature (a possibility of reversibility) of the pseudosenescent state. The term "viable" is difficult to define and as a result, we employ a pragmatic approach to the term, i.e., bacteria that have the ability to grow on media. For example, Kurath and Morita (1983) referred to cells that had the ability to grow on an agar medium as viable, yet they recognize that a subpopulation (10 times greater than the CFUs) that had INT activity did not show up on agar plates. Further evidence for actively metabolizing bacteria in nature that could not form colonies was reported by Hoppe (1976, 1978). These actively metabolizing bacteria that did not have the ability to form CFUs on agar represented the predominant inhabitants of offshore marine regions. The viable but nonculturable state can result not only from nutrient deprivation but also due to temperature, antibiotics, salt concentration, chlorine, etc. For a further discussion of the term, consult Roszak and Colwell (1987a).

Viable but unculturable refers to the fact that the organism cannot be cultured on the medium employed. It is defined by Oliver (1993) as a cell that can be demonstrated to be metabolically active, while being incapable of undergoing the sustained cellular division required for growth in or on a medium normally supporting growth of that cell. In culturing bacteria from nature, Gest (1993) points out what we have known for a long time, that unraveling complex growth requirements and formulating optimal growth media is frequently a very difficult and time-consuming effort. He further points out that "unculturable" assumes that no one will ever be able to grow the organism in question in the laboratory, and obviously this is not a defensible scientific proposition. The term "unculti-vated" is also currently being used in place of "unculturable" (Ward et al., 1990; Kopczynski et al., 1994). The term "unplateable" has also been employed (Oliver, 1993).

When dealing with the viable but nonculturable state, the question is whether we can classify the organisms as dead (lethally injured by starvation or irreversibly injured), injured (sublethally or reversibly injured), and normal (uninjured) cells. In injured cells, does repair occur in dilute medium so that repair can involve de novo synthesis of proteins, rRNA, and DNA? As mentioned previously, food microbiologists have devised many different media for resuscitation of injured bacteria. However, in starved cells, we recognize the loss of lipopolysaccharides so that repair may also require the synthesis and reorganization of the lipopoly-saccharides in the membrane. Other sites of injury due to starvation, such as the cell wall, ribosomes, DNA, and many enzymes, must be studied in unculturable bacteria. By passage through a ligated rabbit ileal loop, pathogenic *E. coli* was resuscitated from the viable but nonculturable state (Grimes and Colwell, 1986). Resuscitation of *Campylobacter jejuni* from the viable but nonculturable state by animal passage was observed by Saha et al. (1991).

Many types of bacteria that have not been cultured have been identified in various environments by probes (eg., Fuhrman et al., 1989, 1993; Liesack and Stackebrandt, 1991; Giovannoni et al., 1990). The mere presence of a particular microorganism does not necessarily imply ecological importance (Karl, 1986).

Nonculturable *Vibrio vulnificus* cells represent an alternative physiological re-sponse to that induced by starvation and can occur in the presence of excess nutrient (Oliver et al., 1991). Xu et al. (1982) found a greater proportion of viable but nonculturable cells after starvation of *E. coli* for 4 days at 25 ‰ salinity at 25°C than at 10°C.

"Viable" and "nonviable" cells of *Micrococcus luteus* can be easily and quan-titatively distinguished in a flow cytometer by the extent to which they accumulate fluorescent dye rhodamine 123 (Kaprelyants and Kell, 1992). Furthermore, the viability of a very low-growing chemostat culture is only about 40–50%, as judged by plate counts, but most of the "nonviable" cells can be resuscitated by incu-bation of the organism in nutrient medium prior to plating. Thus, "nonviable-but-resuscitable" cells can also be determined.

Binnerup et al. (1993) studied the viable but nonculturable *P. fluorescens* DF57 (kanamycin-resistant strain) employing a microcolony epifluorescence technique. After placing the organisms in soil for 40 days, only 0.2 to 0.35% of the initial inoculum was culturable. They found that microcolonies (20% of the initial in-oculum) underwent two or three divisions and stopped. Cells in microcolonies that underwent two or three divisions were also noted by Torrella and Morita

(1981). These microcolonies were considered by the authors to be viable but nonculturable *P. fluorescens*.

The viable but nonculturable state has also been found in *Pseudomonas aeruginosa* (Binnerup and Sorensen, 1993). In all likelihood, the viable but nonculturable state will be found in other species.

For more information concerning nonculturable bacteria, consult Rosak and Colwell (1987a), Oliver (1993), and Colwell and Grimes (1997).

7.11 Starvation Protection (General Cellular Resistance) Against Different Environmental Factors

Starved cells of *E. coli* exhibited enhanced resistance to heat (57°C) or hydrogen peroxide (15 mM), compared to their exponential growing counterpart (Jenkins et al., 1988). This degree of resistance increased with the length of the starvation period. If chloramphenicol was added at the onset of the starvation process, no protection against heat or hydrogen peroxide could be detected, indicating that the starvation protein synthesis must take place before protection can be developed. The 30 subsets of glucose starvation proteins were also synthesized during heat and hydrogen peroxide adaptations. Later, Jenkins et al. (1990) demonstrated that osmotolerance developed during starvation. Five of the 22 polypeptides induced during osmotic shock were also starvation proteins. Starvation proteins of *E. coli* include several heat shock proteins. The induction of heat shock proteins is controlled by the minor sigma factor, σ^{32} (Jenkins, et al., 1991). Upon starvation, the level of σ^{32} increased in wild-type *E. coli,* and the three σ^{32}-controlled heat shock proteins (DnaK, GroEL, and HtpG) were not induced during starvation in an isogenic Δ *proH* strain, which is unable to synthesize σ^{32}. As a result, the σ^{32} plays a role in the induction of these proteins during both heat and starvation. *P. putida* KT2442 developed a pattern of cross-protection when starved, which enabled the cells to survive environmental stresses such as high and low temperature, elevated osmolarity, solvents, and oxidative agents (Givskov et al., 1994a,b).

Starvation-induced cross-protection was also noted with a *Vibrio* sp. (Jouper-Jaan et al., 1992). Maximal stress resistance to autolysis, $CaCl_2$, heat, near UV, and UV resistance occurred at different times during the starvation process in *Vibrio* sp. Starvation protects Ant-300 from the effects of 200 atm of hydrostatic pressure, whereas nonstarved cells cannot tolerate this 200 atm (Novitsky and Morita, 1978a). Due to these results, cells in Antarctic waters, when convergence takes place, are mainly in the starved state. As a result, these cells can tolerate the low temperature, time, and hydrostatic pressure that is associated with the Antarctic Convergence waters so that when upwelling occurs, the cells can resume their metabolic activity. This water mass does not sink below 1500 m (Gordon, 1967; Gordon and Goldberg, 1970) and is assumed to diverge 20 to 30 years later.

Welch et al. (1993) demonstrated that some of the pressure-induced proteins in *E. coli* were also induced by heat shock or cold shock. It is possible that some of the pressure-induced stress proteins are also identical with the starvation-induced proteins.

Preyer and Oliver (1993) subjected starved and unstarved Ant-300 cells to 17°C (4°C above their maximum growth temperature) and the total counts (AODC), direct viable counts (method of Kogure et al., 1979), and plate counts were employed to determine the effect of heat exposure on the cells. At 17°C, viability of unstarved cells was lost within 40 hours as indicated by the plate counts, whereas by the Kogure method, less than 5% was noted. On the other hand, when cells were starved for 1 week prior to heat challenge, significant plateability was maintained for 6 days and the direct viable counts indicated their presence for at least 12 days. Starvation-mediated protection against heat was noted in *Vibrio* sp. DW1 and in *E. coli* K165 and Sc122 by Jouper-Jaan et al. (1992).

Starvation did not protect *Alteromonas espejiana* strain 261 from heat stress (Chan, 1977). The experimental procedure was such that glucose, lysine, or glucose and lysine were added to the cells when heat stressed to permit the organism to activate its enzymatic systems. In this situation, heat stress could definitely have an adverse effect. When glucose, lysine, or glucose and lysine were added to the starved cells 60 minutes before the heat stress, this helped ward off the heat stress.

E. coli AK7, an unsaturated fatty acid auxotroph, supplied with linolenic acid, while appearing normal during logarithic growth, showed a fast decline in CFU during starvation as a result of an osmotic downshift when transferred to standard agar plates unsupplemented with an osmolyte such as 300 mM sucrose of salt (NaCl or KCl) (Munro et al., 1994). Starved cells recovered their osmoresistance after the addition of a carbon and energy source to the starvation medium. Normally, carbon-starved *E. coli* cells are more resistant to osmotic challenges and other stresses as noted above. Munro et al. (1994) also noted that *E. coli* M4100 with the survival gene (proS$^+$) survived in seawater much better than its counterpart (RH 90) without the survival gene. *E. coli,* genetically marked with luminescence, when starved for 54 days at 4° and 30°C became viable but nonculturable at 30°C, but not at 4°C (Duncan et al., 1994). Luminescence of starved cells fell below background levels during starvation, but occasionally increased to detectable levels. However, employing *P. fluorescens* (also genetically marked with luminescence), Duncan et al. (1994) could not detect any nonculturable forms when starved for 54 days at either 4° or 30°C, nor did they show any decrease in luminescence. *Vibrio harveyi,* when starved for 54 days at 4°C, remained viable but nonculturable, but maintained both culturability and viability at 30°C (Duncan et al., 1994).

Bacteria isolated on low-nutrient media from the environment are physiologically more tolerant than those isolated on rich media (Harowitz et al., 1983). It appears that some of the stress proteins that are synthesized during starvation are also produced by other stresses. Thus, we have cross-protection. In oligotrophic environments, most of the bacterial cells are starved when they are exposed to other stresses; they may be better equipped to handle the imposed stress.

7.12 Conclusions

Each species has its own mechanisms to withstand the lack of nutrients and this ability varies widely. We have much to learn about the starvation-survival process and how it affects the microorganism's ability to survive long periods without an energy source. Also, the starvation-survival mechanism is probably very important to the survival of the species, as well as the evolution of the species. Starved cells appear to be more resistant to stresses. As a result, its importance in medical, food, environment, and applied microbiology becomes very evident. Because not all microorganisms in a clone or in the environment survive the starvation-survival process, Darwin's law of the survival of the fittest comes into play.

8

Molecular Genetics of Bacterial Starvation

Paul Blum

They are as sick that surfeit with too much as they that starve with nothing.
—W. Shakespeare, *Merchant of Venice,* Act I, Scene II

8.1 Scope and General Considerations

Studies on the ecology of starving bacteria are at a technical transition that promises to increase greatly our understanding of processes inherent to all niches occupied by starving cells. Identification of key starvation genes and proteins derived from laboratory-based studies should lead to the development of molecular probes for the evaluation of these critical genes in natural samples. Correlating environmental parameters with expression of these genes will provide a novel mechanistic viewpoint to evaluate ecological processes. A major challenge, however, remains to be addressed, consisting of building bridges to cross the disciplinary gap between microbial ecologists and molecular microbiologists. The gap comprises methodological and theoretical differences demanding an interdisciplinary approach to resolve their conflict. This chapter attempts to facillitate this process by providing a physiological viewpoint for much of the recent studies on the molecular biology of starvation.

Stationary phase is operationally defined as the condition of nongrowth elicited by nutrient deprivation or nutrient excess. High cell density and nutrient excess conditions also are considered in this chapter, as much of the molecular research relies on these conditions. However, such conditions are unlikely to be particularly relevant to the environment and necessitate qualification. This chapter is confined to studies employing the "model" organisms typical of molecular genetic starvation studies, all of which exhibit starvation-survival and are members of the Proteobacteria. They include *Escherichia coli, Salmonella typhimurium,* and selected *Vibrio* and *Pseudomonas* species, with continued emphasis on the *E. coli* system due to its highly developed genetics. Although these studies depend almost exclusively on laboratory-based batch culture, application of the results to natural populations and communities can and should be anticipated. It is also important to note that most of the selected studies focus primarily on starvation for carbon; corresponding analyses of the consequence of deprivation for nitrogen, phosphorus, or other nutrients is more limited. For additional information, the reader is referred to earlier reviews on the molecular genetics of the stationary phase (Lowen and Hengge-Aronis, 1994; Kolter et al., 1993; Matin, 1991; Matin et al., 1989).

8.2 Adaptive Mechanisms for Starvation-Survival

8.2.1 Formation of the Starving Cell by Reductive Division

Stationary phase cells produced by nutrient deprivation (Jenkins et al., 1988) or nutrient excess (Elliker and Frazier, 1938) exhibit a distinctive rounded cell shape with a significantly reduced volume (Lange and Hengge-Aronis, 1991b; Maaløe and Kjeldgaard, 1966; Morita, 1993; Östling et al., 1993). This morphological alteration is an adaptive response thought to promote starvation-survival as it results in an increase in the cell surface area to volume ratio and thus the opportunity to more efficiently scavenge passing nutrients. Many natural environments, for example the oceans, are dominated by such cells, which remanifest a rod-like shape upon nutrient amendment. Recovery of such cells from natural environments frequently necessitates the use of filters with surprisingly small porosity providing unambigous evidence for the existence of such cells. The morphological changes accompanying starvation are likely to involve changes in the rate of peptidoglycan synthesis and degradation, which modify the morphology and structure of the bacterial wall. The so-called morphogenes, *bolA* (Aldea et al., 1989) and *ficA* (Kawamukai et al., 1989) may be involved in altered peptidogycan production. Artificially elevated expression of *bolA* creates a rounded cell morphology in growing cells, whereas *bolA* expression itself is normally inversely correlated with growth phase. BolA may be a regulatory protein involved in controlling levels of the penicillin-binding protein PBP6. PBP6 is a carboxypeptidase involved in cell wall septum formation, a process integral to cell wall conformation; thus altered expression of PBP6 by BolA may result in changes in cell wall biosynthesis and consequently cell shape. However, other proteins also contribute to the process of reductive division. The transcription factor (sigma factor), RpoS (σ^S), is present in stationary phase cells under conditions of nutrient excess and is required for reductive division (Lange and Hengge-Aronis, 1991a; Lange and Hengge-Aronis, 1991b). It has been shown to play a critical role in the ex-

pression of a range of stationary phase genes (see section 8.2.2), is phylogeneti-
cally conserved, and thus is known as the "stationary phase" sigma. RpoS may
exert its control over reductive division by controlling *bolA* expression, since
expression of the *bolA* gene has been shown to be compromised in an *rpoS* mutant.
In addition we have recently discovered a role for the molecular chaperone and
heat shock protein, DnaK, in the process of reductive division in starving *E. coli*.
Mutants lacking *DnaK* were found to be unable to complete the process of reductive division and instead retained the rod cell shape of nutrient-replete growing
cells (Rockabrand et al., 1995). Finally the phosphorylated nucleotide, guanosine
tetraphosphate (ppGpp) may play an important role in reductive division as well
(see Chapter 7). Genes involved in cell wall biosynthesis have been shown to
undergo increased expression in response to amino acid starvation (Cashel and
Rudd, 1987). Amino acid starvation elicits the stringent response in *relA*[+] strains
of *E. coli*. Thus the morphology of starving cells may also reflect processes under
the control of ppGpp. The role of this molecule is discussed in greater detail later
(section 8.3.3).

8.2.2 Starvation Proteins and Genes

The now classic work by Neidhardt and coworkers led to the early identification
of proteins whose rates of synthesis were inversely correlated with growth rate
(Pederson et al., 1978). This study utilized the method of two-dimensional SDS
PAGE to generate "maps" of protein synthesized by *E. coli*. An inverse correlation between growth rate and the de novo rate of protein production defined a
protein set that currently is categorized as "stationary phase proteins." This class
of proteins (originally termed class IA) includes many that have been identified
in later studies based on their comigration with purified protein standards
(VanBogelen et al., 1992). These include (the corresponding gene mnemonic is
given in parentheses): ATP synthase-F1 sector β subunit *(atpD);* pyruvate water
dikinase *(pps);* RecA *(recA);* succinate-CoA ligase-(ADP-forming) β *(sucC);* Hns
(hns); dihydrolopamide succinyltransferase *(sucB);* succinate dehydrogenase
(sdhA); ATP synthase-F1 sector, α subunit *(atpA);* formate acetyltransferase *(pfl);*
and citrate-(si)-synthase *(gltA).* The pattern of production of these proteins in
batch culture would result in a diagnostic trend of increasing synthesis with the
approach to the stationary phase. Surprisingly, little attention has been paid to
this important dataset. An identical approach also based on 2-D SDS PAGE was
later used by various investigators to address the question of what proteins might
be synthesized de novo during the early stages of nutrient starvation in *E. coli*
(Groat et al., 1986; Spence et al., 1990), in *S. typhimurium* (Spector et al., 1986),
in *P. putida* (Giskov et al., 1994b), and in a *Vibrio* sp. (Nyström et al., 1990a).

Such studies have indicated that at least for these organisms, starvation elicits either the production of previously undetectable and therefore "new" proteins, elevated synthesis of certain proteins, or the reduction or elimination of synthesis of other proteins. Thus the application of 2-D SDS PAGE has led to the identification of a range of proteins collectively defined as stationary phase proteins. Surprising, however, has been the lack of corresponding information on changes in the absolute concentrations of cell protein composition; that is, nearly all studies have been confined to the synthesis of de novo protein. Since it is well accepted that starvation is accompanied by a significant elevation in cellular rates of proteolysis (Mandelstam, 1963), it is unclear whether mere elevation in the rates of de novo protein synthesis are synonymous with increased absolute protein concentrations. But it seems likely that the absolute level of proteins would be related to their cellular activity.

An alternative approach involved the use of phage Mu *d* (*lac bla*) gene fusions, including both translational and transcriptional types, to identify and clone genes of interest due to their starvation-specific regulatory properties (Blum et al., 1990; Spector and Cubitt, 1992; Spector et al., 1988; Lange and Hengge-Aronis, 1991a). Typically, the promoter of the starvation gene was cloned by virtue of its ability to reconstitute expression of a truncated plasmid marker gene (Wanner, 1987). Cloning of the gene of interest was accomplished by genomic DNA isolation and subsequent ligation into an antibiotic resistant plasmid vector. Lambda phage derivatives (Bremer et al., 1985) have also been used to isolate by phage infection and subsequent lysogenization, genes with stationary phase regulatory patterns (Blum et al., 1990; Weichart et al., 1993). In addition, the use of transposons to produce mutations with stationary phase phenotypes has yielded several interesting genes including a locus required for high cell density stationary phase survival called *surA* (Tormo et al., 1990). However, the utilization of the gene disruption technique for mutant isolation is likely to preclude the identification of many genes that play critical roles in stationary phase physiology; loss of function of such genes would require their maintenance in an exponential growth phase. The failure to ensure use of permissive conditions would likely lead to compensatory changes in gene expression.

Of particular relevance is a uniquely comprehensive approach used to identify starvation-induced genes involving the use of ordered Lambda phage libraries or encyclopedias hybridized to labeled mRNA prepared from starving cells. An encyclopedia is an assignment of all genomic cloned fragments in an ordered dataset. Blattner and coworkers screened a phage DNA encyclopedia of the *E. coli* chromosome for recombinant phages that contained genes preferentially expressed in response to the limitation of specific nutrients (Chuang et al., 1993). The identity of most of these genes is still largely unknown though their availability in cloned form and map location should greatly increase the analysis of their function.

8.2.3   The Starvation Gene Switch: Turning the Starvation Response On and Off

Bacterial genes are regulated at adjacent sequences termed *promoters*. These lie immediately adjacent to the gene itself on the 5' or upstream side. Promoter recognition is a prerequisite for gene expression and is mediated by transcription proteins called *sigma* (σ) *factors*. Sigma factors recognize only their specific promoters; thus sigma factor availability or activity sets the level of expression of genes via action at the promoters. Promoter types include those of genes for housekeeping functions and more specialized roles such as motility, nitrogen metabolism, and heat shock. However, the identity of a true stationary phase promoter is as yet elusive. Efforts to understand the most likely class, those controlled by RpoS, the stationary phase transcription protein, remain unresolved. This class includes genes such as the morphogene (see foregoing) *bol*Ap1 (see foregoing; Aldea et al., 1989), the bacteriocin, microcin B17 gene *mcbA* (Bohannon et al., 1991; Connell et al., 1987), and a gene involved in filament formation, *ftsQ* (Aldea et al., 1990). They are all *rpoS* controlled, share promoter sequences, and are distinct from those for housekeeping functions (Harley and Reynolds, 1987). Despite this, other *rpoS*-regulated starvation genes, including that encoding the DNA binding protein, *dps* (Altuvia et al., 1994), or fibronectin-binding curli *csgBA* (Arnqvist et al., 1994), have promoters with no sequence relationship to the other RpoS-regulated class. Additional conflicting results have been reported (Bohannon et al., 1991; Lange and Hengge-Aronis, 1991b). A requirement for the outer membrane protein, *ompR,* in the stationary phase induction of *mcbA*p may provide a possible signal transduction mechanism between the interior and exterior of the cell (Hernandez-Chico et al., 1986). The most surprising aspect surrounding *rpoS* is the observation that this sigma factor efficiently recognizes σ^{70} (housekeeping) promoters in vitro (Tanaka et al., 1993); thus the search for an *rpoS*-specific sequence motif may be irrelevant. Stationary phase high cell density, nutrient excess conditions were employed for several of these studies, therefore it is also unclear how such promoters respond under starvation-induced conditions. Interestingly the conserved-starvation promoter class share a tendency for their promoter DNA to bend during electrophoretic gel analysis of promoter fragments (Espinosa and Tormo, 1993).

Several factors implicate a more complex view of the action of *rpoS*. These include a crucial role for this protein in growing cells and the widespread existence of *rpoS* nonsense mutations. An essential role for RpoS has been recently established in growing, not stationary phase cells, in response to "osmotic" stress produced by the addition of excess sodium chloride (Hengge-Aronis et al., 1993). This role may be crucial for understanding the osmotic-sensitive phenotype of *rpoS* mutants in high cell density stationary phase conditions (Lange and Hengge-Aronis, 1991b). In addition, studies conducted by Matin and coworkers using stationary phase cells elicited by starvation indicate only a modest effect of loss of function *rpoS* mutations under aerobic conditions. Interestingly a more profound effect was noted under anaerobic conditions (McCann et al., 1991). The significance of this observation is unclear, but it suggests that attention to the physiological conditions used to study this important gene are crucial for understanding its role in the stationary phase.

A large number of laboratory stock strains of *E. coli* harbor loss of function mutations, typically nonsense mutations, in *rpoS* (Kassen et al., 1992; Ivanova et al., 1992; Zambrano et al., 1993). Such mutations are often suppressed at the translational level due to the widespread distribution of tRNA nonsense suppressor alleles in laboratory stocks of *E. coli* strains. This is of particular surprise in light of the "stationary phase" importance of *rpoS*. It is common practice to maintain this organism in laboratories for prolonged periods in rich medium agar (permanents or stabs) at ambient temperatures. If *rpoS* is crucial for stationary phase cells, then under such conditions there should be significant selective pressure to maintain a functional *rpoS* gene. It is surprising, therefore, that *rpoS* nonsense mutations are readily found in standard laboratory strains of *E. coli*. It may be that suppression of premature translation of the *rpoS* genes underlies the "killer" phenotype of aged *E. coli* under nutrient-excess stationary phase conditions (Zambrano et al., 1993). Regardless of these considerations, *rpoS* homologs have been recently reported in other gram-negative bacteria including *P. aeruginosa* (Fujita et al., 1994) and *Salmonellae* (Kowarz et al., 1994). In the latter case, a requirement for *rpoS* has been demonstrated in the expression of plasmid encoded virulence genes (Fang et al., 1992).

The mechanisms governing starvation-induced genes are as yet unclear. Many are likely to require *rpoS* (see section 8.3.3), and as many may be independent of this sigma factor. Several of the heat shock protein genes, including *dnaK,* have been shown to be under the control of the heat shock transcription factor during starvation (Jenkins et al., 1991), indicating that starvation somehow elicits heat shock gene expression. Thus, the heat shock promoters including those of DnaK must be also considered as stationary phase or starvation-inducible promoters. The physiological overlap between thermal stress and starvation may be of particular importance, as it suggests a general mechanism may exist that contributes to starvation-survival. Finally, many genes are controlled by levels of the small molecule, cAMP and ppGpp (see section 8.3.3; Blum et al., 1990). Thus three of the six *E. coli* sigma factors are crucial for starvation gene expression.

8.2.4 Saving the Chromosome with DNA-Binding Proteins

A major class of proteins induced in bacteria following starvation or entry into a high cell density stationary phase are those generally known as the histonelike proteins. These share a highly basic amino acid content and therefore interact electrostatically with the negatively charged polyphosphate backbone of DNA much as do eukaryotic histones. The first reports indicating coregulation of histonelike protein synthesis with the stationary phase were noted originally by Pederson and coworkers in their general analysis of proteins produced in *E. coli* (Pederson et al., 1978). However, the true identity of this as yet unidentified protein was determined later (Spassky et al., 1984). This mode of expression was identified subsequently for other histonelike proteins including *dps* (Almiron et al., 1992) or *pexB* (Lomovskaya et al., 1994). Because *dps* mutants are oxidant-sensitive in the stationary phase, it may be conjectured that the accumulation of at least this histonelike protein may be involved through perhaps an ability to condense the genome, in the normal resistance to oxidizing agents observed with wild-type strains. The H1a (Spassky et al., 1984) or H-NS protein (Dersch et al., 1993) is now well characterized at both the protein and genetic levels. Mutants in the *hns* gene exhibit pleiotropic alterations in the regulation of diverse genes (Higgins et al., 1990). This has led to the idea that H-NS acts in a sequence-independent manner to effect transcription of certain promoters. In addition such mutants exhibit alterations in recombination, transposition, and deletion formation, which also point to a generalized defect in DNA function. Overproduction of H-NS is growth inhibitory (Spurio et al., 1992), and the CspA cold shock protein positively regulates *hns* expression (La Teana et al., 1991). H-NS appears to have an affinity for "bent" DNA (Owen-Hughes et al., 1992). Since *rpoS*-regulated promoters also share a tendency to be bent (Espinosa-Urgel and Tormo, 1993; and see below, Section 8.3) H-NS and RpoS share an affinity for a similar substrate. This may help to explain the apparent overlap between H-NS and RpoS in the expression of the stationary phase–induced CsgBA operon (Arnqvist et al., 1994; Olsen et al., 1993). Interestingly *hns* transcription is enhanced in stationary phase and autoregulates its own synthesis (Dersch et al., 1993).

8.2.5 The Mechanism of Starvation-Induced Environmental Stress Resistance

Stationary phase cells produced by nutrient deprivation (Jenkins et al., 1988) or nutrient excess (Elliker and Frazier, 1938) exhibit a generalized resistance to extremes of heat, oxidizing agents such as hydrogen peroxide (H_2O_2), and sodium chloride (for review see Kolter et al., 1993; Matin et al., 1989). Starvation-induced resistance is dependent upon protein synthesis (Jenkins et al., 1988) and the regulons controlled by the alternative sigma factors, σ^{32} (Jenkins et al., 1991) and σ^S (McCann et al., 1991) encoded by the *rpoH* and *rpoS (katF)* genes respectively (for review see Lowen and Hengge-Aronis, 1994). A requirement for σ^S in the development of stress resistance has also been demonstrated in stationary phase cells under conditions of nutrient excess (Lange and Hengge-Aronis, 1991a). Because stress resistance in starving cells is significantly greater than that produced by the adaptive treatment of growing cells (Jenkins et al., 1988), unique mechanisms may be employed in nongrowing cells to create the stress-resistant state. For example, an important role for trehalose synthesis in nutrient-excess stationary phase thermotolerance has been reported (Hengge-Aronis et al., 1991). In addition, expression of the *htrE* operon, which is required for growth above 43.5°C and is controlled by σ^E (Erickson and Gross, 1989) and *rpoS* (Raina et al., 1993), may play some role in stationary phase thermotolerance, or as discussed below (section 8.3.3) it may reflect the newly recognized growth phase component of *rpoS* regulation (Lowen and Hengge-Aronis, 1994). Stationary phase cells are also typified by an extreme resistance to oxidants such as H_2O_2. An important role for *rpoS* (Lange and Hengge-Aronis, 1991a; McCann et al., 1991) and the *rpoS*-controlled gene, *dps* (Almiron et al., 1992) or *pexB* (Lomovskaya et al., 1994) in this unique physiologic condition is supported by the apparent H_2O_2-sensitive and stationary phase–specific phenotypes of mutants in these genes. *RpoS* is also important for the expression of *katE* encoding the HPII catalase, and *xthA* encoding exonuclease III. In contrast, distinct regulatory systems appear to control H_2O_2 resistance in growing cells (for review see Demple, 1991).

Whatever the molecular mechanisms are for starvation- or stationary phase–induced stress resistance, its temporal aspects are of great importance for starvation-survival in the environment; hostile conditions may well occur in addition to nutrient deprivation. To better understand the basis of starvation-induced thermotolerance, glucose-starved *E. coli* were tested for the ability to retain thermotolerance for prolonged starvation periods. After 45 days of continous starvation, the percentage of the starving cells that retained thermotolerance remained at a high level and close to the level observed in cells at the onset of starvation (Fig. 8.1). Thus starvation-induced stress resistance is a diagnostic characteristic of cells subjected to prolonged starvation.

8.2.6   Stopping Starvation-Induced Protein Damage: The Role of HSPs

The heat shock proteins (HSPs) are highly conserved proteins found in all organisms that appear to share common modes of expression and function. They include regulators of protein stability and structure called *molecular chaperones*. One of these, termed DnaK (HSP70), is associated with starvation-induced stress responses and expression of starvation genes. Studies on DnaK deficiency suggest a role for DnaK in bacterial carbon starvation as well as general carbon metabolism. The *dnaK103* mutation (Jacobson and Gillespie, 1970; Spence et al., 1990) reduces cell survival on a carbon-deficient solid medium and alters the synthesis of a subset of starvation proteins (Spence et al., 1990). Another mutation, *dnaK25*, inhibits mannose and sorbitol metabolism (Burkholder et al., 1994). The *dnaK* gene is highly expressed under normal growth conditions (Pederson et al., 1978)

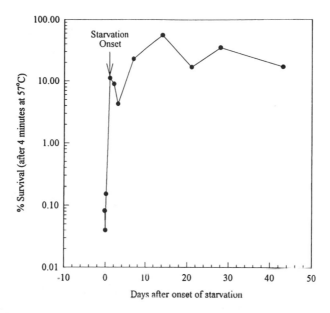

Figure 8.1 Prolonged thermotolerance in glucose-starved *Escherichia coli*. Wild-type *E. coli* K12 (strain PBL500, Rockabrand et al., submitted) was tested for thermotolerance during growth and in response to glucose starvation. Thermotolerance is expressed as the percentage of cells surviving the heat treatment relative to an otherwise identical but unheated control. The onset of starvation is indicated by the arrow.

but retains the capacity for further induction in response to heat (Yamamori and Yura, 1982), oxidation (Morgan et al., 1986), osmotic stress (Meury and Kohiyama, 1991), and starvation (Groat et al., 1986). In *E. coli,* the concentration and rate of synthesis of this protein has been shown to correlate directly with increasing growth rate and is therefore at its lowest level in slowly dividing cells (Pederson et al., 1978). Growth limitation elevates DnaK levels in starving *Vibrio* species (Holmquist et al., 1993) and in sodium chloride–stressed *E. coli* (Meury and Kohiyama, 1991). Artificially elevated DnaK levels are also preferentially toxic in stationary phase *E. coli* in a *dnaJ*-dependent manner (Blum et al., 1992a).

We have discovered a role for the HSP and molecular chaperone, DnaK, in starvation-induced stress resistance and other aspects of the nongrowth state including reductive division (Rockabrand et al., 1995). DnaK levels closely parallel changes in starvation-induced stress resistance. Glucose resupplementation of starving cells resulted in rapid loss of thermotolerance, H_2O_2 resistance and the elevated DnaK levels. A *dnaK* deletion mutant but not an otherwise isogenic wild-type strain, failed to develop starvation-induced thermotolerance or H_2O_2 resistance. When starved for glucose, the nonfilamentous and rod-shaped *dnaK* mutant strain failed to convert into the small spherical form typical of starving wild-type cells. This indicates that DnaK plays an important role in the process of reductive division. Complementation of DnaK deficiency using P_{tac} regulated *dnaK*$^+$ and *dnaK*$^+$*J*$^+$ expression plasmids, confirmed a specific role for the DnaK molecular chaperone in these starvation-induced phenotypes.

Recent analysis of a *dnaK* mutant strain in this laboratory has revealed additional stationary phase mutant phenotypes. Exposure of an isogenic wild-type and *dnaK* mutant strain to 2.5 M sodium chloride after 3 days of glucose starvation indicates that the *dnaK* mutant is preferentially killed by such exposure (Fig. 8.2). Thus DnaK deficiency also results in an inability of such cells to develop starvation-induced osmotic resistance. It remains unknown why DnaK plays this important role in stationary phase physiology. It might be related to an increased capacity of starving cells to refold or process protein substrates. In either case, a specific role for DnaK was observed previously in three of the distinguishing features of starving *E. coli* including thermotolerance, H_2O_2 resistance, and reductive division (Rockabrand et al., 1995).

Interestingly, all three of the *dnaK* mutant starvation phenotypes are shared by *rpoS* mutants (Lange and Hengge-Aronis, 1991a; Lange and Hengge-Aronis, 1991b; McCann et al., 1991). The apparent requirement for a functional *dnaK* gene in the development of starvation-induced stress resistance and reductive division clearly indicates that other genes such as *rpoS* are necessary but are not sufficient for these stationary phase processes.

8.3 Coordinating Genome Expression for Starvation-Survival

Four classes of genes can be distinguished based on their patterns of expression in starvation. Those that are reduced in expression, those that are elevated in expression, those that remain expressed at current levels, and those that previously were not expressed but were induced by nutrient deprivation. These four patterns of expression have been used to select and identify genes that are likely to have a role in the starvation or stationary phase. A wide range of genes have been identified at the molecular level and are considered in the following discussion according to functional class.

8.3.1 Protein Metabolism in Starving Bacteria

A crucial role for protein degradation exists throughout a range of stationary phase processes. The bulk of cell protein consists of ribosomes. Polysome levels drop upon carbon starvation in *E. coli* (Dresden and Hoagland, 1967) in a process recently identified to involve a ribosome modulation factor (Yamagishi et al., 1993). Proteolysis in general increases in rate upon the entry into the stationary

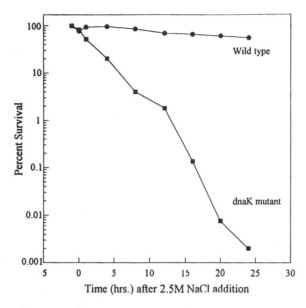

Figure 8.2   Requirement for DnaK in the salt resistance of glucose-starved *E. coli*. Otherwise wild type of *dnaK* mutant *E. coli* strains (PBL500 and PBL501, Rockabrand et al., submitted) were starved for glucose for three days and then treated with 2.5 M sodium chloride. Cell samples were plated after the exposure times indicated on a rich medium lacking elevated sodium chloride. Resistance is indicated as the percentage of surviving cells relative to an otherwise identical but untreated control.

phase (Mandlestam, 1963; Nath and Koch, 1971a). Inhibition of this process using multiply deficient peptidase mutants resulted in a severe reduction in cell viability (Reeve et al., 1984b). Surprisingly, an important and therefore specific role of any particular *E. coli* protease in this process has been elusive (Damerau and St. John, 1993). It is possible that an overlap in proteolytic specificity during substrate selection would genetically result in a phenotype only apparent in the simultaneous absence of all overlapping activities. The specific identification of those proteins that are degraded in stationary phase are unknown. However, one primary target is ribosomes and their proteins. This topic is controversial. Starvation for phosphate, nitrogen, and carbon were reported to elicit severe ribosome degradation (Davis et al., 1986). In contrast, more recent work indicates that starving *E. coli* (Yamagishi et al., 1993), *Pseudomonads* (Givskov et al., 1994a), and *Vibrio* (Albertson et al., 1990b) preferentially retain an excess of ribosomes relative to rRNA per cell. Regardless of the absolute amount of the ribosome pool that is degraded, these data indicate that a significant amount of total cell protein is transformed upon entry into the stationary phase.

Protein synthesis during entry into the stationary phase has received significantly more attention than proteolysis. For example, the identity of a large number of proteins and genes, which are synthesized in a stationary phase-specific manner has been determined. The failure to synthesize such proteins results in reduced survival and an inability to develop the generalized stress-resistant phenotype of stationary phase cells (Jenkins et al., 1988; Jenkins et al., 1991). The magnitude of the reorganization of cell protein may not be as apparent from such studies as they traditionally rely upon the measurement of de novo protein rather than on total changes in protein concentrations.

To gain some additional information concerning this process, we evaluated the bulk incorporation of radioactively labeled methionine (^{35}S-methionine) into carbon-starved *E. coli* over a prolonged carbon starvation regimen (Almiron et al., 1992; Rockabrand et al., 1995). As expected, there is a near-exponential decrease in incorporation rates as starvation proceeds (Fig. 8.3). However, integration of the quantity of de novo protein synthesized throughout the time course and corrected for background indicates that such cells synthesize twice their mass, which indicates that a tremendous degree of protein processing or transformation must occur during the starvation period. In light of this information we examined the pattern of bulk proteins in carbon-starved *E. coli.*

Consistent with the apparent quantity of protein synthesis we consistently find a dramatic change in the pattern of abundant proteins detected by mere coomassie blue dye staining. This pattern becomes most evident between 7 and 9 days of starvation and remains largely unchanged thereafter. The new proteins observed in *E. coli* K12 strain PBL500 had the following apparent molecular masses: 81.5, 70, 55.3, 35.9, 32.5, 22.5, 20.4, and 12.7 kDa. One of these, the 70-kDa protein, has been unambigously identified as the DnaK protein (Rockabrand et al., 1995). Such global changes in cell protein composition are distinct from previously described changes, as they reflect the balance between ongoing synthesis and degradation and not merely the latter process alone. It is likely therefore that the more abundant proteins in starving *E. coli* play important roles in stationary phase. Surprisingly, however, very few of these have as yet been identified, possibly due to the experimental challenges inherent in their investigation.

8.3.2 DNA and RNA in Starving Bacteria

Genome copy number in stationary phase cells must accommodate pregrowth rates of DNA replication. That is, a single but rapidly growing cell can have as many as four replicons. If such a cell is starved, it must complete current rounds of replication, then cease replication and divide twice. This situation presents potential problems, the most notable of which is a situation in which cell division is uncoupled from replication. The mechanism(s) employed to accomplish the uncoupling remain to be defined but is integral to potential applications using

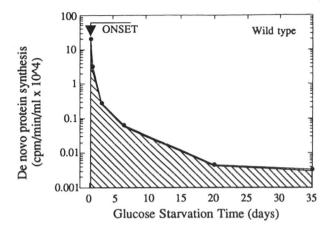

Figure 8.3   De novo protein synthesis in glucose-starved *E. coli*. Wild type *E. coli* K12 (strain PBL500, Rockabrand et al., 1995) was pulse labeled at various times with ^{35}S-methionine during prolonged glucose starvation. Labeled proteins were recovered by trichloroacetic acid precipitation and the degree of incorporation determined by liquid scintillation counting. The results were normalized to the amount of label used, the duration of the labeling period, and the number of cells labeled.

nongrowing cells in a biosynthetic manner such as uncoupling growth from "production" (Blum et al., 1992b). It is also clear that the extent of supercoiling in genomic and plasmid DNA undergoes significant changes in response to carbon starvation (Balke and Gralla, 1987). Such changes may be the result of or contribute to alterations in the levels of DNA-binding or histonelike proteins such as Dps (PexB) or H-NS (see section 8.3.3). Some evidence does in fact exist coupling the supercoiling state to stationary phase expression, in particular for the outer membrane protein, TonB (Dorman et al., 1988), but as yet the role of supercoiling in stationary phase processes is unclear

The role of RNA in microbial ecology is well known, particularly that of ribosomal RNA (rRNA) as a tool to discern taxonomic features in natural samples. However the use of this molecule as an indicator of physiological status in natural samples is controversial. It is well accepted that the levels of rRNA (and ribosomes) are directly proportional to the growth rate for rapidly growing *E. coli* and *S. typhimurium* with doubling times between 24 and 100 minutes (Neidhardt et al., 1990). In more slowly growing cells, the relationship between growth rate and rRNA is uncoupled and this molecule exists in excess of that required for protein biosynthesis (Jinks-Robertson and Nomura, 1987; Neidhardt et al., 1990). It has been argued that a linear relationship between growth rate and rRNA content is preserved in more slowly growing cells, but that the content of rRNA decreases more slowly relative to decreasing growth rate than that observed in more rapidly growing cells (Kemp et al., 1993). In response to nutrient deprivation, however, the fate of the cellular rRNA and ribosome content is less clear and reports concerning their levels in high cell density stationary phase cells have varied between undetectable and low but readily detectable levels (Dresden and Hoagland, 1967; Kaplan and Apirion, 1975a,b). In *E. coli* and *P. cepacia,* it is now recognized that a large excess of rRNA is mantained following nutrient starvation (Yamagishi et al., 1993; Givskov et al., 1994a). Such levels are clearly in great excess of that required to support stationary phase requirements for protein synthesis. Upon the exit of *Vibrio* sp. from stationary phase, the levels of rRNA and ribosomes are also in excess of that required for protein synthesis (Albertson et al., 1990). This latter observation strongly supports the contention that starving bacteria retain an excess of rRNA and ribosomes, because it is unlikely that de novo synthesis of these molecules occurs to any appreciable extent during starvation. It appears likely that at least some portion of the reduction in rRNA concentration may be mediated by increased levels of ppGpp (see section 8.3.3). However, it has been noted that one of the two promoters of the *E. coli* ribosomal RNA genes, *rrn,* is ppGpp-insensitive and remains active in response to carbon starvation (Sarmientos and Cashel, 1983).

In contrast to the levels of rRNA exhibited by bacteria at various growth rates and in response to starvation, tRNA levels appear to vary in an independent fashion. Bulk tRNA concentrations do not correlate with increasing growth rate (Neidhardt et al., 1990). This may reflect in part the variation in patterns of regulation of individual tRNA species. For example, five tRNA species cognate to "major" codons increase both as a relative fraction of total RNA and in absolute concentration with increasing growth rate, while three tRNA species cognate to "minor" codons behave in an inverse fashion (Emilsson and Kurland, 1990). Although it has become clear that individual tRNA species are differentially regulated in response to changing rates of growth, the consequences of starvation or stationary phase on their bulk levels or on the levels of individual species are largely unknown.

8.3.3 The Role of Small Molecules as Cellular Signals for Starvation: ppGpp, cAMP, and Homoserine Lactones

Amino acid starvation in bacteria induces rapid accumulation of the small signal molecule guanosine tetraphosphate (ppGpp), also called *magic spot* (reviewed in Cashel and Rudd, 1987). ppGpp accumulates in an "idling" reaction of ribosomes where ribosomes pause at particular codons during translation elongation. This stalling occurs in response to a codon-specific uncharged tRNA present in the A site of the ribosome and results in the rapid accumulation of millimolar intracellular concentrations (Haseltine and Block, 1973). Nearly all investigations utilized artificial means to produce this condition, typically including either addition of amino acid analogs (Shand et al., 1989b) or conditional heat-sensitive tRNA synthetase mutations. Nonpermissive conditions resulted in reduced levels of charged tRNA and thus signaled amino acid insufficiency. Elevated levels of ppGpp result in a global change in gene expression affecting large numbers of genes and metabolic pathways in a process called the *stringent response* (Cashel and Rudd, 1987). This response restricts synthesis of the protein synthesis apparatus including rRNA and ribosomal proteins, which collectively represent the bulk of the cell protein mass. In addition various biosynthetic pathways are induced. These changes are consistent with a reorientation of metabolism to reduced growth. However, the environmental relationship between the amino acid deprivation-induced stringent response and growth in the natural environment is unclear. Because free-living bacteria are typically prototrophic, amino acid limitation is unexpected. In particular, environmental conditions that might result in amino acid imbalance are largely unknown; however, it is likely that a generalized starvation for any particular nutrient such as carbon or nitrogen could lead to amino acid imbalances. In fact, an important role for ppGpp is evident in response to nutrient starvation.

Cashel and coworkers demonstrated that carbon, nitrogen, or phosphorus deprivation elicited a large increase in the levels of ppGpp (Lazzarini et al., 1971). ppGpp is thought to control the stringent response by mediating RNAP interactions with promoters such that the rate of initiation and elongation is either decreased (negatively regulated) or increased (positively regulated). This reaction depends on a "discriminator" sequence located adjacent to where mRNA synthesis begins (Shand et al., 1989a). Examples of the former case include the major promoter of the rRNA operons, whereas those of the latter include the amino acid biosynthetic operons including the biosynthetic histidine *(his)* or tryptophan *(trp)* of *S. typhimurium* and *E. coli* (Cashel and Rudd, 1987). Gross and coworkers identified a component of the expression of the more common heat shock proteins including DnaK and GroEL called "stringent induction," which occurred during a moderate (4–5°C) increase in growth temperature (Grossman et al., 1985). The component of the observed increase in HSP expression resulting from the imposition of amino acid starvation (due to loss of function of a conditionally mutant amino-acyl synthetase) was subtracted from a total change due also in part to the heat shock response itself. *relA* was required for the stringent component of the increase, further confirming a role for ppGpp. If ppGpp were important for this effect, and knowing that ppGpp levels and HSPs increase in response to starvation (Lazzarini et al., 1971), it would seem that HSPs would be positively regulated by ppGpp during nutrient starvation. However, an interaction between σ^{32}, the heat shock sigma factor, and ppGpp has not been reported. ppGpp is inversely correlated with cell growth rate. In addition, DnaK levels are directly correlated with growth rate (Pederson et al., 1978). Thus if ppGpp plays a role in HSP expression, then under starvation conditions such regulation must be less operative than occurs during growth.

If ppGpp is a general mediator of starvation-induced changes in gene expression, then it is of interest to note that ppGpp displays the sort of global metabolic control that would be necessary to grossly reorient cell metabolism from a growth to a nongrowth state. ppGpp is involved in regulation of the expression of enzyme activities for cell wall biosynthesis, proteolysis, phospholipid synthesis, protein synthesis, and amino acid synthesis (Cashel and Rudd, 1987). However, it has been reported that neither *relA* or the ppGpp-related *spoT* are required for carbon starvation-survival (Reeve et al., 1984a). However, the *relA* mutation does not typically exist in the absence of compensatory mutations directly related to the metabolism of ppGpp. Therefore *relA* strains harbor mutations in the *spoT* locus that may compensate for reduced ppGpp levels caused by *relA*. Because ppGpp accumulates in response to starvation in a *relA*-independent fashion (Lazzarini et al., 1971), the lack of a role for *relA* does not address the issue of the role of ppGpp in the starvation response. Although the unresolved role of ppGpp in the regulation of starvation-induced gene expression remains, this signal nucleotide is important in the exit from starvation (Flärdh and Kjelleberg, 1994).

The importance of cAMP in the expression of a variety of carbon starvation-induced proteins is well established (Blum et al., 1990; Alexander and St. John, 1994; Weichart et al., 1993) and the reader is referred to other sources for additional information (Matin, 1991). Such genes are most likely to play an important role in the escape from starvation and they appear to be largely dispensable for starvation-survival and growth under defined conditions (Blum et al., 1990). It is likely, however, that the commonly used method involving gene disruption may have biased recovery of specific gene classes.

The consequence of specific cell densities in stationary phase cell assemblages must now be considered in the context of particular signal transduction mechanisms governing gene expression. In high cell density conditions, the process referred to as quorum sensing (Meighen, 1991; Choi and Greenberg, 1992) becomes operative. This is typically accomplished by the excretion of a small "signal" molecule that coordinates the expression of specific genes in a cell-density-dependent manner. Quorum sensing requires high cell density conditions and as a result of differential gene expression can influence cell responses to specialized environmental conditions. This process is well documented in *Vibrio* and Pseudomonads (Jones et al., 1993; Pirhonen et al., 1993). In *E. coli,* two distinct signal molecules have been proposed: a homoserine lactone (Huisman and Kolter, 1994) and uridine diphosphoglucose (Bohringer et al., 1995), which play a role in sensing the stationary phase. It is possible that such molecules, particularly the former, may in fact be related to sensing high cell density; however, such speculation remains to be investigated. It is clear, however, that studies concerned with the stationary phase elicited by or because of nutrient excess conditions or because of starvation will provide important opportunities to understand a critical aspect of bacterial biology.

8.4 Conclusions

Important progress has been made in the analysis of the stationary phase including the identification of specific genes, their regulation, and their roles in the global and molecular reorganization essential to bacterial starvation. Reconciliation between the disparate conditions used to examine the stationary phase remains crucial, however, as only in a limited number of niches are microbes confronted with an excess of nutrients. However, it is also clear that nongrowth, regardless of the preceeding growth conditions, comprises a range of states with common molecular aspects including the termination of the dilution of cellular molecules due to growth and the cessation of macromolecular synthesis. With the identification of conserved starvation genes present among diverse members of the proteobacteria, it is likely that molecular probes will be developed to discern the stationary phase state in natural samples. Such procedures would present an important new tool in microbial ecology and provide a direct measure of bacterial dormancy in the environment. Such approaches will help to bridge the gap that has historically existed between the molecular genetics and microbial ecology of bacterial starvation.

9

Effects of Host Starvation on Bacteriophage Dynamics

H. S. Schrader, J. O. Schrader, J. J. Walker,
N. B. Bruggeman, J. M. Vanderloop, J. J. Shaffer, and
T. A. Kokjohn

> Nowadays, in the field of lysogeny, many facts and theoretical views are
> gloriously discovered which were known a long time ago.
>
> —André Lwoff (1953)

9.1 Introduction

Electron micrographic studies have revealed the presence of very large numbers
of viruslike particles in marine and freshwater samples (Torrella and Morita, 1979;
Bergh et al., 1989; Proctor and Fuhrman, 1990; Børsheim et al., 1990; Paul et
al., 1991; Wommack et al., 1992; Fuhrman and Suttle, 1993; Hennes and Simon,
1995). The sheer magnitude of viruslike particles observed in both aquatic and
terrestrial ecosystems suggests the potential importance of bacteriophages and
other viruses in microbial ecology. While the actual origin(s) of these particles
remains undefined, many of these particles superficially resemble bacteriophages
and these findings have prompted additional efforts to better understand the role
of viruses in the ecology of bacteria, cyanobacteria, and algae.

Bacteria have fundamental roles in nutrient cycling in ecosystems. In marine
systems, it is clear that heterotrophic bacteria are critical for the recycling of
dissolved organic carbon (DOC) (Azam et al., 1983). It is now certain that a view
of energy and carbon flow through food webs as a steady progression to ever
larger consumers in higher trophic levels is simplistic. Recent work has revealed
that a "microbial loop" exists, responsible for cycling a substantial portion of
input carbon in a short-circuit fashion (Azam et al., 1983). An interesting and at
present unknown aspect of carbon flow concerns the impact of viruses on these
epicyclic exchanges in the lower trophic levels. Clearly, viral infection in situ will
result in production of DOC on lysis of the host and this DOC will be subject to
the cycling activities of bacteria (Proctor and Fuhrman, 1990; Bratbak et al., 1990;

Suttle et al. 1990; Suttle, 1994; Hennes and Simon, 1995). The exact rate of contribution of viral lysis to the microbial carbon loop is not known and it is uncertain how conditions such as increased carbon dioxide concentrations in the atmosphere will ultimately impact oceanic biogeochemical cycling processes.

The overall significance of the viruses of bacteria could be that these are agents of population control and may influence the activities and distribution of their hosts (Kokjohn et al., 1991). In addition, bacteriophages have the potential to influence the genetic diversity of bacterial populations by imposition of selection for ability to grow in the face of predation and by allowing for the direct exchange of genes via transductional processes. The challenge for the ecologist is to understand and accurately model the effects of bacteriophages on natural assemblages of bacteria.

9.2 The Infective Process

9.2.1 One-Step Growth Curves

The replication of bacteriophages is understood in considerable detail for some systems, both lytic and lysogenic (Hayes, 1968; Ackermann and DuBow, 1987; Kutter et al., 1994). Using the one-step protocol of Ellis and Delbrück (1939), the overall mode of reproduction of bacteriophages that result in the death and lysis of the host cell follows a general and reproducible pattern. Bacteriophages have no independent metabolism and are dependent on the host for provision of ribosomes, precursor pools of macromolecules, and energy to assemble progeny bacteriophages after infection. For many virulent bacteriophages, the host metabolism is completely redirected to produce only bacteriophage components and in some cases, host macromolecules are extensively degraded. For example, bacteriophage T4 of *Escherichia coli* causes the rapid unfolding and depolymerization of the host nucleoid within minutes of infection.

After infection, a series of recognizable states of replication of bacteriophage progeny are detectable (Hayes, 1968). Initially, during the eclipse phase, all trace of infective particles, including intracellular ones, are lost as the nuclear material of the bacteriophage is separated from the capsid and used to direct the formation of new phage particles. The time during which no net increase in bacteriophages is detectable in infected cultures is termed the latent period. During this time bacteriophage components are synthesized but not yet assembled into mature infective particles. The point at which infective particles are detectable in infected cultures marks the end of the latent period and the beginning of the rise period. At this time, the host cell is lysed by the action of viral-encoded enzymes, ending the cycle with a burst (release) of progeny bacteriophage particles.

9.2.2   Potential for Replication

Systematic studies of bacteriophage replication have clearly demonstrated that the kinetics of progeny reproduction and yield of the typical infected host cell are affected by environmental conditions, with maximum rates and yields correlated with optimal growth conditions for the host cells (Ackermann and DuBow, 1987; Kutter et al., 1994). Poorer growth conditions for the host bacterium result in extensions of latent periods and decreased yields of progeny bacteriophage (Ackermann and DuBow, 1987; Kutter et al., 1994).

Studies of microbial ecology have led to the appreciation that nonsporulating bacteria are capable of surviving under conditions in which growth is very slow or impossible (Roszak and Colwell, 1987a; Morita, 1993). It now seems probable that bacteria in natural ecosystems actually spend considerable periods in starvation conditions during which time both cell physiology and structure undergo fundamental alterations (Novitsky and Morita, 1976; Matin et al., 1989; Morita, 1993).

Bacteriophage dynamics have been studied primarily in rapidly growing host cell populations under conditions prevalent for only limited periods, if they occur at all, in natural ecosystems. The classic model of bacteriophage dynamics is virus replication only in growing hosts. For example, studies of bacteriophage T4 have revealed that this virus definitely does not replicate in stationary phase cultures (Delbrück, 1940; Kutter et al., 1994). Indeed, one general method of distinguishing plaques in lawns of indicator cultures of bacteria resulting from bacteriophages from those due to parasitic organisms such as *Bdellovibrio* depends on the observation that usually bacteriophage plaques stop growing after overnight incubation while those plaques due to *Bdellovibrio* parasitism develop much more slowly on lawns that are no longer replicating (Holt and Krieg, 1994).

The results of Delbrück (1940) and Kutter et al. (1994) strongly suggest that bacteriophage replication under nutrient-depleted conditions may be sharply inhibited or precluded. To better understand the potential for bacteriophages to replicate in natural aquatic ecosystems, we have conducted one-step growth experiments using host cells that are in the stationary phase or subsisting under starvation conditions.

We have obtained from our field site at Antelope Creek (Lincoln, Nebraska) a temperate bacteriophage, designated AC-Q, that forms turbid plaques on *Pseudomonas aeruginosa* PAO. Using a modification of the one-step growth protocol of Ellis and Delbrück (1939), we have quantified replication of this bacteriophage in *P. aeruginosa* PAO host cells that are growing in Luria broth (Fig. 9.1) or incubated under starvation conditions in buffer (Fig. 9.2). Under experimental conditions that support rapid growth of the host cells, replication of bacteriophage is rapid; bursts begin to occur after approximately 120 minutes and the average burst size exceeds 200 plaques (Fig. 9.1). When cells are incubated under con-

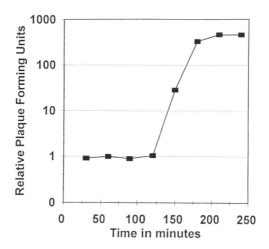

Figure 9.1   One-step growth curves of bacteriophage ACQ in *Pseudomonas aeruginosa* host cells. Actively growing cells. *P. aeruginosa* PAO303 was grown to logarithimic phase (45–55 Klett$_{660}$ units) at 37°C in Luria broth (Kokjohn and Miller, 1985). Cells were harvested by centrifugation and suspended in an equal volume of 10 mM Tris, pH7.5, 10 mM MgCl$_2$,150 mM NaCl (TMN buffer). Suspended cells were infected with bacteriophages at a low multiplicity of infection (MOI), approximately 0.1, and infection was allowed to proceed at room temperature (approximately 22°C) for 30 minutes. The moment of mixture of phage and cells was taken as time zero for the experiments. Infected cells were diluted in Luria broth and incubated at 30°C. Growth tubes were sampled periodically for infectious centers (Kokjohn et al., 1991) to observe the kinetics of progeny production. *P. aeruginosa* PAO303 was used as bacteriophage indicator using soft agar overlay titration procedures (Arber et al., 1983).

ditions in which growth of the host cells is not possible, the kinetics and yield of bacteriophage replication are altered (Fig. 9.2). The time of burst is delayed and the yield of progeny is reduced, but significantly, replication of bacteriophage AC-Q is definitely not prohibited under nongrowth conditions for the host cell. These experiments reveal that while bacteriophage replication is clearly responsive to the environmental conditions of the host, at least some replication is possible in stationary or short-term starved *P. aeruginosa* host cells for this bacteriophage.

How prevalent are bacteriophage with the potential to accomplish at least nominal replication in nongrowing host populations? Several papers published in the peer-reviewed scientific literature have described a bacteriophage of *Achromobacter* capable of replication when the host culture was definitely in the stationary phase (Woods, 1976; Robb et al., 1977; Robb et al., 1978). In addition, studies detailed in this work and with host cells other than *P. aeruginosa* (data not shown)

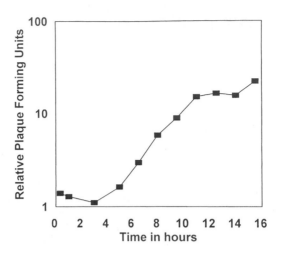

Figure 9.2   One-step growth curve of bacteriophage ACQ in starved *P. aeruginosa* host cells. Starved cells were prepared from LB (Luria broth) cultures of *P. aeruginosa* PAO303 in the early exponential phase of growth (45–55 Klett $_{660}$ units). Cells were harvested by centrifugation and suspended in an equal volume of TMN buffer. Cultures were starved by continuation of incubation at 37°C for a period of 24 hours. Starved cultures were infected and assayed for progeny bacteriophage production as detailed for log phase cultures. *P. aeruginosa* PAO303 served as the bacteriophage indictator using soft agar overlay titration procedures (Arber et al., 1983).

support the hypothesis that bacteriophage competent to replicate in stationary phase or even starved cultures are actually common. If long periods of nongrowth are the true norm for heterotrophic bacteria in nature, it seems that the process of natural selection would favor bacteriophages with at least some ability to use abundant nongrowing host populations.

   While isolating bacteriophages from natural freshwater and marine ecosystems, we have noticed that plaques definitely due to the activities of bacteriophages are often observed to continue expanding even though all growth of the bacterial lawn has ceased. Other investigators have reported similar observations of plaque expansion for certain viruses of *E. coli* (Kutter et al., 1994). While we have experimentally tested only a few of these expanding plaque morphology bacteriophages to date, our results suggest that this phenotype is due to continued, albeit slower, replication of bacteriophages on stationary phase lawns of host cells. This type of plaque morphology forms a high percentage of the total plaques detected on our standard indicator cultures. These observations suggest that not only are bacteriophages capable of replication in stationary phase host cells present in aquatic environments, they may actually be the dominant type of bacteriophages for some hosts.

9.3 Lysis-Lysogeny Pathways

9.3.1 Factors Influencing Lysogenic Development

Certain bacteriophages can interact with host cells in a manner that does not always cause the obligatory lysis and death of the host cell (Lwoff, 1953; Hayes, 1968). These bacteriophages are termed temperate and may establish a state of peaceful coexistence with the host in which most viral functions required for replication of progeny virions are not expressed. In this state, the viral chromosome is replicated precisely in step with the host but without the production of mature virions and lysis of the host cell. Bacterial cells that carry quiescent temperate viruses, termed prophage, are immune to superinfection by the same variety of bacteriophage, because genes on the resident prophage are actively expressed that repress the lytic growth functions of these viruses. Bacteria carrying temperate viruses are termed lysogens because they will cause the lysis of noncarrier (nonimmune) strains.

9.3.2 Effects of Starvation on the Lysis-Lysogeny Decision

Studies of temperate bacteriophages such as lambda of *E. coli* show that a complex interplay of both host and phage genes determine whether the result of an infection will be establishment of lysogeny or the lytic growth of the phage (Herskowitz and Hagen, 1980). In the *E. coli*–phage lambda system, the nutritional status of the host determines the developmental pathway selected by the incoming virus. When the host cell is rapidly growing, cAMP levels are low and lytic growth is favored. In the opposite case where the host cell is growing poorly and cAMP levels are elevated, lysogeny establishment is more frequent. This particuliar response is hypothesized to allow phage replication under those conditions in which it is most likely to be successful, allowing the phage to await better conditions for growth by establishing lysogeny if conditions are poor.

For the lambda–*E. coli* system lysogeny establishment is efficient whenever sufficient repressor of bacteriophage lytic functions is synthesized soon after initiation of infection; this blocks continued phage development (Wulff and Rosenberg, 1983). In addition to the *c*I gene product, which is the repressor of bacteriophage lytic growth, two other bacteriophage proteins are required for efficient production of *c*I and establishment of lysogeny; the spare protein products of the *c*II and *c*III genes of the bacteriophage. Failure to produce any of the three proteins, the *c*I, *c*II, or *c*III gene products, results in the failure of the infecting

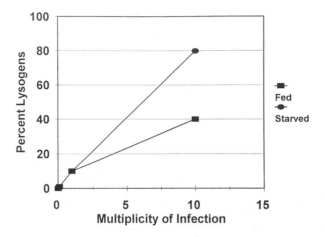

Figure 9.3   Establishment of lysogeny by bacteriophage D3 in log phase and starved cultures of *P. aeruginosa*. Cells were grown in LB at 37°C to 45–55 Klett $_{660}$ units. For log phase experiments, the cells were harvested by centrifugation and suspended in an equal volume of TMN buffer. Cells were immediately infected with bacteriophage at several different MOI. Infected cells were diluted in 0.85% saline and plated on LB agar to enumerate survivors. The number of lysogens among the survivors was quantified by patching onto lawns of phage-sensitive indicator *P. aeruginosa* PAO303. Lysogens were revealed by the production of halos due to the lysis of the sensitive PAO303 lawn.

For experiments using starved cells, harvested cultures were suspended in an equal volume of TMN and incubation continued for approximately 24 hours. At the end of this period, cells were infected with bacteriophage and survivors and lysogens quantified as described for fed cells.

bacteriophage to establish lysogeny. The bacteriophage *c*II gene product is necessary for transcription from the promoters required to express high levels of *c*I to allow for repression of lytic growth and for integration of the prophage genome into the host cell chromosome, functions vital for the establishment of lysogeny. The *c*III gene product extends the functional half-life of the *c*II protein in the infected cell. The host *hflA* gene protein, a putative protease, antagonizes the bacteriophage *c*II protein activities and so decreases probability of establishment of lysogeny. The bacteriophage *c*III protein extends the half-life of functional *c*II protein by interfering with bacterial HflA protein activity and several other cellular functions that tend to cause inactivation of the bacteriophage *c*II gene product. Host cell nutritional state enters in the biochemical interplay of host and bacteriophage genes by influencing the level of expression of the *hflA* and other genes (Herskowitz and Hagen, 1980; Wulff and Rosenberg, 1983). Under conditions in which the host cell is well charged with energy and biochemical precursors for continued growth, cAMP levels are low and expression of lysogeny antagonists such as *hflA* are maximal, leading to election of the lytic pathway of bacteriophage development. Under poorer growth conditions, host cell functions that would otherwise tend to prevent high levels of lysogeny are not effectively expressed and stable lysogens are more frequently the result of infection.

Given the available facts concerning lysogeny and its establishment, does the situation in natural ecosystems mean that election of lysogeny will predominate after infection by temperate phage? If starvation is the norm, or at least frequent for heterotrophic bacterial hosts, the prediction is that if the *E. coli*-lambda model is universal, lysogeny establishment after infection may be prevalent in nutrient-poor natural ecosystems. We have tested lysogeny establishment in *P. aeruginosa* under several conditions of growth with several bacteriophages (Figs. 9.3 and 9.4). Bacteriophage D3 is analogous to phage lambda of *E. coli* in several respects (Miller and Kokjohn, 1987) and seems to follow the same pattern of lysogeny establishment predicted by the *E. coli* model (Fig. 9.3). For this particuliar bacteriophage, the frequency of establishment of lysogeny is definitely enhanced by prior starvation of the host cell. However, studies of bacteriophage F116 lysogeny establishment reveal that this bacteriophage is rather indifferent to the nutritional status of the host with regard to establishment of lysogeny and forms lysogens at approximately the same rate in both fed and starved host cells (Fig. 9.4). Thus, it is immediately evident that the actual situation in situ will be complex with different varieties of bacteriophage exhibiting different tendencies with regard to ability to grow or tendency to establish lysogeny under the prevalent environmental conditions. These experiments clearly reveal that one model for the probability of lysogeny in natural ecosystems will not suffice.

9.4 Prophage Induction

9.4.1 Background

The termination of a stable lysogenic state may be provoked by environmental stress (Lwoff, 1953; Walker, 1984; Ackermann and DuBow, 1987). Cultures of lysogens spontaneously release bacteriophages at a readily detectable level as growth proceeds. Under normal conditions of growth in a small fraction of the bacterial population the prophages are shifted from a repressed state to a condition in which bacteriophage replication is active. This irreversible induction results in the rapid production of progeny virions and death of the host cell, just as if the lytic pathway of development was elected at the time of initial infection of the host cell. Many types of lysogens are inducible to lytic growth at very high rates by application of stress. In general, the common aspect of inducing agents seems to be that most are DNA-damaging agents (Roberts and Devoret, 1983; Ackermann and DuBow, 1987). In these DNA damage–inducible systems, the prophage

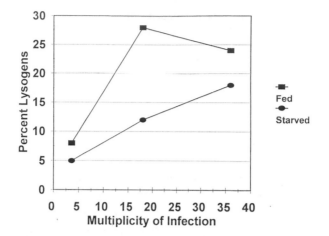

Figure 9.4   Establishment of lysogeny by bacteriophage F116 in log phase and starved cultures of *P. aeruginosa*. Cells were grown in LB at 37°C to 45–55 Klett$_{660}$ units. For log phase experiments, the cells were harvested by centrifugation and suspended in an equal volume of TMN buffer. Cells were immediately infected with bacteriophage at several different MOI. Infected cells were diluted in 0.85% saline and plated on LB agar to enumerate survivors. The number of lysogens among the survivors was quantified by patching onto lawns of phage-sensitive indicator *P. aeruginosa* PAO303. Lysogens were revealed by the production of halos due to the lysis of the sensitive PAO303 lawn.

For experiments using starved cells, harvested cultures were suspended in an equal volume of TMN and incubation continued for approximately 24 hours. At the end of this period, cells were infected with bacteriophage and survivors and lysogens quantified as described for fed cells.

recognizes that the bacterial cell has sustained severe damage and proceeds to execute an irreversible lytic cycle to escape a dying host (Roberts and Devoret, 1983; Ackermann and DuBow, 1987).

The recognition that damage has occurred in the host cell is due to the interception by the prophage of the DNA repair (SOS) system signal generated by certain DNA-damaging agents. In the case of ultraviolet (UV)-C damage to the cell DNA, formation of a high density of pyrimidine-pyrimidine dimers along the host chromosome that interfere with the progression of DNA replication forks activate a protease-promoting activity of the cellular RecA protein (Roberts and Devoret, 1983). The activated RecA protein results in the enhancement in the rate of destruction of the repressor of the SOS regulon, the product of the *lexA* gene (Walker, 1984). If the host cell is not lysogenized by a prophage capable of DNA damage induction, the removal of repression of the SOS system results in enhanced repair of DNA damage and greater chance for cell survival due to repair of lethal lesions. UV-C-inducible bacteriophages possess repressors of vegetative growth that have extensive sequence homology to the host LexA repressor and are inactivated by RecA protein as well (Walker, 1984) when DNA damage is extensive. When the repressor is eliminated, the prophage is activated to lytic growth in an irreversible manner (Roberts and Devoret, 1983; Walker, 1984).

9.4.2 Solar UV as an Inducing Agent

Given the fact that agents such as UV-C radiation will damage DNA and lead to induction of lytic growth of prophages, is the production of viruses in the oceans and other aquatic ecosystems substantially enhanced by solar irradiation? One possible complication becomes immediately apparent; not all lysogens are DNA damage-inducible. While acting as lysogens by most standard criteria, they simply are not induced when the host cell is exposed to UV-C radiation. Examples of such UV-noninducible lysogens include P1 (Yarmolinsky and Sternberg, 1988) and Mu (Kupelian and DuBow, 1986) of *E. coli*. A survey of bacteriophages isolated using *E. coli* as host revealed that about 50% are UV-C inducible (Roberts and Devoret, 1983).

It is important to consider that solar UV radiation, the UV-B wavelengths of 280–320 nm and UV-A wavelengths of 320–400 nm, differs substantially from UV-C radiation in both chromophores and effects on bacterial cells (Jagger, 1985; Eisenstark, 1989). UV-C radiation, commonly produced for experiments with the use of a germicidal lamp (Jagger, 1985), is nearly monochromatic radiation centered on 254 nm. This wavelength is strongly absorbed by bacterial DNA, resulting in the formation of pyrimidine:pyrimidine dimers that are the dominant lethal lesions in irradiated cells. UV-C radiation is absorbed completely by the ozone layer and does not penetrate to the biosphere. In contrast, the UV-B and -A regions of natural sunlight do penetrate the ozone layer. The UV-B radiation present in natural sunlight is absorbed by DNA, but clearly pyrimidine dimers are not the primary lethal lesions for irradiated bacterial cells (Jagger, 1985). In addition, solar UV damage differs from UV-C radiation damage in that it is not photoreactivable, confirming the lack of involvement of dimers in killing effects. UV-A wavelengths may not be directly absorbed by DNA at all, but may exert damaging effects in a complex manner by activation of another chromophore such as 4-thiouridyl-containing tRNA (Jagger, 1985; Eisenstark, 1989). An important aspect of solar UV radiation is that it may be only a poor inducer of bacterial SOS responses (Eisenstark, 1989) and prophages (Turner and Eisenstark, 1984). Studies of the prophage-inducing effects of solar UV induction have revealed that in certain hosts such as *E. coli*, even though pyrimidine dimers are very likely formed in the DNA (Jagger, 1985), induction of some prophages will be inefficient (Turner and Eisenstark, 1984). In the particuliar case of *E. coli*, a key reason seems to be the result of the consequences of irradiation of the thiouridyl-containing tRNA chromophore for UV-A. The overall effect of solar irradiation of these tRNA molecules is a sharp decrease in the ability of the irradiated host to induce any SOS functions, including prophages when present, a decreased efficiency of translation of mRNA, and general impairment of bacteriophage replication (Jagger, 1985; Eisenstark, 1989). Although there is a danger in extrapolating the *E. coli* model too far, clearly, it is not automatically the case that prophage UV-C induction competence will mean that solar UV is likewise an effective inducing agent for that same system.

We have tested lysogens of *E. coli* and *P. aeruginosa* and found that some lysogens definitely are activated to lytic growth by exposure to UV-C radiation (Fig. 9.5) but fail to show efficient induction after exposure to broadband solar UV (Fig. 9.6). These data are consistent with those of Turner and Eisenstark (1984), who have studied induction of lambda lysogens of *E. coli* with both monochromatic solar UV radiation and broadband solar UV radiation.

While some lysogens may not be induced at high levels by exposure to broadband solar UV, we have isolated several that do show at least some induction after exposure to low doses of solar radiation (Figure 9.7). It is important to note that the total dose of solar UV to those lysogens used for the experiment shown in Fig. 9.7 is low and would actually represent only a period of tens of minutes exposure to natural sunlight at temperate latitudes during summer. Thus, the lysogens we have examined to date that have shown solar UV induction potential may simply be those that are very sensitive to even low densities of chromosomal DNA lesions and are quickly activated to lytic growth by minor damage. Typically, prophage induction curves show an optimal dose effect and excess irradiation actually decreases the rate of induction. There is no guarantee that the higher doses of solar UV that would be expected to occur in natural ecosystems with solar exposure will be effective inducing agents in these systems and this point is presently under investigation. The true situation in natural ecosystems with regard to induction of prophage by solar UV exposure may be very complex and dependent on rates of exposure, the precise spectrum of the radiation, and the exact nature of prophage and host irradiated.

9.4.3 Aptitude

Induction of many prophages to lytic growth is definitely dependent on the growth state of the host cell. The ability to support induction, termed "aptitude" by Lwoff (1953), is dependent in a complex way on the environmental conditions of culture. For example, lysogens of *Bacillus megaterium* that are capable of efficient induction by exposure to UV radiation when growing in rich broth medium fail to support such induction when cultures grown in minimal medium are exposed in a similar fashion at similar doses. In addition, studies of *P. aeruginosa* and *E. coli* lysogens reveal that short-term starvation (periods of hours) for nutrients such as

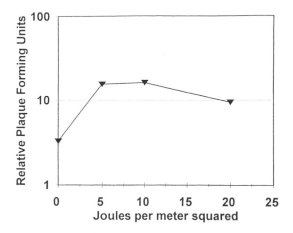

Figure 9.5   Induction of *P. aeruginosa* prophage D3 by ultraviolet (UV-C) radiation. Cells were grown to approximately 50 Klett$_{660}$ units in Luria broth at 37°C, harvested by centrifugation, and suspended in an equal volume of 0.85% saline. Samples of saline-suspended cells were exposed to UV-C radiation. Infectious centers were quantified by plating diluted samples on *P. aeruginosa* PAO303 lawns using the soft agar overlay procedure (Arber et al., 1983).

glucose impaired the induction to lytic growth of prophages by UV irradiation (Lwoff, 1953). Given the fact that it seems likely that bacteria in natural ecosystems will spend considerable time in the starved state, it seems very likely that lysogenic induction may be impossible or weak for many heterotrophic bacterial lysogens.

While induction by DNA damage may be inefficient in natural ecosystems, spontaneous induction of *P. aeruginosa* lysogens does occur at substantial rates even in cultures that have starved for long periods of time. Thus, even though some modes for production of virus may be precluded under starvation conditions, some lysogens present in an environment will be found to continuously produce mature infectious virions. Most interestingly, lysogens of *recA* mutants of *P. aeruginosa* also release mature virions at a substantial rate in contrast to the condition for *E. coli* lysogens of phage lambda (Table 9.1). It is unknown if either these hosts or phage encode antirepressor functions analogous to that of bacteriophage P22 (Susskind and Botstein, 1978; Poteete, 1988), but it is clear that whatever the mechanism of spontaneous induction in *P. aeruginosa,* this induction does continue, even after extensive starvation of the host cell. Thus, again we are led to conclude that the situation in natural ecosystems with regard to induction of lysogens will be complex and dependent on the exact system examined, the growth state of the host cell at the time of irradiation or DNA damage, the ability

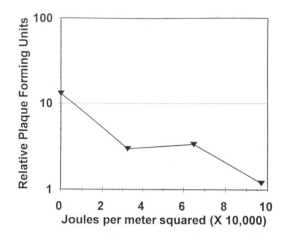

Figure 9.6   Induction of *P. aeruginosa* prophage D3 by solar ultraviolet (UV-A and -B) radiation. Cells were grown to approximately 50 Klett$_{660}$ units in Luria broth at 37°C, harvested by centrifugation, and suspended in an equal volume of 0.85% saline. Samples of saline-suspended cells were exposed to broadband solar UV radiation from an Oriel Corp. (Stratford, Connecticut) solar simulator. Infectious centers were quantified by plating diluted samples on *P. aeruginosa* PAO303 lawns using the soft agar overlay procedure (Arber et al., 1983).

of the host to rapidly repair any DNA damage, and the environmental conditions prevalent before and after inducing treatment.

9.5   Modeling Natural Ecosystems

9.5.1   Laboratory Conditions vs. Natural Aquatic Ecosystems

Most of the knowledge of microbiology and microbial ecology is based on work with microbes that are readily culturable. Often, microbes are cultured under conditions that allow for the most rapid growth and are studied during the exponential phase of reproduction. Such conditions, while convenient for the investigator, may occur only rarely in nature and it is possible that information generated in such studies may not really represent the true situation in situ (Morita, 1993).

Recent discoveries relevant to the natural ecology of bacteria have increased appreciation of the inherent complexities of modeling dynamic microbial communities. The highly significant and now unquestioned demonstration of the existence of the viable, but nonculturable, state of certain heterotrophic bacteria

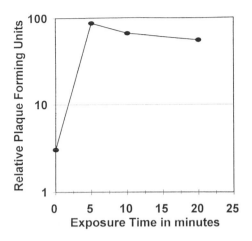

Figure 9.7   Induction of *P. aeruginosa* prophage BLB by solar ultraviolet (UV-A and -B) radiation. Cells were grown to approximately 50 Klett$_{660}$ units in Luria broth at 37°C, harvested by centrifugation, and suspended in an equal volume of 0.85% saline. Samples of saline-suspended cells were exposed to broadband solar UV radiation from an Oriel Corp. (Stratford, Connecticut) solar simulator. Infectious centers were quantified by plating diluted samples on *P. aeruginosa* PAO303 lawns using the soft agar overlay procedure (Arber et al., 1983).

(Roszak and Colwell, 1987a) sharply illustrate problems of predicting microbial activities in situ based on laboratory experiments with bacteria studied under optimal environmental conditions. New studies that have revealed the remarkable changes, both physiologic and genetic, that occur in cultures of bacteria subjected to starvation and selection for growth competence in stationary phase (Zambrano et al., 1993) underscore the suspicion that bacterial populations as they exist in nature may differ substantially from those studied in the laboratory. Other investigators have called into question even the standard methods of study of nutrient-deprived cultures of bacteria and have proposed that conditions used in experiments be developed with more care to truly represent the state of those bacteria as they may reasonably be expected to exist in natural ecosystems (Morita, 1993).

9.5.2   Bacteriophage Enumeration Using Survivors of Long-Term Starvation

It is clear that cells subjected to stresses (starvation, solar UV irradiation, xenobiotic agent exposure, etc.) will be substantially altered in physiology from their exact counterparts maintained under optimal laboratory conditions. Moreover, cultures maintained for long periods in the laboratory inevitably change and adapt to laboratory conditions by periodic selection (Kubitschek, 1970). Does the ability

Table 9.1 Prophage induction in wild-type and *recA* mutants of *Escherichia coli* and *Pseudomonas aeruginosa*.

Strain	PFU/ml (Spontaneous)	PFU/ml (Induced)
E. coli AB1157 (RecA$^+$)	10^5/ml	10^8/ml
E. coli AB1157 (*recA56*)	< 10/ml	< 10/ml
P. aeruginosa PAO1 (RecA$^+$)	10^6/ml	10^8/ml
P. aeruginosa PAO1 (*recA7*)	10^6/ml	10^6/ml

of cells selected for survival under long-term starvation conditions to support the replication of bacteriophages differ from standard laboratory strains? During the course of studies conducted in this laboratory directed toward understanding the potential for bacteriophage replication in natural aquatic environments, we have made several observations that have led us to a new hypothesis concerning bacteriophage replication in natural ecosystems. We reasoned that some varieties of bacteriophage must exist that are adapted to the infection and lysis of starved cells because such hosts would be frequently encountered in nature. Initial attempts to detect enhanced levels of bacteriophages in environmental samples by plating on indicators obtained from short-term starved host cell cultures failed. However, using the survivors of very long-term starvation experiments (over 4 years in duration) on *P. aeruginosa* PAO1, now designated *P. aeruginosa* strain PAO1-S, as bacteriophage indicators, we have observed strikingly elevated bacteriophage plaque counts from the same environmental samples that produce very few or no plaques using our standard (isogenic) indicators. A representative experiment with water samples obtained from Antelope Creek, Lincoln, Nebraska, is shown in Table 9.2. At the time of writing, our experiments have revealed that the use of these novel indicator cells results in a nominal 100-fold enhancement in bacteriophage detection ability. These counts are almost certainly *underestimates* of total bacteriophages because samples were treated with chloroform prior to plating and thus only free bacteriophages resistant to chloroform contribute to the plaque count in this determination. Moreover, these are initial experiments; no attempts to optimize phage plating as yet have been conducted. Remarkably, we are detecting bacteriophages during those times of the year (December and January) in which we routinely detected no plaques due (we formerly hypothesized) to cold water temperatures and low metabolic activity of potential *P. aeruginosa* host cells. Taken together, these data suggest that survivors of long-term starvation have stable alterations that also substantially improve their ability to act as indicators of the bacteriophages present in natural environmental samples.

Can simple starvation result in the enhanced bacteriophage plaquing activity observed with the PAO1-S strain? Experiments conducted by Sano et al. (1990)

Table 9.2   Bacteriophage titer of Antelope Creek water samples.

Date Plaques	Indicator	Sample Dilution	Mean Number of
1-5-95	*P. aeruginosa* PAO1	10^0	none
1-5-95	*P. aeruginosa* PAO1 (starved 72 hours)	10^0	none
1-5-95	*P. aeruginosa* PAO1-S (survivor of 4 years starvation)	10^0	97

have revealed that growth under iron-limited conditions enhances killing activity of certain bacteriophage tail-like aerugocins and perhaps all that we are observing is the expression of a receptor that is induced during extended incubation in stationary phase or under starvation conditions. Simply starving *P. aeruginosa* PAO1 for brief periods prior to use as phage indicator is not sufficient to provoke enhanced bacteriophage detection (Table 9.1). This is a critical control for it lends support to the hypothesis that the enhanced bacteriophage detection ability is the result of genetic alteration(s) of the original bacterial host strain. Is a loss of restriction ability the explanation for the enhanced bacteriophage detection phenotype? Attempts to detect increased numbers of bacteriophages using restrictionless varieties of *P. aeruginosa* were not successful and plating samples on *P. aeruginosa* PAO1 that had the restriction system inactivated by high temperature growth (Rolfe and Holloway, 1966) likewise did not reveal increased bacteriophage numbers (H. S. Schrader and T. A. Kokjohn, unpublished observations).

The *P. aeruginosa* PAO1-S strain is altered in several phenotypes in comparison to isogenic parental strains. In addition to enhanced phage plating characteristics this strain exhibits greater resistance to ultraviolet radiation (UV-A, -B, and -C regions of the spectrum) and enhanced hydrogen peroxide resistance. Significantly, these cells retain sensitivity to our standard reference bacteriophages, F116 and UT1. Based on the data available, our working hypothesis is that *P. aeruginosa* strain PAO1-S has undergone mutation and selection during starvation and is altered in a way that leads to the stable expression of the enhanced bacteriophage plating ability phenotype. These survivors of long-term starvation may be more representative of *P. aeruginosa* cells that are present in natural aquatic environments and are more avidly infected and lysed by those bacteriophages present in those ecosystems.

9.6   Conclusions

The one thing that seems inescapable when attempting to understand, predict, or model bacteriophage replication in natural ecosystems is the complexity of the undertaking. It is clear that the potential for replication of bacteriophages under the nutritive and physicochemical conditions of natural environments depends on the exact phage-host system being examined. In some cases, potential for replication in even stationary phase hosts seems limited (Delbrück, 1940; Kutter et al., 1994). However, our experiments reveal that some replication is possible for other bacteriophages in hosts that are actually tolerating starvation conditions. Observations are rapidly accumulating that suggest or are consistent with effective replication of cyanophages and bacteriophages as the norm in aquatic environments (Proctor and Fuhrman, 1990; Proctor et al., 1993; Suttle, 1994; Hennes and Simon, 1995) despite the often harsh conditions of those environments.

How do bacteriophages such as T4 that replicate very poorly or not at all when the host cell is in stationary phase manage not to become extinct? One possibility is that the infection process is efficient even with nongrowing hosts and that the bacteriophage is retained in this intracellular nonreplicative state for extended periods. There are some reports and discussions in the scientific literature of the existence of states that are neither true lytic growth nor classic lysogeny (Zinder, 1958; Kutter et al., 1994). In addition, in heterogeneous natural environments it must be considered that a subset of the heterotrophic bacteria will be experiencing growth conditions even though the typical cell is not reproducing. These growing cells may provide enough reservoir for production of bacteriophages to ensure their continued existence.

Temperate bacteriophages present in natural ecosystems do not universally enter into stable lysogeny on infection of the host cell. Our experiments reveal that some temperate bacteriophages are rather indifferent to the growth state of the host cell and may elect to grow lytically even in cells that have been starved for short periods. It seems probable based on our observations that many lysogens will not be efficiently induced by exposure to solar UV radiation due to the effects of nutrient limitation and a concomitant lack of aptitude (Lwoff, 1953) to support prophage induction. This effect of cell nutrition on prophage induction may confound efforts to quantify the frequency of lysogens present in natural aquatic ecosystems.

At the present time, the hypothesis that many bacteriophages in natural ecosystems preferentially recognize host cells that are either not growing or are actually starved is not proven and we are investigating the nature of the mutation(s) that account for enhanced plating characteristics of the bacterial survivors of long-term starvation experiments. It is important to recognize that an enhancement of bacteriophage plating ability may come about due to several trivial mechanisms and these must be eliminated as the cause of the enhanced plating effects we have observed to bolster the case for bacteriophage preference for starved hosts.

If bacterial cells express a set of conserved genes in response to starvation stress, might there be present on the surface of starved cells conserved structures that are exploited by bacteriophages as receptors? Taking this idea a step further, would bacteriophages capable of recognizing such hypothetical evolutionarily conserved receptors have an extended host range and be capable of infecting any host bacterium that displayed such receptors? Examples of bacteriophages with extraordinarily broad host range, limited only by the presence of suitable receptors, are known (Olsen et al., 1974). It will be interesting to determine if bacteriophages that replicate preferentially in starved hosts can actually be isolated and to determine the characteristics and host ranges of those viruses.

10

Energy of Maintenance

> First, the enormous success of classical microbiology and microbial genet-
> ics has to a considerable degree depended on selection and utilization of
> strains of organisms that behave in a reliable and consistent manner on
> laboratory media and usually bore little or no resemblance to the natural
> environment. Furthermore, the whole objective of microbial research was
> to discern order in a seemingly chaotic Nature; thus, elements of disorder
> were, by and large, perforce ignored. These two aspects of microbiological
> research perhaps conspired to impose a regularity and predictability upon
> Nature which may be misleading if we wish to understand how microor-
> ganisms behave outside the confines of the laboratory test tube or petri dish.
> —Terzaghi and O'Hara, 1990

10.1 Introduction

Do microorganisms employ energy of maintenance to help them survive starvation? Pirt (1965) proposed the concept of energy of maintenance for a growing
culture of *Escherichia coli* in order to explain how an organism maintains itself
when substrate becomes limiting. It was applied to *E. coli* growing in broth culture. Energy of maintenance is that energy required by all organisms to replace
the unstable cell constituents and is independent of growth (Marr et al., 1963).
No new biomass is formed. The energy that is expended is for osmotic regulation,
maintenance of intracellular pH, turnover of macromolecules, and motility, which
must be supplied by endogenous metabolism (Dawes, 1976, 1984; Dawes and
Ribbons, 1963, 1964a,b). Recognizing that turnover of proteins and nucleic acids
may occur in growing cultures, Marr et al. (1963) stated that most of the energy
of maintenance in starved cells would be for this turnover. Difficulties are encountered in determining the proportions of energy utilized for growth or maintenance in an actively growing population (Marr et al., 1963; Babiuk and Paul,
1970). Pirt (1966) also states that the maintenance energy for growing cells is
likely to be greater than for organisms in the stationary phase. Do bacteria in
nature require energy of maintenance? There does not appear to be sufficient
energy in ecosystems (see Chapter 3) to provide the indigenous bacteria in ecosystems with sufficient energy for the energy of maintenance. Furthermore, there
is not sufficient energy in soil to provide fungi with energy of maintenance (Lockwood, 1977; Lockwood and Filonow, 1981).

Lamanna and Mallette (1953) reviewed the problem associated with attempts to observe the existence of the energy of maintenance and its measurement in bacteria. They challenged the concept. There is no energy of maintenance in cells in the cryptobiotic state and the maintenance and growth are under control of the organism (Hinton, 1968). The process of cryptobiosis increases the life span enormously but not the duration of its active life. On the other hand, McGrew and Mallette (1962) demonstrated an energy of maintenance in *E. coli,* later confirmed by Marr et al. (1963). However, when energy of maintenance was first demonstrated, one should realize that the organisms were starved only a few hours (not days, weeks, months, or years) and investigators worried about the viability of the entire cell population. Foster (1947) stated that endogenous metabolism occurs simply because the organisms cannot help it, and therefore has no relationship, direct or otherwise, to the period of survival. Experimentally one cannot separate the energy used for endogenous respiration from the energy needed for maintenance for microorganisms in the environment (see section 7.5.4 in Chapter 7).

Monod (1942) examined the problem of energy of maintenance using *Bacillus subtilis* and *E. coli* as the test organisms and came to the conclusion that energy for maintenance was zero. Bauchop and Elsden (1961) came to the same conclusion a year earlier using growth-yield experiments. Lamanna (1963) further stated that the concept of energy of maintenance should be viewed with suspicion. The metabolic activity of the starving cell could be the expression of an unreasonable but natural compulsion for an enzyme to work with no necessary productive end in sight (Lamanna, 1963). Bacteria rapidly use any surplus and then the starvation process begins again. After the onset of endogenous metabolism, viability is lost, the time depending on the organism and its nutritional status; but the organism remains osmotically intact, metabolizing entities after the loss of viability (Postgate and Hunter, 1962).

Fildes (1940) demonstrated that growth of *Bacterium coli* could be completely stopped when exposed to $HgCl_2$ at 8×10^{-6} M. However, this growth could be reinstated if thioglycolic acid, cysteine, or glutathione (all containing the -C-SH group) was added. Even after an exposure for 4 days, the addition of -SH compounds will still reinstate growth. During this time, endogenous metabolism cannot occur due to the inactivation of the -SH enzymes or coenzymes. Some of the early texts such as Thimann (1955) state that energy of maintenance may not be required by cells, citing the data of Fildes (1940). Battley (1987) maintained that "the maintenance of the fabric of microbial cells is not intrinsically necessary." However, if the organism's environment is hostile, then maintenance energy is required depending on the vicissitudes of the environment.

The first step in the starvation-survival process of a clone is to use the energy from endogenous metabolism. This situation has been the focus of many studies and it should be stressed that these studies are usually conducted over short-term periods of hours or a few days. Rarely do you see experiments carried out for a week. During this initial period of starvation, reserve energy supplies are used to delay the death of most cells within a clone. Nevertheless, within the clone a few cells do survive. A clone of bacteria deprived of nutrients generally decreases in cell numbers, as determined by plate or spread plate counts, but there is often a fluctuation in viable cells with time as the cells are preparing themselves for longer-term survival against energy deprivation. In cells that have a long doubling time, the cell is faced with the problem of extracting the maximum yield of ATP. This energy demand in slow-growing cells must also take into consideration the increased need for energy of maintenance that can require 70% or more of the cell's energy flow (Van Verseveld et al., 1984a,b; Stouthamer et al., 1990).

Preservation of a functional transport system that will permit the concentration of substrates against a concentration gradient is vital. This brings up the subject of an energized membrane system. As previously mentioned, *Vibrio cholerae*'s membrane loses most of its phospholipids upon starvation and permits the transport of substrates across the membrane without an energy-consuming transport mechanism.

Endogenous metabolism has been studied by many investigators, generally on *E. coli,* initially on laboratory media and for short periods of time. However, not all the energy must go to maintain the cell. As can be seen from the chemotactic behavior of starved cells, some of the energy may be directed to flagellar movement. There is now evidence to suggest that the rapid metabolism of endogenous substrates, which generate energy at a rate greatly in excess of the maintenance requirements and which must, therefore, involve energy-splitting reactions, accelerates the death of starved microorganisms; whereas prolonged survival is associated with a low rate of endogenous metabolism more closely attuned to the maintenance energy requirements. Evolution has favored the development of species capable of both storing reserve energy and of utilizing it for the preservation or replacement of essential cellular components. Ribbons and Dawes (1963) stated that processes change qualitatively or quantitatively as starvation proceeds and glycogen in *E. coli* disappears during starvation. In all probability, endogenous metabolism does play an important part in the first phase of the starvation-survival process where there is a fluctuation in protein, RNA, and DNA content of the cells and generally an increase in cell numbers, especially as shown in stage 1 of Figure 7.9.

Many reviews deal with endogenous metabolism and its relationship to starvation (e.g., Dawes, 1976, 1984; Pirt, 1982; Dawes and Ribbons, 1963, 1964a,b). Endogenous metabolism provides the energy for starving cells in order to make the internal cell adjustment necessary for survival. However, in this chapter, I wish to deal with the subject in terms of whether there is an energy of maintenance.

10.2 Energy of Maintenance in the Environment

10.2.1 Chemostat Studies

Working with mixed culture of a *Spirillum* sp. and a *Pseudomonas* sp. in a chemostat at low dilution rate, Matin and Veldkamp (1978) demonstrated that the *Spirillum* had the advantage because it had a more active and/or efficient respiratory chain, a lower energy of maintenance, and a lower minimal growth rate. Competitiveness under nutrient limitation is related to a maintenance energy requirement (Dow and Morgan, 1986).

10.2.2 Energy of Maintenance in Soil

The earlier studies on energy of maintenance were performed on pure cultures in the laboratory. Generally the cells were taken from the log growth phase. When the values of the energy of maintenance published in the literature are applied to natural ecosystems, then extrapolations on a per annum basis most often surpass the values known for the average C input (Gray and Williams, 1971; Barber and Lynch, 1977; Lynch 1979; Lynch and Panting, 1980; Anderson and Domsch 1985a,b). The adopted maintenance values for the soil biomass are one order of magnitude lower than known values for single-cell cultures and are in the range of 0.001 and 0.003/hour. These figures poorly meet or exceed the C input values (Babiuk and Paul, 1970; Shields et al., 1973; Behera and Wagner, 1974; Paul and Voroney, 1980). Jenkinson and Ladd (1981) calculated maintenance values from the unmanured Rothamsted soil and found it to be three orders of magnitude smaller than the value of 0.041/hour for *Aerobacter cloacae*. Thus, the old concept of energy of maintenance applies only to in vitro conditions where the viability of the clone is addressed. In nature, we address survival and, in long-term survival (stage 3, Fig. 7.9), the microbial cells that remain viable do not need an energy of maintenance because the survival mechanism itself permits the cell to remain viable. Various assumptions have been made concerning the energetic efficiency of growth in soil. Some have taken into account the recycling of microbial cells and some have used maintenance requirements, which suggest little or no energy is available for growth (Lynch and Panting, 1980).

Babiuk and Paul (1970) first examined the maintenance energy requirement of the soil population. Later Gray and Williams (1971), Shields et al. (1973), Hunt (1977), Barber and Lynch (1977), and Paul and van Veen (1978) all agreed that specific maintenance rates must, at most, be very much smaller than those observed for organisms grown in vitro. Lynch (1979) and Mason et al. (1986) emphasize the fact that almost nothing is known about the maintenance energy requirements of bacterial populations growing at very slow rates. The first calculation of maintenance energy of the soil population was made by Babiuk and Paul (1970) who employed a figure of 0.04/hour. Using a more appropriate figure of 0.001/hour, Barber and Lynch (1977) recalculated the maintenance energies of several investigations. These revised figures indicate that only the results of Gray and Williams (1971) showed that the energy input was greater than the requirement for the energy of maintenance. When the data were revised by Gray and Williams (1971), they demonstrated that there was only sufficient energy for transient growth. All calculations show that the soil population does not receive sufficient substrate energy even to maintain itself (Lynch, 1979).

Calculations of energy of maintenance derived from actively growing cultures are clearly inappropriate to the mixed populations in soil (Williams, 1985). Although the value of 0.04/hour was calculated by Pirt (1965) for actively growing cultures, the value of 0.001/hour has been suggested (Babiuk and Paul, 1970). This value has been used by Gray and Williams (1971), Gray et al. (1973), and Parkinson et al. (1978). A value of 0.001/hour for the maintenance rate would utilize half of the carbon entering the soil (Parkinson et al., 1978). A value of 0.002/hour has been calculated by Shields et al. (1973) whereas Behara and Wagner (1974) suggested a value of 0.003/hour. According to Paul and Voroney (1980), it does not take a great deal of calculation to show that the population of natural systems, although alive and resistant to decomposition, is largely dormant, with only a small proportion of the population actively growing or even having maintenance energy rates comparable to those in the laboratory. *Arthrobacter globiformis* is known to be able to survive many weeks in culture. Chapman and Gray (1981) found that, following starvation, the specific maintenance rate fell even further, which could have been due to a shutting down of cell processes in the whole population. The specific maintenance rate has been shown not to exceed 0.002/hour when the maximum yield factor was about 0.6 g/g; while in a starving population of *A. globiformis*, it was estimated to be between 0.0008 and 0.00005/hour at 25°C and approximately one-quarter of these values at 10°C (Chapman and Gray, 1981).

Recognizing that the supply of readily utilizable carbon in soil is severely limited (Clark, 1965; Lockwood, 1977), Lockwood and Filonow (1981) tabulated the figures in Table 10.1, based on the Monod growth equation (Marr et al., 1963), which permits one to calculate the amount of substrate required for cell maintenance. This equation is as follows:

$$dx/dt + ax = Y(ds/dt)$$

where a is the specific maintenance constant (per hour), x the concentration of cells in g, Y the yield coefficient (efficiency of conversion of substrate to cells), t is time, and s the amount of substrate required for maintanance. Subtracting s from the total substrate input gives the amount available for growth. The number of generations of microbial growth per year is given by the equation

Table 10.1 Predicted number of generations of microbial cells per year and maintenance energy requirements from estimates of substrate inputs, yield coefficients, biomass, and specific maintenance constants in several soils.

Reference	Substrate Input (S)	Yield Coefficient (Y)	Biomass (x)	Specific Maintenance Constant (a)	Maintenance Energy Requirements (s)	Number of Generations (R)
Clarholm and Rosswall (1980)	441 g/m²	0.5[a]	35 g/m²	0.001[a]	613 g/m²	<1
Lynch and Panting (1980)	3540 kg C/ha	0.5[a]	200 kg C/ha	0.002[a]	3504 kg C/ha	<1
Gray and Williams (1971)	764 g/m²	0.35	46 g/m²	0.001	1156 g/m²	<1
Babiuk and Paul (1970)	500 g/m²	0.35	55 g/m²[b]	0.001	1364 g/m²	<1
Shields et al. (1973)	937 µg C/g	0.60	165 µg C/g	0.002	300 µg C/g	5.8
Flanagan and Van Cleve (1977)	398 g/m²	0.40	5.7 g/m²	0.00055	22.6 g/m²	21
Flanagan and Bunnell (1976)	10 g/m²[c]	0.54	0.45 g/m²[d]	0.00032	1.2 g/m²	23
Behara and Wagner (1974)	1040 kg/ha	0.39	8.4 kg/ha	0.0036	265 kg/ha	36
Gray et al. (1974)	706 g/m²	0.35	3.7 g/m²[b]	0.001	92 g/m²	87
Gray et al. (1974)	706 g/m²	0.35	11.1 g/m²[e]	0.001	277 g/m²	20

Source: Reprinted by permission from *Advances in Microbial Ecology* 5:1–61, J. L. Lockwood and A. B. Filonow, 1981; copyright 1981, Plenum Publishing Corporation.

[a] Value supplied or substituted for that in reference.

[b] Bacteria only.

[c] Amount available for microbial growth of annual input of 70 g/m².

[d] Fungi only.

[e] Assuming fungal biomass = 2 × bacterial biomass.

$$Y(S + xR) = xR$$

where Y is the yield coefficient, x is the concentration of cells in grams, Sp is the substrate available for growth, and R is the number of generations per year (Gray and Williams, 1971). The striking data are the number of generations per year and the amount of maintenance energy requirements by the bacteria in soil (Table 10.1).

Values of maintenance energy requirements, in g glucose/g biomass dry wt/hour for various species are as follows: 0.011 for *Arthrobacter globimormis* (Klotz, see Anderson and Domsch, 1985a), 0.094 for *Enterobacter cloacae* (Pirt, 1965), 0.005 for *E. coli* (McGrew and Mallette, 1962), 0.005 for *Mycobacterium phlei* (Klotz, see Anderson and Domsch, 1985a), and 0.085 for *Pseudomonas putida* (Klotz, see Anderson and Domsch, 1985a). These values were obtained from in vitro experiments employing glucose as the substrate. Marr et al. (1963) observed the specific maintenance requirements of bacteria ranged from 0.025 to 0.028/hour where glucose was added to a batch culture. A recalculation based on protein yield gave a specific maintenance of 0.016/hour; whereas, based on protein turnover, a value less than 0.001/hour was obtained by Koch and Levy (1955). When an efficiency estimate of 60% was used, the figure of 0.002/hour was obtained by Shields et al. (1973) for microbes under field conditions. This 60% was obtained from the original amount of labeled C stabilized in the soil as microbial tissue and metabolites. This figure agrees with Payne (1970) for the theoretical growth yields of aerobic cultures grown where the energy supply limits growth. If an efficiency factor of 0.35% was employed, the specific maintenance value would be 0.001/hour. The factor of 0.35% has been used by Babiuk and Paul (1970) and Gray and Williams (1971). These specific maintenance values calculated for viable mixed populations under field conditions are much less than those determined in laboratory studies.

Laboratory cultures of actively multiplying bacteria are capable of producing many times their dry weight of CO_2 in 24 hours; whereas in stationary phase, the production of CO_2 for 24 hours is approximately the equivalent to the dry weight of the cells (Clark and Paul, 1970). This would represent a microbial biomass of 200 g/m^2 in soil to a depth of 30 cm. The amount of CO_2 produced, according to Domsch (1986), Monteith et al. (1964), Tamm and Krzysch (1965), and Clark and Paul (1970), would represent 35–70 times as much weight of CO_2 as is usually respired from field soils daily. The soil fauna and plant roots also help the production of CO_2 in the soil. Thus, it can be seen that the CO_2 produced in the laboratory represents a great discrepancy between laboratory studies and field studies. Shields et al. (1973) demonstrated that the amount of carbonaceous substrate (energy) consumed by microorganisms in soil is not directly related to growth, mainly because a portion of the energy is required by living organisms to replace unstable cell constituents. The concept that total biomass in soil at any given time reflects the microbial activity is encountered in the maintenance energy calculations by Pirt (1965) and Veldkamp (1968), whereas Oginsky and Umbreit (1955) used total cell nitrogen to calculate the maintenance energy. Clark and Paul (1970) made some calculations concerning the data of Oginsky and Umbreit (1955) who stated that a gram of bacterial cell nitrogen would require 50 kcal/day of cell nitrogen to maintain their life functions. Thus, 60 grams of bacteria, containing 6 g of cell nitrogen, would require 300 kcal/day. The inclusion of 140 g of fungal biomass would raise the requirement to 1000 kcal/day. The net productivity in Matador Grassland soil communities rarely exceeds 5000 kcal/year. This falls short of annual maintenance energy requirements for 200 g of microbial biomass (fungi and bacteria). Clark and Paul (1970) showed that 1 g dry weight of cells requires 1.4 g of dextrose, hence 200 g of microbial biomass would require 280 g of dextrose (ca. 1100 kcal) for each generation of microbial cells produced. This would indicate that the soil microflora at the Matador plant community site, as well as that in soil generally, could reproduce only a very few times per year. Definitely, laboratory data should not be used to present field studies.

Anderson and Domsch (1985a) determined the maintenance carbon requirements of the dormant microbial biomass of two agricultural soils (labeled I, II) and a forest soil (labeled III). They found that the amount of carbon needed for preventing microbial-C loss during incubation expressed as coefficient m (mg glucose C/mg biomass C/hour) was 0.00031, 0.00017, and 0.00017/hour at 28°C and 0.000043, 0.000034, and 0.000016/hour at 15°C for soils I, II, and III, respectively. Hence, when these soils are incubated there is a loss of microbial biomass-C with time. These authors define "maintenance" as the amount of carbon necessary to keep a microbial population at a relatively constant level by preventing net microbial-C loss. Anderson and Domsch (1985a) showed the response of dormant microbial biomass to weekly glucose-C amendments in three soils at 15 and 28°C. Even when glucose amendments approximating the maintenance value were used, the biomass decreased. Naturally when the glucose amendments below this maintenance level were used, the microbial biomass decreased even more. The determined m values of the dormant population, depending on the temperature, were two to three orders of magnitude below known values from pure cultures or m values of metabolically activated biomass under in situ conditions. Because of their results, they posed the question of how much carbon is necessary to maintain the level of carbon that constitutes the living microbial biomass.

In another set of experiments Anderson and Domsch (1985b) measured the CO_2-C rate resulting from the added glucose to an actively-metabolizing in situ soil microbial population. In this case, soil samples were treated with increasing amounts of glucose to determine not only the maximal initial respiratory response but also where the CO_2 rate remained constant over an 8–13 hour period. A decrease in CO_2 production after 8–13 hours was directly related to glucose exhaustion. The concentration of glucose-C which kept the CO_2-C rate at the previous level was regarded as the maintenance level. For the three soils mentioned in the previous paragraph, the values for the maintenance coefficient m (mg glucose-C/mg biomass-C) was 0.012/hour for the two agricultural soils and 0.03/hour for the forest soil. They also determined that the metabolic qCO_2 and the maintenance values were identical for all soils, although the glucose-C maintenance level was different. No growth took place when the maintenance level of glucose was used. When comparing the output of CO_2 to the bacterial population in soil, the only plausible conclusion is that most of the soil bacteria are in a dormant state and bacteria are in a chronic state of starvation (Clark, 1965). In soil microbes, no relationship was found between starvation-survival and endogenous respiration (Chen and Alexander, 1972). For further information concerning the energy of maintenance in soil, consult Lockwood (1981) and Lockwood and Filonow (1981).

10.2.3 Energy of Maintenance in the Marine Environment

The marine bacterium, *Vibrio natriegens,* probably has a low requirement for energy of maintenance (Stouthamer and Bettenhaussen, 1973; Neijssel and Tempest, 1976).

Employing a chemostat, Jones and Rhodes-Roberts (1981) calculated the maintenance energy of a marine *Pseudomonas* as 0.042 g glucose/g dry weight/hour. They further added that it was not possible to relate any metabolic properties or maintenance requirements to the remarkable capacity for prolonged survival under starvation conditions observed for two marine bacteria. Maintenance energy is a concept that emerged from chemostat growth studies of bacteria, but as yet has not been amenable to measurement in the bacterioplankton (Wright, 1988a). Furthermore, Wright (1988a) states that "in nature, of course, we do not expect the steady state to be all that steady."

10.3 Conclusions

Maintenance of the starved state was termed "envirotolerant" by MacDonell and Hood (1984). Novitsky and Morita (1978b) argued that, during evolutionary selection for the ability to maintain viability during nutrient stress, reduction in endogenous respiration may have emerged as the means by which macromolecular degradation ceases.

I would like to introduce a new term, energy of survival. It would apply to the "Amazons" of a species in the starvation-survival state, especially as in stage 3 of Figure 7.9. This energy of survival would provide enough energy to the cell to metabolize to take care of the destruction of amino acids (racemization) and depurination of the DNA so that the cell survives until it meets more optimal conditions that would permit growth and reproduction to occur. Where might this energy come from? Geological material and the atmosphere do contain a very low concentration of organic matter. Experimentally, the concept of energy of survival would be very difficult to demonstrate, if not impossible.

The term "viability" applies to populations, not to individuals (Postgate, 1976). Accordingly, the term "energy of maintenance" should apply to a clone or population. More specifically, it would apply to a single species and not a mixed population that occurs in nature. There are differences between cells within a clone (Koch, 1987). In my opinion, the concept of energy of maintenance should be reserved for pure cultures in the laboratory where we are dealing with a clone and its ability to maintain its colony-forming units and should not be applied to organisms in the environment. When energy is no longer available to the cell, endogenous metabolism takes place and this endogenous metabolism prepares the cell for prolonged starvation. Endogenous metabolism is most rapid in stage 1 (Fig. 7.9) and extremely low, if any, in stage 3 (Fig. 7.9). Very few organisms are active at any given time in soil (Jenkinson and Ladd, 1981) or in oligotrophic environments. When cells reach the metabolic arrest stage (stage 3, Fig. 7.9), little or no energy is expended. The metabolic arrest or precursor growth cell stage helps explain why bacteria in the marine oligotrophic environment can remain viable in spite of the organic matter being old (Williams et al., 1969), its resistance to decomposition (Barber, 1968), and its residence time (Broecker, 1963), and the presence of bacteria in ancient material. The concept of energy of maintenance has no bearing on the long-term starvation-survival of bacteria, where we are addressing the survival of a species and not a clone.

Clearly, it can be seen that in most natural environments there is insufficient organic matter (energy) for the heterotrophic bacteria present to provide for the so-called energy of maintenance. In addition, the fungi and other microorganisms in natural environments also need the organic matter. This fact, plus the data from Chapter 3, definitely shows that most ecosystems are oligotrophic, and therefore, the main mode of existence for microbes is the starvation-survival lifestyle.

References

Abelson, P. H. 1959. Paleobiochemistry and organic geochemistry. *Fortschr. Chem. Org. Naturstoffe* **17**:379–403.

Ackermann, H.-W., and M. S. DuBow. 1987. *Viruses of Prokaryotes*. CRC Press, Boca Raton, FL.

Adler, J. Chemotaxis in bacteria. *J. Supromolec. Struct.* **4**:305–317.

Adler, H. I., W. D. S. Fisher, A. Cohen, and A. A. Gardigree. 1967. Miniature *Escherichia coli* cells deficient in DNA. *Proc. Nat. Acad. Sci.* **57**:321–326.

Adler, J., G. L. Hazelbauer, and M. M. Dahl. 1973. Chemotaxis towards sugars in *Escherichia coli*. *J. Bacteriol.* **115**:824–847.

Aida, W., T. Muraoka, and H. Seki. 1988. Effect of rapid oligotrophication by an aquatic treatment pilot plant on the microbial community of a mesotrophic bog. *Water, Air, Soil Pollu.* **42**:433–438.

Akagi, Y., and N. Taga. 1980. Uptake of D-glucose and L-proline by oligotrophic and heterotrophic marine bacteria. *Can. J. Microbiol.* **26**:454–459.

Akagi, Y., N. Taga, and U. Simidu. 1977. Isolation and distribution of oligotrophic marine bacteria. *Can. J. Microbiol.* **23**:981–987.

Akagi, Y., U. Simidu, and N. Taga. 1980. Growth responses of oligotrophic and heterotrophic marine bacteria in various substrate concentrations, and taxonomic studies on them. *Can. J. Microbiol.* **26**:800–806.

Albertson, N. A., G. W. Jones, and S. Kjelleberg. 1987. The detection of starvation-specific antigens in two marine bacteria. *J. Gen. Microbiol.* **133**:2225–2231.

Albertson, N. A., T. Nyström, and S. Kjelleberg. 1990a. Exoprotease activity of two marine bacteria during starvation. *Appl. Environ. Microbiol.* **56**:218–223.

Albertson, N. A., T. Nyström, and S. Kjelleberg. 1990b. Starvation-induced modulations in binding protein-dependent glucose transport by the marine *Vibrio* sp. S14. *FEMS Microbiol. Lett.* **70**:205–210.

Albertson, N. A., T. Nyström, and S. Kjelleberg. 1990c. Functional mRNA half-lives in the marine Vibrio sp. S14 during starvation and recovery. *J. Gen. Microbiol.* **136**:2195–2199.

Albertson, N. A., T. Nyström, and S. Kjelleberg. 1990d. Macromolecular synthesis during recovery of the marine *Vibrio* sp. S14 from starvation. *J. Gen. Microbiol.* **136**:2201–2207.

Albrechtsen, H.-J. 1994. Distribution of bacteria, estimated by a viable count method, and heterotrophic activity in different size fractions of aquifer sediment. *Geomicrobiol. J.* **12**:253–264.

Aldea, M., T. Garrido, C. Hernandez-Chico, M. Vincente, and S. R. Kushner. 1989. Induction of a growth-phase-dependent promoter triggers transcription of *bolA,* an *Escherichia coli* morphogene. *EMBO J.* **8**:3923–3931.

Aldea, M., T. Garrido, J. Pla, and M. Vicente. 1990. Division genes in *Escherichia coli* are expressed coordinately to cell septum requirements by gearbox promoters. *EMBO J.* **9**:3787–3794.

Alexander, D. M., and A. C. St. John. 1994. Characterization of the carbon starvation-inducible and stationary phase-inducible gene sip encoding an outer membrane lipoprotein in *Escherichia coli. Mol. Microbiol.* **11**:1059–1071.

Alexander, M. 1965. Biodegradation: Problems of molecular recalcitrance and microbial fallibility. *Adv. Appl. Microbiol.* **7**:35–76.

Alexander, M. 1971. Biochemical ecology of microorganisms. *Ann. Rev. Microbiol.* **15**:361–392.

Alexander, M. 1973. Nonbiodegradable and other recalcitrant molecules. *Biotech. Bioeng.* **15**:611–647.

Alexander, M. 1979. Recalcitrant molecules, fallible micro-organisms. *In Microbial Ecology: A Conceptual Approach,* J. M. Lynch and N. J. Poole (eds.), pp. 246–253. Blackwell Scientific Publications, Oxford.

Alexander, M., R. J. Wagenet, P. C. Baveye, J. T. Gannon, U. Mingelgrin, and Y. Tan. 1991. *Movement of Bacteria Through Soil and Aquifer Sand.* EPA Project Summary. EPA/600/S2-91/010 June, 1991.

Alldredge, A. L. 1976. Discarded appendicularian houses as sources of food, surface habitats, and particulate organic matter in planktonic environments. *Limnol. Oceanogr.* **21**:14–23.

Alldredge, A. L. 1979. The chemical composition of macroscopic aggregates in two neretic seas. *Limnol. Oceanogr.* **24**:855–866.

Alldredge, A. L. 1989. The significance of suspended detrital aggregates of marine snow as microhabitats in the pelagic zone of the ocean. *In Recent Advances in Microbial Ecology,* T. Hattori, Y. Ishida, Y. Maruyama, R. Y. Morita, and A. Uchida (eds.), pp. 108–112. Japan Scientific Societies Press, Tokyo.

Alldredge, A. L., and Y. Cohen. 1987. Can microscale chemical patches persist in the Sea? Microelectrode study of marine snow, fecal pellets. *Science* **235**:689–691.

Alldredge, A. L., and C. C. Gotschalk. 1988. *In situ* settling behavior of marine snow. *Limnol. Oceanogr.* **33**:339–351.

Alldredge, A. L., and M. W. Silver. 1988. Characteristics, dynamics and significance of marine snow. *Prog. Oceanogr.* **20**:41–82.

Alldredge, A. L., J. J. Cole, and D. A. Caron. 1986. Production of heterotrophic bacteria inhabiting macroscopic organic aggregates (marine snow) from surface waters. *Limnol. Oceanogr.* **31**:68–78.

Allen, H. L. 1976. Dissolved organic matter in lake water: characteristics of molecular weight size fractions and ecological implications. *Oikos* **27**:64–74.

Allen, J. A. 1979. The adaptation and radiation of deep-sea bivalves. *Sarsia* **64**:19–27.

Allen-Austin, D., B. Austin, and R. R. Colwell. 1984. Survival of *Aeromonas salmonicida* in river water. *FEMS Microbiol. Lett.* **21**:143–146.

Aller, R. C., and J. Y. Yingst. 1978. Biogeochemistry of tube dwellings: a study of the sedentary polychaete *Amphitrite ornata* (Leidy). *J. Mar. Res.* **36**:201–254.

Allweiss, B., J. Dostal, K. E. Carey, T. F. Edwards, and R. Freter. 1977. The role of chemotaxis in the ecology of bacterial pathogens of mucosal surfaces. *Nature* **260**:448–450.

Almiron, M., A. J. Link, D. Furlong, and R. Kolter. 1992. A novel DNA-binding protein with regulatory and protective role in starved *Escherichia coli*. *Genes Devel.* **6**:2646–2654.

Alongi, D. M. 1985. Effect of disturbance on population dynamics and trophic interactions among microbes and meiofauna. *J. Mar. Res.* **43**:351–364.

Alongi, D. M., and R. B. Hansen. 1985. Effect of detritus supply on trophic relationships within experimental benthic food webs. II. Microbial responses, fate, and composition of decompositing detritus. *J. Exp. Mar. Biol. Ecol.* **88**:167–182.

Altuvia, S., M. Almiron, G. Huisman, R. Kolter, and G. Storz. 1994. The *dps* promoter is activated by OxyR during growth and by IHF and σ^S in stationary phase. *Mol. Microbiol.* **13**:265–272.

Amador, J. A., M. Alexander, and R. G. Zika. 1989. Sequential photochemical and microbial degradation of organic molecules bound to humic acid. *Appl. Environ. Microbiol.* **55**:2843–2849.

Amako, K., S. Shimodori, T. Imoto, S. Miake, and A. Umeda. 1987. Effects of chitin and its solubility derivatives on survival of *Vibrio cholerae* O1 at low temperature. *Appl. Environ. Microbiol.* **53**:603–605.

Ames, G. F., and J. Lever. 1970. Components of histidine transport: histidine binding proteins and hisP protein. *Proc. Natl. Acad. Sci. (USA)* **66**:1096–1103.

Ammerman, J. W., and F. Azam. 1981. Dissolved cyclic adenosine monophosphate (cAMP) in the sea and uptake of cAMP by marine bacteria. *Mar. Ecol. Prog. Ser.* **5**:85–89.

Ammerman, J. W., and F. Azam. 1987. Characteristics of cyclic AMP transport by marine bacteria. *Appl. Environ. Microbiol.* **53**:203–206.

Ammerman, J. W., and F. Azam. 1992. Uptake of cyclic AMP by natural populations of marine bacteria. *Appl. Environ. Microbiol.* **43**:869–876.

Ammerman, J. W., J. A. Fuhrman, A. Hagström, and F. Azam. 1984. Bacterioplankton growth in seawater: 1. Growth kinetics and cellular characteristics in seawater cultures. *Mar. Ecol. Prog. Ser.* **18**:31–39.

Amthor, K. 1989. The activity of extracellular enzymes in starved and unstarved cells of a psychrophilic vibrio. M.S. Thesis. Oregon State University, Corvallis.

Amy, P. S., and R. Y. Morita. 1983a. Starvation-survival patterns of sixteen freshly isolated open-ocean bacteria. *Appl. Environ. Microbiol.* **45**:1109–1115.

Amy, P. S. and R. Y. Morita. 1983b. Protein patterns of growing and starved cells of a marine *Vibrio* sp. *Appl. Environ. Microbiol.* **45**:1685–1690.

Amy, P. S., C. Pauling, and R. Y. Morita. 1983a. Starvation-survival processes of a marine *Vibrio. Appl. Environ. Microbiol.* **45**:1041–1048.

Amy, P. S., C. Pauling, and R. Y. Morita. 1983b. Recovery from nutrient starvation by a marine *Vibrio* sp. *Appl. Environ. Microbiol.* **45**:1685–1690.

Amy, P. S., B. A. Caldwell, A. H. Soeldner, R. Y. Morita, and L. J. Albright. 1987. Microbial activity and ultrastructure of mineral-based marine snow from Howe Sound, British Columbia. *Can. J. Fish. Aquat. Sci.* **44**:1135–1142.

Amy, P. S., C. Durham, D. Hall, and D. L. Haldeman. 1993. Starvation-survival of deep subsurface isolates. *Current Microbiol.* **26**:345–352.

Amy, P. S., D. L. Haldeman, D. Ringelberg, D. H. Hall, and C. Russell. 1992. Comparison of identification systems for classification of bacteria isolated from water and endolithic habitats within deep subsurface. *Appl. Environ. Microbiol.* **58**:3367–3373.

Anderson, G. 1958. Identification of derivatives of deoxyribonucleic acid in humic acid. *Soil Sci.* **86**:169–174.

Anderson, J. I. W., and W. P. Heffernan. 1965. Isolation and characterization of filterable marine bacteria. *J. Bacteriol.* **90**:1713–1718.

Anderson, J. M. 1987. Interactions between invertebrates and microorganisms: noise or necessity for soil processes. *Symp. Soc. Gen. Microbiol.* **41**:125–145.

Anderson, J. M. 1988. Invertebrate-mediated transport processes in soils. *In Biological Interactions in Soil,* C. A. Edwards, B. R. Stinner, D. Stinner, and S. Babatin (eds.), pp. 6–19. Elsevier, Amsterdam.

Anderson, T.-H., and K. H. Domsch. 1975. Measurement of bacterial and fungal contributions to respiration of selected agricultural and forest soils. *Can. J. Microbiol.* **21**:314–322.

Anderson, T.-H., and K. H. Domsch. 1978. A physiological method for the quantitative measurement of microbial biomass in soils. *Soil Biol. Biochem.* **10**:215–221.

Anderson, T.-H., and K. H. Domsch. 1985a. Maintenance carbon requirements of actively metabolizing microbial populations under *in situ* conditions. *Soil Biol. Biochem.* **17**:197–203.

Anderson, T.-H., and K. H. Domsch. 1985b. Determination of ecophysiological maintenance carbon requirements of soil microorganisms in a dormant state. *Biol. Fert. Soils* **1**:81–89.

Anderson, T.-H., and K. H. Domsch. 1986. Carbon assimilation and microbial activity in soil. *Z. Pflanzenernachr. Bodenk.* **149**:457–486.

Andersson, A., U. Larsson, and A. Hagström. 1986. Size-selective grazing by a microflagellate on pelagic bacteria. *Mar. Ecol. Prog. Ser.* **33**:51–57.

Andrew, M. H. E., and A. D. Russell. 1984. *The Revival of Injured Microbes.* Academic Press, London.

Andrews, J. H. 1984. Relevence of r- and K-theory to the ecology of plant pathogens. *In Current Perspective in Microbial Ecology,* M. J. Klug and C. A. Reddy (eds.), pp. 1–7. American Society of Microbiology, Washington, DC.

Anonymous. 1979. Physics examines bacteria's world. *Science News* **113**:214.

Anraku, Y. 1967. The reduction and restoration of galactose transport in osmotically shocked cells of *Escherichia coli. J. Biol. Chem.* **242**:2123–2127.

Ansback, J., and T. H. Blackburn. 1980. A method for the analysis of acetate turnover in a coastal marine sediment. *Microb. Ecol.* **5**:253–264.

Arber, W., L. Enquist, B. Hohn, N. E. Murray, and K. Murray. 1983. Experimental methods for use with lambda. In *Lambda II,* R. W. Hendrix, J. W. Roberts, F. W. Stahl, and R. A. Weisberg (eds.), pp. 433–466. Cold Spring Harbor Laboratory Press, Cold Spring Harbor, New York.

Arnqvist, A., A. Olsen, and S. Normark. 1994. σ^S-dependent growth-phase induction of the *csgBA* promoter in *Escherichia coli* can be achieved in vivo by σ^{70} in the absence of the nucleoid-associated protein H-NS. *Mol. Microbiol.* **13**:1021–1032.

Arons, A. B., and H. Stommel. 1967. On the abyssal circulation of the world oceans. *Deep-Sea Res.* **14**:441–457.

Arrhenius, G. 1952. Sediment cores from the East Pacific, Pt. 1. Properties of the sediment and their distribution. *Report on the Swedish Deep-Sea Expedition 5.* 91 pp.

Atkinson, D. E. 1971. Adenine nucleotides as stoichiometric coupling agents in metabolism as regulatory modifiers: the adenylate energy charge. *In Metabolic Regulation,* H. J. Vogel (ed.), pp. 1–21. Academic Press, New York.

Atkinson, D. E. 1977. *Cellular Energy Metabolism and Its Regulation.* Academic Press, New York.

Atkinson, D. E., and G. M. Walton. 1967. Adenosine triphosphate conservation in metabolic regulation: rate liver citrate cleavage eyzyme. *J. Biol. Chem.* **242**:3239–3241.

Atlas, R. M. 1984. *Petroleum Microbiology.* Macmillian, New York.

Atlas, R. M. 1991. Response of microbial populations to environmental disturbance. *Microb. Ecol.* **22**:249–256.

Atlas, R. M., and R. Bartha. 1981. *Microbial Ecology: Fundamentals and Applications.* Addison Wesley, Reading, MA.

Aumen, N. G., P. J. Bottomley, G. M. Ward, and S. V. Gregory. 1983. Microbial decomposition of wood in streams: distribution of microflora and factors affecting [^{14}C]lignocellulose mineralization. *Appl. Environ. Microbiol.* **46**:1409–1416.

Aumen, N. G., P. J. Bottomley, and S. V. Gregory. 1985. Impact of nitrogen and phosphorus on [^{14}C]lignocellulose decomposition by stream wood microflora. *Appl. Environ. Microbiol.* **49**:1113–1118.

Austin, B. 1988. *Marine Microbiology.* Cambridge Univ. Press, Cambridge.

Austin, B., and D. A. Austin. 1987. *Bacterial Fish Pathogens—Diseases in Farmed and Wild Fish.* Ellis Horwood Ltd., Chichester.

Austin, B., and J. N. Rayment. 1985. Epizootiology of *Renibacterium salmoninarum*, the causal agent of bacterial kidney disease in salmonid fish. *J. Fish Dis.* **8**:505–509.

Ayanaba, A., and M. Alexander. 1972. Changes in nutritional types in bacteria successions. *Can. J. Microbiol.* **18**:1427–1430.

Ayanaba, A., S. B., Tuckwell, and D. S. Jenkinson. 1976. The effects of clearing and cropping on the organic reserves and biomass of tropical forest soils. *Soil Biol. Biochem.* **8**:519–525.

Azam, F., and J. W. Ammerman. 1984a. Mechanisms of organic matter utilization by marine bacterioplankton. *In Marine Phytoplankton and Productivity*, O. Holm-Hansen, L. Bolis, and R. Giles (eds.), pp. 45–54. Springer-Verlag, Berlin.

Azam, F., and J. W. Ammerman. 1984b. Cycling of organic matter by bacterioplankton in pelagic marine ecosystems: microenvironmental considerations. *In Flow and Energy and Materials in Marine Ecosystems*, M. J. Fashman (ed.), pp. 345–360. Plenum Publishing, New York.

Azam, F., and B. C. Cho. 1987. Bacterial utilization of organic matter in the sea. *Symp. Soc. Gen. Microbiol.* **41**:261 281.

Azam, F., and J. A. Fuhrman. 1984. Measurement of bacterioplankton growth in the sea and its regulation by environmental conditions. *In Heterotrophic Activity in the Sea*, J. E. Hobbie and P. J. leB. Williams (eds.), pp. 179–196. Plenum Press, New York.

Azam, F., and R. E. Hodson. 1977a. Size distribution and activity of marine microheterotrophs. *Limnol. Oceanogr.* **22**:492–501.

Azam, F., and R. E. Hodson. 1977b. Dissolved ATP in the sea and its utilization by marine bacteria. *Nature* **267**:696–697.

Azam, F., and R. E. Hodson. 1981. Multiphasic kinetics for D-glucose uptake by assemblages of natural marine bacteria. *Mar. Ecol Prog. Ser.* **6**:213–222.

Azam, F., T. Fenchel, J. G. Field, J. S. Gray, L. A. Meyer-Reil, and T. F. Thingstad. 1983. The ecological role of water-column microbes in the sea. *Mar. Ecol. Prog. Ser.* **10**:257-263.

Bååth, E. 1994. Thymidine and leucine incorporation in soil bacteria with different cell size. *Microbiol. Ecol.* **27**:267–278.

Babiuk, L. A., and E. A. Paul. 1970. The use of fluorescein isothiocyanate in the determination of bacterial biomass of grassland soil. *Can. J. Microbiol.* **16**:57–62.

Bacon, J. S. D. 1968. The chemical environment of bacteria in soil. *In The Ecology of Soil Bacteria*, T. R. G. Gray and D. Parkinson (eds.), pp. 25–43. Univ. Toronto Press, Toronto.

Bada, J. L. 1985. Racemization of amino acids. *In Chemistry and Biochemistry of Amino Acids*, G. C. Barrett (ed.), pp. 399–414. Chapman & Hall, London.

Bada, J. L., and C. Lee. 1977. Decomposition and alteration of organic compounds in sea water. *Mar. Chem.* **5**:523–534.

Bada, J. L., E. Hoopes, and M.-S. Ho. 1982. Combined amino acids in Pacific waters. *Earth Planet. Sci. Lettr.* **58**:276–284.

Bae, H. C., and L. E. Casida, Jr. 1973. Responses of indigenous microorganisms to soil incubation as viewed by transmission electron microscopy of cell thin sections. *J. Bacteriol.* **113**:1462–1473.

Bae, H. C., E. H. Cota-Robles, and L. E. Casida, Jr. 1972. Microflora of soil as viewed by transmission electron microscopy. *Appl. Microbiol.* **23**:637–648.

Bain, N., W. Hodgkiss, and J. M. Shewan. 1968. The bacteriology of salt used in fish curing. *In The Microbiology of Fish and Meat Curing Brines,* B. P. Eddy (ed.), pp. 1–11. Her Majesty's Stationary Office, London.

Baker, R. M., F. L. Singleton, and M. A. Hood. 1983. Effect of nutrient deprivation on *Vibrio cholerae. Appl. Environ. Microbiol.* **46**:930–940.

Baker, S. A., M. H. B. Hayes, R. G. Simmonds, and M. Stacey. 1967. Studies on soil polysaccharides I. *Carbohydr. Res.* **5**:12–24.

Bakhrouf, A., M. Jeddi, A. Bouddabous, and M. J. Gauthier. 1989. Evolution of *Pseudomonas aeruginosa* towards a filterable state in seawater. *FEMS Microbiol. Lett.* **59**:187–190.

Bakken, L. R. 1985. Separation and purification of bacteria from soil. *Appl. Environ. Microbiol.* **49**:1482–1487.

Bakken, L. R., and R. A. Olsen. 1986. Dwarf cells in soil—A result of starvation of "normal" bacteria, or separate population. *In Perspectives in Microbial Ecology,* F. Megusar and M. Ganthar (eds.), pp. 561–566. Slovene Society for Microbiology, Ljubljana, Yugoslavia.

Bakken, L. R., and R. A. Olsen. 1987a. The relationship between cell size and viability of soil bacteria. *Microb. Ecol.* **13**:103–114.

Bakken, L. R., and R. A. Olsen. 1987b. Viability of soil bacteria: Optimization of plate-counting technique and comparison between total counts and plate counts within different size groups. *Microb. Ecol.* **13**:59–74.

Bakken, L. R., and R. A. Olsen. 1989. DNA-content of soil bacteria of different cell size. *Soil Biol. Biochem.* **21**:789–793.

Balba, M. T., and D. B. Nedwell. 1982. Microbial metabolism of acetate, propionate and butyrate in anoxic sediment from Colne Point saltmarsh, Essex, U.K. *J. Gen. Microbiol.* **128**:1415–1422.

Balke, V. L., and J. D. Gralla. 1987. Changes in the linking number of supercoiled DNA accompany growth transitions in *Escherichia coli. J. Bacteriol.* **169**:4499–4506.

Balkwill, D. L. 1989. Numbers, diversity, and morphology of aerobic, chemoheterotrophic bacteria in deep subsurface sediments from a site in South Carolina. *Geomicrob. J.* **7**:33–51.

Balkwill, D. L. 1990. Density and distribution of aerobic, chemoheterotrophic bacteria in deep southeast coastal plain sediment at the Savannah River site. *In Proceedings of the First International Symposium on Microbiology of the Deep Subsurface,* C. B. Fliermans and T. C. Hazen (eds.), Section 3, pp. 3–10. W. S. R. C., Information Services Section, Publication Group, Savannah, SC.

Balkwill, D. L., D. P. Labeda, and L. E. Casida, Jr. 1975. Simplified procedures for releasing and concentrating microorganisms from soil for transmission electron microscopy viewing as thin-sectioned and frozen-etch preparation. *Can. J. Microbiol.* **21**:252–262.

Balkwill, D. L., T. E. Rucinsky, and L. E. Casida, Jr. 1977. Release of microorganisms from soil with respect to transmission electron microscopy viewing and plate counts. *Antonie van Leeuwenhoek J. Microbiol. Serol.* **43**:73–87.

Ball, W. J., and D. E. Atkinson. 1975. Adenylate energy charge in *Sacchromycetes cerevisiae* during starvation. *J. Bacteriol.* **121**:975–982.

Banse, K. 1990. New views on the degradation and disposition of organic particles as collected by sediment traps: the open sea. *Deep Sea Res.* **37**:1177–1195.

Banoub, M. W., and P. J. LeB. Williams. 1973. Seasonal changes in the organic forms of carbon, nitrogen and phosphorus in sea water at E_1 in the English Channel during 1968. *J. Mar. Biol. Assoc. U.K.* **53**:695–703.

Barash, H., and Y. S. Halpern. 1971. Glutamate-binding protein and its relation to glutamate transport in *Escherichia coli* K-12. *Biochem. Biophys. Res. Commun.* **45**:681–699.

Barber, D. A., and J. M. Lynch. 1977. Microbial growth in the rhizosphere. *Soil Biol. Biochem.* **9**:305–308.

Barber, R. T. 1966. Interaction of bubbles and bacteria in the formation of organic aggregates in seawater. *Nature.* **211**:257–258.

Barber, R. T. 1968. Dissolved organic carbon from deep waters resists microbial oxidation. *Nature* **220**:274–275.

Barcina, I., J. M. Gonzalez, J. Iriberi, and L. Egea. 1990. Survival strategy of *Escherichia coli* and *Enterococcus faecalis* in illuminated fresh and marine systems. *J. Appl. Bacteriol.* **68**:189–198.

Barghoorn, E. S., and J. W. Schopf. 1966. Microorganisms three billion years old from the Precambrian of South Africa. *Science* **152**:758–763.

Barghoorn, E. S., and S. A. Tyler. 1963. Fossil organisms from Precambrian sediments. *Ann. New York Acad. Sci.* **108**:451–452.

Bartholomew, J. W., and G. Paik. 1966. Isolation and identification of obligate thermophilic sporeforming bacilli from ocean basin cores. *J. Bacteriol.* **92**:635–638.

Basaraba, J., and R. L. Starkey. 1966. Effect of plant tannins on decomposition of organic substances. *Soil Sci.* **101**:17–23.

Bastin, E. S. 1926. The problem of the natural reduction of sulphates. *Bull. Amer. Assoc. Petrol. Geol.* **10**:1270–1299.

Bastin, E. S., and F. E. Greer. 1930. Additional data on sulphate-reducing bacteria in soils and waters of Illinois oil field. *Bull. Assoc. Petrol. Geol.* **14**:153–159.

Bateman, J. B., and F. E. White. 1963. Relative humidity and the killing of bacteria: the survival of *Serratia marcescens* dehydrated by concentrated glycerol and sucrose solutions. *J. Bacteriol.* **85**:918–926.

Bate-Smith, E. C., and T. Swain. 1962. Flavonoid compounds. *In Comparative Biochemistry,* vol. III, M. Florkin and H. S. Mason (eds.), pp. 755–809. Academic Press, New York.

Batoosingh, E., G. A. Riley, and B. Keshwar. 1969. An analysis of experimental methods for producing particulate organic matter in seawater by bubbling. *Deep-Sea Res.* **16**:213–216.

Battley, E. H. 1987. *Energetics of Microbial Growth.* Wiley-Interscience, New York.

Bauchop, T. L., and S. R. Elsden. 1961. The growth of microorganisms in relation to their energy supply. *J. Gen. Microbiol.* **23**:457–469.

Bauer, J. E., P. M. Williams, and E. R. M. Druffel. 1992. ^{14}C activity of dissolved organic carbon fractions in the north-central Pacific and the Sargasso Sea. *Nature* **357**:670.

Baxter, M., and J. McN. Sieburth. 1984. Metabolic and ultrastructural response to glucose of two eurytrophic bacteria isolated from seawater at different enriching concentrations. *Appl. Environ. Microbiol.* **47**:31–38.

Baylor, E. F., and W. H. Sutcliffe, Jr. 1963. Dissolved organic matter in seawater as a source of particulate food. *Limnol. Oceanogr.* **8**:369–371.

Beauchamp, E. G., J. T. Trevors, and J. W. Paul. 1989. Carbon sources for bacterial denitrification. *Adv. Soil Sci.* **10**:113–142.

Becker, R. E., and C. M. Volkman. 1961. A preliminary report on the bacteriology of permafrost in the Fairbanks area. *Proceeding of the Alaskan Science Conference* **12**:188.

Beerstecker, E. 1954. *Petroleum Microbiology.* Elsevier Press, Houston, TX.

Behara, B., and G. H. Wagner. 1974. Microbial growth rate in glucose-amended soil. *Soil Sci. Soc. Amer. Proc.* **38**:591–594.

Beijerinck, M. W., and A. Van Delden. 1903. Über eine farblose Bacterie, deren C-Bedarf aus der atmosphärischen Luft herrührt. *Zbl. Bakt. II. Abt.* **10**:33–47.

Bell, C. R., and L. J. Albright. 1982. Attached and free-floating bacteria in a diverse selection of water bodies. *Appl. Environ. Microbiol.* **43**:1227–1237.

Bell, R. T., G. M. Ahlgren, and I. Ahlgren. 1983. Estimating bacterioplankton production by measuring ^{3}H-thymidine incorporation in a eutrophic Swedish lake. *Appl. Environ. Microbiol.* **45**:1709–1721.

Bell, W., and R. Mitchell. 1972. Chemotactic and growth response of marine bacteria to algal extracellular products. *Biol. Bull.* **143**:265–277.

Beloin, R. M., J. L. Sinclair, and W. C. Ghiorse. 1988. Distribution and activity of microorganisms in subsurface sediments of a pristine site in Oklahoma. *Microb. Ecol.* **16**:85–97.

Bengtsson, G. 1991. Bacterial exopolymer and PHB production in fluctuating groundwater habitats. *FEMS Microb. Ecol.* **86**:15–24.

Benner, R., J. D. Pakulski, M. McCarthy, J. L. Hedges, and P. G. Hatcher. 1992. Bulk characteristics of dissolved organic matter in the ocean. *Science* **255**:1561–1564.

Benoit, R. E., and R. L. Starkey. 1968. Enzyme inactivation as a factor in the inhibition of the decomposition of organic mattter by tannins. *Soil Sci.* **105**:203–208.

Bergh, Ø., K. Y. Børsheim, G. Bratbak, and M. Heldal. 1989. High abundance of viruses found in aquatic environments. *Nature* **340**:467-468.

Berland, T. A., and L. R. Bakken. 1991. Microbial growth and nitrogen immobilization in the root zone of barley (*Hordeum vulgare* L.), Italian ryegrass (*Lolium multiforum* Lam.), and white clover (*Trifolium repens* L.). *Biol. Fert. Soils* **12**:154–160.

Betzer, P. R., W. J. Showers, E. A. Lawa, C. D. Winn, G. R. DiTullio, and P. M. Kroopnick. 1984. Primary productivity and particle fluxes on a transect of the equator at 153°W in the Pacific Ocean. *Deep-Sea Res.* **31**:1–11.

Beveridge, T. J. 1989. The structure of bacteria. In *Bacteria in Nature*, vol. 3, J. S. Poindexter and E. R. Leadbetter (eds.), pp. 1–65. Plenum Press, New York.

Bewley, R. J. F., and D. Parkinson. 1985. Bacterial and fungal activity in sulphur dioxide polluted soils. *Can. J. Microbiol.* **31**:13–15.

Bhatnagar, T. 1975. Lombriciens et humifiction un aspect nouveau de l'incorporation microbienne d'azote induite par les vers de terre. In *Biodégradation et Humification*, G. Kilbertus, O. Reisinger, A. Mourey, and J. P. Cancela da Fonseca (eds.) pp. 169–182. Sarreguemines, Pierron.

Bianchi, T. S., and J. S. Levinton. 1981. Nutrition and food limitation of deposit-feeders. II. Differential effects of *Hydrobia toteni* and *Ilyanassa obsoleta* on the microbial community. *J. Mar. Res.* **39**:547–556.

Bigger, J. W. 1937. The growth of coliform bacilli in water. *J. Path. Bacteriol.* **44**:167–211.

Billen, G. 1984. Heterotrophic utilization and regeneration of nitrogen. In *Heterotrophic Activity in the Sea*, J. E. Hobbie and P. J. leB. Williams (eds.), pp. 313–356. Plenum Press, New York.

Binnerup, S. J., D. F. Jensen, H. Thordal-Christensen, and J. Sørensen. 1993. Detection of viable, but non-culturable *Pseudomonas fluorescens* DF57 in soil using a microcolony epifluorescence technique. *FEMS Microbiol. Ecol.* **12**:97–105.

Binnerup, S. J., and J. Sørensen. 1993. Long-term oxidant deficiency in *Pseudomonas aeruginosa* PAO303 results in cells which are non-culturable under aerobic conditions. *FEMS Microb. Ecol.* **13**:79–84.

Bioardi, J. L., N. Moreni, and M. L. Galar. 1988. Survival and infectivity of a *Rhizobium meliloti* strain maintained in water and buffer suspensions. *J. Appl. Bacteriol.* **65**:189–193.

Birch, H. F. 1958. The effect of soil drying on humus decomposition and nitrogen availability. *Plant Soil* **10**:9–31.

Birch, H. F. 1959. Further observations on humus decompostion and nitrification. *Plant Soil* **11**:262–286.

Birch, H. F. 1960. Soil drying and soil fertility. *Tropical Agric.* **37**:3–10.

Birch, H. F., and M. T. Friend. 1961. Resistance of humus to decomposition. *Nature* **191**:731–732.

Bird, D. F., and D. M. Karl. 1988. Microbial incorporation of exogenous thymidine and glutamic acid in surface waters of the Bransfield Strait: a RACER analysis. *Antarctic J. U.S.* **23**:119–120.

Birge, E. A., and C. Juday. 1934. Particulate and dissolved organic matter in inland lakes. *Ecol. Monogr.* **4**:440–474.

Bishop, J. O. 1969. Interpretation of DNA-RNA hydridization data. *Nature* **224**:600–603.

Bisset, K. A. 1932. *Bacteria.* B. and S. Livington, Edinburgh.

Bitton, G., and K. C. Marshall. 1980. *Adsorption of Microorganisms to Surfaces.* John Wiley & Sons, New York.

Bjørnsen, P. K. 1986a. Automatic determination of bacterioplankton biomass by image analysis. *Appl. Environ. Microbiol.* **51**:1199–1204.

Bjørnsen, P. K. 1986b. Bacterioplankton growth yield in continuous seawater cultures. *Mar. Ecol. Prog. Ser.* **30**:191–196.

Bjørnsen, P. K., and J. Kuparinen. 1991. Determination of bacterioplankton biomass, net production and growth efficiency in the Southern Ocean. *Mar. Ecol. Prog. Ser.* **71**:185–194.

Blanchard, D. C., and L. D. Syzdek. 1970. Mechanism for the water-to-air transfer and concentrations of bacteria. *Science* **170**:626–628.

Bloem, J., F. M. Ellenbrook, M.-J. B. Bar-Gilissen, and T. E. Capapenberg. 1989. Protozoan grazing and bacterial production in stratified Lake Vechten estimated with fluorescently labeled bacteria and by thymidine incorporation. *Appl. Environ. Microbiol.* **55**:1787–1795.

Blum, P. H., S. B. Jovanovich, M. P. McCann, J. E. Schultz, S. A. Lesley, R. R. Burgess, and A. Matin. 1990. Cloning and in vivo and in vitro regulation of cyclic AMP-dependent carbon starvation genes from *Escherichia coli. J. Bacteriol.* **172**:3813–3820.

Blum, P., J. Ory, J. Bauernfeind, and J. Krska. 1992a. Physiological consequences of DnaK and DnaJ overproduction in *Escherichia coli. J. Bacteriol.* **174**:7436–7444.

Blum, P., M. Velligan, N. Lin, and A. Matin. 1992b. DnaK-mediated alterations in human growth hormone protein inclusion bodies. *Bio/Technol.* **10**:301–304.

Boethling, R. S., and M. Alexander. 1979. Effect of concentration of organic chemicals on their biodegradation by natural microbial communities. *Appl. Environ. Microbiol.* **37**:1211–1216.

Boetius, A., and K. Lochte. 1994. Regulation of microbial enzymatic degradation of organic matter in deep-sea sediments. *Mar. Ecol. Prog. Ser.* **104**:299–207.

Bohannon, D. E., N. Connell, J. Keener, A. Tormo, M. Espinosa-Urgel, M. M. Zambrano, and R. Kolter. 1991. Stationary-phase-inducible "gearbox" promoters: differential effects of *katF* mutations and role of σ^{70}. *J. Bacteriol.* **173**:4482–4492.

Bohringer, J., D. Fischer, G. Mosler, and R. Hengge-Aronis. 1995. UDP-glucose is a potential intracellular signal molecule in the control of expression of σ^{S}-dependent genes in *Escherichia coli. J. Bacteriol.* **177**:413–422.

Boiardi, J. L., N. Moreni, and M. L. Galar. 1988. Survival and infectivity of *Rhizobium meliloti* strain maintained in water and buffer suspensions. *J. Appl. Bacteriol.* **65**:189–193.

Bolin, B., and R. B. Cook. 1983. *Scope Report 21. The Major Biogeochemical Cycles and Their Interactions.* John Wiley & Sons, Chichester.

Bolin, I., L. Norlander, and H. Wolf-Waltz. 1982. Temperature-inducible outer membrane protein of *Yersinia pseudotuberculosis* and *Yersina enterocolitica* associated with the virulence plasmid. *Infect. Immun.* **37**:506–512.

Bollen, W. B. 1977. Sulfur oxidation and respiration in 54-year soil samples. *Soil Biol. Biochem.* **9**:405–410.

Boltjes, T. 1934. Onderzoekingen over Nitrificeerende Bacterien. Thesis. Techn. Hoogeschool, Delft, The Netherlands (see Dow and Lawrence, 1980).

Bone, T. L., and D. L. Balkwill. 1988. Morphological and cultural comparison of microorganisms in surface and subsurface sediments at a pristine study site in Oklahoma. *Microb. Ecol.* **16**:49–64.

Boon, J. J., J. W. de Leeuw, and A. L. Burlingame. 1978. Organic geochemistry of Walvin Bay diatomaceous ooze. III. Structural analysis of the monoenoic and polycyclic fatty acids. *Geochim. Cosmochim. Acta* **42**:631–644.

Borek, E., and A. Ryan. 1958. Studies on a mutant of *Escherichia coli* with unbalanced ribonucleic acid synthesis. II. The concomitance of ribonucleic acid synthesis with resumed protein synthesis. *J. Bacteriol.* **75**:72–76.

Børsheim, K. Y., G. Bratbak, and M. Heldal. 1990. Enumeration and biomass estimation of planktonic bacteria and viruses by transmission electron microscopy. *Appl. Environ. Microbiol.* **56**:352-356.

Bosco, G. 1960. Studio della sensibilita, in vitro algi antiotics de parte di microorganismi isolate in epoca preantibiotica. *Nouvi. Ann. Ingiuno Microbiol.* **11**:227 240.

Bove, E., and A. Lobner-Olesen. 1991. Bacterial growth control studied by flow cytometry. *Res. Microbiol.* **142**:131–135.

Bowden, W. B. 1977. Comparison of two direct count techniques for enumerating aquatic bacteria. *Appl. Environ. Microbiol.* **33**:1229–1232.

Boyaval, P., E. Boyaval, and M. J. Desmazeaud. 1985. Survival of *Brevibacterium linens* during nutrient starvation and intracellular changes. *Arch. Microbiol.* **141**:128–132.

Boyd, S. A., and M. M. Mortland. 1990. Enzyme interactions with clays and clay-organic matter complexes. *Soil Biochem.* **6**:1–28.

Boyd, W. L., and J. W. Boyd. 1963. Viability of coliform bacteria in antarctic soil. *J. Bacteriol.* **85**:1121–1123.

Boyd, W. L., and J. W. Boyd. 1964. The presence of bacteria in permafrost of the Alaskan Arctic. *Can. J. Microbiol.* **10**:917–919.

Boylen, C. W., and J. C. Ensign. 1970a. Long-term starvation-survival of rod and spherical cells of *Arthrobacter crystallopoites*. *J. Bacteriol.* **103**:569–577.

Boylen, C. W., and J. C. Ensign. 1970b. Intracellular substrates for endogenous metabolism during long-term starvation of rod and spherical cells of *Arthrobacter crystallopoietes*. *J. Bacteriol.* **103**:578–587.

Boylen, C. W., and M. H. Mulks. 1978. The survival of coryneform bacteria during nutrient starvation. *J. Gen. Microbiol.* **105**:523–334.

Boylen, C. W., and J. L. Pate. 1973. Fine structure of *Arthrobacter crystallopoietes* during long-term starvation of rod and spherical stage cells. *Can. J. Microbiol.* **19**:1–5.

Bradley, W. H. 1963. Unmineralized fossil bacteria. *Science* **141**:919–921.

Bratbak, G. 1985. Bacterial biovolume and biomass estimations. *Appl. Environ. Microbiol.* **49**:1488–1493.

Bratbak, G., M. Heldal, S. Norland, and T. F. Thingstad. 1990. Viruses as partners in spring bloom microbial trophodynamics. *Appl. Environ. Microbiol.* **56**:1400–1405.

Brdar, B., E. Kos, and M. Drakulic. 1965. Metabolism of nucleic acids and protein in starving bacteria. *Nature* **208**:303–304.

Bremer, E., T. J. Silhavy, and G. M. Weinstock. 1985. Transposable lambda *plac* Mu bacteriophages for creating *lacZ* operon fusions and kanamycin resistance insertion in *Escherichia coli. J. Bacteriol.* **162**:1092–1099.

Bremmer, J. H. 1965. Organic nitrogen in soils. *In Soil Nitrogen*, W. V. Barthlomew and F. C. Clark (eds.), pp. 93–149. American Society for Agronomy, Madison, WI.

Bretz, H. W. 1962. Simple method for estimating slide culture survival. *J. Bacteriol.* **84**:1115–1116.

Brewer, P. G., Y. Nozaki, D. W. Spencer, and A. P. Fleer. 1980. Sediment trap experiments in the deep North Atlantic: Isotope and elemental fluxes. *J. Mar. Res.* **38**:703–728.

Breznak, J. A., C. J. Potrikus, N. Pfennig, and J. C. Ensign. 1978. Viability and endogenous substrates used during starvation survival of *Rhodospirillum rubrun. J. Bacteriol.* **134**:381–386.

Bright, J. J., and M. Fletcher. 1983. Amino acid assimilation and electron transport system activity in attached and free living bacteria. *Appl. Environ. Microbiol.* **45**:818–825.

Brinkhurst, R. O., K. E. Chua, and E. Batoosingh. 1971. The free amino acids in the sediments of Toronto Harbor. *Limnol. Oceanogr.* **16**:555–559.

Brinkkerhoff, L. A., and G. B. Fink. 1964. Survival and infectivity of *Xanthomonas malvacearum* in cotton plant debris and soil. *Phytopathology* **54**:1198–1201.

Brisbane, P. G., and J. N. Ladd. 1968. Interactions between soil humic acids and bacteria. *Trans. 9th. Congr. Soil Sci.* **4**:309–317.

Brittain, A. M., and D. M. Karl. 1990. Catabolism of tritiated thymidine by aquatic microbial communities and incorporation of tritium into RNA and protein. *Appl. Environ. Microbiol.* **56**:1245–1254.

Broadbent, F. E. 1964. The characterization of soil humus. *In Microbiology and Soil Fertility*, C. M. Gilmour and O. N. Allen (eds.), pp. 59–76. Oregon State University Press, Corvallis.

Broadbent, F. E. 1965. Effect of fertilizer nitrogen on release of soil nitrogen. *Proc. Soil Sci. Soc. Amer.* **29**:692–696.

Broadbent, F. E., and A. G. Norman. 1947. Some factors affecting the availability of the organic nitrogen in soil, a preliminary report. *Proc. Soil Sci. Soc. Amer.* **11**:264–279.

Brock, T. D. 1967. Bacterial growth rate in the sea: direct analysis by thymidine autoradiography. *Science* **155**:81–83.

Brock, T. D. 1971. Microbial growth rates in nature. *Bacteriol. Rev.* **35**:39–58.

Brockman, F. J., T. L. Kieft, J. K. Frederickson, B. N. Bjornstad, W. Spangenburg, and P. E. Long. 1992. Microbiology of vadose zone paleosols in south-central Washington State. *Microb. Ecol.* **23**:279–301.

Broecker, W. 1963. Radioisotopes and large-scale organic mixing. *In The Sea,* vol. 2, M. N. Hill (ed.), pp. 88–108. Interscience Publishers, Inc., New York.

Broecker, W. S., R. Harowitz, T. Takahashi, and Y. H. Li. 1968. Factors influencing the CaCO₃ compensation levels in the ocean, abstr. *In Abstract of the Annual Meetings of the Geological Society of America,* Mexico City, Mexico.

Brookes, P. C. 1986. The adenylate energy charge ratio and ATP concentration of the soil microbial biomass. *In Perspectives in Microbial Ecology,* F. Megusar, and M. Gantar (eds.), pp. 673–678. Slovene Society for Microbiology, Ljubljana, Yugoslavia.

Brookes, P. C., K. R. Tate, and D. S. Jenkinson. 1983. The adenylate energy charge of the soil biomass. *Soil Biol. Biochem.* **15**:9–16.

Brookes, P. C., D. S. Powelson, and D. S. Jenkinson. 1984. Phosphorous in the soils of different textures and drainage classes. *Soil Biol. Biochem.* **16**:169–175.

Brophy, J. E., and D. J. Carlson. 1989. Production of biologically refractory dissolved organic carbon by natural seawater microbial populations. *Deep-Sea Res.* **36**:497–507.

Brown, C. M., D. C. Ellwood, and J. R. Hunter. 1977. Growth of bacteria at surfaces: Influence of nutrient concentrations. *FEMS Microb. Lett.* **1**:1630–1636.

Brown, M. R. W., and P. Williams. 1985. The influence of environment on envelope properties affecting survival of bacteria in infections. *Ann. Rev. Microbiol.* **39**:527–556.

Brumbach, M. K. 1972. Factory cleanliness for the manufacture of microelectronic devices. *Solid State Technol.* **15**:46–50.

Bryan, B. A. 1981. Physiology and biochemistry of denitrification. *In Denitrification, Nitrification and Atmospheric Nitrous Oxide,* C. C. Delwiche (ed.), pp. 67–84. John Wiley & Sons, New York.

Buck, J. D. 1974. Effects of medium composition on the recovery of bacteria from sea-water. *J. Exp. Mar. Biol. Ecol.* **15**:25–34.

Buck, J. D. 1979. The plate count in aquatic microbiology. In *Native Aquatic Bacteria: Enumeration Activity and Ecology,* J. W. Costerton and R. R. Colwell (eds.), pp. 19–28. American Society for Testing Materials, Philadelphia.

Burchill, S., M. H. B. Hayes, and D. J. Greenland. 1981. Adsorption. *In The Chemistry of Soil Processes,* D. J. Greenland and M. H. B. Hayes (eds.), pp. 221–400. John Wiley & Sons, Chichester.

Burges, N. A., H. M. Hurst, and B. Walkden. 1964. The phenolic constituents of humic acid and their relation to lignin of the plant cover. *Geochim. Geochim. Acta* **28**:1547–1554.

Burke, V. 1936. Effect of starvation medium on bacterial variations. *J. Bacteriol.* **32**:467–468.

Burke, V., and L. Taschner. 1936. Growth of *Escherichia coli* in a small amount of organic matter. *J. Bacteriol.* **32**:124–125.

Burke, V., and A. J. Wiley. 1937. Bacteria in coal. *J. Bacteriol.* **34**:475–481.

Burkholder, W. F., C. A. Panagiotidis, S. J. Silverstein, A. Cegielska, M. E. Gottesman and G. A. Gaitanaris. 1994. Isolation and characterization of an *Escherichia coli* DnaK mutant with impaired ATPase activity. *J. Molec. Biol.* **242**:364–377.

Burleigh, D. J., and C. S. Martens. 1988. Biogeochemical cycling in an organic rich basin: the role of amino acids in sedimentary carbon and nitrogen recycling. *Geochim. Cosmochim. Acta* **52**:1571–1584.

Burleigh, I. G., and E. A. Dawes. 1967. Studies on the endogenous metabolism and senescence of starved *Sarcina lutea. Biochem. J.* **102**:236–250.

Burnison, B. K., and R. Y. Morita. 1974. Heterotrophic potential of amino acid uptake in a naturally eutrophic lake. *Appl. Microbiol.* **27**:488–495.

Burns, R. G. 1978. *Soil Enzymes.* Academic Press, London.

Burns, R. G. 1980. Microbial adhesion to soil surfaces: consequences for growth and enzyme activities. *In Microbial Adhesion to Surfaces,* C. R. W. Berkeley, J. M. Lynch, J. Melling, P. R. Rutter, and B. Vincent (eds.), pp. 249–262. Ellis Harwood, Chichester.

Burns, R. G. 1982. Microbial interactions and community structure. *In Microbial Interactions and Communities,* A. T. Bull and J. H. Slater (eds.), pp. 13–44. Academic Press, London.

Burns, R. G. 1983. Extracellular enzyme-substrate interactions in soil. *Symp. Soc. Gen. Microbiol.* **34**:249–298.

Burns, R. G. 1986. Interaction of enzymes with soil mineral and organic colloids. *In Interactions of Soil Minerals with Natural Organics and Microbes,* P. M. Huang and M. Schnitzer (eds.), pp. 429–451. Soil Science Society of American, Madison, WI.

Burns, R. G., and L. J. Audus. 1970. Distribution and breakdown of paraquat in soil. *Weed Res.* **10**:49–58.

Burns, R. G., M. M. El-Sayed, and A. D. McLaren. 1972a. Extraction of an urease-active organo-complex from soil. *Soil Biol. Biochem.* **4**:107–108.

Burns, R. G., A. H. Pukite, and A. D. McLaren. 1972b. Concerning the location and persistence of soil urease. *Soil Sci. Soc. Amer. Proc.* **36**:308–311.

Burton, S. D., and R. Y. Morita. 1964. Effect of catalase and cultural conditions on *Beggiatoa. J. Bacteriol.* **88**:1755–1761.

Bush, P. L., and W. Strumm. 1968. Chemical interactions in the aggregation of bacteria: bioflocculation in water treatment. *Environ. Sci. Technol.* **2**:49–53.

Busta, F. F. 1978. Introducton to injury and repair of microbial cells. *Adv. Appl. Microbiol.* **23**:195–201.

Butkevich, N. V., and V. S. Butkevich. 1936. Multiplication of sea bacteria depending on the composition and on temperature. *Microbiology* (Moscow) **5**:1005–1021.

Butman, C. A. 1986. Sediment trap biases in turbulent flow: results from a laboratory flume study. *J. Mar. Res.* **44**:645–693.

Butman, C. A., W. D. Grant, and K. D. Stolzenbach. 1986. Predictions of sediment trap biases in turbulent flow: a theoretical analysis based on observations from the literature. *J. Mar. Res.* **44**:601–644.

Butterfield, C. T. 1933. Observations on changes in number of bacteria in polluted water. *Sewage Works J.* **5**:600–622.

Buttner, M. P. 1989. Survival of ice nucleation-active and genetically engineered inactive strains of *Pseudomonas syringae*. M. S. Thesis. University of Nevada at Las Vegas, Las Vegas.

Button, D. K. 1978. On the theory of nutrient concentration control of microbial growth kinetics. *Deep-Sea Res.* **25**:1163–1177.

Button, D. K. 1983. Differences between the kinetics of nutrient uptake by micro-organisms, growth and enzyme kinetics. *Trends Biochem. Sci.* **8**:121–124.

Button, D. K., and B. R. Robertson. 1989. Kinetics of bacterial processes in natural aquatic systems based on biomass as determined by high resolution flow cytometry. *Cytometry* **10**:558–563.

Button, D. K., F. Schut, P. Quang, R. Martin, and B. R. Robertson. 1993. Viability and isolation of marine bacteria by dilution culture: Theory, procedures, and initial results. *Appl. Environ. Microbiol.* **59**:881–891.

Cabral, J. P. S. 1992. Starved *Pseudomonas syringae* cells release strong Cu^{2+}-complexing compounds. *Chem. Speci. Bioavail.* **4**:105–107.

Cabrera, M. L., and D. E. Kissel. 1988. Potentially mineralizable nitrogen in disturbed and undisturbed soil samples. *Soil Sci. Soc. Am. J.* **52**:1010–1015.

Calcott, P. H., and J. R. Postgate. 1971. Substrate-accelerated death: Role of recovery medium and prevention by $3'$–$5'$ cyclic AMP. *J. Gen. Microbiol.* **66**:i.

Calhoun, D. H. 1985. Bacterial longevity in salt-free medium: responses. *ASM News* **51**:1.

Calleja, G. B. 1984. *Microbial Aggregation.* CRC Press Inc., Boca Raton.

Cameron, R. E., and F. A. Morelli. 1974. Viable microorganisms from ancient Ross Island and Tayler Valley drill core. *Antarctic J. of the U.S.* **9**:113–116.

Campbell, A. 1981. Evolutionary significance of accessory DNA elements in bacteria. *Ann. Rev. Microbiol.* **35**:55–83.

Campbell, C. A., and V. O. Biederbeck. 1976. Soil bacterial changes as affected by growing season weather conditions. A field and laboratory study. *Can. J. Soil Sci.* **56**:293–310.

Campbell, J. J. R., A. F. Gronlund, and M. G. Duncan. 1963. Endogenous metabolism of *Pseudomonas*. *Ann. N. Y. Acad Sci.* **102**:669–677.

Campbell, C. A., E. A. Paul, D. A. Rennie, and K. J. McCallum. 1967. Applicability of the carbon-dating method of analysis to soil humus studies. *Soil Sci.* **104**:217–224.

Cano, R. J., and M. K. Boruchi. 1995. Revival and identification of bacterial spores in 25- to 40-million-year-old Dominican Amber. *Science* **268**:1060–1064.

Cano, R. J., M. Borucki, H. N. Poinar, and G. O. Poinar. 1993. Isolation and nucleotide sequencing of *Bacillus*-sp. rDNA from the 25–40 million year old bee *Proplebeia-dominicana* in Dominican amber. *Abstract of the General Meeting of the American Society for Microbiology,* abstr., **93**:207.

Cano, R. J., M. K. Borucki, M. H. Higby-Schweitzer, H. N. Poinar, G. O. Poinar, Jr., and K. J. Pollard. 1994. *Bacillus* DNA in fossil bees: an ancient symbiosis. *Appl. Environ. Microbiol.* **60**:2164–2167.

Carlile, M. J. 1980. Positioning mechanisms—the role of motility, taxis and tropism in the life of microorganisms. *In Contemporary Microbial Ecology,* D. C. Ellwood, J. N. Hedger, M. J. Latham, J. M. Lynch and J. H. Slater, (eds.), pp. 55–74. Academic Press, London.

Carlson, D. J., L. M. Mayer, M. L. Braun, and T. H. Mague. 1985a. Molecular weight distribution of dissolved organic materials in seawater as determined by ultrafiltration: a reexamination. *Mar. Chem.* **16**:155–171.

Carlson, D. J., L. M. Mayer, M. L. Brann, and T. H. Mague. 1985b. Binding of monomeric compounds to macromolecular dissolved organic matter in seawater. *Mar. Chem.* **16**:141–153.

Carlucci, A. F. 1974. Nutrients and microbial response to nutrients in seawater. *In Effect of the Ocean Environment on Microbial Activities,* R. R. Colwell and R. Y. Morita (eds.), pp. 245–248. University Park Press, Baltimore.

Carlucci, A. F., and P. M. Bowes. 1970. Production of vitamin B_{12}, thiamine and biotin by phytoplankton. *J. Physiol.* **6**:351–357.

Carlucci, A. F., and D. Pramer. 1957. Factors influencing the plate method for determining abundance of bacteria in sea water. *Proc. Soc. Exp. Biol. Med.* **96**:392–394.

Carlucci, A. F., and S. L. Shimp. 1974. Isolation and growth of a marine bacterium in low concentrations of substrate. *In Effect of the Ocean Environment on Microbial Activities,* R. R. Colwell and R. Y. Morita (eds.), pp. 363–367. University Park Press, Baltimore.

Carlucci, A. F., and P. N. Williams. 1978. Simulated in situ growth of pelagic marine bacteria. *Naturwissenschaften* **65**:541–542.

Carlucci, A. F., S. L. Shimp, and D. B. Craven. 1986. Growth characteristics of low-nutrient bacteria from the north-east and central Pacific Ocean. *FEMS Microbiol. Ecol.* **38**:1–10.

Carlucci, A. F., S. L. Shimp, P. A. Jumars, and H. W. Paerl. 1976. In situ morphologies of deep-sea and sediment bacteria. *Can. J. Microbiol.* **22**:1667–1671.

Carman, K. R., F. C. Dobbs, and J. B. Guckert. 1988. Consequences of thymidine catabolism for estimates of bacteria production: an example from a coastal marine sediment. *Limnol. Oceanogr.* **33**:1595–1606.

Caron, D. A., P. G. Davis, L. P. Mardin and J. McN. Sieburth. 1982. Heterotrophic bacteria and bacteriovorous protozoa in oceanic macroaggregates. *Science* **218**:795–797.

Carter, M. R. 1986. Microbial biomass and mineralizable nitrogen in solonetzic soils: Influence of gypsum and lime admendments. *Soil. Biol. Biochem.* **18**:531–537.

Carter, P. W. 1978. Adsorption of amino acid-containing organic matter by calcite and quartz. *Geochim. Cosmochim. Acta* **42**:1239–1242.

Carter, P. W., and R. M. Mitterer. 1978. Amino acid composition of organic matter associated with carbonate and non-carbonate sediments. *Geochim. Cosmochim. Acta* **42**:1231–1238.

Cash, H. A. 1985. Bacterial longevity in salt-free medium: responses. *ASM News* **51**:2.

Cashel, M., and K. E. Rudd. 1987. The stringent response. *In Escherichia coli and Salmonella typhimurium: Cellular and Molecular biology,* Vol. 2, F. C. Neidhardt, J. L. Ingraham, K. B. Low, B. Magasanik, M. Schaechter, and H. E., Umberger (eds.), pp. 1410–1438. American Society for Microbiology, Washington DC.

Casida, L. E., Jr. 1977. Small cells in pure culture of *Agromyces ramosus* in natural soil. *Can. J. Microbiol.* **23**:214–216.

Casida, L. E., Jr. 1980. Death of *Micrococcus luteus* in soil. *Appl. Environ. Microbiol.* **39**:1031–1034.

Cauwet, G. 1978. Organic chemistry of sea water particulates: concepts and developments. *Oceanol. Acta* **1**:99–105.

Cavalier-Smith, T. 1981. The origin and early evolution of the eukayotic cell. *Symp. Soc. Gen. Microbiol.* **32**:33–84.

Cerri, C. C., and D. S. Jenkinson. 1981. Formation of microbial biomass during the decomposition of ^{14}C labelled ryegrass in soil. *J. Soil Sci.* **32**:619–626.

Certes, A. 1884. Sur la culture, à l'abri des germes atmosphérique, des eaux et des sediments rapportés par les expéditions du Travailleur et du Talisman, 1882–1883. *C. R. hebd. Séanc, Acad. Sci. Paris.* **98**:690.

Chahal, K. S., and G. H. Wagner. 1965. Decomposition of organic matter in Sanborn Field soils amended with ^{14}C glucose. *Soil Sci.* **100**:96–103.

Chalvignac, M. A., and J. Mayaudon. 1971. Extraction and study of soil enzymes metabolizing tryptophan. *Plant Soil* **34**:25–31.

Chan, K. 1977. Responses of marine bacteria to nutrient addition and secondary heat stress in relation to starvation. *Microbios Letters* **6**:137–144.

Chapelle, F. H. 1993. *Ground-Water Microbiology and Geochemistry.* John Wiley & Sons, New York.

Chapelle, F. H., and D. R. Lovely. 1990. Rates of microbial metabolism in deep coastal plain aquifers. *Appl. Environ. Microbiol.* **56**:1865–1874.

Chapelle, F. H., J. L. Zelibor, Jr., D. J. Grimes, and L. L. Knobel. 1987. Bacteria in deep coastal plain sediments of Maryland: a possible source of CO_2 to ground water. *Water Resour. Res.* **23**:1625–1632.

Chapman, A. G., and D. E. Atkinson. 1977. Adenine nucleotide concentrations and turnover rates. Their correlation with biological activity in bacteria and yeasts. *Adv. Microb. Physiol.* **15**:253–306.

Chapman, A. G., L. Fall, and D. E. Atkinson. 1971. Adenylate energy change in *Escherichia coli* during growth and starvation. *J. Bacteriol.* **108**:1072–1086.

Chapman, S. J., and T. R. G. Gray. 1981. Endogenous metabolism and macromolecular composition of *Arthrobacter globiformis. Soil Biol. Biochem.* **13**:11–18.

Chapman, S. J., and T. R. G. Gray. 1986. Importance of cryptic growth, yield factors and maintenance energy in models of microbial growth in soil. *Soil Biol. Biochem.* **18**:1–4.

Chave, K. E. 1965. Carbonates: Association with organic matter in surface seawater. *Science* **148**:1723–1724.

Chave, K. E., and E. Suess. 1967. Suspended minerals in seawater. *Trans. N.Y. Acad. Sci. Ser. II* **29**:991–1000.

Chebsch, A. J. 1960. Tritium age of ground water at the Nevada Test Site, Nye County, Nevada. Preliminary report TEL\I-763. U.S. Geological Survey, Denver.

Chen, M., and M. Alexander. 1972. Resistence of soil microorganisms to starvation. *Soil Biol. Biochem.* **4**:283–288.

Cheng, C.-N. 1975. Extraction and desalting amino acids from soils and sediments: evaluation of methods. *Soil Biol. Biochem.* **7**:319–322.

Chesbro, W. 1988. The domains of slow bacterial growth. *Can. J. Microbiol.* **34**:427–435.

Cheshire, M. V. 1977. Origins and stability of soil polysaccharides. *J. Soil. Sci.* **28**:1–10.

Chet, I., and R. Mitchell. 1976. The relationship between chemical structure of attractants and chemotaxis by a marine bacterium. *Can. J. Microbiol.* **22**:1206–1208.

Chibata, I., and T. Tosa. 1977. Transformations of organic compounds by immobilized cells. *Adv. Appl. Microbiol.* **22**:1–22.

Chin-Leo, G., and D. L. Kirchman. 1990. Unbalanced growth in natural assemblages of marine bacterioplankton. *Mar. Ecol. Prog. Ser.* **63**:1–8.

Chisholm, S. W., R. J. Olson, E. R. Zettler, R. Goericke, J. B. Waterbury, and N. A. Welschmeyer. 1988. A novel free-living apochlorophyte abundant in the oceanic euphotic zone. *Nature* **334**:340–343.

Cho, B. C., and F. Azam. 1988. Heterotrophic bacterioplankton production measurement by the tritiated thymidine incorporation method. *Arch. Hydrobiol. Beih. Ergeb. Limnol.* **31**:153–162.

Chocair, J. A., and L. A. Albright. 1981. Heterotrophic activities of bacterioplankton and bacteriobenthos. *Can. J. Microbiol.* **27**:259–266.

Choi, S. H., and E. P. Greenberg. 1992. Genetic dissection of DNA binding and luminescence gene activation by the *Vibrio fischeri* LuxR protein. *J. Bacteriol.* **174**:4064–4069.

Christensen, D., and T. H. Blackburn. 1980. Turnover of tracer (^{14}C,^{3}H labelled) alanine in inshore sediments. *Mar. Biol.* **58**:97–103.

Christensen, D., and T. H. Blackburn. 1982. Turnover of ^{14}C-labelled acetate in marine sediments. *Mar. Biol.* **71**:113–119.

Christensen, H. 1993. Conversion factors for the thymidine incorporation technique estimated with bacteria in pure culture and on seeding roots. *Soil Biol. Biochem.* **25**:1085–1096.

Christensen, M. 1969. Soil microfungi of dry to mesic conifer-hardwood forest in northern Wisconsin. *Ecology* **50**:9–27.

Christian, R. R., R. B. Hanson, and S. Y. Newell. 1982. Comparison of methods for measurement of bacteria growth rates in mixed batch culture. *Appl. Environ. Microbiol.* **43**:1160–1165.

Christian, R. R., R. L. Wetzel, S. M. Harlan, and D. W. Stanley. 1986. Growth and decomposition in aquatic microbial systems: Alternate approaches in simple models. *In*

Perspectives in Microbial Ecology, F. Megusar and M. Gantar (eds.), pp. 38–45. Slovene Society for Microbiology, Ljubljana, Yugoslavia.

Chróst, R. J. 1981. The composition and bacterial utilization of DOC released by phytoplankton. *Kieler Meeresforsch. Sonderh.* **5**:325–332.

Chróst, R. J. 1988. Phosphorus and microplankton development in an eutrophic lake. *Acta Microbiol. Polonca* **37**:205–225.

Chróst, R. J. 1989. Characterization and significance of ß-glucosidase activity in lake water. *Limnol. Oceanogr.* **34**:660–672.

Chróst, R. J. 1990. Microbial ectoenzymes in aquatic environments. *In Aquatic Microbial Ecology: Biochemical and Molecular Approach,* J. Overbeck and R. J. Chróst (eds.), pp. 47–78. Springer-Verlag, New York.

Chróst, R. J. 1991. *Microbial Enzymes in Aquatic Environments.* Springer-Verlag, New York.

Chróst, R. J. 1992. Significance of bacterial ectoenzymes in aquatic environments. *Hydrobiologia* **243**:61–70.

Chróst, R. J., and M. A. Faust. 1983. Organic carbon release by phytoplankton: its composition and utilization by bacterioplankton. *J. Plankt. Res.* **11**:223–242.

Chróst, R. J., and J. Overbeck. 1987. Kinetics of alkaline phosphatase activity and phosphorus availability for phytoplankton and bacterioplankton in Lake Plusssee (north German eutrophic lake). *Microb. Ecol.* **13**:229–248.

Chróst, R. J. and H. Rai. 1993. Ectoenzyme activity and bacterial secondary production in nutrient-impoverished and nutrient-enriched freshwater mesocosms. *Microb. Ecol.* **25**:131–150.

Chróst, R. J., and B. Riemann. 1994. Storm-stimulated enzymatic decomposition of organic matter in benthic/pelagic coastal microcosms. *Mar. Ecol. Prog. Ser.* **108**:185–192.

Chróst, R. J., J. Overbeck, and R. Wcislo. 1988. Evaluation of the [^{3}H]thymidine method for estimating bacterial growth rates and production in lake water: re-examination and methodological comments. *Acta Microbiol. Polonica* **37**:95–112.

Chrzanowski, T. H., and K. Simek. 1990. Prey-size selection by freshwater flagellated protozoa. *Limnol. Oceanogr.* **35**:1429–1436.

Chu, M., and R. Stephen. 1967. A study of free-living and root-surface fungi in cultivated and fallow soils in Hong Kong. *Nova Hedwigia* **14**:301–311.

Chuang, S-E., D. L. Daniels, and F. R. Blattner. 1993. Global regulation of gene expression in *Escherichia coli. J. Bacteriol.* **175**:2026–2036.

Chuang, T. Y., and W. H. Ko. 1983. Propagule size: Its relation to longevity and reproductive capacity. *Soil Biol. Biochem.* **15**:269–274.

Clarholm, M., and T. Rosswall. 1980. Biomass and turnover of bacteria in a forest soil and a peat. *Soil Biol. Biochem.* **12**:49–57.

Clark, F. E. 1965. The concept of competition in microbiology. *In Ecology of Soil-Borne Plant Pathogens,* K. F. Baker, and W. C. Snyder (eds.), pp. 339–347. University of California Press, Berkeley.

Clark, F. E. 1967. Bacteria in soil. *In Soil Biology,* A. Burges and F. Raw (eds.), pp. 15–49. Academic Press, New York.

Clark, F. E. 1968. The growth of bacteria in soil. *In The Ecology of Soil Bacteria,* T. R. G. Gray and D. Parkinson (eds.), pp. 441–457. University of Toronto Press, Toronto.

Clark, F. E., and E. A. Paul. 1970. The microflora of grassland. *Adv. Agron.* **22**:375–435.

Clark, M . E., G. A. Jackson, and W. J. North. 1972. Dissolved free amino acids in southern California coastal waters. *Limnol. Oceanogr.* **17**:749–758.

Clarke, P. H. 1974. The evolution of enzymes for the utilization of novel substrates. *Symp. Soc. Gen. Microbiol.* **24**:183–217.

Clarke, P. H., and L. N. Ornston. 1975. Metabolic pathways and regulation. *In Genetics and Biochemistry of Pseudomonas,* P. H. Clarke and M. H. Richmond (eds.), pp. 191–340. John Wiley & Sons, New York.

Clifton, C. E. 1966. Aging of *Escherichia coli. J. Bacteriol.* **92**:905–912.

Clifton, C. E. 1967. Aging of *Pseudomonas aeruginosa. J. Bacteriol.* **94**:2077–2078.

Cloud, P. E., Jr. 1965. Significance of the gunflint (Precambrian) microflora. *Science* **148**:27–35.

Coffin, R. B. 1989. Bacterial uptake of dissolved free and combined amino acids in estuarine waters. *Limnol. Oceanogr.* **34**:531–542.

Coffin, R. B., and J. H. Sharp. 1987. Microbial trophodynamics in the Delaware Estuary. *Mar. Ecol. Prog. Ser.* **41**:253–266.

Coffin, R. B., J. P. Connolly, and P. S. Harris. 1993. Availability of dissolved organic carbon to bacterioplankton examined by oxygen utilization. *Mar. Ecol. Prog. Ser.* **101**:9–22.

Cohen, B. 1922. Disinfection studies. The effect of temperature and hydrogen ion concentration upon the viability of *Bact. coli* and *Bact. typhosum* in water. *J. Bacteriol.* **7**:183–230.

Cohen, S. S., and H. D. Barner. 1954. Studies on unbalanced growth in *Escherichia coli. Proc. Natl. Acad. Sci. (USA)* **40**:179–185.

Colbert, S. F., T. Isakeit, M. Ferri, A. R. Weinhold, M. Hendson, and M. N. Schroth. 1993a. Use of an exotic carbon source to selectively increase metabolic activity and growth of *Pseudomonas putida* in soil. *Appl. Environ. Microbiol.* **59**:2056–2063.

Colbert, S. F., M. N. Schroth, A. R. Weinhold, and M. Hendson. 1993b. Enhancement of population densities of *Pseudomonas putida* PpG7 in agricultural ecosystems by selective feeding with the carbon source salicylate. *Appl. Environ. Microbiol.* **59**:2064–2070.

Cole, J. J., and N. F. Caraco. 1993. The pelagic microbial food web of oligotrophic lakes. *In Aquatic Microbiology,* T. E. Ford (ed.), pp. 101–111. Blackwell Scientific Publications, Boston.

Cole, J. J., N. F. Caraco, and D. L. Strayer. 1989. A detailed organic carbon budget as an ecosystem-level calibration of bacteria respiration in an oligotrophic lake during midsummer. *Limnol. Oceanogr.* **34**:286–296.

Cole, J. J., S. Findlay, and M. L. Pace. 1988. Bacterial production in fresh and saltwater ecosystems: a cross-system overview. *Mar. Ecol. Prog. Ser.* **43**:1–10.

Cole, J. J., G. E. Likens, and D. L. Strayer. 1982. Photosynthetically produced dissolved organic carbon: an important carbon source for planktonic bacteria. *Limnol. Oceanogr.* **27**:1080–1090.

Cole, J. J., W. H. McDowell, and G. E. Likens. 1984. Sources and molecular weight of "dissolved" organic carbon in an oligotrophic lake. *Oikos* **43**:1–9.

Coleman, D. C., J. M. Oades, and G. Uehara. 1989. *Dynamics of Soil Organic Matter in Tropical Ecosystems*. University of Hawaii Press, Honolulu.

Collins, V. G., and L. G. Willoughby. 1962. The distribution of bacteria and fungal spores in Blelham Tarn with particular reference to an experimental turnover. *Arch. Mikrobiol.* **43**:294–307.

Colwell, F. S. 1989. Microbiological comparison of surface soil and unsaturated subsurface soil from a semiarid high desert. *Appl. Environ. Microbiol.* **55**:2420–2423.

Colwell, R. R., and D. J. Grimes. 1997. *Nonculturable Microorganisms in the Environment*. Chapman & Hall, New York.

Confer, D. R., and B. E. Logan. 1991. Increased bacterial uptake of macromolecular substrates with fluid shear. *Appl. Environ. Microbiol.* **57**:3093–3100.

Conn, H. J. 1948. The most abundant groups of bacteria in soil. *Bacteriol. Rev.* **12**:257–273.

Connell, J., Z. Han, F. Moreno, and R. Kolter. 1987. An *E. coli* promoter induced by the cessation of growth. *Molec. Microbiol.* **1**:195–201.

Connolly, J. P., R. B. Coffin, and R. E. Landeck. 1992. Modeling carbon utilization by bacteria in natural water systems. *In Modeling the Metabolic and Physiologic Activities of Microorganisms*, C. J. Hurst (ed.), pp. 249–276. John Wiley & Sons, New York.

Conrad, R., and W. Seiler. 1980a. Photooxidative production and microbial consumption of carbon monoxide in seawater. *FEMS Microbiol. Lett.* **9**:61–64.

Conrad, R., and W. Seiler. 1980b. Role of microorganisms in the consumption and production of carbon monoxide by soil. *Appl. Environ. Microbiol.* **40**:437–445.

Conrad, R., and W. Seiler. 1982. Utilization of traces of carbon monoxide by aerobic, oligotrophic microorganisms in oceans, lakes, and soil. *Arch. Microbiol.* **132**:41–46.

Conway, P. L., J. Maki, R. Mitchell, and S. Kjelleberg. 1986. Starvation of marine flounder, squid, and laboratory mice and its effect on the intestinal microbiota. *FEMS Microbiol. Ecol.* **38**:187–195.

Cook, A. M., and B. A. Willis. 1958. The use of stored suspension of *Escherichia coli* I in the evaluation of bactericidal action. *J. Appl. Bacteriol.* **21**:180–187.

Corpe, W. A. 1964. Factors influencing growth and polysaccharide formation by strains of *Chromobacterium violaceum*. *J. Bacteriol.* **88**:1433–1441.

Corpe, W. A. 1970a. An acidic polysacchride produced by a primary film forming marine bacterium. *Develop. Industr. Microbiol.* **11**:1433–1441.

Corpe, W. A. 1970b. Attachment of marine bacteria to solid surfaces. *In Adhesions of Microorganisms in Biological Systems*, R. S. Manley (ed.), pp. 73–87. Academic Press, New York.

Corpe, W.A. 1974. Periphytic marine bacteria and the formation of microbial films on solid surfaces. *In Effect of Ocean Environment on Microbial Activities*, R. R. Colwell and R. Y. Morita (eds.), pp. 397–417, University Park Press, Baltimore, MD.

Corpe, W. A. 1980. Microbial surface components involved in adsorption of microorganisms to surfaces. *In Adsorption of Microorganisms to Surfaces*, G. Bitton and K. C. Marshall (eds.), pp. 105–144. John Wiley and Sons, Inc., New York.

Costerton, J. W. 1979. Summary. *In Native Aquatic Bacteria Enumeration, Activity, and Ecology*, W. J. Costerton and R. R. Colwell (ed.), STP 695, pp. 207–211. American Society for Testing and Materials, Philadelphia.

Costerton, J. W., K.-J. Cheng, G. G. Geesey, T. I. Ladd, J. C. Nickel, M. Dasgupta, and T. J. Marrie. 1987. Bacterial biofilms in nature and disease. *Annu. Rev. Microbiol.* **41**:435–464.

Costerton, J. W., and R. R. Colwell. 1979. *Native Aquatic Bacteria Enumeration, Activity, and Ecology*. STP 695 American Society for Testing Materials. Philadelphia.

Costerton, J. W., J. M. Ingram, and K.-J. Cheng. 1974. Structure and function of the envelope of gram negative bacteria. *Bacteriol. Rev.* **38**:87–110.

Costerton, J. W., R. T. Irvin, and K-J. Cheng. 1981. The bacterial glycocalyx in nature and disease. *Ann. Rev. Microbiol.* **35**:299–324.

Coveney, M. F., and R. G. Wetzel. 1992. Effects of nutrients on specific growth rate of bacterioplankton in oligotrophic lake water cultures. *Appl. Environ. Microbiol.* **58**:150–156.

Cozzone, A. J. 1981. How do bacteria synthesize proteins during amino acid starvation? *Trends Biochem. Sci.* **6**:108–110.

Craig, H. 1971. The deep metabolism: oxygen consumption in abyssal ocean water. *J. Geophys. Res.* **76**:5078–5091.

Craswell, E. T., and S. A. Waring. 1972. Effect of grinding on the decomposition of soil organic matter. I. The mineralization of organic nitrogen in relation to soil type. *Soil Biol. Biochem.* **4**:427–433.

Craven, D. G., and D. M. Karl. 1984. Microbial RNA and DNA synthesis in marine sediments. *Mar. Biol.* **83**:128–139.

Crist, D. K., R. E. Wyza, K. K. Mills, W. D. Bauer, and W. R. Evans. 1984. Preservation of *Rhizobium* viability and symbiotic infectivity by suspension in water. *Appl. Environ. Microbiol.* **47**:895–900.

Cronan, J. E., and E. P. Gelmann. 1975. Physical properties of membrane lipids: biological relevance and regulation. *Bacteriol. Rev.* **39**:232–256.

Crosse, J. E. 1968. Plant pathogens in soil. *In The Ecology of Soil Bacteria*, T. R. G. Gray and D. Parkinson (eds.), pp. 552–572. University of Toronto Press, Toronto.

Cummins, K. W. 1974. Structure and function of steam ecosystems. *Bioscience* **24**:631–641.

Cummins, K. W., G. W. Minshall, J. R. Sedell, C. E. Cushing, and R. C. Petersen. 1984. Stream ecosystem theory. *Verh. Internat. Vereine. Limnol.* **22**:1842–1846.

Cunningham, H. W., and R. G. Wetzel. 1989. Kinetic analysis of protein degradation by a freshwater wetland sediment community. *Appl. Environ. Microbiol.* **55**:1963–67.

Curds, C. R. 1982. The ecology and role of protozoa in aerobic sewage treatment process. *Ann. Rev. Microbiol.* **36**:27–46.

Cutler, D. W., and L. M. Crump. 1920. Daily periodicity in the numbers of active soil flagellates: with a brief note on the relations of trophic amoebae and bacterial numbers. *Ann. Appl. Biol.* **7**:11–24.

Dagley, S., and J. Skyes. 1957. Effect of "starvation" upon the constitution of bacteria. *Nature* **179**:1249–1250.

Daley, R. J., and J. E. Hobbie. 1975. Direct count of aquatic bacteria by a modified epifluorescent technique. *Limnol. Oceanogr.* **20**:875–881.

Dale, N. G. 1975. Bacteria in intertidal sediments: factors relation to their distribution. *Limnol. Oceanogr.* **19**:509–518.

Dalhbäck, B., M. Hermansson, S. Kjelleberg, and B. Norkrans. 1981. The hydrophobicity of bacteria: an important factor in their initial adhesion at the air-water interface. *Arch. Microbiol.* **128**:267–270.

Dallenberg, J. W., and G. Jager. 1981. Priming effect of small glucose additions to ^{14}C-labelled soil. *Soil Biol. Biochem.* **13**:219–223.

Damerau, K., and A. C. St. John. 1993. Role of Clp protease subunits in degradation of carbon starvation proteins in *Escherichia coli. J. Bacteriol.* **175**.53–63.

Darling, C. A., and P. A. Siple. 1941. Bacteria of Antarctica. *J. Bacteriol.* **42**:83–98.

Dashman, T., and G. Stotzky. 1986. Microbial utilization of amino acids and a peptide bound on homoionic montmorillonite and kaolinite. *Soil Biol. Biochem.* **18**:5–14.

Daumas, R., and M. Bianchi. 1984. Bioturbation and microbial activity. *Arch. Hydrobiol. Beih. Ergebn. Limnol.* **19**:289–294.

Davis, B. D., R. Delbecco, N. Eisen, H. S. Ginsberg, and W. B. Wood. 1967. Microbiology, p. 149. Harper & Row, New York.

Davis, B. D., S. M. Luger, and P. C. Tai. 1986. Role of ribosome degradation in the death of starved *Escherichia coli. J. Bacteriol.* **166**:439–445.

Davis, C. L. 1985. Physiological and ecological studies of mannitol utilizing marine bacteria. PhD. Thesis University of Cape Town, Cape Town, South Africa.

Davis, C. L. 1989. Uptake and incorporation of thymidine by bacterial isolates from an upwelling environment. *Appl. Environ. Microbiol.* **55**:1267–1272.

Davis, C. L., and F. T. Robb. 1985. Maintenance of different mannitol uptake systems during starvation in oxidative and fermentative marine bacteria. *Appl. Environ. Microbiol.* **50**:543–748.

Davis, D. G., A. M. Chakrabarty, and G. G. Geesey. 1993. Exopolysaccharide production in biofilms: Substratum activation of alginate gene expression by *Pseudomonas aeruginosa. Appl. Environ. Microbiol.* **59**:1181–1186.

Davis, J. B. 1967. *Petroleum Microbiology.* Elsevier Publ. Co., Amsterdam.

Davis, W. M., and D. C. White. 1988. Fluorometric determination of adenosine nucleotide derivatives as measures of the microfouling, detrital and sedimentary microbial biomass and physiological status. *Appl. Environ. Microbiol.* **40**:539–548.

Dawes, E. Λ. 1976. Endogenous metabolism and survival of starved prokaryotes. *Symp. Soc. Gen. Microbiol.* **26**:19–53.

Dawes, E. A. 1984. Stress of unbalanced growth and starvation in microorganisms. *In The Revival of Injured Microbes,* M. H. E. Andrew and A. D. Russell (eds.), pp. 19–43. Academic Press, London.

Dawes, E. A. 1989. Growth and survival of bacteria. *In Bacteria in Nature,* Vol. 3, J. S. Poindexter and E. R. Leadbetter (eds.), pp. 67–187. Plenum Press, New York.

Dawes, E. A., and W. H. Holmes. 1958. Metabolism of *Sarcina lutea.* III. Endogenous metabolism. *Biochem. Biophys. Acta* **30**:278–293.

Dawes, E. A., and P. J. Large. 1970. Effect of starvation on the viability and cellular constituents of *Zymomonas anerobia* and *Zymomonas mobilis. J. Gen. Microbiol.* **60**:31–42.

Dawes, E. A., and D. W. Ribbons. 1962. The endogenous metabolism of microorganisms. *Ann. Rev. Microbiol.* **16**:241–264.

Dawes, E. A., and D. W. Ribbons. 1963. Endogenous metabolism and survival of *Escherichia coli* in aqueous suspensions. *J. Appl. Bacteriol.* **26**:vi.

Dawes, E. A., and D. W. Ribbons. 1964a. Some aspects of the endogenous metabolism of *Escherichia coli. Biochem. J.* **95**:332–343.

Dawes, E. A., and D. W. Ribbons. 1964b. Some aspects of the endogenous metabolism of bacteria. *Bacteriol. Rev.* **28**:126–149.

Dawes, E. A., and D. W. Ribbons. 1965. Some aspects of the endogenous metabolism of bacteria. *Biochem. J.* **95**:332–343.

Dawes, E. A. and P. J. Senior. 1973. The role and regulation of energy reserve polymers in micro-organisms. *Adv. Microb. Physiol.* **10**:135–264.

Dawson, M. P., B. A. Humphrey, and K. C. Marshall. 1981. Adhesion: a tactic in the survival strategy of a marine *Vibrio* during starvation. *Curr. Microbiol.* **6**:195–199.

Dawson, R. and K. Gocke. 1978. Heterotrophic activity in comparison to the free amino acid concentrations in Baltic sea samples. *Oceanol. Acta* **1**:45–54.

Dean, G. A. 1963. The iodine content of some New Zealand drinking water with a note on the contribution from sea spray to the iodine in rain. *New Zealand J. Sci.* **6**:208–214.

Dean, A. C. R., and C. Hinshelwood. 1966. *Growth, Function and Regulation in Bacterial Cells.* Oxford University Press, London.

DeBaar, H. J. W., J. W. Farrington, and S. G. Wakeham. 1983. Vertical flux of fatty acids in the North Atlantic Ocean. *J. Mar. Res.* **41**:19–41.

Decho, A. W. 1990. Microbial exopolymer excretions in ocean environments: their role(s) in food webs and marine processes. *Oceanogr. Mar. Biol. Annu. Rev.* **28**:73–153.

Degens, E. T. 1970. Molecular nature of nitrogenous compounds in sea water and recent sediments. *In Organic Matter in Natural Waters,* D. W. Hood (ed.), pp. 77–106. Inst. Mar. Sci. Publication No. 1, University of Alaska, College, AK.

Degens, E. T., and J. Matheja. 1971. Formation of organic polymers on minerals and vice versa. *In Organic Compounds in Aquatic Environments,* S. D. Faust and J. V. Hunter (eds.), pp. 29–40. Marcel Dekker, New York.

Degens, E. T., J. M. Hunt, H. Reuter, and W. E. Reed. 1964. Data on the distribution of amino acids and oxygen isotopes in petroleum brine waters of various geological ages. *Sedimentology* **3**:199.

Degens, E. T., R. R. L. Guillard, W. M. Sackett, and J. A. Hellebust. 1968. Metabolic fractionation of carbon isotopes in marine plankton. I. Temperature and respiration experiments. *Deep Sea Res.* **15** :1–10.

Delaquis, P. J., D. E. Caldwell, J. R. Lawrence, and A. R. McCurdy. 1989. Detachment of *Pseudomonpas fluorescens* from biofilms on glass surfaces in response to nutrient stress. *Microb. Ecol.* **18**:199–210.

De Ley, J., K. Kersters, and I. W. Park. 1966. Molecular-biological and taxonomic studies on *Pseudomonas halocrenaea,* a bacterium from Permian salt deposits. *Antonie van Leeuwenhoek J. Microbiol. Serol.* **32**:315–331.

Delbrück, M. 1940. The growth of bacteriophage and lysis of the host. *J. Gen. Physiol.* **23**:643–651.

DeLong, E. F. 1992. High-pressure habitats. *In Encyclopedia of Microbiology,* vol. 2, J. Lederberg (ed.), pp. 405–417. Academic Press, San Diego.

DeLong, E. F., G. S. Wickman, and N. R. Pace. 1989. Phylogenetic stains: ribosomal RNA-based probes for the identification of single cells. *Science* **243**:1360–1363.

Demers, S., J. C. Theriault, E. Bourget, and A. Bah. 1987. Resuspension in the shallow sublittoral zone of a macrotidal estuarine environment: wind influence. *Limnol. Oceanogr.* **32**:327–339.

Deming, J. W., and J. A. Baross. 1993. The early diagenesis of organic matter: Bacterial activity. *In Organic Geochemistry,* M. Engel and S. Macko (eds.), pp. 119–144. Plenum Press, New York.

Deming, J. W., and P. L. Yager. 1992. Natural bacterial assemblages in deep-sea sediments: towards a global view. *In Deep-sea Food Chains and the Carbon Cycle,* G. T. Rowe and V. Pariente (eds.), pp. 11–27. Kluwer Academic Publications, Dordrecht, The Netherlands.

Deming, J. W., P. S. Taber, and R. R. Colwell. 1981. Barophilic growth of bacteria from intestinal tracts of deep-sea invertebrates. *Microbial Ecol.* **7**:85–94.

Demple, B. 1991. Regulation of bacterial oxidative stress genes. *Annu. Rev. Genet.* **25**:315–337.

Dempsey, M. 1981. Marine bacterial fouling: a scanning electron microscope study. *Mar. Biol.* **61**:305–315.

Derenbach, J. B., and P. J. leB. Williams. 1974. Autotrophic and bacterial production: fractionation of plankton populations by differential filtration of samples from the English Channel. *Mar. Biol.* **25**:263–269.

Dersch, P., K. Schmidt, and E. Bremer. 1993. Synthesis of the *Escherichia coli* K-12 nucleoid-associated DNA-binding protein H-NS is subjected to growth-phase control and autoregulation. *Molec. Microbiol.* **8**:875–889.

DeVay, J. E., and W. C. Schnathorst. 1963. Single cell isolation and preservation of bacterial cultures. *Nature* **199**:775–777.

Dickey, R. S. 1961. Relation of some edaphic factors to *Agrobacterium tumefaciens*. *Phytopathology* **51**:607–614.

Dilling, W. L., N. B. Tefertiller, and G. J. Kallos. 1975. Evaporation rates of methylene chloride, chloroform, 1,1,1-trichloroethane, trichloroethylene, tetrachloroethylene, and other chlorinated compounds in dilute aqueous solutions. *Environ. Sci. Technol.* **9**:833–838.

Dische, Z. 1955. Color reaction of nucleic acid components. *Nucleic Acids* **1**:285–305.

Dobbs, F. C., J. B. Guckert, and K. R. Carman. 1989. Comparison of three techniques for administering radiolabeled substrates to sediments for trophic studies: Incorporation of microbes. *Microb. Ecol.* **17**:237–250.

Dombrowski, H. 1963. Bacteria from paleozoic salt deposits. *Ann. New York Acad. Sci.* **108**:453–460.

Domsch, K. H. 1986. Influence of management on microbial communities in soil. *In Microbial Communities in Soil*, V. Jensen, A. Kjøller, and L. H. Sørensen (eds.), pp. 355–367. Elsevier Applied Science, London.

Donache, W. D., K. J. Begg, and M. Vicenie. 1976. Cell length, cell growth, and cell division. *Nature* **264**:328–333.

Dorman, C. J., G. C. Barr, N. N. Bhirain, and C. F. Higgins. 1988. DNA supercoiling and the anaerobic and growth phase regulation of *tonB* gene expression. *J. Bacteriol.* **170**:2816–2826.

Dortch, Q., R. L. Roberts, J. R. Clayton, Jr., and S. I. Ahmed. 1983. RNA/DNA ratios and DNA concentrations as indicators of growth rate and biomass in planktonic marine organisms. *Mar. Ecol. Prog. Ser.* **13**:61–71.

Douglas, D. J., J. A. Novitsky, and R. O. Fournier. 1987. Microautoradiography-based enumeration of bacteria with estimates of thymidine-specific growth and production rates. *Mar. Ecol. Prog. Ser.* **36**:91–99.

Dow, C. S., and A. Lawrence. 1980. Microbial growth and survival in oligotrophic freshwater environments. *In Microbial Growth and Survival in Extremes of Environment*, G. W. Gould and J. E. L. Corry (eds.), pp. 1–20. Academic Press, London.

Dow, C. S., and P. Morgan. 1986. Prosthecate bacteria—are they uniquely adapted to low nutrient ecosystems? *In Perspectives in Microbial Ecology*, F. Megusar and M. Gantar (eds.), pp. 159–162. Slovene Society for Microbiology, Ljubljana, Yugoslavia.

Dow, C. S., and R. Whittenbury. 1980. Prokaryotic form and function. *In Contemporary Microbial Ecology*, D. C. Ellwood, J. N. Hedger, M. J. Latham, J. M. Lynch, and J. H. Slater (eds.), pp. 391–417. Academic Press, London.

Dow, C. S., R. Whittenbury, and N. G. Carr. 1983. The "shut down" or "growth precursor" cell—an adaptation for survival in a potentially hostile environment. *Symp. Soc. Gen. Microbiol.* **34**:187–247.

Draycott, A. P., and P. J. Last. 1971. Some effects of partial sterilization on mineral nitrogen in a light soil. *J. Soil Sci.* **22**:152–157.

Dresden, M. H., and M. B. Hoagland. 1967. Polyribosome of *Escherichia coli*. Breakdown during glucose starvation. *J. Biol. Chem.* **242**:1065–1068.

Drew, G. W. 1912. Report of investigation on marine bacteria carried on at Andros Island. *Yearbook Carnegie Inst. Wash.* **11**:136–154.

Druffel, E. R. M., P. M. Williams, P. M. Brauer, and J. R. Ertel. 1992. Cycling of dissolved and particulate organic matter in the open ocean. *J. Geophys. Res.* **97**:15639–15659.

Druilhet, R. E., and J. M. Sobek. 1976. Starvation survival of *Salmonella enteritidis*. *J. Bacteriol.* **125**:119–124.

Duce, R. A. 1983. Biochemical cycles and the air-sea exchange of aerosols. *In* Scope report 21. *The Major Biogeochemical Cycles and Their Interactions.* B. Bolin and R. B. Cook (eds.), pp. 427–456. John Wiley and Sons, Chichester.

Ducklow, H. W., and C. A. Carlson. 1992. Oceanic bacterial production. *Adv. Microb. Ecol.* **12**:113–182.

Ducklow, H. W., and F.-K. Shiah. 1993. Bacterial production in estuaries. *In Aquatic Microbiology,* T. E. Ford (ed.), pp. 261–287. Blackwell Scientific Publications, Boston.

Ducklow, H. W., D. A. Burdie, P. J. LeB. Williams, and J. M. Davies. 1986. Bacterioplankton: a sink for carbon in a coastal marine plankton community. *Science* **232**:865–867.

Ducklow, H. W., D. L. Kirchman, and G. T. Rowe. 1982. Production and vertical flux of attached bacteria in the Hudson River Plume of the New York Bight as studied with sediment traps. *Appl. Environ. Microbiol.* **43**:769–776.

Ducklow, H. W., S. M. Hill, and W. D. Gardner. 1985. Bacterial growth and the decomposition of particulate organic carbon collected in sediment traps. *Continental Shelf Res.* **4**:445–464.

Duncan, S., A. Glover, K. Killham, and J. I. Prosser. 1994. Luminescence-based detection of activity of starved and viable but nonculturable bacteria. *Appl. Environ. Microbiol.* **60**:1308–1316.

Dunn, J. E., G. E. Windom, K. L. Hansen, and R. J. Seidler. 1974. Isolation and characterization of temperature sensitive mutants of host-dependent *Bdellovibrio bacteriovorus* 109D. *J. Bacteriol.* **117**:1341–1349.

Dussault, H. P. 1958. The fate of red halophilic bacteria in solar salt during storage. *In The Microbiology of Fish and Meat Curing Brines,* B. P. Eddy (ed.), pp. 13–19. Her Majesty's Stationary Office, London.

Duursma, E. K. 1961. Dissolved organic carbon, nitrogen and phosphorus in the sea. *Netherl. J. Sea Res.* **1**:1–147.

Duursma, E. K. 1963. The production of dissolved organic matter in the sea, as related to the primary gross production of organic matter. *Netherl. J. Sea Res.* **2**:85–93.

Duursma, E. K. 1965. The dissolved organic constituents of seawater. *In Chemical Ocean-ography,* vol. 1, J. P. Riley and G. Skirrow (eds.), pp. 433–475. Academic Press, London.

Duursma, E. K., and R. Dawson. 1981. *Marine Organic Chemistry.* Elsevier Science Publications Co., Amsterdam.

Duxbury, J. M., M. S. Smith, and J. W. Doran. 1989. Soil organic matter as a source and a sink of plant nutrients. *In Dynamics of Soil Organic Matter in Tropical Ecosystems,* D. C. Colman, J. M. Oades, and G. Uehara (eds.), pp. 33–67. University of Hawaii Press, Honolulu.

Eagon, R. G. 1962. *Pseudomonas natriegens,* a marine bacterium with a generation time of less than 10 minutes. *J. Bacteriol.* **83**:736–737.

Eastermann, E. F., and A. D. McLaren. 1959. Stimulation of bacterial proteolysis by adsorbents. *J. Soil Sci.* **10**:64–78.

Eastermann, E. F., G. H. Peterson, and A. D. McLaren. 1959. Digestion of clay-protein, lignin-protein and silica-protein complexes by enzymes and bacteria. *Proc. Soil. Sci. Soc. Am.* **23**:31–36.

Eberhart, J. L., R. P. Griffiths, J. S. Rohovec, and R. Y. Morita. 1992. Survival of *Aeromonas salmonicida* and *Renibacterium salmoninarium* in the aquatic environment. *Abstract of the Workshop of the Eastern Fish Health and the American Fish Society,* abstr., p. 34. Workshop (June 16–19). Auburn, AL.

Ehrlich, H. L. 1990. *Geomicrobiology,* 2nd ed. Marcel Dekker, Inc., New York.

Ehrlich, H. L., W. C. Ghiorse, and G. L. Johnson II. 1972. Distribution of microbes in manganese nodules from the Atlantic and Pacific Oceans. *Dev. Ind. Microb.* **13**:57–65.

Eigener, U. 1976. Adenine nucleotide pool variation in intact *Nitrobacter winogradskyi* cells. *Arch. Microbiol.* **102**:233–240.

Eisenbach, M. 1990. Functions of the flagellar modes of rotation in bacteria motility and chemotaxis. *Molec. Microbiol.* **4**:161–167.

Eisenstark, A. 1989. Bacterial genes involved in the response to near-ultraviolet radiation. *Adv. Genet.* **26**:99–147.

Eguchi, M., and Y. Ishida. 1986. An ecological study of oligotrophic bacteria in the Antarctic Ocean. *Mem. Natl. Inst. Polar Res. Spec. Issue* **40**:413.

Eguchi, M., and Y. Ishida. 1990. Oligotrophic properties of heterotrophic bacteria and *in situ* heterotrophic activity in pelagic seawaters. *FEMS Microbiol. Ecol.* **73**:23–30.

Eguchi, M., T. Nishikawa, K. MacDonald, R. Civicchioli, J. C. Gottschal, and S. Kjelleberg. 1996. Responses to stress and nutrient availability by the marine ultramicrobacterium *Sphingomonas* sp. strain RB2256. *Appl. Environ. Microbiol.* **62**:1287–1294.

Elliker, P. R., and W. C. Frazier. 1938. Influence of time and temperature of incubation on heat resistance of *Escherichia coli. J. Bacteriol.* **36**:83–97.

Ellis, D. 1915. Fossil microorganisms from the Jurassic and Cretaceous rocks from Great Britain. *Proc. Roy. Soc. Edinburgh, B,* **35**:110–132.

Ellis, E. L., and M. Delbrück. 1939. The growth of bacteriophage. *J. Gen Physiol.* **22**:365–384.

Ellwood, D. C., C. W. Keevil, P. D. Marsch, C. M. Brown and J. N. Wardell. 1982. Surface-associated growth. *Phil. Trans. Roy. Soc. London B* **297**:517–532.

Ellwood, D. C., J. Melling, and P. Rutter. 1980. *Adhesion of Microorganisms on Surfaces*. Academic Press, London.

Emejuaiwe, S. O., and C. Okafor. 1989. Storage and preservation of stock cultures of some local yeasts in sterile distilled water: An economically useful technique for Culture Collections in a developing country such as Nigeria. *Abstract of the Fifth International Symposium on Microbial Ecology*, abstr., p. 183. Kyoto.

Emilsson, V., and C. G. Kurland. 1990. Growth rate dependence of transfer RNA abundance in *Escherichia coli*. *EMBO J.* **9**:4359–4366.

Engel, M. H., and S. A. Macko. 1993. *Organic Geochemistry Principles and Applications*. Plenum Press, New York.

Engel, A. E. J., B. Nagy, L. A. Nagy, C. G. Engel, G. O. Kremp, and C. M. Drew. 1968. Alga-like forms in Onverwacht series, South Africa: oldest recognized lifelike forms on Earth. *Science* **161**:1005–1008.

Engelking, H. M., and R. J. Seidler. 1974. The involvement of extracellular enzymes in the metabolism of *Bdellovibrio*. *Arch. Microbiol.* **95**:293–304.

Engelking, H. M., and R. J. Seidler. 1975. Application of the deoxyribonucleic acid/ribonucleic acid hybridization technique in *Bdellovibrio* as a model for studying ribonucleic acid turnover in hot-parasite systems. *Appl. Microbiol.* **30**:97–102.

Enger, Ø. 1992a. Microbial ecology of marine fish farms, with special emphasis on the fish pathogenic bacteria *Vibrio salmonicida* and *Aeromonas salmonicida*. Ph. D. Thesis. University of Bergen, Bergen, Norway.

Enger, Ø. 1992b. Possible ecological implication of the high cell surfaces hydrophobacity of the fish pathogen *Aeromonas salmonicida*. *Can. J. Microbiol.* **38**:1025–1052.

Enger, Ø., B. Gunlaugsdottir, B. K. Thorsen, and B. Hjeltnes. 1992. Infectious load of *Aeromonas salmonicida* subsp. salmonicida during the initial phase of a cohabitant infection with Atlantic salmon, *Salmo solar* L. *J. Fish. Dis.* **15**:425–430.

Enger, Ø., H. Nygaard, M. Solberg, G. Schei, J. Nielsen, and I. Dundas. 1987. Characterization of *Aeromonas denitrificans* sp. nov. *Int. J. Syst. Bacteriol.* **37**:416–421.

Enger, Ø., K. A. Hoff, G. H. Schei, and I. Dundas. 1990. Starvation survival of the fish pathogenic bacteria *Vibrio anguillarum* and *Vibrio salmonicida* in marine environments. *FEMS Microb. Ecol.* **74**:215–220.

English, V. C., and A. T. McManus. 1985. Bacterial longevity in salt-free medium: responses. *ASM News* **51**:1.

Ensign, J. C., and R. W. Wolfe. 1964. Nutritional control of morphogenesis in *Arthrobacter crystallopoietes*. *J. Bacteriol.* **87**:924–932.

Eppley, R. W., E. H. Renger, and P. R. Betzer. 1983. The residence time of particulate organic carbon in the surface layer of the ocean. *Deep Sea Res.* **30**:281–323.

Erdman, J. G., E. M. Marlett, and W. E. Hanson. 1956. Survival of amino acids in marine sediments. *Science* **124**:1026.

Erickson, J. W., and C. A. Gross. 1989. Identification of the σ^E subunit of *Escherichia coli* RNA polymerase: a second alternate σ factor involved in high-temperature gene expression. *Genes Devel.* **3**:1462–1471.

Espinosa-Urgel, M., and A. Tormo. 1993. σ^S-dependent promoters in *Escherichia coli* are located in DNA regions with intrinsic curvature. *Nucl. Acids Res.* **21**:3667–3670.

Evelyn, T. P. T. 1987. Bacterial kidney disease in British Columbia, Canada: comments on its epizootiology and on methods for its control on fish farms. Talk on Aqua Nor Meetings, Trondheim, Norway.

Ezekiel, D. H. 1964. Intracellular charging of soluble ribonucleic acid in *Escherichia coli* subjected to isoleucine starvation and chloramphenicol treatment. *Biochem. Biophys. Res. Comm.* **14**:64–68.

Faegri, A., V. L. Torsvik, and J. Goksoyr. 1977. Bacterial and fungal activities in soil: separation of bacteria and fungi by a rapid fractionated centrifuge technique. *Soil Biol. Biochem.* **9**:105–112.

Fallon, R. D., and S. Y. Newell. 1986. Thymidine incorporation by the microbial community of standing dead *Spartina alterniflora*. *Appl. Environ. Microbiol.* **52**:1206–1208.

Fallon, R. D., S. Y. Newell, and G. S. Hopkinson. 1983. Bacterial production in marine sediments: will cell-specific measures agree with whole system metabolism? *Mar. Ecol. Prog. Ser.* **11**:117–119.

Fan, C. L., and V. W. Rodwell. 1975. Physiological consequences of starvation in *Pseudomonas putida*: degradation of intracellular proteins and loss of activity of the inducible enzyme of L-arginine catabolism. *J. Bacteriol.* **124**:1302–1311.

Fang, F. C., S. J. Libby, N. A. Buckmeier, P. C. Lowen, J. Switala, J. Harwood, and D. G. Guiney. 1992. The alternative σ factor KatF (RpoS) regulates *Salmonella* virulence. *Proc. Natl. Acad. Sci. (USA)* **89**:11978–11982.

Faquin, W. C., and J. D. Oliver. 1984. Arginine uptake by psychrophilic marine *Vibrio* sp. during starvation-survival induced morphogenesis. *J. Gen. Microbiol.* **130**:1331–1335.

Farrell, M. A., and H. G. Turner. 1932. Bacteria in anthracite coal. *J. Bacteriol.* **23**:155–162.

Faust, S. J., and J. V. Hunter. 1971. *Organic Compounds in Aquatic Environments*. Marcel Dekker, New York.

Favero, M. S., L. A. Carson, W. W. Bond, and N. J. Peterson. 1971. *Pseudomonas aeruginosa*: growth in distilled water from hospitals. *Science* **173**:836–838.

Federle, T. W., D. C. Dobbins, J. R. Thornton-Manning, and D. D. Jones. 1986. Microbial biomass, activity and community structure in subsurface soils. *Ground Water* **24**:365–374.

Feeny, P. 1976. Plant appparency and chemical defense. *In Biochemical Interactions between Plants and Insects. Recent Advances in Photochemistry,* vol. 10, J. W. Wallace and R. L. Mansell (eds.), pp. 1–40. Plenum Press, New York.

Fenchel, T. M. and T. H. Blackburn. 1979. *Bacteria and Mineral Cycling*. Academic Press, London.

Ferek, R. J., and M. O. Andreae. 1984. Photochemical production of carbonyl sulphide in marine surface waters. *Nature* **307**:148–150.

Ferguson, R. L., and P. Rublee. 1976. Contribution of bacteria to standing crop of coastal plankton. *Limnol. Oceanogr.* **21**:141–145.

Ferguson, R. L., and W. G. Sunda. 1984. Utilization of amino acids by planktonic marine bacteria: Importance of clean techniques and low substrate additions. *Limnol. Oceanogr.* **29**:258–274.

Ferguson, R. L., E. N. Buckley, and A. V. Palumbo. 1984. Response of marine bacterioplankton to differential filtration and confinement. *Appl. Environ. Microbiol.* **47**:49–55.

Fildes, P. 1940. The mechanism of the anti-bacterial action of mercury. *Brit. J. Exp. Path.* **21**:67–73.

Filip, Z. 1973. Clay minerals as a factor influencing the biochemical activity of soil microorganisms. *Folia Microbiol.* **18**:56–74.

Filip, Z. 1975. Wechselbeziehungen zwischen Mikroorganismen und Tonmineralen und ihre Auswirkung auf die Bodendynamik. Habilitationsschrift. Justus Liebie Universitat, Giessen, West Germany.

Filip, Z. 1978. Effect of solid particles on growth and metabolic activities of microorganisms. *In Microbial Ecology,* M. W. Loutit and J. A. R. Miles (eds.), pp. 102–104. Springer-Verlag, New York.

Findlay, R. H., and D. C. White. 1983. Polymeric beta-hydroxy alkanoates from environmental samples and *Bacillus megatherium. Appl. Environ. Microbiol.* **45**:71–78.

Findlay, R. H., P. C. Pollard, D. J. W. Moriarty, and D. C. White. 1985. Quantitative determination of microbial activity and community nutritional status in estuarine sediments: evidence for a disturbance artifact. *Can. J. Microbiol.* **31**:493–498.

Findlay, R. H., M. B. Trexler, J. B. Gukert, and D. C. White. 1990a. Laboratory study of disturbance in marine sediments; Response of a microbial community. *Mar. Ecol. Prog. Ser.* **62**:121–133.

Findlay, R. H., M. B. Trexler, and D. C. White. 1990b. Response of a benthic microbial community to biotic disturbance. *Mar. Ecol. Prog. Ser.* **62**:135–148.

Findlay, S. E. G., J. L. Meyer, and R. T. Edwards. 1984. Measuring bacterial production via rate of incorporation of [^{3}H]thymidine into DNA. *J. Microb. Methods* **2**:57–72.

Fischer, B. 1894. Die Bakterien des Meeres nach den Untersuchungen der Plankton-Expedition unter Fleichzeitiger einiger alterer und nauerer Untersuchungen. *Ergebnisse der Plankton-Expedition der Humbold-Stiftung* **4**:1–83.

Fitzwater, S. E., G. A. Knauer, and J. H. Martin. 1982. Metal contamination and its effect on primary production measurements. *Limnol. Oceanogr.* **27**:544–551.

Flanagan, P. W., and F. L. Bunnell. 1976. Decomposition models based on climatic variables, substrate variables, microbial respiration and production. *In The Role of Terrestrial and Aquatic Organisms in Decomposition Processes,* J. M. Anderson and A. Mac Fadyn (eds.), pp. 437–457. Blackwell, Oxford.

Flanagan, P. W., and K. Van Cleve. 1977. Microbial biomass, respiration and nutrient cycling in a black spruce taiga ecosystem. *In Soil Organisms as Components of Ecosystems,* U. Lohm and T. Persson (eds.), pp. 261–273. *Proc. 6th Coll. Soil Zool., Ecol. Bull (Stockholm).* Swedish Natural Science Research Council, Stockholm.

Flärdh, K., N. H. Albertson, L. Adler, and S. Kjelleberg. 1991. Stability of ribosomes in the marine *Vibrio* sp S14 during multiple nutrient starvation. *Abstract from the General Meetings of the American Society for Microbiology,* abstr., p. 200. American Society for Microbiology, Washington, DC.

Flärdh, K., P. S. Cohen, and S. Kjelleberg. 1992. Ribosomes exist in large excess over the apparent demand for protein synthesis during carbon starvaton in marine *Vibrio* sp. strain CCUG 15956. *J. Bacteriol.* **174**:6780–6788.

Flärdt, K., and S. Kjelleberg. 1994. Glucose upshift of carbon-starved marine *Vibrio* sp. strain S14 causes amino acid starvation and induction of the stringent response. *J. Bacteriol.* **176**:5897–5903.

Fletcher, M. 1977. The effect of culture concentration and age, time and temperature on bacterial attachment to polystyrene. *Can. J. Microbiol.* **23**:1–6.

Fletcher, M. 1979a. The attachment of bacteria to surfaces in aquatic environments. *In Adhesion of Microorganisms to Surfaces,* D. C. Ellwood, J. Melling, and P. Rutter (eds.), pp. 87–108. Academic Press, London.

Fletcher, M. 1979b. A microautoradiographic study of the activity of attached and free-living bacteria. *Arch. Microbiol.* **122**:271–279.

Fletcher, M. 1986. Measurement of glucose utilization by *Pseudomonas fluorescens* that are free-living and that are attached to surfaces. *Appl. Environ. Microbiol.* **52**:672–676.

Fletcher, M., and G. D. Floodgate. 1973. An electron microscope demonstration of an acidic polysaccharide involved in the adhesion of a marine bacterium to solid surfaces. *J. Gen. Microbiol.* **74**:325–334.

Fletcher, M., and G. I. Loeb. 1979. Influence of substratum characteristics on the attachment of a marine pseudomonad to solid surfaces. *Appl. Environ. Microbiol.* **37**:67–72.

Fletcher, M., and K. C. Marshall. 1982. Are solid surfaces of ecological significance to aquatic bacteria? *Adv. Microb. Ecol.* **6**:199–236.

Fliermans, C. B., and T. C. Hazen. 1990. *Proceedings of the First International Symposium on Microbiology of the Deep Subsurface.* W. S. R. C., Information Services Section, Publication Group, Savannah River, SC.

Flint, K. P. 1987. The long-term survival of *Escherichia coli* in river water. *J. Appl. Bacteriol.* **63**:261–270.

Flynn, K. J. 1988a. The concept of "primary production" in aquatic ecology. *Limnol. Oceanogr.* **33**:1215–1216.

Flynn, K. J. 1988b. Some practical aspects of measurements of dissolved free amino acids in natural waters and within microalgae by use of HPLC. *Chem. Ecol.* **3**:259–293.

Flynn, K. J. 1989. Nutrient limitation of marine microbial production: Fact or artefact? *Chem. Ecol.* **4**:1–13.

Flynn, K. J., and J. Fielder. 1989. Changes in intracellular and extracelluar amino acids during predation of the chlorophyte *Dunaliella primolecta* by the heterotrophic dino-flagellate *Oxyrrhis marina* and the use of the glutamine/flutamate ratio as an indicator of nutrient status in mixed populations. *Mar. Ecol. Prog. Ser.* **53**:117–127.

Fogg, G. E. 1983. The ecological significance of extracellular products of phytoplankton photosynthesis. *Bot. Mar.* **26**:3–14.

Folger, T. 1992. Oldest living bacteria tell all. *Discover* **13**:30–31.

Fonden, R. 1968. A yeast-extract peptone agar for the determination of heterotrophic bacteria in lakes. *Vatten* **2**:161–166.

Fonselius, S. 1954. Amino acids in rain water. *Tellus* **6**:90.

Forest, W. W. 1969. Energetic aspects of microbial growth. *Symp. Soc. Gen. Microbiol.* **19**:65–86.

Forrest, W. W. 1965. Adenosine triphosphate pool during the growth of *Streptococcus facaelis*. *J. Bacteriol.* **90**:1013–1016.

Foster, J. W. 1947. Some introspections of mold metabolism. *Bacteriol. Rev.* **11**:166–191.

Foster, P., and A. W. Morris. 1971a. The seasonal variation of dissolved ionic and organically associated copper in the Menai Straits. *Deep-Sea Res.* **18**:231–236.

Foster, P., and A. W. Morris. 1971b. The use of ultraviolet absorption measurements for the estimation of organic pollution in inshore sea waters. *Water Res.* **5**:19–27.

Foster, J. W., D. M. Kinney, and A. G. Moat. 1979. Pyridine nucleotide cycle of *Salmonella typhimurium*: isolation and characterization of *pncA*, *pncB*, and *pncC* mutants and utilization of exogenous nicotinamide dinucleotide. *J. Bacteriol.* **137**:1165–1175.

Fowden, L. 1962. Amino acids and proteins. *In Physiology and Biochemistry of Algae*, R. A. Lewin (ed.), pp. 189–209. Academic Press, N.Y.

Fowler, S. W., and L. F. Small. 1972. Sinking rates of euphausid fecal pellets. *Limnol. Oceanogr.* **17**:293–296.

Fraps, G. S., and A. J. Sterges. 1932. Causes of low nitrification capacity of soil. *Soil Sci.* **34**:353–363.

Fred, E. B., F. C. Wilson, and A. Davenport. 1924. The distribution and significance of bacteria in Lake Mendota. *Ecology* **5**:322–339.

Fredricks, A. D., and W. M. Sackett. 1970. Organic carbon in the Gulf of Mexico. *J. Geophys. Res.* **75**:2199–2106.

Fredrickson, J. K., D. L. Balkwill, J. M. Sachara, S. W. Li, F. J. Brockman, and M. A. Simmons. 1991. Physiological diversity and distribution of heterotrophic bacteria in deep cretaceous sediments of the Atlantic Coastal Plain. *Appl. Environ. Microbiol.* **57**:402–411.

Freter, R. 1981. Mechanisms of association of bacteria with mucosal surfaces in adhesion and microbial pathogenicity. *CIBA Found. Symp.* **80**:36–55.

Frimmel, F. H., and R. F. Christman. 1988. Humic substances and their role in the environment. John Wiley & Sons, New York.

Fry, J. C. 1982. The analysis of microbial interactions and communities *in situ*. *In Micro-*

bial Interactions and Communities, A. T. Bull and J. H. Slater (eds.), pp. 103–152. Academic Press, London.

Fry, J. C. 1990a. Oligotrophs. *In Microbiology of Extreme Environments,* C. Edwards (ed.), pp. 93–116. McGraw-Hill, New York

Fry, J. C. 1990b. Direct methods and biomass estimations. *In Methods in Microbiology,* vol. 22, R. Grigorova and J. R. Norris (eds.), pp. 40–85. Academic Press, London.

Fry, J. C., and A. R. Davis. 1985. An assessment of methods for measuring volumes of planktonic bacteria, with particular reference to television image analysis. *J. Appl. Bacteriol.* **58**:105–112.

Fry, J. C., and T. Zia. 1982a. Viability of heterotrophic bacteria in freshwater. *J. Gen. Microbiol.* **128**:2841–2850.

Fry, J. C., and T. Zia. 1982b. A method for estimating viability of aquatic bacteria by slide culture. *J. Appl. Bacteriol.* **53**:189–198.

Führ, F., and D. R. Sauerbeck. 1968. Decomposition of wheat straw in the field as influenced by cropping and rotation. *In Isotopes and Radiation in Soil Organic Matter Studies,* pp. 241–250. International Atomic Energy Agency, Vienna.

Fuhrman, J. A. 1981. Influence of method on the apparent size distribution of bacterioplankton cells: epifluorescent microscopy compared to scanning electron microscopy. *Mar. Ecol. Prog. Ser.* **5**:103–106.

Fuhrman, J. A. 1987. Close coupling between release and uptake of dissolved free amino acids in seawater studied by an isotope dilution approach. *Mar. Ecol. Prog. Ser.* **37**:45–52.

Fuhrman, J. A. 1990. Dissolved free amino acid cycling in an estuarine outflow plume. *Mar. Ecol. Prog. Ser.* **66**:197–203.

Fuhrman, J. A., and F. Azam. 1980. Bacterioplankton secondary production estimates for coastal waters of British Columbia, Antarctica and California. *Appl. Environ. Microbiol.* **39**:1085–1095.

Fuhrman, J. A., and F. Azam. 1982. Thymidine incorporation as a measure of heterotrophic bacterioplankton production in marine surface waters: Evaluation of field results. *Mar. Biol.* **66**:109–120.

Fuhrman, J. A., and T. M. Bell. 1985. Biological considerations in the measurement of dissolved free amino acids in seawater and implications for chemical and microbiological studies. *Mar. Ecol. Prog. Ser.* **25**:13–21.

Fuhrman, J. A., and R. L. Ferguson. 1986. Nanomolar concentrations and rapid turnover of dissolved free amino acids in seawater: Agreement between chemical and microbiological measurements. *Mar. Ecol. Prog. Ser.* **33**:237–242.

Fuhrman, J. A., and G. B. McManus. 1984. Do bacteria-sized eukaryotes consume significant bacterial production? *Science* **224**:1257–1260.

Fuhrman, J. A., and C. A. Suttle. 1993. Viruses in marine planktonic systems. *Oceanography* **6**:50–62.

Fuhrman, J. A., H. W. Ducklow, D. L. Kirchman, and G. B. McManus. 1986. Adenine and total microbial production: a reply. *Limnol. Oceanogr.* **13**:1395–1400.

Fuhrman, J. A., T. D. Sleeter, C. A. Carlson, and L. M. Proctor. 1989. Dominance of bacterial biomass in the Sargasso Sea and its ecological implications. *Mar. Ecol. Prog. Ser.* **57**:207–342.

Fuhrman, J. A., K. McCallum, and A. A. Davis. 1993. Phylogenetic diversity of subsurface marine microbial communities from the Atlantic and Pacific Oceans. *Appl. Environ. Microbiol.* **59**:1294–1302.

Fujita, M., K. Tanaka, H. Takahashi, and A. Amemura. 1994. Transcription of the principal sigma-factor genes, *rpoD* and *rpoS,* in *Pseudomonas aeruginosa* is controlled according to the growth phase. *Molec. Microbiol.* **13**:1071–1077.

Fulton, H. R. 1920. Decline of *Pseudomonas citrii* in the soil. *J. Agric. Res.* **19**:207–223.

Gagosian, R. B., and C. Lee. 1981. Processes controlling the distribution of biogenic compounds in seawater. *In Marine organic chemistry,* E. K. Duursma and R. Dawson (eds.), pp. 91–123. Elsevier, Amsterdam.

Gagosian, R. B., and D. H. Stuermer. 1977. The cycling of biogenic compounds and their diagenetically transformed products in seawater. *Mar. Chem.* **5**:605–632.

Gahl, R., and B. Anderson. 1928. Sulphate reducing bacteria in California soil waters. *Centrabl. f. Bakteriol. Abt. 2.* **73**:331–338.

Gaill, F. 1979. Digestive structure of abyssal tunicates. *Sarsia* **64**:97–102.

Galippe, V. 1920. Recherches sur la résistance de microzymes à l'action du temps et leur survivance dans l'ambre. *C. R. Acad. Sci. Paris* **170**:856–858.

Galippe, V., and G. Souffland. 1921. Recherches sur la présence dans les météorites, etc., les cendres et les laves volcaniques, d'organites susceptibles de reviviscence et sur leur resistance aux hautes temperératures. *C. R. Acad. Sci. Paris* **172**:1252–1254.

Gallant, J. 1979. Stringent control in *Escherichia coli. Annu. Rev. Genet.* **13**:393–415.

Galle, E. 1910–11. Über Selbstentzündung der Steinkohle. *Centrab. f. Bakteriol. Abt. 2.* **28**:461–473.

Gamble, T. N., M. R. Betlach, and J. M. Tiedje. 1977. Nutritionally dominant denitrifying bacteria from world soils. *Appl. Environ. Microbiol.* **33**:926–939.

Gandy, E. L., and D. C. Yoch. 1988. Relationship between nitrogen-fixing sulfate reducers and fermenters in salt marsh sediments and roots of *Spartina alternifora. Appl. Environ. Microbiol.* **54**:2031–2036.

Garbosky, A. J., and N. Giambiagi. 1962. The survival of nitrifying bacteria in soil. *Plant Soil* **17**:271–278.

García-Lara, J., J. Martímez, M. Vilamu, and J. Vives-Rego. 1993. Effect of previous growth conditions on the starvation-survival of *Escherichia coli* in seawater. *J. Gen. Microbiol.* **139**:1425–1431.

Gardner, W. S., J. F. Chandler, G. A. Laird, and D. Scavia. 1986. Microbial response to amino acid additions in Lake Michigan: Grazer control and substrate limitation of bacterial population. *J. Great Lakes Res.* **12**:161–174.

Gardner, W. S., J. F. Chandler, and G. A. Laird. 1989. Organic nitrogen mineralization and substrate limitation of bacteria in Lake Michigan. *Limnol. Oceanogr.* **34**:478–485.

Garrett, S. D. 1950. Ecology of the root-inhabiting fungi. *Biol. Rev.* **25**:220–254.

Garrett, W. D. 1967. The organic chemical composition of the ocean surface. *Deep-Sea Res.* **14**:221–227.

Garvie, E. I. 1955. The growth of *Escherichia coli* in buffer substrate and distilled water. *J. Bacteriol.* **69**:393–398.

Gauthier, M. J., P. M. Munro, and V. A. Breittmayer. 1989. Influence of prior growth conditions on low nutrient response of *Escherichia coli* in seawater. *Can. J. Microbiol.* **35**:379–383.

Gauthier, M. J., P. M. Munro, and S. Mohajer. 1987. Influence of salts and sodium chloride on the recovery of *Escherichia coli* from seawater. *Curr. Microbiol.* **15**:5–10.

Gay, F. P. 1936. *Agents of Disease and Host Resistance.* Charles Thomas, Baltimore.

Gee, A. H. 1932. Lime deposition and the bacteria. I. Estimate of bacterial activity at the Florida Keys. *Papers from the Tortugase Lab., Carnegie Inst. Wash.* **28**:67–82.

Geesey, G. G. 1982. Microbial exopolymers: ecological and economic considerations. *ASM News* **48**:9–14.

Geesey, G. G. 1987. Survival of microorganisms in low nutrient waters. *In Biological Fouling of Industrial Water Systems: A Problem Solving Approach,* M. W. Mittelman and G. G. Geesey (eds.), pp. 1–23. Water Micro Associates, San Diego.

Geesey, G. G., and R. Y. Morita. 1979. Capture of arginine at low concentrations by a marine psychrophilic bacterium. *Appl. Environ. Microbiol.* **38**:1092–1097.

Geller, A. 1983a. Growth of bacteria in inorganic medium at different levels of airborne organic substances. *Appl. Environ. Microbiol.* **46**:1258–1262.

Geller, A. 1983b. Degradability of dissolved organic lake water compounds in culture of natural bacterial communities. *Arch. Hydrobiol.* **99**:60–79.

Geller, A. 1985a. Light-induced conversion of refractory, high molecular weight lake water constituents. *Schweiz. Z. Hydrol.* **47**:21–26.

Geller, A. 1985b. Degradation and formation of refractory DOM by bacteria during simultaneous growth on labile substrates and persistent lake water constituents. *Schweiz. Z. Hydrol.* **47**:27–44.

Geller, A. 1986. Comparison of mechanisms enhancing biodegradability of refractory lake water constituents. *Limnol. Oceanogr.* **31**:755–764.

George, R. Y., and R. P. Huggins. 1979. Eutrophic hadal benthic community in the Puerto Rico Trench. *Ambio Spec. Rept.* **64**:51–58.

Gerard, J. F., and G. Stotzky. 1975. Smectite-protein complexes vs non–complexed proteins as energy and carbon sources for bacteria. *Agronomy Abstr.,* p. 91.

Gerba, C. P., and J. S. MacLoed. 1976. Effects of sediments on survival of *Escherichia coli* in marine waters. *Appl. Environ. Microbiol.* **32**:114–120.

Gest, H. 1987. *The World of Microbes.* Science Technical Publications, Madison, WI.

Gest, H. 1993. Bacterial growth and reproduction in nature and in the laboratory. *ASM News* **59**:542–543.

Ghiorse, W. C., and D. L. Balkwill. 1983. Enumeration and morphological characterization of bacteria indigenous to subsurface environments. *Dev. Ind. Microbiol.* **24**:213–224.

Ghiorse, W. C., and J. T. Wilson. 1988. Microbial ecology of the terrestrial subsurface. *Adv. Appl. Microbiol.* **33**:107–172.

Gibbon, G. R., R. J. Parkes, and R. A. Herbert. 1989. Biological availability and turnover rate of acetate in marine and estuarine sediments in relation to dissimilatory sulphate reduction. *FEMS Microbiol. Ecol.* **62**:303–306.

Gibbons, R. J. 1977. Adherence of bacteria to host tissue. *In Microbiology, 1977,* D. Schlessinger (ed.), pp. 385–496. American Society for Microbiology, Washington, DC.

Gibbons, R. J., and B. Kapsimalis. 1967. Estimates of the overall rate of growth of the intestinal microflora of hamsters, guinea pigs, and mice. *J. Bacteriol.* **93**:510–512.

Gibbs, W. M. 1919. Isolation and study of nitrifying bacteria. *Soil Sci.* **8**:427–482.

Gillan, F. T., and R. W. Hogg. 1984. A method for the estimation of bacteria biomass and community structure in mangrove-associated sediments. *J. Microbiol. Methods* **2**:275–295.

Gilles, R. 1975. Mechanism of ion and osmoregulation. *In Marine Ecology,* vol. 2, pt. 1, O. Kinne (ed.), pp. 259–348. John Wiley & Sons, London.

Gillespie, P. A., R. Y. Morita, and L. P. Jones. 1976. The heterotrophic activity for amino acids, glucose, and acetate in Antarctic waters. *J. Oceanogr. Soc. Jpn.* **32**:74–82.

Ginsburg-Karagitscheva, T. L. 1933. Microflora or oil waters and oil-bearing formations and biochemical processes caused by it. *Bull. Amer. Assoc. Petrol. Geol.* **17**:52–65.

Ginter, R. L. 1930. Causative agents of sulphate reduction in oil-well waters. *Bull. Amer. Assoc. Petrol. Geol.* **14**:139–152.

Giovannoni, S. J., T. B. Britschgi, C. L. Moyer, and K. G. Field. 1990. Genetic diversity in Sargasso Sea bacterioplankton. *Nature* **345**:60–63.

Givskov, M., L. Eberl, and S. Molin. 1994a. Responses to nutrient starvation in *Pseudomonas putida* KT2442: analysis of general cross-protection, cell shape, and macromolecular content. *J. Bacteriol.* **176**:7–14.

Givskov, M., L. Eberl, and S. Molin. 1994b. Responses to nutrient starvation in *Pseudomonas putida* KT2442: two-dimensional and stress-induced proteins. *J. Bacteriol.* **176**:4816–4824.

Glathe, H., K. H. Knoll, and A. A. M. Makawi. 1963. Die Lebensfähigkeit von *Escherichia coli* in verschiedenen Bodennarten. *Z. Pflanzenerhaehr. Düng. Bodenk.* **100**:142–150.

Glick, M. A. 1980. Substrate capture, uptake, and utilization of some amino acids by starved cells of a psychrophilic marine vibrio. M.S. Thesis. Oregon State University, Corvallis, OR.

Gloxhuber, C. 1974. Toxicological properties of surfactants. *Arch. Toxicol.* **32**:245–270.

Gochenauer, S. E. 1981. Response of soil fungal communities to disturbance. *In The fungal community,* D. T. Wicklow and G. C. Caroll (eds.), pp. 459–479. Marcel Dekker, New York.

Gocke, K., R. Dawson, and G. Liebezeit. 1981. Availability of dissolved free glucose to heterotrophic microorganisms. *Mar. Biol.* **62**:209–216.

Godschalk, G. I., and R. G. Wetzel. 1978. Decomposition of aquatic angiosperms. III. *Zostera marina* L. and a conceptual model of decomposition. *Aquat. Bot.* **5**:329–354.

Gohn, G. S. 1988. Late Mesozoic and early Cenozoic geology of the Atlantic Coastal Plain: North Carolina to Florida. *In The Geology of North America, Vol. 1, The Atlantic Continental Margin,* R. E. Sheridan and J. A. Grow (eds.), pp. 107–130. U. S. Geological Society of America, Denver.

Goksöyr, J. 1975. Decomposition, microbiology and ecosystem analysis. *In Fennoscandian tundra ecosystems,* Pt. 1: *Plants and Microorganisms,* F. E. Wielgolaski (ed.), pp. 230–238. Springer-Verlag, Berlin.

Goldberg, A. L., and A. C. St. John. 1976. Intracellular protein degradation in mammalian and bacterial cells. *Ann. Rev. Biochem.* **45**:747–903.

Goldberg, A. L., K. Olden, and W. F. Prouty. 1973. Studies of the mechanisms and selectivity of protein degradation in *Escherichia coli. In Intracellular Protein Turnover,* R. T. Schimke and N. Katunuma (eds.), pp. 17–55. Academic Press, New York.

Goldman, J. C. 1984. Oceanic nutrient cycles. *In Flows of Energy and Nutrients in Marine Ecosystems: Theory and Practice,* M. J. Fashman (ed.), NATO Conferences series IV, Marine Sciences 13, pp. 137–170. Plenum Press, New York.

Goldman, J. C., and M. R. Dennett. 1985. Susceptibility of some marine phytoplankton species to cell breakage during filtration and post-filtration rinsing. *J. Expt. Mar. Biol. Ecol.* **86**:47–58.

Goldman, J. C., A. D. Caron, and M. R. Dennett. 1987a. Nutrient cycling in a microflagellate foodchain: IV. Phytoplankton-microflagellate interactions. *Mar. Ecol. Prog. Ser.* **24**:231–242.

Goldman, J. C., A. D. Caron, and M. R. Dennett. 1987b. Regulation of gross growth efficiency and ammonium regeneration in bacteria by substrate C:N ratio. *Limnol. Oceanogr.* **32**:1239–1252.

Goldstein, G. 1991. Isolation of *Enterobacter cloacae* from the intestinal remains of 11,000 year old mastodon. *Abstract of the General Meeting of the American Society for Microbiology,* abstr., p. 298. American Society for Microbiology, Washington, DC.

Gomes, S. L., and L. Shapiro. 1984. Differential expression and positioning of chemotaxis methylation proteins in *Caulobacter. J. Molec. Biol.* **178**:551–568.

González, J. M., E. B. Sheer, and B. F. Sheer. 1990. Size-selective grazing on bacteria by natural assemblages of estuarine flagellates and ciliates. *Appl. Environ. Microbiol.* **56**:583–589.

Gooday, A. J., and C. M. Turley. 1990. Responses by benthic organisms to inputs of organic material to the ocean floor: A review. *Philos. Trans. R. Soc. London A.* **331**:119–138.

Goodfellow, M., and C. H. Dickinson. 1985. Delineation and description of microbial populations using numerical methods. *In Computer-Assisted Bacterial Systematics,* M. Goodfellow, D. Jones, and F. G. Priest (eds.), pp. 165–225. Academic Press, London.

Goodrich, T. D. and R. Y. Morita. 1977. Low temperature inhibition on binding, transport, and incorporation of leucine, arginine, methionine and histidine in *Escherichia coli. Ztschr. F. Allg. Mikrobiol.* **17**:91–97.

Gordon, A. L. 1967. Structure of Antarctic waters between 20°W and 170°W. Antarctic Map Folio Ser. 6. Folio American Geographical Society of New York, New York.

Gordon, A. L., and R. D. Goldberg. 1970. Circumpolar characteristics of Antarctic waters. Antarctic Map Folio Ser. 13. Folio American Geographical Society of New York, New York.

Gordon, A. S., and F. J. Millero. 1985. Adsorption mediated decrease in the biodegradation rate of organic compounds. *Microb. Ecol.* **11**:289–298.

Gordon, A. S., S. M. Gerchakov, and I. R. Udey. 1981. The effect of polarization on the attachment of marine bacteria to copper and platinum. *Can. J. Microbiol.* **27**:698–703.

Gordon, A. S., S. M. Gerchakov, and F. J. Millero. 1983. Effects of inorganic particles on metabolism by a periphytic marine bacterium. *Appl. Environ. Microbiol.* **45**:411–417.

Gordon, D. C. 1970. Some studies on the distribution and composition of particulate organic matter in the North Atlantic Ocean. *Deep-Sea Res.* **18**:233–243.

Gorini, L. 1960. Antagonism between substrate and repressor in controlling the formation of a biosynthetic enzyme. *Proc. Natl. Acad. Sci. (USA)* **46**:682–690.

Gottesman, S. 1984. Bacterial regulation: global regulatory networks. *Annu. Rev. Genet.* **18**:415–441.

Gottschal, J. C. 1985. Some reflections on microbial competitiveness among heterotrophic bacteria. *Antonie Van Leeuwenhoek J. Microbiol. Serol.* **51**:473–494.

Gottschal, J. C. 1990. Phenotypic response to environmental changes. *FEMS Microb. Ecol.* **74**:93–102.

Gottschal, J. D. 1992. Substrate capturing and growth in various ecosystems. *J. Appl. Bacteriol. Symp. Suppl.* **72**:39S–48S.

Goulder, R. 1977. Attached and free bacteria in an estuary with abundant suspended solids. *J. Appl. Bacteriol.* **43**:399–405.

Gough, A., R. W. Attwell, D. F. D. Hardy, and R. Caldwell. 1986. *Solid State Technology* **29**:139–142.

Governal, R. A., M. T. Yahya, C. P. Gerba, and F. Shadman. 1992. Comparison of assimilable organic carbon and UV-oxidizable carbon for evaluation of ultrapure-water systems. *Appl. Environ. Microbiol.* **58**:724–726.

Gray, T. R. G. 1976. The survival of vegetative microbes in soil. *Symp. Soc. Gen. Microbiol.* **26**:327–364.

Gray, T. R. G., R. Hissett, and T. Duxbury. 1974. Bacterial populations of litter and soil in a deciduous woodland. II. Numbers, biomass, and growth rates. *Rev. Ecol. Biol. Soil* **11**:15–26.

Gray, T. R. G., and G. Parkinson. 1968. *The Ecology of Soil Bacteria.* University of Toronto Press, Toronto.

Gray, T. R. G., and J. R. Postgate. 1976. Editor's Preface. *Symp. Soc. Gen. Microbiol.* **26**:ix–x.

Gray, T. R. G., and S. T. Williams. 1971. Microbial productivity in soil. *Symp. Soc. Gen. Microbiol.* **21**:255–286.

Greaves, J. M., and L. W. Jones. 1944. The influence of temperature on the microflora of the soil. *Soil Sci.* **58**:377–387.

Greenland, D. J. 1971. Interactions between humic and fulvic acids and clays. *Soil Sci.* **111**:34 41.

Greenwood, D. J. 1968. Measurement of microbial metabolism in soil. *In The Ecology of Soil Bacteria,* T. R. G. Gray and D. Parkinson. (eds.), pp. 138–157. Liverpool University Press, Liverpool.

Gregory, P. H. 1973. *The Microbiology of the Atmosphere,* 2nd ed. John Wiley & Sons, New York.

Griffin, D. M. 1972. *Ecology of Soil Fungi.* Syracuse University Press, Syracuse, New York.

Griffin, G. J., and D. A. Roth. 1979. Nutritional aspects of soil mycostasis. *In Soil-borne Plant Pathogens,* B. Schippers and W. Gams (eds.), pp. 79–96. Academic Press, London.

Griffith, P. C., D. J. Douglas, and S. C. Wainright. 1990. Metabolic activity of size-fractioned microbial plankton in estuarine, nearshore, and continental shelf water of Georgia. *Mar. Ecol. Prog. Ser.* **59**:263–270.

Griffiths, E., and R. G. Burns. 1972. Interaction between phenolic substances and microbial polysaccharides in soil aggregation. *Plant Soil.* **36**:599–612.

Griffiths, R. P., and R. Y. Morita. 1973. Applicability of the reverse-flow filter technique to marine microbial studies. *Appl. Environ. Microbiol.* **26**:687–691.

Griffiths, R. P., J. A. Baross, F. J. Hanus, and R. Y. Morita. 1974. Some physical and chemical parameters affecting the formation and retention of glutamate pools in a marine psychrophilic bacterium. *Ztschr. f. Allg. Mikrobiol.* **14**:359–369.

Grimes, D. J., and R. R. Colwell. 1986. Viability and virulence of *Escherichia coli* suspended by membrane chamber in semitrophical ocean water. *FEMS Microb. Lettr.* **34**:161–165.

Groat, R. G., and A. Matin. 1986. Synthesis of unique polypeptides at the onset of starvation in *Eschericha coli. J. Ind. Microbiol.* **1**:69–73.

Groat, R. G., J. E. Schultz, E. Zychlinsky, A. Bockman, and A. Matin. 1986. Starvation proteins in *Escherichia coli:* kinetics of synthesis and role in starvation survival. *J. Bacteriol.* **168**:486–493. (Erratum 169:3866, 1987).

Gronlund, A. F., and J. J. R. Campbell. 1963. Nitrogenous substrates of endogenous respiration in *Pseudomonas aeruginosa. J. Bacteriol.* **86**:58–66.

Gronlund, A. F., and J. J. R. Campbell. 1965. Enzymatic degradation of ribosomes during endogenous respiration of *Pseudomonas aeruginosa. J. Bacteriol.* **90**:1–7.

Grossman, A. D., W. E. Taylor, Z. F. Burton, R. R. Burgess, and C. A. Gross. 1985. Stringent response in *Escherichia coli* induces expression of heat shock proteins. *J. Molec. Biol.* **186**:357–365.

Grossman, N., and E. Z. Ron. 1989. Apparent minimal size required for cell division in *Escherichia coli. J. Bacteriol.* **171**:80–82.

Gruner, J. W. 1917. Organic matter and the origin of the Biwabick iron-bearing formation. *Econ. Geol.* **17**:407–460.

Guckert, A., M. Valla, and F. Jacquin. 1975. Adsorption of humic acids and soil polysaccharides to montmorillonite. *Pochvovedenye* **2**:41–47.

Guckert, J. B., and D. C. White. 1986. Phospholipid, ester-linked fatty acid analysis in microbial ecology. *In Perspectives in Microbial Ecology,* F. Megusar and M. Gantar (eds.), pp. 455–459. Slovene Society for Microbiology, Ljubljana, Yugoslavia.

Guckert, J. B., C. P. Antworth, P. D. Nichols, and D. C. White. 1985. Phospholipid ester-linked fatty acid profiles as reproducible assays for changes in prokaryotic community structure of estuarine sediments. *FEMS Microbiol. Ecol.* **31**:147–158.

Guckert, J. B., M. A. Hood, and D. C. White. 1986. Phospholipid ester-linked fatty acid profile changes during nutrient deprivation of *Vibrio cholerae:* Increase in *cis/trans* ratio and porportions of cyclopropyl fatty acids. *Appl. Environ. Microbiol.* **52**:794–801.

Güde, H. 1984. Test for validity of different radioisotopes activity measurements by microbial pure and mixed cultures. *Arch. Hydrobiol.* **19**:257–266.

Gunkel, W. 1972. Organic substances. *In Marine Ecology,* vol. 1, pt. 3, O. Kinne (ed.), pp. 1533–1549. Wiley-Interscience, London.

Gurijala, K. R., and M. Alexander. 1990. Explanation for the decline of bacteria introduced into lake water. *Microb. Ecol.* **20**:231–244.

Gustafson, L., and J. Paleos. 1971. Interactions responsible for the selective adsorption of organics on organic surfaces. *In Organic Compounds in Aquatic Environments,* S. D. Faust and J. V. Hunter (eds.), pp. 213–238. Marcel Dekker, New York.

Haan, H. de. 1977. Effect of benzoate on microbial decompostion of fulvic acid in Tjeukemeer (The Netherlands). *Limnol. Oceanogr.* **22**:38–44.

Hagström, A. 1984. Aquatic bacteria: Measurements and significance of growth. *In Current Perspectives in Microbial Ecology,* M. J. Klug and C. A. Reddy (eds.), pp. 495–501. American Society for Microbiology, Washington, DC.

Hagström, A., F. Azam, A. Andersson, J. Wikner, and F. Rassoulzadegan. 1988. Microbial loop in an oligotrophic pelagic marine ecosystem: possible roles of cyanobacteria and nanoflagellates in the organic fluxes. *Mar. Ecol. Prog. Ser.* **49**:171–178.

Hagström, A., J. W. Ammerman, S. Henrichs, and F. Azam. 1984. Bacterioplankton growth in seawater. II. Organic matter utilization during steady-state growth in seawater cultures. *Mar. Ecol. Prog. Ser.* **18**:41–48.

Hagström, A., U. Larsson, P. Hörstedt, and S. Normark. 1979. Frequency of dividing cells, a new approach to the determination of bacteria growth rates in aquatic environments. *Appl. Environ. Microbiol.* **37**:805–812.

Hahn, J. 1980. Organic constituents of natural aerosols. *Ann. N.Y. Acad. Sci.* **338**:359–376.

Haider, K. M., and J. P. Martin. 1968. The role of microorganisms in the formation of humic acids. *In Isotopes and Radiation in Soil Organic-Matter Studies,* pp. 189–196. IAEA Proc. Series, Vienna.

Haidler, K. 1992. Problems related to the humification processes in soils of temperate climates. *Soil Biochem.* **7**:55–94.

Haight, R. D., and R. Y. Morita. 1966. Thermally induced leakage from *Vibrio marinus,* an obligately psychrophilic marine bacterium. *J. Bacteriol.* **92**:1388–1393.

Haldeman, D. L., and P. S. Amy. 1993a. Diversity within a colony morphotype: Implications for ecological research. *Appl. Environ. Microbiol.* **59**:933–935.

Haldeman, D. L., and P. S. Amy. 1993b. Bacterial heterogenicity in deep subsurface tunnels at Rainier Mesa, Nevada Test Site. *Microbl. Ecol.* **25**:183–194.

Haldeman, D. L., P. S. Amy, D. Ringelberg, and D. C. White. 1993. Characterization of microbiology within a 21 m³ section of rock from the deep subsurface. *Microb. Ecol.* **26**:145–159.

Haldeman, D. L., B. J. Pitonzo, S. P. Story, and P. S. Amy. 1994. Comparison of the microbiota recovered from surface and deep subsurface rock, water, and soil along an elevational gradient. *Geomicrobiology* **12**:99–111.

Hale, J. M., and W. V. Halversen. 1940. The enumeration of ammonia-oxidizing bacteria in forest litter. *J. Bacteriol.* **39**:100–101.

Hall, G. H., J. G. Jones, R. W. Pickup, and B. M. Simon. 1990. Methods to study the bacterial ecology of freshwater environment. *In Methods in Microbiology,* vol. 22, R. Grigorova and J. R. Norris (eds.), pp. 181–209. Academic Press, London.

Hallum, M. J., and W. V. Bartholomew. 1953. Influence of rate of plant residue addition in accelerating the decomposition of soil organic matter. *Proc. Soil Sci. Soc. Am.* **17**:365–368.

Hamilton, R. D., and A. F. Carlucci. 1966. Use of ultra-violet-irradiated seawater in the preparation of culture media. *Nature* **211**:483–484.

Hamilton, R. D., and O. Holm-Hansen. 1967. Adenosine triphosphate content of marine bacteria. *Limnol. Oceanogr.* **12**:319–324.

Hamilton, R. D., K. M. Morgan, and J. D. H. Strickland. 1966. The glucose uptake kinetics of some marine bacteria. *Can. J. Microbiol.* **12**:995–1003.

Hanson, R. B., and H. K. Lower. 1983. Nucleic acid synthesis in oceanic microplankton from Drake Passage, Antarctica: An evaluation of the steady state. *Mar. Biol.* **73**:79–89.

Hanson, R. B., and W. S. Gardner. 1978. Uptake and metabolism of two amino acids by anaerobic microorganisms in four diverse salt-marsh soils. *Mar. Biol.* **46**:101–107.

Hanson, R. B., and K. R. Tenore. 1981. Microbial metabolism and incorporation by the polychaete *Capitella capitata* of aerobically and anaerobically decomposed detritus. *Mar. Ecol. Prog. Ser.* **6**:299–307.

Harada, T., and R. Hayashi. 1968. Studies on the organic nitrogen becoming decomposable through the effect of drying a soil. *Soil Sci. Plant Nutri.* **14**:13–19.

Harder, W., and M. M. Atwood. 1978. Biology, physiology and biochemistry of Hyphomicrobia. *Adv. Microb. Physiol.* **16**:303–356.

Harder, W., and L. Dijkhuizen. 1982. Strategies of mixed substrate utilization in microorganisms. *Philos. Trans. R. Soc. London Ser. B* **297**:459–479.

Harder, W., and L. Dijkhuizen. 1983. Physiological responses to nutrient limitation. *Ann. Rev. Microbiol.* **37**:1–23.

Harder, W., and H. Veldkamp. 1971. Competition of marine psychrophilic bacteria at low temperature. *Antonie von Leeuwenhoek J. Microbiol. Serol.* **37**:51–63.

Harder, W., L. Dijkhuizen, and H. Veldkamp. 1984. Environmental regulation of microbial metabolism. *Symp. Soc. Gen. Microbiol.* **36** (pt. 2):51–96.

Hargrave, B. T. 1970. The effect of a deposit-feeding amphipod on the metabolism of benthic microflora. *Limnol. Oceanogr.* **15**:21–30.

Harley, C. B., and R. P. Reynolds. 1987. Analysis of the *E. coli* promoter sequences. *Nucleic Acids Res.* **15**:2343–2361.

Harley, J. L. 1971. Fungi in ecosystems. *J. Ecol.* **59**:653–668.

Harold, F. M. 1986. *The Vital Force: A Study of Bioenergetics.* W. H. Freeman and Co., New York.

Harowitz, A., M. I. Krichevsky, and R. M. Atlas. 1983. Characteristics and diversity of subarctic marine oligotrophic, stenoheterotrophic, and euryheterotrophic bacterial populations. *Can. J. Microbiol.* **29**:527–535.

Harris, C. M. and D. B. Kell. 1985. The estimation of microbial biomass. *Biosensors* **1**:17–84.

Harris, N. D. 1963. The influence of the recovery medium and the incubation temperature on the survival of damaged bacteria. *J. Appl. Bacteriol.* **26**:387–397.

Harris, R. H., and R. Mitchell. 1973. The role of polymers in microbial aggregation. *Ann. Rev. Microbiol.* **27**:27–50.

Harrison, A. F., D. D. Harkness, and P. J. Bacon. 1989. The use of bomb-^{14}C for studying organic matter and N and P dynamics in a woodland soil. *In Nutrient Cycling in Terrestrial Ecosystems. Field Methods, Appliction and Interpretation,* A. F. Harrison, P. Ineson and O. W. Heal (eds.), pp. 246–258. Elsevier Applied Science, London.

Harrison, A. P., Jr. 1960. The response of *Bacterium lactis aerogenes* when held at growth temperature in the absence of nutrients: An analysis of survival curves. *Proc. Roy. Soc. B* **152**:418–428.

Harrison, A. P., Jr. 1961. Ageing and decline of bacteria in phosphate buffer. *Bacteriol. Proc. A* **10**:53.

Harrison, A. P., Jr., and F. R. Lawrence. 1963. Phenotypic, genotypic, and chemical changes in starving populations of *Aerobacter aerogenes J. Bacteriol.* **85**:742–750.

Harrison, F. C., and M. E. Kennedy. 1922. The red discoloration of cured codfish. *Trans. Roy. Soc. Can.* **15**:101–152.

Hartke, A., S. Bouche, X. Gansel, P. Boutibonnes, and Y. Auffray. 1994. Starvation-induced stress resistance in *Lactococcus lactis* subsp. *lactis* IL1403. *Appl. Environ. Microbiol.* **60**:3474–3478.

Harvey, R. H., and D. A. Boran. 1985. Geochemistry of humic substances in seawater. *In Humic Substances in Soil, Sediment and Water,* G. R. Aiken, D. M. McKnight, R. L. Wershaw, and P. MacCarthy (eds.), pp. 233–247. John Wiley & Sons, New York.

Harwood, C. S., and E. Canale-Parola. 1981. Branched-chain amino acid fermentation by a marine spirochete: Strategy for starvation survival. *J. Bacteriol.* **148**:109–116.

Haseltine, W. A., and R. Block. 1973. Synthesis of guanosine tetraphosphate and pentaphosphate requires the presence of a codon-specific uncharged transfer ribonucleic acid in the acceptor site of ribosomes. *Proc. Natl. Acad. Sci. (USA)* **70**:1564–1568.

Hashimoto, T., and T. Hattori. 1989. Grouping of soil bacteria by analysis of colony formation on agar plates. *Biol. Fertil. Soil* **7**:198–201.

Hassett, D. J., M. S. Bisesi, and R. Hartenstein. 1987. Bactericidal action of humic acids. *Soil Biol. Biochem.* **19**:111–113.

Hattori, T. 1973. *Microbial Life in the Soil.* Marcel Dekker, New York.

Hattori, T. 1976. Plate count of bacteria in soil on a diluted nutrient broth as a culture medium. *Rept. Inst. Agric. Res. Tohoku University* **27**:23–30.

Hattori, T. 1980. A note on the effect of different types of agar on plate count of oligotrophic bacteria in soil. *J. Gen. Appl. Microbiol.* **26**:373–374.

Hattori, T. 1984. Physiology of soil oligotrophic bacteria. *Microbiol. Sci.* **1**:102–104.

Hattori, T. 1986. Method of isolation of microorganisms in different physiological states. *In Microbial Communities,* V. Jensen, A. Kjøller, and L. H. Sørensen (eds.), pp. 163–176. Elsevier Applied Science Publication, London.

Hattori, T., and C. Furusaka. 1960. Chemical activities of *E. coli* adsorbed on a resin. *J. Biochem.* **48**:831–837.

Hattori, T., and C. Furusaka. 1961. Chemical activities of *Azotobacter agile* adsorbed on a resin. *J. Biochem.* **50**:312–315.

Hattori, R. and T. Hattori. 1963. Effect of a liquid-solid interface on the life of microorganisms. *Ecological Rev.* **16**:63–70.

Hattori, R., and T. Hattori. 1976. The physical environment in soil microbiology: an attempt to extend principles of microbiology to soil microorganisms. *Crit. Rev. Microbiol.* **4**:423–461.

Hattori, R., and T. Hattori. 1987. Interaction of microorganisms with a charged surface— a model experiment. *Rept. Inst. Agricul. Res., Tohoku University* **36**:21–67.

Hayashi, R., and T. Harada. 1969. Characterization of the organic nitrogen becoming decomposable through the effect of drying of a soil. *Soil Sci. Plant Nutri.* **15**:226–234.

Hayes, W. 1968. *The Genetics of Bacteria and Their Viruses,* 2nd ed. Blackwell Scientific Publication, London.

Hazelbauer, G. L., and J. Adler. 1971. Role of galactose binding protein in chemotaxis of *Escherichia coli* toward galactose. *Nat. New Biol.* **230**:101–104.

Hazelbauer, G. L., and J. S. Parkinson. 1977. Bacterial chemotaxis. *In Receptors and Recognition,* ser. B., vol. 3, J. S. Reissig (ed.), pp. 59–98. Chapman and Hall, London.

Hazen, T. C., L. Jiménez, G. L. de Victoria, and C. B. Fliermans. 1991. Comparison of bacteria from deep subsurface sediment and adjacent groundwater. *Microb. Ecol.* **22**:293–304.

Hazen, T. C., M. I. Raker, and G. W. Esch. 1979. Chemotaxis of *Aeromonas hydrophila.* *Abstr. Annu. Mtg. Soc. Amer. Microbiol.* I–26.

Hedges, J., C. Lee, and P. Wangersky. 1993. Comments from the editors on the Suzuki statement. *Mar. Chem.* **41**:289–290.

Hedges, J. I. 1987. Organic matter in sea water. *Nature* **330**:205–206.

Hedges, J. I., and P. E. Hare. 1987. Amino acid adsorption by clay minerals in distilled water. *Geochim. Cosmochim. Acta* **51**:255–259.

Hedrick, D. B., A. Vass, B. K. Richards, W. J. Jewell, J. B. Guckert, and D. C. White. 1991a. Starvation and overfeeding stress on microbial activities in high-solids high–yield methanogenic digesters. *Biomass Bioenergy* **1**:75–82.

Hedrick, D. B., B. Richards, W. Jewell, J. B. Guckert, and D. C. White. 1991b. Disturbance, starvation, and overfeeding stresses detected by microbial lipid biomarkers in high-solid high-yield methanogenic reactors. *J. Ind. Microbiol.* **8**:91–98.

Heinmets, F. 1953. Reactivation of ultraviolet inactivated *Escherichia coli* by pyruvate. *J. Bacteriol.* **66**:455–457.

Heinmets, F., J. J. Lehman, W. W. Taylor, and R. H. Kathan. 1954a. The study of factors which influence metabolic reactivation of the ultraviolet inactivated *Escherichia coli. J. Bacteriol.* **67**:511–522.

Heinmets, F., W. W. Taylor, and J. J. Lehman. 1954b. The use of metabolites in the restoration of the viability of heat and chemically inactivated *Escherichia coli. J. Bacteriol.* **67**:5–14.

Heise, S., and W. Reichardt. 1991. Anaerobic starvation survival of marine bacteria. *Kiel. Meeresforsch.* **8**:97–101.

Hellebust, J. A. 1965. Excretion of some organic compounds by marine phytoplankton. *Limnol. Oceanogr.* **10**:192–206.

Helmke, E., and H. Weyland. 1991. Effect of temperature on extracellular enzymes occurring in permanently cold marine environments. *Kieler Meeresforsch. Sonderh.* **8**:198–204.

Hendricks, C. W. 1967. Multiplication and growth of selected enteric bacteria in clear mountain stream water. *Water Res.* **1**:567–576.

Hendriksen, K., J. I. Hansen, and T. H. Blackburn. 1980. The influence of benthic infauna on exchange rates of inorganic nitrogen between sediment and water. *Ophelia* (Suppl. 1):249–256.

Hengge-Aronis, R. 1993. Survival of hunger and stress: the role of *rpoS* in early stationary phase gene regulation in *E. coli. Cell* **72**:165–168.

Hengge-Aronis, R., W. Klein, R. Lange, M. Rimmele, and W. Boos. 1991. Trehalose synthesis genes are controlled by the putative σ factor encoded by *rpoS* and are involved in stationary-phase thermotolerance in *Escherichia coli. J. Bacteriol.* **173**:7918–7924.

Hengge-Aronis, R., R. Lange, N. Henneberg and D. Fischer. 1993. Osmotic regulation of *rpoS*-dependent genes in *Escherichia coli. J. Bacteriol.* **175**:259–265.

Henis, Y., K. R. Gurijala, and M. Alexander. 1989. Factors involved in multiplication and survival of *Escherichia coli* in lake water. *Microb. Ecol.* **17**:171–180.

Hennes, K. P., and M. Simon. 1995. Significance of bacteriophages for controlling bacterioplankton growth in a mesotrophic lake. *Appl. Environ. Microbiol.* **61**:333-340.

Henrichs, S. M. 1980. Biogeochemistry of dissolved amino acids in marine sediment.

Ph.D. Thesis. Massachusetts Institute of Technology and Woods Hole Institution. WHOI Report No 80–39.

Henrichs, S. M., and J. W. Farrington. 1979. Amino acids in interstitial waters of marine sediments. *Nature* **279**:319–322.

Henrici, A. T. 1928. *Morphological Variation and the Rate of Growth of Bacteria.* Charles C. Thomas, Springfield, IL.

Henrici, A. T., and D. E. Johnson. 1935. Studies on freshwater bacteria II. Stalked bacteria; a new order of Schizomycetes. *J. Bacteriol.* **30**:61–93.

Herbert, D. 1976. Stoichiometric aspects of microbial growth. *In Continuous Culture 6: Applications and New Fields,* A. C. Denn, D. C. Ellwood, C. G. T. Evans, and J. Melling (eds.), pp. 1–30. Ellis Harwood, Chichester.

Herbert, R. A. 1990. Methods for enumerating microorganisms and determining biomass in natural environments. *In Methods in Microbiology,* vol. 22, R. Grigorova and J. R. Norris (eds.), pp. 1–39. Academic Press, New York.

Herman, D. C., and J. W. Costerton. 1993. Starvation-survival of a *p*-nitrophenol-degrading bacterium. *Appl. Environ. Microbiol.* **59**:340–343.

Hermansson, M., and K. C. Marshall. 1985. Utilization of surface-localized substrate by non-adhesive marine bacteria. *Microb. Ecol.* **11**:91–105.

Hernandez-Chico, C., J. L. San Millan, R. Kolter, and F. Moreno. 1986. Growth phase and OmpR regulation of transcription of microcin B17 genes. *J. Bacteriol.* **167**:1058–1065.

Herskowitz, I., and D. Hagen. 1980. The lysis-lysogeny decision of phage lambda: explicit programming and responsiveness. *Ann. Rev. Genet.* **14**:399–445.

Hespell, R. B., and M. Mertens. 1978. Effect of nucleic acid compounds on viability and cell composition of *Bdellovibrio bacteriovorus* during starvation. *Arch. Microb.* **116**:151–159.

Hespell, R. B., M. R. Thomashow, and S. C. Rittenberg. 1974. Changes in cell composition and viability of *Bdellovibrio bacteriovorus* during starvation. *Arch. Microbiol.* **97**:313–327.

Heukelekian, H., and A. Heller. 1940. Relation between food concentration and surface for bacterial growth. *J. Bacteriol.* **40**:547–558.

Hicks, R. J., and J. K. Fredrickson. 1989. Aerobic metabolic potential of microbial populations indigenous to deep subsurface environments. *Geomicrob. J.* **7**:67–78.

Higgins, C. F., J. C. D. Hinton, C. S. J. Julton, T. Owen-Hughes, G. D. Pavitt, and A. Sierafi. 1990. Protein H1: a role for chromatin structure in the regulation of bacterial gene expression and virulence. *Mol. Microbiol.* **4**:2007–2012.

Hines, M. E., and G. E. Jones. 1985. Microbial biogeochemistry and bioturbation in the sediments of Great Bay, New Hampshire. *Estuar. Coast Shelf Sci.* **20**:719–742.

Hines, M. E., W. H. Orem, W. B. Lyons, and G. E. Jones. 1982. Microbial activity and bioturbation-induced oscillations in pore water chemistry of estuarine sediments in spring. *Nature* **299**:433–435.

Hinga, K. R., J. McN. Sieburth, and G. R. Heath. 1979. The supply and use of organic material at the deep sea floor. *J. Mar. Res.* **37**:557–579.

Hinton, H. E. 1968. Reversible suspension of metabolism and the origin of life. *Proc. Roy. Soc. (London) Series B* **171**:43–56.

Hippe, H. 1965. Wiederverwertung von Poly-ß-hydroxybuttersäure durch Knallgasbakterien. *Zentr. Bakteriol. Parasitenk. Abt. I.* **198**:321–323.

Hirsch, P. 1964. Oligocarbophilie Wachstum auf Koisten von Luftverunreinigungen bei Mycobakterien und einigen ihnen nahestehenden Actinomyceten. *Zentralbl. Bakteriol. Parasitenkd. Infekionskr. Abt. I.* **194**:70–82.

Hirsch, P. 1968. Gestielte und knospende Bakterien. Spezialisten für C_1-Stoffwechsel an nährstoffarmen Standorten. *Mitt. Int. Ver. Limnol.* **14**:52–63.

Hirsch, P., and E. Rades-Rohkohl. 1983. Microbial diversity in a groundwater aquifer in northern Germany. *Dev. Ind. Microbiol.* **24**:183–200.

Hirsch, P., and S. F. Conti. 1964a. Biology of budding bacteria. I. Enrichment, isolation and morphology of *Hyphomicrobium* spp. *Arch. Mikrobiol.* **48**:339–357.

Hirsch, P., and S. F. Conti. 1964b. Biology of budding bacteria. II. Growth and nutrition of *Hyphomicrobium* spp. *Arch. Mikrobiol.* **48**:358–367.

Hirsch, P., S. S. Cohen, J. C. Ensign, W. H. Jannasch, A. L. Koch, K. C. Marshall, A. Matin, J. S. Poindexter, S. C. Rittenberg, D. C. Smith, and H. Veldkamp. 1979. Life under conditions of low nutrients, Group Report. *In Strategies of Microbial Life in Extreme Environments,* M. Shilo (ed.), pp. 357–372. Verlag Chemie, Weinheim and New York

Hirte, W. 1962. Einige Untersuchungen zue Methodik der mikrobiol. Bodenverarbeitung. *Zentral. Bakt. und Hygiene, II. Abt.* **115**:394–403.

Hissett, R., and T. G. R. Gray. 1976. Microsites and time changes in soil microbe ecology. *In The Role of Terrestrial and Aquatic Organisms in Decomposition Processes,* J. M. Anderson and A. MacFadyen (eds.), pp. 23–39. Blackwell, Oxford.

Hiura, K., T. Hattori, and C. Furusaka. 1976. Bacteriological studies on the mineration of organic nitrogen in paddy soils. I. Effect of mechanical distruption of soils on ammonification and bacterial number. *Soil Sci. Plant Nutr. (Tokyo)* **22**:459–465.

Ho, K. P., and W. J. Payne. 1979. Assimilation efficiency and energy content of prototrophic bacteria. *Biotechnol. Bioeng.* **21**:787–802.

Hobbie, J. E., and T. E. Ford. 1993. A perspective on the ecology of aquatic microbes. *In Aquatic Microbiology: An Ecological Approach,* T. E. Ford (ed.), pp. 1–14. Blackwell Scientific Publications, Boston.

Hobbie, J. E., R. J. Daley, and S. Jasper. 1977. Use of nucleopore filters for counting bacteria by fluorescent microscopy. *Appl. Environ. Microbiol.* **33**:1225–1228.

Hobot, J. A., E. Carleman, W. Villiger, and E. Kellenberger. 1984. Periplasmic gel: new concept resulting from reinvestigation of bacterial cell envelope ultrastructure by new methods. *J. Bacteriol.* **160**:143–152.

Hodson, R. E., F. Azam, A. Carlucci, J. A. Fuhrman, D. M. Karl, and O. Holm-Hansen. 1981a. Microbial uptake of dissolved organic matter in McMurdo Sound, Antarctica. *Mar. Biol.* **61**:89–94.

Hodson, R. E., A. E. Maccubbin, and L. R. Pomeroy. 1981b. Dissolved adenosine tri-phosphate utilization by free-living and attached bacterioplankton. *Mar. Biol.* **64**:43–51.

Hoff, K. A. 1989. Survival of *Vibrio anguillarum* and *Vibrio salmonicida* at different salinities. *Appl. Environ. Microbiol.* **55**:1775–1786.

Höfle, M. F. 1988. Bacterial community structure and dynamics after large-scale release of nonindigenous bacteria as revealed by low-molecular-weight-RNA analysis. *Appl. Environ. Microbiol.* **58**:3387–3394.

Höfle, M. G. 1982. Glucose uptake of *Cytophage johnsonae* studied in batch and chemostat culture. *Arch. Microb.* **133**:289–294.

Höfle, M. G. 1983. Long-term changes in chemostat cultures of *Cytophaga johnsonae*. *Appl. Environ. Microbiol.* **46**:1045–1053.

Höfle, M. G. 1984. Transient responses of glucose-limited cultures of *Cytophaga john-sonae* to nutrient excess and starvation. *Appl. Environ. Microbiol.* **47**:356–362.

Hogg, R. W., and E. Englesberg. 1969. L-arabinose binding protein from *Escherichia coli* B/r. *J. Bacteriol.* **100**:423–432.

Holden, J. T. 1958. Degradation of intracellular nucleic acid and leakage of fragments by *Lactobacillus arabinosus*. *Biochim. Biophys. Acta* **29**:667–668.

Hollibough, J. T. 1976. The biological degradation of arginine and glutamic acid in sea-water in relation to the growth of phytoplankton. *Mar. Biol.* **36**:303–312.

Hollibough, J. T. 1978. Nitrogen regeneration during the degradation of several amino acids by plankton communities collected near Halifax, Nova Scotia, Canada. *Mar. Biol.* **45**:191–201.

Hollibough, J. T. 1979. Metabolic adaptation in natural bacterial populations supplemented with selected amino acids. *Estuar. Coastal Mar. Sci.* **9**:215–230.

Hollibough, J. T. 1988. Limitations of the [^{3}H]thymidine method for estimating bacterial productivity due to thymidine metabolism. *Mar. Ecol. Prog. Ser.* **43**:19–30.

Holmes, R. W., P. M. Williams, and R. W. Eppley. 1967. Red water in La Jolla Bay, 1964–1966. *Limnol. Oceanogr.* **12**:503–512.

Holm-Hansen, O., and H. W. Paerl. 1972. The applicability of ATP determination for estimation of microbial biomass and metabolic activity. *Mem. Ist. Ital. Idrobiol.* **29** Suppl.:149–168.

Holmquist, L., Å. Jouper-Jaan, D. Weichart, D. R. Nelson, and S. Kjelleberg. 1993. The induction of stress proteins in three marine *Vibrio* during carbon starvation. *FEMS Microbiol. Ecol.* **12**:185–194.

Holt, J. G, and N. R. Krieg. 1994. Enrichment and isolation. In *Methods for General and Molecular Bacteriology*, P. Gerhardt (ed.), pp. 179–215. American Society for Micro-biology, Washington, D.C.

Honeyman, B. D., and P. H. Santschi. 1989. A Browning pumping model for oceanic trace metal scavenging: evidence from Th isotopes. *J. Mar. Res.* **47**:951–992.

Honjo, S. 1978. Sedimentation of materials in the Sargasso Sea at a 5367 m deep station. *J. Mar. Sci.* **36**:469–492.

Honjo, S. 1990. Ocean particles and fluxes of material to the interior of the deep ocean: the Azoic theory 120 years later. *In Facets of Modern Biogeochemistry*, V. Ittekkot, S. Kempe, W. Michaelis and A. Spitzy (eds.), pp. 62–73. Springer-Verlag, Berlin.

Honjo, S., and M. R. Roman. 1978. Marine copepod fecal pellets: Production preservation and sedimentation. *J. Mar. Res.* **36**:45–57.

Hood, D. W. 1970. *Organic Matter in Natural Waters*. Institute of Marine Science, Occasional Publications, No. 1. University of Alaska, College, AK.

Hood, D. W., and T. C. Loder. 1973. Microbial conversion of dissolved organic carbon compounds in sea water. *In The Aquatic Environment: Microbial Transformations and Water Management Implications*, L. J. Guarraia and R. K. Ballentine (eds.). pp. 211–234. Superintendent of Documents, U. S. Government Printing Office. Washington, D. C.

Hood, M. A., and M. T. MacDonell. 1987. Distribution of ultramicrobacteria in a Gulf Coast Estuary and induction of ultramicrobacteria. *Microb. Ecol.* **14**:113–127.

Hood, M. A., and G. E. Ness. 1982. Survival of *Vibrio cholerae* and *Escherichia coli* in estuarine waters and sediments. *Appl. Environ. Microbiol.* **43**:578–584.

Hood, M. A., J. B. Guckert, D. C. White, and F. Deck. 1986. Effect of nutrient deprivation on lipid, carbohydrates, DNA, RNA, and protein levels in *Vibrio cholerae*. *Appl. Environ. Microbiol.* **52**:788–793.

Hopkinson, C. C., Jr., B. Sherr, and W. J. Wiebe. 1989. Size fractioned metabolism of coastal microbial plankton. *Mar. Ecol. Prog. Ser.* **51**:155–166.

Hoppe, H.-G. 1976. Determination and properties of actively metabolizing heterotrophic bacteria in the sea, investigated by means of autoradiography. *Mar. Biol.* **36**:291–302.

Hoppe, H.-G. 1978. Relations between active bacteria and heterotrophic potential in the sea. *Netherlands J. Sea Res.* **12**:78–98.

Hoppe, H.-G. 1984. Attachment of bacteria: advantage or disadvantage for survival in the aquatic ecosystem. *In Microbial Adhesion and Aggregation*, K. C. Marshall (ed.), pp. 283–301. Dahlem Konferenzen, Springer-Verlag.

Hoppe, H.-G., H. Ducklow, and B. Karrasch. 1993. Evidence of dependency of bacterial growth on enzymatic hydrolysis of particulate organic matter in the mesopelagic ocean. *Mar. Ecol. Prog. Ser.* **93**:277–283.

Hoppe, H.-G., S.-J. Kim, and K. Gocke. 1988. Microbial decomposition in aquatic environments: combined process of extracellular enzyme activity and substrate uptake. *Appl. Environ. Microbiol.* **54**:784–790.

Hopton, J. W., U. Melchiorri-Santolini, and Y. I. Sorokin. 1972. Enumeration of viable cells of micro-organisms by plate count technique. *In Techniques for the Assessment of Microbial Production and Decomposition in Fresh Waters*, IBP Handbook No. 23, Y. I. Sorokin and H. Kadota (eds.), pp. 59–63. Blackwell Scientific Publications, Oxford.

Horan, N. J., M. Midgley, and E. A. Dawes. 1981. Effect of starvation on transport, membrane potential and survival of *Staphylococcus epidermidis* under anaerobic conditions. *J. Gen. Microbiol.* **127**:223–230.

Horowitz, A., M. I. Krichevsky, and R. M. Atlas. 1983. Characteristics and diversity of subarctic marine oligotrophic, stenoheterotrophic, and euryheterotrophic bacterial population. *Can. J. Microbiol.* **29**:527–535.

Horvath, R. 1972. Microbial co-metabolism and degradation of organic compounds in nature. *Bacteriol. Rev.* **36**:147–155.

Howarth, R. W. 1988. Nutrient limitation of net primary production in marine ecosystems. *Ann. Rev. Ecol.* **19**:89–110.

Howarth, R. W., and J. M. Teal. 1979. Sulfate reduction in a New England salt marsh. *Limnol. Oceanogr.* **24**:999–1013.

Houwink, A. L. 1955. Caulobacter—its morphologenesis, taxonomy and parasitism. *Antonie van Leeuwenhoe J. Microbiol. Serol.* **21**:49–64.

Hughes, M. N., and R. K. Poole. 1989. *Metals and Micro-organisms.* Chapman & Hall, London.

Huisman, G. W., and R. Kolter. 1994. Sensing starvation: a homoserine lactone-dependent signalling pathway in *Escherichia coli. Science.* **265**:537–539.

Humphrey, B. A., and K. C. Marshall. 1984. The triggering effect of surfaces and surfactant on heat output, oxygen consumption and size reduction of a starving marine *Vibrio. Arch. Microbiol.* **140**:166–170.

Humphrey, B. A., M. R. Dickson, and K. C. Marshall. 1979. Physicochemical and in situ observations on the adhesion of gliding bacteria to surfaces. *Arch. Microbiol.* **120**:231–238.

Humphrey, B. A., S. Kjelleberg, and K. C. Marshall. 1983. Responses of marine bacteria under starvation conditions at a solid-water interface. *Appl. Environ. Microbiol.* **45**:43–47.

Hunt, H. W. 1977. A simulation model for decomposition in grasslands. *Ecology* **58**:469–484.

Hunt, H. W., C. V. Cole, and E. T. Elliott. 1985. Models for growth of bacteria inoculated in sterilized soil. *Soil. Sci.* **139**:156–165.

Hussong, D., R. R. Colwell, M. O'Brian, A. D. Weiss, A. D. Pearson, R. M. Weiner, and W. D Burge. 1987. Viable *L. pneumonphila* not detectable by culture on agar media. *Biotech.* **5**:947–950.

Hutchinson, G. E. 1973. Eutrophication, past and present. *In Eutrophication: causes, consequences, corrections,* pp. 17–26. National Academy of Science, Washington, D.C.

Hylleberg, J., and K. Henriksen. 1980. The central role of bioturbance in sediment mineralization and element re-cycling. *Ophelia* (Suppl. 1):1–16.

Iacobellis, N.S., and J. E. DeVay. 1986. Long-term storage of plant-pathogenic bacteria in sterile distilled water. *Appl. Environ. Microbiol.* **52**:388–389.

Il'in, N. P., and D. S. Orlov. 1973. Photochemical destruction of humic acids. *Soviet Soil Sci.* **5**:73–81.

Ingraham, J. L., O. Maaloe, and F. C. Neidhardt. 1983. *Growth of the Bacterial Cell.* Sinauer Associates, Inc., Sunderland, MA.

Iriberri, J., M. Unannue, I. Barcina, and L. Egea. 1987. Seasonal variation in population density and heterotrophic activity of attached and free-living bacteria in coastal waters. *Appl. Environ. Microbiol.* **53**:2308–2314.

Ishida, Y., and H. Kadota. 1974. Ecological studies on bacteria in the sea and lake waters polluted with organic substances. I. Responses of bacteria to different concentrations of organic substances. *Bull. Jpn. Soc. Sci. Fish.* **40**:999–1005.

Ishida, Y., and H. Kadota. 1975. A comparison between viable count and direct count of bacteria in polluted sea water. *Bull. Jpn. Soc. Sci. Fish.* **41**:271.

Ishida, Y., and H. Kadota. 1977. Distribution of oligotrophic bacteria in Lake Mergozzo. *Bull. Jpn. Soc. Sci. Fish.* **43**:885–892.

Ishida, Y., and H. Kadota. 1979. A new method for enumeration of oligotrophic bacteria in lake water. *Arch. Hydrobiol. Beih.* **12**:77–85.

Ishida, Y., and H. Kadota. 1981a. Growth patterns and substrate requirements of naturally occurring obligate oligotrophs. *Microb. Ecol.* **7**:123–130.

Ishida, Y., and H. Kadota. 1981b. Obligately oligotrophic bacteria in Lake Biwa. *Verh. Internat. Verin. Limnol.* **21**:552–555.

Ishida, Y., A. Uchida, and H. Kadota. 1977. Ecological studies on bacteria in the sea and lake waters polluted with organic substances. IV. Determination of bacteria; degradable organic matter in aquatic environments. *Bull. Jpn. Soc. Sci. Fish.* **43**:885–892.

Ishida, Y., K. Shibahara, H. Uchida, and H. Kadota. 1980. Distribution of obligately oligotrophic bacteria in Lake Biwa. *Bull. Jpn. Soc. Sci. Fish.* **46**:1151–1158.

Ishida, Y., M. Eguchi, and H. Kadota. 1986. Existence of obligately oligotrophic bacteria as dominant population in the south China Sea and the west Pacific Ocean. *Mar. Ecol. Prog. Ser.* **30**:197–203.

Ishida, Y., K. Fukami, M. Eguchi, and I. Yoshinaga. 1989. Strategies for growth of oligotrophic bacteria in the pelagic environment. *In Recent Advances in Microbial Ecology,* Hattori, T., Y., Ishida, Y. Maruyama, R. Y. Morita, and A. Uchida (eds.), pp. 89–93. Japan Scientific Societies Press, Tokyo.

Issatchenko, V. 1940. On the microorganisms of the lower limits of the biosphere. *J. Bacteriol.* **40**:379–381.

Ittekkot, V. 1988. Global trends in the nature of organic matter in river suspensions. *Nature* **332**:436–438.

Ittekkot, V., and B. Haake. 1990. The terrestrial link in the removal of organic carbon in the sea. *In Facets of Modern Biogeochemistry,* V. Ittekkot, S. Kempe, W. Michaelis, and A. Spitzy (eds.), pp. 318–325. Springer-Verlag, Berlin.

Ittekkot, V., U. Brockmann, W. Michaelis, and E. T. Degans. 1981. Dissolved free and combined carbohydrates during a phytoplankton bloom in the northern North Sea. *Mar. Ecol. Prog. Ser.* **4**:299–305.

Iturriaga, R., and H.-G. Hoppe. 1977. Observations of heterotrophic activity on photoassimilated organic matter. *Mar. Biol.* **40**:100–108.

Iturriaga, R., and A. Zsolnay. 1981. Transformation of some dissolved organic compounds by a natural heterotrophic population. *Mar. Biol.* **62**:125–129.

Ivanova, A., M. Renshae, R. V. Guntaka, and A. Eisenstark. 1992. DNA base sequence variability in *katF* (putative sigma factor) gene of *Escherichia coli. Nucl. Acids Res.* **20**:5479–5480.

Iwashima, A., A. Matsumura, and Y. Nose. 1971. Thiamine-binding of *Escherichia coli. J. Bacteriol.* **108**:1419–1421.

Jackson, G. A. 1987a. Physical and chemical properties of aquatic environments. *Symp. Soc. Gen. Microbiol.* **41**:213–233.

Jackson, G. A. 1987b. Simulating chemosensory responses of marine microorganisms. *Limnol. Oceanogr.* **32**:1253–1266.

Jackson, G. A. 1989. Stimulation of bacterial attraction and adhesion to falling particles in an aquatic environment. *Limnol. Oceanogr.* **34**:514–530.

Jackson, G. A. 1990. A model of the formation of marine algal flocs by physical coagulation processes. *Deep-Sea Res.* **38**:1197–1211.

Jackson, T. A. 1975. Humic matter in natural waters and sediments. *Soil Sci.* **119**:56–64.

Jacobson, A., and D. Gillespie. 1968. Metabolic events occurring during recovery from prolonged glucose starvation in *Escherichia coli. J. Bacteriol.* **95**:1030–1039.

Jacobson, A., and D. Gillespie. 1970. An RNA polymerase mutant defective in ATP initiations. *Cold Spring Harbor Symp. Quant. Biol.* **35**:85–93.

Jagger, J. 1985. *Solar UV Actions on Living Cells.* Praeger Publishers, New York.

James, N. 1958. Soil extract in microbiology. *Can. J. Microbiol.* **4**:363.

James, N., and M. I. Sutherland. 1942. Are there living bacteria in permanently frozen subsoil? *Can. J. Res. C.* **20**:228–235.

Jannasch, H. W. 1958. Studies on planktonic bacteria by means of a direct membrane filter method. *J. Gen. Microbiol.* **18**:609–620.

Jannasch, H. W. 1967. Growth of marine bacteria at limiting concentrations of organic carbon in seawater. *Limnol. Oceanogr.* **12**:264–271.

Jannasch, H. W. 1968. Growth characteristics of heterotrophic bacteria in seawater. *J. Bacteriol.* **95**:722–723.

Jannasch, H. W. 1969. Estimation of bacterial growth rates in natural waters. *J. Bacteriol.* **99**:156–160.

Jannasch, H. W. 1974. Steady state and the chemostat in ecology. *Limnol. Oceanogr.* **19**:716–720.

Jannasch, H. W. 1977. Growth kinetics of aquatic bacteria. *In Aquatic Microbiology,* F. A. Skinner and J. M. Shewan (eds.), pp. 55–68. Academic Press, London.

Jannasch, H. W. 1979. Microbial ecology of aquatic low nutrient habitats. *In Strategies of Microbial Life in Extreme Environments,* M. Shilo (ed.), pp. 243–260. Verlag Chemie, Weinheim.

Jannasch, H. W., and C. O. Wirsen. 1973. Deep-sea microorganisms: an in situ response to nutrient enrichment. *Science* **180**:641–643.

Jannasch, H. W., and C. O. Wirsen. 1979. Chemosynthetic primary production at East Pacific sea floor spreading centres. *BioScience* **29**:592–598.

Jannasch, H. W., and G. E. Jones. 1959. Bacterial populations in seawater as determined by different methods of enumeration. *Limnol. Ocenoagr.* **4**:128–139.

Jannasch, H. W., and P. H. Pritchard. 1972. The role of inert particulate matter in the activity of aquatic microorganisms. *Memorie dell'Instituto Italiano di Idrobioilogia* **29**:289–308.

Jannasch, H. W., K. Eimhjellen, K. Wirsen, and A. Farmanfarmaian. 1971. Microbial degradation of organic matter in the deep sea. *Science* **171**:672–675.

Jánsson, S. K. 1960. On the establishment and use of tagged microbial tissue in soil organic matter research. *Trans. 7th Int. Congr. Soil Sci.* **2**:635–642.

Jayne-Williams, D. J. 1963. Report of a discussion on the effect of the diluent on the recovery of bacteria. *J. Appl. Bacteriol.* **26**:398–404.

Jeffrey, L. M. 1968. Lipids of the marine waters. *In Organic Matter in Natural Waters,* D. W. Hood (ed.), pp. 55–76. University of Alaska, College, AK.

Jeffrey, L. M., and D. W. Hood. 1958. Organic matter in sea water: an evaluation of various methods for isolation. *J. Mar. Res.* **17**:247–271.

Jeffrey, W. H., and J. H. Paul. 1986. Activity measurements of planktonic microbial and microfouling communities in a eutrophic estuary. *Appl. Environ. Microbiol.* **51**:157–162.

Jeffrey, W. H., and J. H. Paul. 1988. Effect of 5-fluoro-2'-deoxyuridine on [3H]thymidine incorporation by bacterioplankton in the waters of southwest Florida. *Appl. Environ Microbiol.* **54**:3165–3168.

Jeffrey, W. H., and J. H. Paul. 1990. Thymidine uptake, thymidine incorporation, and thymidine kinase activity in marine bacterium isolates. *Appl. Environ. Microbiol.* **56**:1367–1372.

Jenkins, D. E., E. A. Auger, and A. Matin. 1991. Role of *rpoH,* a heat shock regulator protein, in *Escherichia coli* carbon starvation protein synthesis and survival. *J. Bacteriol.* **173**:1992–1996.

Jenkins, D. E., J. E. Schultz, and A. Matin. 1988. Starvation-induced cross protection against heat or H_2O_2 challenge in *Escherichia coli.* *J. Bacteriol.* **170**:3910–3914.

Jenkins, D. E., S. A. Chaisson, and A. Matin. 1990. Starvation-induced cross protection against osmotic challenge in *Escherichia coli.* *J. Bacteriol.* **172**:2779–2781.

Jenkinson, D. S. 1965. Studies on the decomposition of plant material in soil. I. Losses of carbon from [14]carbon labelled ryegrass inoculated with soil in the field. *J. Soil Sci.* **16**:104–115.

Jenkinson, D. S. 1966a. Studies on the decomposition of plant material in soil. II. *J. Soil Sci.* **17**:280–302.

Jenkinson, D. S. 1966b. The priming action. *In The Use of Isotopes in Soil Organic Matter Studies,* FAO/IAEA Technical Meeting, 1963, pp. 199–208. Pergamon Press, Oxford.

Jenkinson, D. S. 1971. Studies on the decompositon of C[14]labelled organic matter in soil. *Soil Sci.* **11**:64–70.

Jenkinson, D. S. 1976. The effect of biocidal treatments on metabolism in soil. IV. The decomposition of fumigated organisms in soil. *Soil Biol. Biochem.* **8**:203–208.

Jenkinson, D. S., and J. N. Ladd. 1981. Microbial biomass in soil, measurement and turnover. *Soil Biochem.* **5**:415–471.

Jenkinson, D. S., and D. S. Powelson. 1976. The effects of biocidal treatments on metabolism in soil. I. Fumigation with chloroform. *Soil Biol. Biochem.* **8**:167–177.

Jensen, H. L. 1961. Survival of *Rhizobium meliloti* in soil culture. *Nature* **192**:682–683.

Jensen, L. M., K. Sand-Jensen, S. Marcher, and M. Hansen. 1990. Plankton community respiration along a nutrient gradient in a shallow Danish estuary. *Mar. Ecol. Prog. Ser.* **61**:75–85.

Jensen, V. 1968. The plate count technique. *In The Ecology of Soil Bacteria,* T. R. G. Gray and D. Parkinson (eds.), pp. 158–170. University of Toronto Press, Toronto.

Jensen, V. 1974. Decomposition of angiosperm tree leaf litter. *In Biology of Plant Litter Decomposition,* C. H. Dickinson and G. J. F. Pugh (eds.), pp. 69–104. Academic Press, London.

Jiménez, L., G. L. de Victoria, J. Wear, C. B. Fliermans, and T. C. Hazen. 1990. Molecular analysis of deep subsurface bacteria. *In Proceedings of the First International Symposium on Microbiology of the Deep Subsurface,* C. G. Fliermans and T. Hazen (eds.), pp. 2-97–2-113. W.S.R.C., Information Services Section. Publication Group, Savannah River.

Jinks-Robertson, S., and M. Nomura. 1987. Ribosomes and tRNA. *In Escherichia coli* and *Salmonella typhimurium: Cellular and Molecular Biology,* vol. 2, F. C. Neidhardt, J. L. Ingrapam, K. B. Low, B. Magasanik, M. Schaechter, H. E. Umbarger (eds.), pp. 1358–1385. American Society for Microbiology, Washington, DC.

Johannes, R. E., and M. Satomi. 1966. Composition and nutritive value of fecal pellets of a marine crustacean. *Limnol. Oceanogr.* **11**:191–197.

Johnson, B. D., and P. E. Kepkay. 1992. Colloid transport and bacterial utilization of oceanic DOC. *Deep-Sea Res.* **39**:885–869.

Johnson, B. D., and P. J. Wangersky. 1986. Surface coagulation in seawater. *Netherlands J. Sea Res.* **20**:201–210.

Johnston, C. G., and J. R. Vestal. 1991. Photosynthetic carbon incorporation and turnover in Antarctic cryptoendolithic microbial communities: Are they the slowest-growing communities on earth? *Appl. Environ. Microbiol.* **57**:2308–2311.

Johnson, D. M., R. R. Peterson, D. R. Lycan, J. W. Sweet, M. E. Neuhaus, and A. L. Schaedel. 1985. *Atlas of Oregon Lakes.* Oregon State University Press, Corvallis, OR.

Johnston, J. R., and H. Oberman. 1979. Yeast genetics in industry. *Prog. Ind. Microbiol.* **15**:151–205.

Johnstone, B. H., and R. D. Jones. 1988a. Physiological effects of long-term energy-source deprivation on the survival of a marine chemolithotrophic ammonium oxidizing bacterium. *Mar. Ecol. Prog. Ser.* **49**:295–303.

Johnstone, B. H., and R. D. Jones. 1988b. Recovery of a marine chemolithotrophic ammonium-oxidizing bacterium from long-term energy-source deprivation. *Can. J. Microbiol.* **34**:1347–1350.

Johnstone, B. H., and R. D. Jones. 1988c. Effects of light and CO on the survival of a marine ammonium-oxidizing bacterium during energy deprivation. *Appl. Environ. Microbiol.* **54**:2890–2893.

Johnstone, B. H., and R. D. Jones. 1989. A study on the lack of [methyl-^{3}H]thymidine uptake and incorporation by chemolithotrophic bacteria. *Microb. Ecol.* **18**:73–77.

Joint, I. R., and R. J. Morris. 1982. The role of bacteria in the turnover of organic matter in the sea. *Ann. Rev. Oceanogr. Mar. Biol.* **20**:65–118.

Jones, B. Yu, N. J. Bainton, M. Birdsall, W. Bycroft, S. R. Chhabra, A. J. R. Cox, P. Colby, P. J. Reeves, S. Stephens, M. K. Winson, G. P. C. Salmond, G. S. A. B. Stewart, and P. Williams. 1993. The lux autoinducer regulates the production of exoenzyme virulence determinants in *Erwinia carotovora* and *Pseudomonas aeruginosa*. *EMBO J.* **12**:2477–2482.

Jones, D., R. H. Deibel, and C. F. Niven. 1964. Catalase activity of two *Streptococcus faecalis* and its enhancement by aerobiosis and added cations. *J. Bacteriol.* **88**:602–610.

Jones, K. L., and M. E. Rhodes-Roberts. 1981. The survival of marine bacteria under starvation conditions. *J. Appl. Bacteriol.* **50**:247–258.

Jones, R. D., and R. G. Prahl. 1985. Lipid composition of a marine ammonium oxidizer grown at 5 and 25°C. *Mar. Biol. Progr. Ser.* **26**:157–159.

Jones, R. D., and R. Y. Morita. 1985. Survival of a marine ammonium oxidizer under energy source deprivation. *Mar. Ecol. Prog. Ser.* **21**:175–179.

Jørgensen, B. B. 1977. The sulfur cycle of a coastal marine sediment (Limfjorden, Denmark). *Limnol. Oceanogr.* **5**:814–832.

Jørgensen, B. B. 1983. Processes at the sediment water interface. In *The Major Biogeochemical Cycles and Their Interactions*, B. Bolin and R. B. Cook (eds.), Scope Report 21, pp. 477–509. John Wiley, Chichester.

Jørgensen, C. B. 1976. August Pütter, August Krogh and modern ideas on the use of dissolved organic matter in the aquatic environment. *Biol. Rev.* **51**:291–328.

Jørgensen, K. S., and J. M. Tiedje. 1993. Survival of denitrifiers in nitrate-free anaerobic environments. *Appl. Environ. Microbiol.* **59**:3297–3305.

Jørgensen, N. O. G. 1979. Annual variation of dissolved free primary amines in estuarine water and sediment. *Oecologia* **40**:207–217.

Jørgensen, N. O. G., N. Kroer, R. B. Coffin, X.-A. Yang, and C. Lee. 1993. Dissolved free amino acids, combined amino acids, and DNA as sources of carbon and nitrogen to marine bacteria. *Mar. Ecol. Prog. Ser.* **98**:135–148.

Jørgensen, N. O. G., P. Lindroth, and K. Mopper. 1981. Extraction and distribution of free amino acids and ammonium in sediment interstitial waters from the Limfjord, Denmark. *Oceanol. Acta* **4**:465–474.

Jørgensen, N. O. G., K. Mopper, and P. Lindroth. 1980. Occurrence, origin, and assimilation of free amino acids in an estuarine environment. *Ophelia*, Suppl. **1**:179–192.

Jouper-Jaan, Å., B. Dahllöf, and S. Kjelleberg. 1986. Changes in the protein composition of three bacterial isolates from marine waters during short term energy and nutrient deprivation. *Appl. Environ. Microbiol.* **52**:1419–1421.

Jouper-Jaan, Å., A. E. Goodman, and S. Kjelleberg. 1992. Bacteria starved for prolonged periods increased protection against lethal temperatures. *FEMS Microbiol. Ecol.* **101**:229–236.

Jumars, P. A., D. L. Penry, J. A. Baross, M. J. Perry, and B. W. Frost. 1989. Closing the microbial loop: Dissolved organic carbon pathway to heterotrophic bacteria from incomplete ingestion, digestion, and absorption in animals. *Deep-Sea Res.* **36**:483–495.

Kaiser, J.-P., and J.-M. Bollag. 1990. Microbial activity in the terrestrial subsurface. *Experientia* **46**:797–806.

Kalckar, H. M. 1971. The periplasmic galactose binding protein of *Escherichia coli. Science* **174**:557–565.

Kalinenko, V. O. 1957. Multiplication of heterotrophic bacteria in distilled water. *Mikrobiologiya* **26**:148–153.

Kalle, K. 1966. The problem of Gelbstoff in the sea. *Oceanogr. Mar. Biol. Ann. Rev.* **4**:91–104.

Kaneko, T., and R. R. Colwell. 1975. Adsorption of *Vibrio parahaemolyticus* to chitin and copepods. *Appl. Environ. Microbiol.* **29**:269–274.

Kaplan, R., and D. Apirion. 1975a. The fate of ribosomes in *Escherichia coli* cells starved for carbon. *J. Biol. Chem.* **250**:1854–1863.

Kaplan, R., and D. Apirion. 1975b. Decay of ribosomal ribonucleic acid in *Escherichia coli* cells starved for various nutrients. *J. Biol. Chem.* **250**:3174–3178.

Kaprelyants, A. S., J. C. Gottschal, and D. B. Kell. 1993. Dormancy in non-sporulating bacteria. *FEMS Microb. Rev.* **104**:271–286.

Kaprelyants, A. S., and D. B. Kell. 1992. Rapid assessment of bacterial viability and vitality using rhodomine 123 and flow cytometry. *J. Appl. Bacteriol.* **74**:410–422.

Kaprelyants, A. S., and D. B. Kell. 1993. Dormancy in stationary-phase culture of *Micrococcus luteus:* Flow cytometric analysis of starvation and resuscitation. *Appl. 781 Environ. Microbiol.* **59**:3187–3196.

Karl, D. M. 1979. Measurement of microbial activity and growth in the ocean by rates of stable ribonucleic acid synthesis. *Appl. Environ. Microbiol.* **38**:850–860.

Karl, D. M. 1980. Cellular nucleotide measurements and application to the microbial loop. *Microbiol. Rev.* **44**:739–796.

Karl, D. M. 1982. Selected nucleic acid precursors in studies of aquatic microbial ecology. *Appl. Environ. Microbiol.* **44**:891–902.

Karl, D. M. 1986. Determination of in situ microbial biomass, viability, metabolis, and growth. *In Bacteria in nature,* vol. 2, J. S. Poindexter and E. R. Leadbetter (eds.), pp. 85–176. Plenum Press, New York.

Karl, D. M. 1993. Microbial processes in the Southern Ocean. *In Antarctic Microbiology,* E. I. Friedmann (ed.), pp. 1–63. Wiley-Liss, New York.

Karl, D. M. 1994. Accurate estimation of microbial loop processes and rates. *Microb. Ecol.* **28**:147–150.

Karl, D. M., and C. D. Winn. 1986. Does adenine incorporation into nucleic acids measure total microbial production?: a response to comments by Fuhrman et al. *Limnol. Oceanogr.* **31**:1384–1394.

Karl, D. M., and G. A. Knauer. 1984. Vertical distribution, transport, and exchange of carbon in the northeast Pacific Ocean: evidence for multiple zones of biological activity. *Deep-Sea Res.* **31**:221–243.

Karl, D. M., and O. Holm-Hansen. 1978. Methology and measurements of adenylate energy charge ratios in environmental samples. *Mar. Biol.* **48**:185–197.

Karl, D. M., G. A. Knauer, and J. H. Martin. 1988. Downward flux of particulate organic matter in the ocean: a particle decomposition paradox. *Nature* **332**:438–441.

Karl, D. M., O. Holm-Hansen, G. T. Taylor, G. Tien, and D. F. Bird. 1991. Microbial biomass and productivity in the western Bransfield Strait, Antartica during the 1986–87 austral summer. *Deep-Sea Res.* **38**:1029–1055.

Kassen, I., P. Falkenberg, O. B. Styrvold, and A. R. Strom. 1992. Molecular cloning and physical mapping of the *otsBA* genes, which encodes the osmoregulatory trehalose pathway of *Escherichia coli*: evidence that transcription is activated by *katF* (AppR). *J. Bacteriol.* **174**:889–898.

Katznelson, H. 1940. Survival of microorganisms introduced into soil. *Soil Sci.* **49**:283–293.

Kawamukai, M., H. Matsuda, W. Fuji, R. Utsumi, and T. Komano. 1989. Nucleotide sequences of *fic* and *fic-1* gene involved in cell filamentation induced by cyclic AMP in *Escherichia coli. J. Bacteriol.* **171**:4525–4529.

Kefford, B., B. A. Humphrey, and K. C. Marshall. 1986. Adhesion: a possible survival strategy for Leptospires under starvation conditions. *Current Microbiol.* **13**:247–250.

Keil, R. G., and D. L. Kirchman. 1991a. Dissolved combined amino acids in marine waters as determined by a vapor-phase hydrolysis method. *Mar. Chem.* **33**:243–259.

Keil, R. G., and D. L. Kirchman. 1991b. Contribution of dissolved free amino acids and ammonium to the nitrogen requirements of heterotrophic bacterioplankton. *Mar. Ecol. Prog. Ser.* **73**:1–10.

Keilin, D. 1959. The problem of anabiosis or latent life: history and current concept. *Proc. Roy. Soc. of London, Ser. B.* **150**:149–191.

Kelman, A. 1954. The relationship of pathogenicity in *Pseudomonas solanacearum* to colony appearance on a tetrazolium medium. *Phytopathology* **44**:693–695.

Kelman, A. 1956. Factors influencing viability and variation in cultures of *Pseudomonas solanacearum. Phytopathology* **46**:16–17.

Kelman, A., and J. H. Jensen. 1951. Maintaining virulence in isolates of *Pseudomonas solanacearum. Phytopathology* **41**:185–187.

Kemp, A. L. W., and A. Mudrochova. 1973. The distribution and nature of amino acids and other nitrogen-containing compounds in Lake Ontario surface sediments. *Geochim. Cosmochim. Acta* **37**:2191–2206.

Kemp, P. F. 1988. Bacterivory by benthic ciliates: Significance as a carbon source and impact on sediment bacteria. *Mar. Ecol. Prog. Ser.* **49**:163–169.

Kemp, P. F. 1990. The fate of benthic bacterial production. *Rev. Aquat. Sci.* **2**:109–124.

Kemp, P. F. 1994. A philosophy of methods development: The assimilation of new methods and information into aquatic microbial ecology. *Microb. Ecol.* **28**:159–162.

Kemp, P. F., S. Lee, and J. LaRoche. 1993. Estimating the growth rate of slowly growing marine bacteria from RNA content. *Appl. Environ. Microbiol.* **59**:2594–2601.

Kendall, A. I., T. E. Friedemann, and M. Ishikawa. 1930. Methods for the study of "resting" bacteria. *J. Infect. Diseases* **47**:186–193.

Kennedy, M., S. L. Reader, and L. M. Swierczynski. 1994. Preservation records of microorganisms: evidence of the tenacity of life. *Microbiology* **140**:2543–2559.

Kennell, D. 1968. Titration of the gene sites on DNA-RNA hybridization. II. The *Escherichia coli* chromosome. *J. Molec. Biol.* **34**:85–103.

Kenney, M. J., and S. L. Reader. 1932. Ancient microbe database. *Nature* **360**:634–635.

Kepkay, P. E. 1994. Particle aggregation and the biological reactivity of colloids. *Mar. Ecol. Prog. Ser.* **109**:293–304.

Kepkay, P. E., and B. D. Johnson. 1988. Microbial response to organic particle generation by surface coagulation in seawater. *Mar. Ecol. Prog. Ser.* **48**:193–198.

Kepkay, P. E., and B. D. Johnson. 1989. Coagulation on bubbles allows the microbial respiration of oceanic dissolved organic carbon. *Nature* **385**:63–65.

Kepkay, P. E., and M. L. Wells. 1992. Dissolved organic carbon in North Atlantic surface waters. *Mar. Ecol. Prog. Ser.* **80**:275–283.

Kepkay, P. E., W. G. Harrison, and B. Irwin. 1990a. Surface coagulation, microbial respiration and primary production in the Sargasso Sea. *Deep Sea Res.* **37**:145–155.

Kepkay, P. E., D. K. Muschenheim, and D. B. Johnson. 1990b. Surface coagulation and microbial respiration at a tidal front on Georges Bank. *Continental Shelf Res.* **10**:573–588.

Kerr, R. A., and J. G. Quinn. 1975. Chemical studies on the dissolved organic matter in seawater. I. Isolation and fractionation. *Deep-Sea Res.* **22**:107–116.

Khailov, K. M., and Z. Z. Finenko. 1968. Organic macromolecular components dissolved in sea water and their inclusion into food chains. *Proc. Symp. Marine Food Chains.* Arrhus, Denmark.

Khaylov, K. M. 1968. Dissolved organic macromolecules in seawater. *Geochim. Inst.* **5**:497–503.

Khlebnikova, G. M., D. A. Gilichinskii, D. G. Fedorov-Davydov, and E. A. Vorob'eva. 1990. Quantitative evaluation of microorganisms in permafrost deposits and buried soils. *Microbiology* (Moscow) **59**:106–112.

Kieft, T. L., and L. R. Rosacker. 1991. Application of respiration- and adenylate-based soil microbiological assays to deep subsurface terrestrial sediments. *Soil Biol. Biochem.* **23**:563–568.

Kieft, T. L., D. B. Ringelberg, and D. C. White. 1994. Changes in ester-linked phospholipid fatty acid profiles of subsurface bacteria during starvation and desiccation in a porous medium. *Appl. Environ. Microbiol.* **60**:3292–3299.

Kieft, T. L., L. R. Rosacker, D. Willcox, and A. J. Franklin. 1990. Water potential and starvation stress in deep subsurface microorganisms. *In Proceedings of the First International Symposium on Microbiology of the Deep Subsurface*, C. G. Fliermans and T. C. Hazen (eds.), pp. 4-99–4-111. W.S.R.C., Information Services Section, Publications Group. Savannah River.

Kikuchi, E. 1986. Contribution of the polychaete, *Neanthes japonica* (Izuka) to the oxygen uptake and carbon dioxide production of an intertidal mud-flat of the Nanakita estuary, Japan. *J. Exp. Mar. Biol. Ecol.* **97**:81–93.

Kim, J., and C. E. ZoBell. 1974. Occurrence and activities of cell-free enzymes in oceanic environments. *In Effect of the Ocean Environment on Microbial Activities*, R. R. Colwell and R. Y. Morita (eds.), pp. 368–385. University Park Press, Baltimore.

King, G. M. 1983. Sulfate reduction in Georgia salt marsh soils: an evaluation of pyrite formation by ^{35}S and ^{55}Fe tracers. *Limnol. Oceanogr.* **28**:987–995.

King, G. M. 1992. Measurement of acetate concentrations in marine pore waters by using an enzymatic approach. *Appl. Environ. Microbiol.* **57**:3476–3481.

King, W. I., and A. Hurst. 1963. A note on the survival of some bacteria in different diluents. *J. Appl. Bacteriol.* **26**:504–506.

Kirchman, D. 1983. The production of bacteria attached to particles suspended in freshwater pond. *Limnol. Oceanogr.* **28**:858–872.

Kirchman, D. L. 1990. Limitation of bacterial growth by dissolved organic matter in the subarctic Pacific. *Mar. Ecol. Prog. Ser.* **62**:47–54.

Kirchman, D. L. 1993. Particulate detritus and bacteria in the marine environments. *In Aquatic Microbiology: An Ecological Approach*, T. E. Ford (ed.), pp. 321–341. Blackwell Scientific, Boston.

Kirchman, D. L., and R. E. Hodson. 1986. Metabolic regulation of amino acid uptake in marine waters. *Limnol. Oceanogr.* **31**:339–350.

Kirchman, D., and R. Mitchell. 1982. Contribution of particle-bound bacteria to total microheterotrophic activity in five ponds and two marshes. *Appl. Environ. Microbiol.* **43**:200–209.

Kirchman, D. L., H. W. Ducklow, and R. Mitchell. 1982. Estimates of bacterial growth from changes in uptake rates and biomass. *Appl. Environ. Microbiol.* **44**:1206–1307.

Kirchman, D., E. K'Ness, and R. Hodson. 1985. Leucine incorporation and its potential as a measure of protein synthesis by bacteria in natural aquatic systems. *Appl. Environ. Microbiol.* **49**:599–607.

Kirchman, D. L., Y. Suzuki, C. Garside, and H. W. Ducklow. 1991. Bacterial oxidation of dissolved organic carbon in the north Atlantic Ocean during the spring bloom. *Nature* **352**:612–614.

Kjelleberg, S. 1993. *Starvation in Bacteria*. Plenum Press, New York.

Kjelleberg, S., and B. Dahback. 1984. ATP level of a starving surface-bound and free-living marine *Vibrio* sp. *FEMS Lett.* **24**:214–219.

Kjelleberg, S., K. B. G. Flärdh, T. Nyström, and D. J. W. Moriarty. 1993. Growth limitation and starvation of bacteria. *In Aquatic Microbiology: An Ecological Approach,* T. E. Ford (ed.), pp. 289–320. Blackwell Scientific, Boston.

Kjelleberg, S., and M. Hermansson. 1984. Starvation-induced effects on bacterial surface characteristics. *Appl. Environ. Microbiol.* **48**:497–503.

Kjelleberg, S., and M. Hermansson. 1987. Short term responses to energy fluctuation by marine heterotrophic bacteria. *In Microbes in the Sea,* M. A. Sleigh (ed.), pp. 203–219. Ellis Horwood, Ltd., Chichester.

Kjelleberg, S., M. Hermansson, P. Marden, and G. W. Jones. 1987. The transient phase between growth and nongrowth of heterotrophic bacteria, with emphasis on the marine environment. *Ann. Rev. Microbiol.* **41**:25–49.

Kjelleberg, S., B. A. Humphrey, and K. C. Marshall. 1982. Effect of interfaces on small, starved marine bacteria. *Appl. Environ. Microbiol.* **43**:1166–1172.

Kjelleberg, S., B. A. Humphrey, and K. C. Marshall. 1983. Initial phases of starvation and activity of bacteria at surfaces. *Appl. Environ. Microbiol.* **46**:978–984.

Kjelleberg, S., K. C. Marshall, and M. Hermansson. 1985. Oligotrophic and copiotrophic marine bacteria-observations related to attachment. *FEMS Microb. Ecol.* **31**:89–96.

Kjelleberg, S., T. A. Strenström, and G. Odham. 1979. Comparative study of different hydrophobic devices for sampling lipid surface films and adherent micro-organisms. *Mar. Biol.* **53**:21–25.

Klein, D. A., and L. E. Casida, Jr. 1968. *Escherichia coli* dieout from normal as related to nutrient availability and the indigenous microflora. *Can. J. Microbiol.* **13**:1461–1470.

Klein, H. 1987. Benthic storms, vortices, and particle dispersion in the deep West European Basin. *Dt. Hydrogr. Z.* **40**:87–102.

Kluyver, A. J., and J. K. Baars. 1932. On some physiological artefacts. *Proc. Roy. Soc. Amsterdam* **35**:370–378.

Knauer, G. A., J. H. Martin, and K. W. Bruland. 1979. Fluxes of particulate carbon nitrogen and phosphorus in the upper water column in the northeast Pacific. *Deep-Sea Res.* **26**:97–108.

Knoll, A. H., and S. M. Awramik. 1983. Ancient microbial systems. *In Microbial Geochemistry,* W. E. Krumbein (ed.), pp. 287–315. Blackwell Scientific, Oxford.

Knowles, C. J. 1977. Microbial metabolic regulation by adenine nucleotide pools. *Symp. Soc. Gen. Microbiol.* **27**:241–283.

Knowles, C. J., and L. Smith. 1970. Measurements of ATP levels of intact *Azotobacter vinelandii* under different conditions. *Biochim. Biophys. Acta* **197**:152–160.

Koch, A. L. 1971. The adaptive responses of *Escherichia coli* to a feast and famine existence. *Adv. Microb. Physiol.* **6**:147–217.

Koch, A. L. 1976. How bacteria face recession and depression. *Perspect. Biol. Med.* **40**:44–63.

Koch, A. L. 1979. Microbial growth in low concentrations of nutrients, *In Strategies of Microbial Life in Extreme Environments,* M. Shilo (ed.), pp. 261–279. Verlag Chemie. Weinheim and New York.

Koch, A. L. 1985. The macroeconomics of bacterial growth, *In Bacteria in Their Natural Environments,* M. Fletcher and G. D. Floodgate, (eds.), pp. 1–42. Academic Press, London.

Koch, A. L. 1987. The variability and individuality of the bacterium. *In Escherichia coli and Salmonella typhimurium. Cellular and Molecular Biology,* F. C. Neidhardt, J. L. Ingraham, K. Brooks, B. Magasanik, M. Schaechter, and H. E. Umberger (eds.), pp. 1606–1614. American Society for Microbiology, Washington, DC.

Koch, A. L. 1990. Diffusion—The Crucial process in many aspects of the biology of bacteria. *Adv. Microb. Ecol.* **11**:37–70.

Koch, A. L. 1991. Efffective growth by the simplest means: The bacterial way. *ASM News* **57**:633–637.

Koch, A. L., and R. Coffman. 1970. Diffusion, permeation, or enzyme limitation: a probe for the kinetic of enzyme induction. *Biotechn. Bioeng.* **12**:651–677.

Koch, A. L., and H. R. Levy. 1955. Protein turnover in growing cultures of *Escherichia coli. J. Biol. Chem.* **217**:947–957.

Kogure, K., and I. Koike. 1987. Particle counter determination of bacterial biomass in seawater. *Appl. Environ. Microbiol.* **53**:274–277.

Kogure, K., U. Simidu, and N. Taga. 1979. A tentative direct count microscopic method for counting living marine bacteria. *Can. J. Microbiol.* **24**:415–420.

Kogure, K., U. Simidu, and N. Taga. 1980. Distribution of viable marine bacteria in neritic seawater around Japan. *Can. J. Microbiol.* **26**:318–823.

Kogure, K., K. Fukami, U. Simidu, and N. Taga. 1986. Abundance and production of bacterioplankton in the Antarctic. *Memoirs Natl. Inst. Polar Res. Spec. Issue* **40**:414–422.

Koike, I., and A. Hattori. 1974. Growth yields of a denitrifying bacterium. *Pseudomonas denitrificans,* under aerobic and anaerobic conditions. *J. Gen. Microbiol.* **88**:1–10.

Koike, I., S. Hara, K. Terauchi, and K. Kogure. 1990. Role of sub-micrometre particles in the ocean. *Nature* **345**:242–244.

Kokjohn, T. A., and R. V. Miller. 1985. Molecular cloning and characterization of the *recA* gene of *Pseudomonas aeruginosa* PAO. *J. Bacteriol.* **163**:568–572.

Kokjohn, T. A., G. S. Sayler, and R. V. Miller. 1991. Attachment and replication of *Pseudomonas aeruginosa* bacteriophages under conditions simulating aquatic environments. *J. Gen. Microbiol.* **137**:661–666.

Kolter, R. 1992. Life and death in stationary phase. *ASM News* **58**:75–79.

Kolter, R., D. A. Siegele, and A. Tormo. 1993. The stationary phase of the bacterial life cycle. *Ann. Rev. Microbiol.* **47**:855–874.

Kong, S., and D. L. Johnstone. 1994. Toxicity of toluene and xylene to *Acinetobacter calcoaceticus* in starvation/survival mode. *Biotech. Lett.* **16**:1217–1220.

Kong, S., D. L. Johnstone, D. R. Yonge, J. N. Peterson, and T. M. Brouns. 1992. Comparison of chromium adsorption to starved and fresh subsurface bacterial consortium. *Biotechnol. Techniques* **6**:143–148.

Konings, W. N., and H. Veldkamp. 1980. Phenotypic responses to environmental changes. *In Contemporary Microbial Ecology,* D. C. Ellwood, J. N. Hedger, M. J. Lynch, and J. H. Slater (eds.), pp. 161–191. Academic Press, New York.

Konings, W. N., and H. Veldkamp. 1983. Energy transduction and solute transport mechanisms in relation to environments occupied by microorganisms. *Symp. Soc. Gen. Microbiol.* **34**:153–186.

Kononova, M. M. 1966. *Soil Organic Matter,* 2nd ed. Pergamon Press, Oxford.

Kopczynski, E. D., M. M. Bateman, and D. M. Ward. 1994. Recognition of chimeric small-subunit ribosomal DNAs composed of genes from uncultured microorganisms. *Appl. Environ. Microbiol.* **60**:746–749.

Korkhonen, L. K., and P. J. Martikainen. 1991a. Comparison of the survival of *Campylobacter jejuni* and *Campylobacter coli* in culturable form in surface water. *Can. J. Microbiol.* **37**:530–533.

Korkhonen, L. K., and P. J. Martikainen. 1991b. Survival of *Eschrichia coli* and *Campylobacter jejuni* in untreated and filtered lake water. *J. Appl. Bacteriol.* **71**:379–382.

Koshland, D. E. Jr. 1977. Bacterial chemotaxis and some energy enzymes in energy metabolism. *Symp. Soc. Gen. Microbiol.* **27**:317–331.

Köster, M., O. Charfreitag, and L.-A. Meyer-Reil. 1991. Availability of nutrients to a deep-sea benthic microbial community: results from a ship-board experiment. *Kieler Meeresforsch., Sonderh.* **8**:127–133.

Kostiw, L. L., C. W. Boylen, and B. J. Tyson. 1972. Lipid composition of growing and starving cells of *Arthrobacter crystallopoietes. J. Bacteriol.* **111**:103–111.

Kotzias, D., K. Hustert, and A. Weiser. 1987. Formation of oxygen species and their reactions with organic chemicals in aqueous solution. *Chemosphere* **16**:505–511.

Kowarz, L., C. Coynaul, B. Robbe-Saule, and F. Norel. 1994. The *Salmonella typhimurium katF* (*rpoS*) gene: cloning nucleotide sequence, and regulation of *spvR* and *spvABCD* virulence plasmid genes. *J. Bacteriol.* **176**:6852–6860.

Kragelund, L., and O. Nybroe. 1994. Culturability and expression of outer membrane proteins during carbon, nitrogen, or phosphorus starvation of *Pseudomonas fluorescens* DF57 and *Pseudomonas putida* DF14. *Appl. Environ. Microbiol.* **60**:2944–2944.

Krambeck, C. 1979. Applicability and limitations of the Machaelis-Menton equation in microbial ecology. *Arch. Hydrobiol. Beih. Ergebn Limnol.* **12**:64–76.

Krambeck, C., H.-J. Krambeck, and J. Overbeck. 1981. Micro-computer-assisted biomass determination of plankton bacteria on scanning electron micrographs. *Appl. Environ. Microbiol.* **42**:142–149.

Kramer, J. G., and F. L. Singleton. 1992. Variations in rRNA content of marine *Vibrio* spp. during starvation-survival and recovery. *Appl. Environ. Microbiol.* **58**:201–207.

Kramer, J. G., and F. L. Singleton. 1993. Measurement of rRNA variations in natural communities of microorganisms on the southeastern U.S. continental shelf. *Appl. Environ. Microbiol.* **59**:2430–2436.

Kranck, K., and T. G. Milligan. 1988. Macroflocs from diatoms: in situ photography of particles in Bedford Basin, Nova Scotia. *Mar. Ecol. Prog. Ser.* **44**:183–189.

Krasil'nikov, N. A., and S. S. Belyaev. 1967. Distribution of *Caulobacter* in certain soils. *Mikrobiologiya* **36**:1083–1086.

Kriss, A. E. 1963. *Marine Microbiology.* Oliver and Boyd, Edinburgh.

Kriss, A. E., and E. Maarkianovich. 1959. Observations on the rate of reproduction of microorganisms in the seas. *Microbiology* (Moscow) **23**:551.

Kriss, A. E., I. E. Mishustina, N. Mitskevich, and E. V. Zemtsova. 1967. *Microbial Populations of the Oceans and Seas.* Edward Arnold (Publ.) Ltd., London (translated by K. Syers and edited by G. E. Fogg).

Kristensen, E., and T. H. Blackburn. 1987. The fate of organic carbon and nitrogen in experimental marine sediment systems: influence of bioturbation and anoxia. *J. Mar. Res.* **45**:231–257.

Kroer, N. 1993. Bacterial growth efficiency on natural dissolved organic matter. *Limnol. Oceanogr.* **38**:1282–1290.

Kroer, N. 1994. Relationships between biovolume and carbon and nitrogen content of bacterioplankton. *FEMS Microb. Ecol.* **13**:217–224.

Krogh, A. 1934a. Conditions of life in the ocean. *Ecol. Monogr.* **4**:421–429.

Krogh, A. 1934b. Conditions of life at great depths in the ocean. *Ecol. Monogr.* **4**:421–429.

Krska, J., T. Elthon, and P. Blum. 1993. Monoclonal antibody recognition and function of a DnaK (HSP70) epitope found in gram-negative bacteria. *J. Bacteriol.* **175**:6433–6440.

Krumholz, R., and J. M. Suflita. 1996. Ecological interactions among sulfate reducing and other bacteria living in subsurface cretaceous rocks. *In Abstracts of the 96th General Meeting of the American Society for Microbiology, 1996,* abstr. N-56, p. 331. American Society for Microbiology, Washington, DC.

Kubitschek, H. E. 1970. *Introduction to Research with Continuous Cultures.* Prentice-Hall, Englewood Cliffs, NJ.

Kuenen, J. G., J. Boonstra, H. G. J. Schroder, and H. Veldkamp. 1977. Competition for inorganic substrates among chemoorganotrophic bacteria. *Microb. Ecol.* **3**:119–130.

Kuenen, Ph. H. 1950. *Marine Geology.* John Wiley & Sons, New York.

Kunicki-Goldfinger, W., and W. J. H. Kunicki-Goldfinger. 1972. Semicontinuous culture of bacteria on membrane filters. I. Use for the bioassay of inorganic and organic nutrients in aquatic environments. *Acta Microbiol. Pol. Ser. B* **4**:49–60.

Kupelian, A., and M. S. DuBow. 1986. The effect of gamma-irradiation on Mu DNA transposition and gene expression. *Mutat. Res.* **160**:1–10.

Kurath, G. 1980. Some physiological bases for survival of a marine bacterium during nutrient starvation. M.S. Thesis. Oregon State University, Corvallis, OR.

Kurath, G., and R. Y. Morita. 1983. Starvation-survival physiological studies of a marine *Pseudomonas* sp. *Appl. Environ. Microbiol.* **45**:1206–1211.

Kushner, D. J. 1993. Microbial life in extreme environments. *In Aquatic Microbiology: An Ecological Approach,* T. E. Ford (ed.), pp. 383–407. Blackwell Scientific, Boston.

Kutter, E., E. Kellenberger, K. Carlson, S. Eddy, J. Neitzel, L. Messinger, J. North, and B. Guttman. 1994. Effects of bacterial growth conditions and physiology on T4 infection. *In Molecular Biology of Bacteriophage T4*, J. D. Karam (ed.), pp. 406–418. American Society for Microbiology, Washington, DC.

Kuznetsov, S. I. 1970. Distribution of bacteria in lakes. *In The Microbiology of Lakes and Its Geochemical Activity*, S. I. Kuznetzov (ed.), pp. 116–147. University Texas Press, Austin and London (edited by C. H. Oppenheimer).

Kuznetsov, S. I., G. A. Dubinina, and N. A. Lapteva. 1979. Biology of oligotrophic bacteria. *Ann. Rev. Microbiol.* **33**:377–387.

Kuznetsov, S. I., M. V. Ivanov, and N. N. Lyalikova. 1963. *Introduction to Geological Microbiology*. McGraw-Hill, New York.

Kuznetsova, M. A., O. V. Orlova, and A. B. Sabinov. 1984. Aggregates of organic substances in freshwater and their role on feeding of filter-feeding crustaceans. *Hydrobiol. J.* **20**:1–7.

Kuzuwa, G., K. Bronwell, and G. Gurloff. 1971. The phenylalanine-binding protein of *Comamonas* sp. (ATCC 11299a). *J. Biol. Chem.* **246**:6371–6380.

Labeda, D. P., K.-C. Liu, and L. E. Casida, Jr. 1976. Colonization of soil by *Arthrobacter* and *Pseudomonas* under varying conditions of water and nutrient availability as studied by plate counts and transmission electron microscopy. *Appl. Environ. Microbiol.* **31**:551–561.

Ladd, J. N., and J. H. A. Butler. 1966. Comparison of some properties of soil humic acids and synthetic phenolic polymers incorporating amino acid derivatives. *Aust. J. Soil Res.* **4**:41–54.

Ladd, J. N., and J. H. A. Butler. 1975. Humus-enzyme systems and synthetic, organic polymers analogs. *In Soil Biochemistry*, vol. 4, E. A. Paul and A. D. McLaren (eds.), pp. 143–194. Marcel Dekker, New York.

Ladd, J. N., and P. G. Brisbane. 1967. Release of amino acids from soil humic acids by proteolytic enzymes. *Aust. J. Soil. Res.* **5**:161–171.

Ladd, J. N., J. M. Oades, and M. Amato. 1981. Microbial biomass formed [14]-, [15]-C labelled plant material decomposition in soils in the field. *Soil Biol. Biochem.* **13**:119–126.

Ladd, T. I., R. M. Ventullo, P. M. Wallis, and J. W. Costerton. 1982. Heterotrophic activity and biodegradation of labile and refractory compounds in groundwater and stream microbial populations. *Appl. Environ. Microbiol.* **44**:321–329.

Ladd, J. N., R. B. Jackson, M. Amato, and J. H. A. Butler. 1983. Decomposition of plant material in Australian soils. I. The effect of quantity added on decomposition and on residual microbial biomass. *Aust. J. Soil Res.* **21**:563–570.

Ladd, J. N., M. Amato, and J. M. Oades. 1985. Decomposition of plant material in Australian soils. III. Residual organic and microbial biomass C and N from isotope-labelled legume material and soil organic matter, decomposing under field conditions. *Aust. J. Soil Res.* **23**:603–611.

Lamanna, C. 1963. Studies on endogenous metabolism in bacteriology. *Ann. N.Y. Acad. Sci.* **102**:517–520.

Lamanna, C., and M. F. Mallette. 1953. *Basic Bacteriology.* Williams & Wilkins, Baltimore.

La Marre, A. G., S. G. Straley, and S. F. Conti. 1977. Chemotaxis toward amino acids by *Bdellovibrio bacteriovorus. J. Bacteriol.* **114**:201–207.

Lande, R., and A. M. Wood. 1987. Suspension times of particles in the upper ocean. *Deep-Sea Res.* **34**:61–72.

Landry, M. R., L. W. Hass, and V. L. Fagerness. 1984. Dynamics of microbial plankton communities: experiments in Kaneohe Bay, Hawaii. *Mar. Ecol. Prog. Ser.* **16**:127–133.

Lang, G. A., J. D. Early, G. C. Martin, and R. L. Darnell. 1987. Endo-, para-, and eco-dormancy: physiological terminology and classification for dormancy research. *Hort-Science* **22**:371–377.

Lange, R., and R. Hengge-Aronis. 1991a. Growth phase-regulated expression of *bolA* and morphology of stationary-phase *Escherichia coli* cells are controlled by the noval sigma factor σ^{S}· *J. Bacteriol.* **173**:4474–4481.

Lange, R., and R. Hengge-Aronis. 1991b. Identification of a central regulator of stationary phase gene expression in *E. coli. Mol. Microbiol.* **5**:49–59.

Lange, W. 1976. Speculations on the possible essential function of the gelatinous sheath of blue-green algae. *Can. J. Microbiol.* **22**:1181–1185.

LaPat-Polasko, L. T., P. L. McCarty, and A. J. B. Zehnder. 1984. Secondary substrate utilization of methylene chloride by an isolated strain of *Pseudomonas* sp. *Appl. Environ. Microbiol.* **47**:825–830.

Lappin-Scott, H. M., and J. W. Costerton. 1990. Starvation and penetration of bacteria in soils and rocks. *Experientia* **46**:807–812.

Lappin-Scott, H. M., F. Cusack, and J. W. Costerton. 1988a. Nutrient resuscitation and growth of starved cells in sandstone cores: a novel approach to enhanced oil recovery. *Appl. Environ. Microbiol.* **54**:1273–1382.

Lappin-Scott, H. M., F. Cusack, A. MacLeod, and J. W. Costerton. 1988b. Starvation and nutrient resuscitation of *Klebsiella pneumoniae* isolated from oil well waters. *J. Appl. Bact.* **64**:541–549.

Lasker, R. 1966. Feeding, growth, respiration, and carbon utilization of a euphausiid crustacean. *J. Fish. Res. Bd. Can.* **23**:1291–1317.

La Teana, A., A. Brandi, M. Falconi, R. Spurio, C. L. Pon, and C. O. Gualerzi. 1991. Identification of a cold shock transcriptional enhancer of the *Escherichia coli* gene encoding nucleoid protein H-NS. *Proc. Natl. Acad. Sci. (USA)* **88**:10907–10911.

Lauffenburger, D. A. 1991. Quantitative studies of bacteria chemotaxis and microbial population dynamics. *Microb. Ecol.* **22**:175–185.

Law, A. T., and D. K. Button. 1977. Multiple-carbon-source-limited growth kinetics of a marine coryneform bacteria. *J. Bacteriol.* **129**:115–123.

Law, E. A. 1983. Plots of turnover times versus added substrate concentrations provide only upper bounds to in situ substrate concentration. *J. Theore. Biol.* **101**:147–159.

Lawrence, J. R., and D. E. Caldwell. 1987. Behavior of bacterial stream populations within the hydrodynamic boundary layers of surface microenvironments. *Microb. Ecol.* **14**:15–27.

Lawrence, J. R., and D. R. Korber. 1993. Aspects of microbial surface colonization behavior. *In Trends in Microbial Ecology,* R. Guerrero and C. Pedrós-Alió (eds.), pp. 113–118. Spanish Society for Microbiology, Barcelona.

Lawrence, J. R., D. R. Korber, and D. E. Caldwell. 1992. Behavioral analysis of *Vibrio parahaemolyticus* variants in high- and low-viscosity microenvironments by use of digital image processing. *J. Bacteriol.* **174**:5732–5739.

Lazzarini, R. A., M. Cashel, and J. Gallant. 1971. On the regulation of guanosine tetraphosphate levels in stringent and relaxed strains of *Escherichia coli. J. Biol. Chem.* **246**:4381–4385.

Lebaron, P., and F. Joux. 1994. Flow cytometric analysis of the cellular DNA content of *Salmonella typhimurium and Alteromonas haloplanktis* during starvation and recovery in seawater. *Appl. Environ. Microbiol.* **60**:4345–4350.

Lebedjantzev, A. N. 1924. Drying of soil as one of the natural factors in maintaining soil fertility. *Soil Sci.* **18**:419–447.

Lee, C., and C. J. Cronin. 1984. Particulate amino acids in the sea: Effects of primary productivity and biological decomposition. *J. Mar. Res.* **42**:1075–1097.

Lee, C., and J. L. Bada. 1975. Amino acids in equatorial Pacific Ocean water. *Earth Planet. Sci. Lett.* **26**:61–68.

Lee, H. A. 1920. Behaviour of the citrus canker organisms in the soil. *J. Agric. Res.* **19**:189–206.

Lee, C. C., R. F. Harris, J. D. Williams, J. K. Syers, and D. E. Armstrong. 1971. Adenosine triphosphate in lake sediments. II. Origin and significance. *Soil Sci. Soc. Proc.* **35**:86–91.

Lee, S., and J. A. Fuhrman. 1987. Relationships between biovolume and biomass of naturally derived marine bacterioplankton. *Appl. Environ. Microbiol.* **53**:1298–1303.

Lee, S., C. Malone, and P. F. Kemp. 1993. Use of multiple 16S rRNA-targeted fluorescent probes to increase signal strength and measure cellular RNA from natural planktonic bacteria. *Mar. Ecol. Prog. Ser.* **101**:193–201.

Leechman, F. 1961. *The Opel Book.* Ure Smith, Sydney, Australia.

Leenheer, J. A., R. L. Malcolm, P. W. McKinley, and L. A. Eccles. 1974. Occurrence of dissolved organic carbon in selected ground-water samples in the United States. *J. Res. U.S. Geol. Sur.* **2**:361–369.

Le Fèbre, J. 1986. Aspects of the biology of frontal systems. *Adv. Mar. Biol.* **23**:163–299.

Leff, L. G., and J. L. Meyer. 1991. Biological availability of dissolved carbon along the Ogeechee River. *Limnol. Oceanogr.* **36**:315–323.

Leifson, E. 1962a. The bacterial flora of distilled and stored water. I. General observations, techniques and ecology. *Int. Bull. Bact. Nomencl. Taxon.* **12**:133–153.

Leifson, E. 1962b. The bacterial flora of distilled and stored water. II. Caulobacter vibrioides Henrici and Johnson 1935 in distilled water. *Int. Bull. Bact. Nomencl. Taxon.* **12**:155–159.

Lelliot, R. A. 1965. The preservation of plant pathogenic bacteria. *J. Appl. Bacteriol.* **28**:181–193.

Lemaire, J. M., and B. Jovan. 1966. Modifications microbiologiques entrainees par la mise en culture de sols nouvellement defriches: Incidences sur l'installation de l'*Ophiobolus graminis* Sacc: (= *Linmoicarpon caricetdi B. et Br.*) *et du Streptomyces scabies* (Thaxt.) Waksman et Henrici. *Ann. Epiphyt.* **17**:313–333.

Leps, W. T., and J. C. Ensign. 1979a. Adenosine triphosphate pool levels and endogenous metabolism in *Arthrobacter crystallopoietes* during growth and starvation. *Arch. Microbiol.* **122**:61–67.

Leps, W. T., and J. C. Ensign. 1979b. Adenylate nucleotide levels and energy charge in *Arthrobacter crystallopoietes* during growth and starvation. *Arch. Microbiol.* **122**:69–76.

Lewis, A. G., A. Ramnarine, and M. S. Evans. 1971. Natural chelators—an indication of activity with the calanoid copepod *Euchaeta japonica. Mar. Biol.* **11**:1–4.

Lewis, D. L. 1991. Quiescent/senescent/moribund/nutrient-starved/dormant . . . ? *ASM News* **57**:342.

Lewis, D. L., and D. K. Gattie. 1991. The ecology of quiescent microbes. *ASM News* **57**:27–32.

Li, W. K. W. 1983. Consideration of errors in estimating kinetic parameters based on Michaelis-Menton formalism in microbial ecology. *Limnol. Oceanogr.* **28**:185–190.

Li, W. K. W., and M. Wood. 1988. Vertical distribution of North Atlantic ultraplankton: Analysis by flow cytometry and epifluorescence microscopy. *Deep-Sea Res.* **35**:1615–1638.

Li, W. K. W., and P. M. Dickie. 1985a. Metabolic inhibition of size-fractionated marine plankton radiolabeled with amino acids, glucose, bicarbonate, and phosphate in the light and dark. *Microb. Ecol.* **11**:11–24.

Li, W. K. W., and P. M. Dickie. 1985b. Growth of bacteria in seawater filtered through 0.2μm Nuclepore membranes: implication for dilution experiments. *Mar. Ecol. Prog. Ser.* **26**:245–252.

Li, W. K. W., and P. M. Dickie. 1987. Temperature characteristics of photosynthetic and heterotrophic activities: Seasonal variations in temperate microbial plankton. *Appl. Environ. Microbiol.* **53**:2282–2295.

Li, W. K. W., D. V. Subba Roa, W. G. Harrison, J. C. Smith, J. J. Cullen, B. Irwin, and T. Platt. 1983. Autotrophic picoplankton in the tropical ocean. *Science* **219**:292–295.

Liebezeit, G., M. Bolter, I. F. Brown, and R. Dawson. 1980. Dissolved free amino acids and carbohydrates at pycnoline and boundaries in the Sargasso Sea and related microbial activity. *Oceanol. Acta* **3**:357–362.

Liesack, W., and E. Stackebrandt. 1991. Potential risks of gene amplification by PCR as determined by 16S rDNA analysis of mixed-culture of strict bacrophilic bacteria. *Microb. Ecol.* **21**:191–198.

Lieske, R. 1932. Über das vorkommen van bakterien in kohlenflozen. *Biochem. Zeitsch.* **250**:339–351.

Lieske, van R., and E. Hofmann. 1929. Untersuchungen über den Bakteriengehalt der Erde in grossen Tiefen. *Centralb. f. Bakterio,* Abt. 2 **77**:305–309.

Lignell, R. 1990. Excretion of organic carbon by phytoplankton: Its relation to algal biomass, primary productivity and bacterial secondary productivity in the Baltic Sea. *Mar. Ecol. Prog. Ser.* **68**:85–89.

Lim, C.-H., and K. P. Flint. 1989. The effects of nutrient on the survival of *Escherichia coli* in lake water. *J. Appl. Bacteriol.* **66**:559–569.

Lindahl, T., and A. Andersson. 1972. Rate of chain breakage at apurinic sites in double-stranded deoxyribonucleic acid. *Biochemistry* **11**:3618–3623.

Lindahl, T., and B. Nyberg. 1972. Rate of depurination of native deoxyribonucleic acid. *Biochemistry* **11**:3610–3618.

Linder, K., and J. D. Oliver. 1989. Membrane fatty acid and virulence changes in the viable but nonculturable state of *Vibrio vulnificus. Appl. Environ. Microbiol.* **55**:2837–2842.

Lindquist, R., and G. Bengtsson. 1991. Dispersal dynamics of ground water bacteria. *Microb. Ecol.* **21**:49–72.

Linley, E. A. S., and R. C. Newell. 1984. Estimates of bacterial growth yields based upon plant detritus. *Bull. Mar. Sci.* **35**:426–439.

Lipman, C. B. 1930. Artificial bacilli. *Science* **71**:418–419.

Lipman, C. B. 1931. Living organisms in ancient rocks. *J. Bacteriol.* **22**:183–196.

Lipman, C. B. 1932. Are there living bacteria in stony meteorites? *Amer. Mus. Novit., No.* **588**:1–19.

Lipman, C. B. 1934. Further evidence on the amazing longevity of bacteria. *Science* **79**:230–231.

Lipman, C. B. 1935. Bacteria in travertine from the Yellowstone. *J. Bacteriol.* **29**:3.

Lipman, C. B. 1937. Bacteria in coal. *J. Bacteriol.* **34**:483–488.

Liss, P. S. 1983. The exchange of biogeochemically important gases across the air-sea interface. *In* Scope Report 21. *The Major Biogeochemical Cycles and Their Interactions,* B. Bolin and R. B. Cook (eds.), pp. 411–426. John Wiley & Sons, Chichester.

Litchfield, C. D., A. L. S. Munro, L. C. Massie, and G. D. Floodgate. 1974. Biochemistry and microbiology of some Irish Sea sediments: I. Amino acid analysis. *Mar. Biol.* **26**:249–260.

Liu, P. V. 1984. Bacterial longevity in salt-free medium. *ASM News* **50**:471.

Lloyd, B. 1931. Muds of the Clyde Sea area. II. Bacterial content. *J. Mar. Biol. Assoc.* **17**:751–765.

Lloyd, B. 1937. Bacteria in stored water. *J. Roy. Tech. College, Glascow* **4**:173–177.

Lochte, K. 1992. Bacterial standing stock and consumption of organic carbon in the benthic boundary layer of the abyssal North Atlantic. *In Deep-sea Food Chains and the Carbon Cycle,* R. T. Rowe and V. Pariente (eds.), pp. 1–10. Kluwer Academic Publ., Dordrecht, The Netherlands.

Lock, M. A., and T. E. Ford. 1986. Metabolism of dissolved organic matter by attached microorganisms in rivers. *In Perspectives in Microbial Ecology,* F. Megusar and M. Gantar (eds.), pp. 22–29. Slovene Society for Microbiology, Ljubljana, Yugoslavia.

Lockwood, J. L. 1964. Soil fungistasis. *Ann. Rev. Phytopath.* **2**:341–362.

Lockwood, J. L. 1968a. The fungal environment of soil bacteria. *In The Ecology of Soil Bacteria,* T. R. G. Gray and D. Parkinson (eds.), pp. 44–65. University of Toronto Press, Toronto.

Lockwood, J. L. 1968b. Discussion on the environment of soil bacteria. *In The Ecology of Soil Bacteria,* T. R. G. Gray and D. Parkinson (eds.), pp. 89–94. University of Toronto Press, Toronto.

Lockwood, J. L. 1977. Fungistasis in soil. *Biol. Rev. Cambridge Philos. Soc.* **52**:1–43.

Lockwood, J. L. 1981. Exploitation competition. *In The Fungal Community,* D. T. Wicklow and G. C. Carroll (eds.), pp. 319–349. Marcel Dekker, New York.

Lockwood, J. L., and A. B. Filonow. 1981. Responses of fungi to nutrient-limiting conditions and to inhibitory substances in natural habitats. *Adv. Microb. Ecol.* **5**:1–61.

Logan, B. E., and A. L. Alldredge. 1989. The increased potential for nutrient uptake by flocculating diatoms. *Mar. Biol.* **101**:443–450.

Logan, B. E., and J. W. Dettmer. 1990. Increased mass transfer to microorganisms with fluid motion. *Biotechnol. Bioeng.* **35**:1135–1144.

Logan, B. E., and J. R. Hunt. 1987. Advantages to microbes of growth in permeable aggregates in marine systems. *Limnol. Oceanogr.* **32**.1034–1048.

Logan, B. E., and Q. Jiang. 1990. A model for determining molecular size distribution of DOM. *J. Environ. Eng. Div. Amer. Soc. Civ. Eng.* **116**:1046–1062.

Logan, B. E., and D. K. Kirchman. 1991. Increased uptake of dissolved organic by marine bacteria as a function of fluid motion. *Mar. Biol.* **111**:175–181.

Lomovskaya, O. L., J. P. Kidwell, and A. Matin. 1994. Characterization of the σ^{38}-dependent expression of a core *Escherichia coli* starvation gene, *pexB. J. Bacteriol.* **176**:3928–3935.

Lopez, J. G., and G. R. Vela. 1981. True morphology of Azotobacteraceae-filterable bacteria. *Nature* **289**:588–590.

Loquet, M., T. Bhatnagar, and M. B. Bouché. 1977. Essai' d'estimation de l'influence écologique des lombriciens sur les microorganisms. *Pedobiologia* **17**:400–417.

Love, L. G. 1957. Micro-organisms and the presence of syngenetic pyrite. *Quart. J. Geol. Mining Met. Soc. London* **113**:429–440.

Love, L. G., and D. O. Zimmerman. 1961. Bedded pyrite and micro-organisms from the Mount Isa shale. *Econ. Geol.* **56**:873–896.

Lovett, S. 1964. Effect of deuterium on starving bacteria. *Nature* **203**:429–430.

Lovley, D. R., and M. J. Klug. 1982. Intermediary metabolism of organic matter in the sediments of a eutrophic lake. *Appl. Environ. Microbiol.* **43**:522–560.

Lowe, W. E., and T. R. G. Gray. 1972. Ecological studies on coccoid bacteria in a pine forest soil. I. Classification. *Soil Biol. Biochem.* **4**:459–467.

Lowen, P. C., and R. Hennge-Aronis. 1994. The role of the sigma factor σ^S (*katF*) in bacterial global regulation. *Ann. Rev. Microbiol.* **48**:53–80.

Lowendorf, H. S. 1983. Factors affecting survival of *Rhizobium* in soil. *Adv. Microb. Ecol.* **4**:87–124.

Lucas, C. E. 1955. External metabolites in the sea. *Deep-Sea Res.* **3**:139–148.

Lund, V., and J. Goksoyr. 1980. Effects of water fluctuations on microbial biomass and activity in soil. *Microb. Ecol.* **6**:115–123.

Lundgren, B., and B. Söderstöm, 1983. Bacterial numbers in a pine forest soil in relation to environmental factors. *Soil Biol. Biochem.* **15**:625–630.

Luscombe, B. M., and T. R. G. Gray. 1974. Characteristics of *Arthrobacter* grown in continuous culture. *J. Gen. Microbiol.* **82**:213–222.

Lwoff, A. 1953. Lysogeny. *Bacteriol. Rev.* **17**:269–337.

Lyakin, Y. I. 1968. Calcium carbonate saturation of Pacific water. *Oceanology* **8**:44–53.

Lynch, D., and C. Lynch. 1958. Resistance of protein-lignin complexes, lignin and humic acids to microbial attack. *Nature* **181**:1478–1479.

Lynch, D. L., and L. J. Cotnoir. 1956. The influence of clay minerals on the breakdown of certain organic substances. *Proc. Soil. Sci. Soc. Am.* **20**:367–370.

Lynch, J. M., and L. M. Panting. 1980. Cultivation and the soil biomass. *Soil Biol. Biochem.* **12**:29–33.

Lynch, J. M. 1979. The terrestrial environment. In *Microbial Ecology: A Conceptual Approach,* J. M. Lynch and N. J. Poole (eds.), pp. 67–91. Blackwell Scientific Publications, Oxford.

Lynch, J. M. 1990. Longevity of bacteria: considerations in environmental release. *Curr. Microbiol.* **20**:387–389.

Lytle, C. R., and E. M. Perdue. 1981. Free, proteinaceous, and humic-bound amino acids in river water containing high concentrations of aquatic humus. *Environ. Sci. Technol.* **15**:224–228.

Maaløe, O. 1960. The nucleic acids and the control of bacterial growth. *Symp. Soc. Gen. Microbiol.* **10**:272–293.

Maaløe, O., and N. O. Kjeldgaard. 1966. Control of Macromolecular Synthesis. W. A. Benjamin, Inc., New York.

Macdonald, R. M. 1980. Cytochemical demonstration of catabolism in soil microorganisms. *Soil Biol. Biochem.* **12**:419–423.

MacDonell, M. T., and M. A. Hood. 1982. Isolation and characterization of ultramicrobacteria from a Gulf Coast Estuary. *Appl. Environ. Microbiol.* **43**:566–571.

MacDonell, M. T., and M. A. Hood. 1984. Ultramicrovibrios in Gulf Coast estuarine waters: isolation, characterization and incidence. In *Vibrios in the Environment,* John R. R. Colwell (ed.), pp. 551–562. Wiley & Sons, New York.

MacDonell, M. T., R. M. Baker, F. L. Singleton, and M. A. Hood. 1984. Effects of surface association and osmolarity on seawater microcosm populations of an environmental isolate of *Vibrio cholerae*. *In Vibrios in the Environment*, R. R. Colwell (ed.), pp. 535–549. John Wiley & Sons, New York.

MacIntyre, W. G., and R. F. Platford. 1968. Dissolved $CaCO_3$ in the Labrador Sea. *J. Fish. Res. Bd. Can.* **21**:1475–1480.

MacKelvie, R. M., J. J. R. Campbell, and A. F. Gronlund. 1968. Survival and intracellular changes of *Pseudomonas aeruginosa* during prolonged starvation. *Can. J. Microbiol.* **14**:639–644.

MacLeod, F. A., H. M. Lappin-Scott, and J. W. Costerton. 1988. Plugging of a model system by using starved bacteria. *Appl. Environ. Microbiol.* **54**:1365–1372.

Macrae, R. M., and J. F. Wilkinson. 1958. Poly-ß-hydroxybutyrate metabolism in washed suspensions of *Bacillus cereus* and *Bacillus megaterium*. *J. Gen. Microbiol.* **19**:210–222.

Macurak, J., J. Szolnoki, and V. Vancura. 1963. Decomposition of glucose in soil. *In Soil organisms*, T. Docksen and J. van der Drift (eds.), North Holland, Amsterdam.

Maeda, M., and N. Taga. 1979. Chromogenic assay method of lipopolysaccharide (LPS) for evaluating bacterial standing crop in seawater. *J. Appl. Bacteriol.* **47**:175–182.

Maeda, M., and N. Taga. 1983. Comparisons of cell size of bacteria isolated from four marine localities. *La Mer* **21**:207–210.

Magasanik, B. 1976. Classical and postclassical modes of regulation of the synthesis of degradative bacteria enzymes *Prog. Nucleic Acid Res. Mol. Biol.* **17**.99–115.

Mallory, L. M., B. Austin, and R. R. Colwell. 1977. Numerical taxonomy and ecology of oligotrophic bacteria isolated from the estuarine environment. *Can. J. Microbiol.* **23**:733–750.

Mallory, L. M., C. S. Yuk, L. N. Liang, and M. Alexander. 1983. Alternative prey: a mechanism for elimination of bacterial species by protozoans. *Appl. Environ. Microbiol.* **46**:1073–1079.

Malmcrona-Friberg, K., A. Tunlid, P. Måden, S. Kjelleberg, and G. Odham. 1986. Chemical changes in cell envelope and poly-ß-hydroxybutyrate during short term starvation of a marine bacterial isolate. *Arch. Microbiol.* **144**:340–345.

Malmcrona-Friberg, K., B. B. Blainley, and K. C. Marshall. 1990a. Chemostatic responses of a marine bacterium towards products of an insoluble substrate. *FEMS Microb. Ecol.* **85**:199–206.

Malmcrona-Friberg, K., A. E. Goodman, and S. Kjelleberg. 1990b. Chemotactic responses of the marine *Vibrio* sp. strain CCUG 15956 to low molecular weight substrates under starvation and recovery conditions. *Appl. Environ. Microbiol.* **56**:3699–3704.

Malone, T. C. 1980. Algal size. *In The Physiological Ecology of Phytoplankton*, I. Morris (ed.), pp. 433–463. University of California Press, Berkeley.

Mandelstam, J. 1958a. The free amino acids in growing and non-growing populations of *Escherichia coli*. *Biochem. J.* **69**:103–109.

Mandelstam, J. 1958b. Turnover of protein in growing and non-growing populations of *Escherichia coli*. *Biochem. J.* **69**:110–119.

Mandelstam, J. 1960. The intracellular turnover of protein and nucleic acids and its role in biochemical differentiation. *Bacteriol. Rev.* **24**:289–308.

Mandelstam, J. 1963. Protein turnover and its function in the economy of the cell. *Ann. New York Acad. Sci.* **102**:621–636.

Mandelstam, J., and H. Halvorson. 1960. Turnover of protein and nucleic acid in soluble and ribosome fractions of non-growing *Escherichia coli*. *Biochim. Biophys. Acta* **40**:43–49.

Maniatis, T., E. Fritsch, and J. Sambrook. 1982. *Molecular Cloning: A Laboratory Manual.* Cold Spring Harbor Laboratory, Cold Spring Harbor, New York.

Mantoura, R. F. C., and E. M. S. Woodward. 1983. Conservative behavior of reverine dissolved organic carbon in the Severn Estuary, chemical and geochemical implications. *Geochim. Cosmochim. Acta* **47**:1293–1309.

Mården, P., A. Tunlid, K. Malmcrona-Friberg, G. Odham, and S. Kjelleberg. 1985. Physiological and morphological changes during short term starvation of bacterial isolates. *Arch. Microbiol.* **142**:326–332.

Mården, P., T. Nystrom, and S. Kjelleberg. 1987. Uptake of leucine by a marine Gram-negative heterotrophic bacterium during exposure to starvation conditions. *FEMS Microb. Ecol.* **45**:223–241.

Mården, P., M. Hermansson, and S. Kjelleberg. 1988. Incorporation of tritiated thymidine by bacterial isolates undergoing a starvation-survival response. *Arch. Microbiol.* **149**:427–432.

Marion, G. M., and P. C. Miller. 1982. Nitrogen mineralization in a tussock tundra soil. *Arct. Alp. Res.* **14**:287–293.

Marr, A. G., E. H. Nilsen, and D. J. Clark. 1963. The maintenance requirements of *Escherichia coli*. *Ann. N. Y. Acad. Sci.* **103**:536–548.

Marshall, K. C. 1964. Survival of root-nodule bacteria in dry soils exposed to high temperature. *Aust. J. Agric. Res.* **15**:273–281.

Marshall, K. C. 1975. Clay mineralogy in relation to survival of soil bacteria. *Ann. Rev. Phytopathol.* **13**:357–373.

Marshall, K. C. 1976. *Interfaces in Microbial Ecology.* Harvard University Press, Cambridge.

Marshall, K. C. 1979. Growth at interfaces. *In Strategies of Microbial Life in Extreme Environments,* M. Shilo (ed.), pp. 281–290. Verlag Chemie, Weinheim.

Marshall, K. C. 1980. Bacterial adhesion in natural environments. *In Microbial Adhesion to Surfaces,* R. C. W., Berkeley et al (ed.), pp. 187–196. Ellis Harwood, Chichester,

Marshall, K. C., J. S. Whiteside, and M. Alexander. 1960. Problems in the use of agar for the enumeration of soil microorganisms. *Soil Sci. Soc. Am. Proc.* **24**:61.

Marshman, N. A., and K. C. Marshall. 1981. Bacterial growth on proteins in the presence of clay minerals. *Soil Biol. Biochem.* **13**:127–134.

Martell, A. E. 1971. Principles of complex formation. *In Organic Compounds in Aquatic Environments,* S. D. Faust and J. V. Hunter (eds.), pp. 239–264. Marcel Dekker, New York.

Martens, R. 1985. Estimation of the adenylate energy charge in unamended and amended agricultural soils. *Soil Biol. Biochem.* **17**:765–772.

Martin, A. Jr. 1963. A filterable *Vibrio* from fresh water. *Proc. Pa. Acad. Sci.* **36**:174–178.

Martin, E. M., C. D. Yegian, and G. S. Stent. 1963. The intracellular conditions of soluble ribonucleic acid in *Escherichia coli* to amino acid starvation. *Biochem. J.* **88**:46P.

Martin, J. H., G. A. Knauer, D. M. Karl, and W. W. Broenkow. 1987. VERTEX: carbon cycling in the northeast Pacific. *Deep-Sea Res.* **34**:267–285.

Martin, J. P., J. O. Erwin, and R. A. Shepherd. 1966. Decomposition of the iron, aluminum, zinc, and copper salts or complexes of some microbial and plant polysaccharides in soil. *Soil Sci. Soc. Amer. Proc.* **30**:322–327.

Martin, P., and R. A. MacLeod. 1984. Observations on the distinction between oligotrophic and eutrophic marine bacteria. *Appl. Environ. Microbiol.* **47**:1017–1022.

Martinez, L. M., W. Silver, J. M. King, and A. L. Alldredge. 1983. Nitrogen fixation by floating diatom mats: A source of new nitrogen to oligotrophic ocean waters. *Science* **221**:152–154.

Marumoto, T., and Y. Yamada. 1977. Significance and characteristics of readily decomposable organic matter in paddy soils. *Bull. Fac. Agric., Yamaguchi University* **28**:71–81.

Marumoto, T., T. Kai, H. Yoshida, and T. Harada. 1977. Drying effect of mineralization of microbial cells and their cell walls in soil and contribution of microbial cell walls as source of decomposable soil organic matter due to drying. *Soil Sci. Plant Nutri.* **23**:9–19.

Marumoto, T., J. P. E. Anderson, and K. H. Domsch. 1982a. Decomposition of ^{14}C- and ^{15}N-labelled microbial cells in soil. *Soil Biol. Biochem.* **14**:461–467.

Marumoto, T., J. P. E. Anderson, and K. H. Domsch. 1982b. Mineralization of nutrient from soil microbial biomass. *Soil Biol. Biochem.* **16**:469–475.

Maruyama, H. B., and S. Okamura. 1972. Ribosome degradation and the degradation products in starved *Escherichia coli*. V. Ribonucleoprotein particles from glucose-starved cells. *J. Bacteriol.* **110**:442–446.

Maruyama, Y., T. Komano, H. Fujita, T. Muroyama, T. Ando, and T. Ogawa. 1977. Synchronization of bacterial cells by glucose starvation. *In Growth and Differentiation,* T. Ishikawa, Y. Maruyama and H. Matsumoto (eds.), pp. 77–93. University Park Press, Baltimore.

Marxsen, J. 1988. Investigations into the number of respiring bacteria in groundwater from sandy and gravelly deposits. *Microb. Ecol.* **16**:65–72.

Mason, C. A., G. Hamer, and J. D. Bryers. 1986. The death and lysis of microorganisms in environmental processes. *FEMS Microb. Rev.* **39**:373–401.

Massa, E. M., A. L. Vinals, and R. N. Farias. 1988. Influence of unsaturated fatty acids membrane components on sensitivity of an *Escherichia coli* fatty acid auxotroph to conditions of nutrient depletion. *Appl. Environ. Microbiol.* **54**:2107–2111.

Mathur, S. P. 1971. Characterization of soil humus through enzymatic degradation. *Soil Sci.* **111**:147–157.

Matin, A. 1979. Microbial regulatory mechanisms at low nutrient concentrations as studied in a chemostat. *In Strategies of Microbial Life in Extreme Environments,* M. Shilo (ed.), pp. 323–339. Verlag Chemie, Weinheim.

Matin, A. 1981. Regulation of enzyme synthesis as studied in continuous culture. *In Continuous Cultures of Cells,* vol. 2, P. H. Calcott (ed.), pp. 69–97. CRC Press, Boca Raton, FL.

Matin, A. 1990. Molecular analysis of the starvation stress in *Escherichia coli. FEMS Microb. Ecol.* **74**:186–196.

Matin, A. 1991. The molecular basis of carbon-starved induced general resistance in *Escherichia coli. Mol. Microbiol.* **5**:3–10.

Matin, A., and H. Veldkamp. 1978. Physiological basis of the selective advantage of a *Spirillum* sp. in a carbon-limited environment. *J. Gen Microbiol.* **105**:187–197.

Matin, A., and J. C. Gottschal. 1976. Influence of dilution rate on NAD(P)H concentrations and ratios in a *Pseudomonas* sp. grown in continuous culture. *J. Gen. Microbiol.* **94**:332–341.

Matin, A., A. Grootjans, and H. Hogenhuis. 1976. Influence of dilution rate on enzymes of intermediary metabolism in two freshwater bacteria grown in continuous culture. *J. Gen. Microbiol.* **94**:323–332.

Matin, A., C. Veldhuis, V. Stegeman, and M. Veenhuis. 1979. Selective advantage of a *Spirillum* sp. in a carbon-limited environment. Accumulation of poly-ß-hydroxybutyric acid and its role in starvation. *J. Gen. Microbiol.* **112**:349–355.

Matin, A., E. A. Auger, P. H. Blum, and J. E. Schultz. 1989. Genetic basis of starvation survival in nondifferentiating bacteria. *Ann. Rev. Microbiol.* **43**:293–316.

Maurer, L. G., and P. L. Parker. 1972. The distribution of dissolved organic matter in nearshore waters of the Texas coast. *Contr. Mar. Sci., University Texas* **16**:109–124.

Mayaudon, J., and P. Simonart. 1958. Study of the decomposition of organic matter in soil by means of radioactive carbon. II. The decomposition of radioactive glucose in soil and distribution of radioactivity in the humus fractions of soil. *Plant Soil* **9**:376–394.

Mayaudon, J., and P. Simonart. 1963. Humification des microorganismes marque par ^{14}C dan le sol. *Ann. Inst. Pasteur* **105**:257–266.

Mayaudon, J., and P. Simonart. 1965. Humification dans le sol d'un complexe polysaccharidique C^{14} d'origine microbienne. *Meded. LandHogesch. Ppzoekstnts Gent.* **30**:941–955.

Mayer, G. H., M. B. Morrow, and O. Wyss. 1962. Viable micro-organisms in a fifty year-old yeast preparation in Antarctica. *Nature* **196**:598.

Mayer, L. M., L. L. Schick, and F. W. Setchell. 1986. Measurement of protein in nearshore sediments. *Mar. Ecol. Prog. Ser.* **30**:159–165.

McBee, R. H., and V. H. McBee. 1956. The incidence of thermophilic bacteria in Arctic soils and waters. *J. Bacteriol.* **71**:182–185.

McCann, M. P., J. P. Kidwell, and A. Matin. 1991. The putative σ factor *katF* has a central role in development of starvation-mediated general resistance in *Escherichia coli. J. Bacteriol.* **173**:4188–4194.

McCann, M. P., J. Kidwell, and A. Matin. 1992. Microbial starvation survival, genetics. *In Encyclopedia of Microbiology,* vol. 3, J. Lederberg (ed.), pp. 159–170. Academic Press, San Diego.

McCarthy, D. H. 1977. Some ecological aspects of bacteria fish pathogen-*Aeromonas salmonicida. Aquat. Microb.* **6**:299–32.

McCarthy, J. J., and J. C. Goldman. 1979. Nitrogen nutrition of marine phytoplankton in nutrient depleted waters. *Science* **203**:670–672.

McCarty, P. L. 1971. Energetics and bacterial growth. *In Organic Compounds in Aquatic Environments,* S. D. Faust and J. V. Hunter (eds.), pp. 495–531. Marcel Dekker, New York.

McCave, I. N. 1975. Vertical flux of particles in the ocean. *Deep-Sea Res.* **22**:491–502.

McCave, I. N. 1984. Size spectra and aggregation of suspended particles in the deep ocean. *Deep-Sea Res.* **31**:329–352.

McCorkle, D. C., and S. R. Emerson. 1988. The relationship between pore water carbon isotopic composition and bottom oxygen concentration. *Geochim. Cosmochim. Acta* **52**:1169–1178.

Mcfadyen, A. 1970. Soil metabolism in relation to ecosystem energy flow and primary and secondary production. *In Methods for the Study of Production and Energy Flow in Soil Communities,* J. Phillipson (ed.), pp. 167–172. UNESCO, Paris.

McFeters, G. A. 1990. Enumeration, occurrence, and significance of injured indicator bacteria in drinking water. *In Drinking Water Microbiology,* G. A. McFeters (ed.), pp. 478–495 Springer Verlag, New York.

McGill, W. B., and E. A. Paul. 1976. Fractionation of soil and ^{15}N nitrogen to separate the organic and clay interactions of immobilized N. *Can. J. Soil Sci.* **56**:203–212.

McGill, W. B., E. A. Paul, J. A. Shields, and W. E. Lowe. 1973. Turnover of microbial populations and their metabolites in soil. *Bull. Ecol. Res. (Stockholm)* **17**:293–301.

McGrew, S. B., and M. F. Mallette. 1962. Energy of maintenance in *Escherichia coli. J. Bacteriol.* **83**:844–850.

McKinley, V. L., J. W. Costerton, and D. C. White. 1988. Microbial biomass, activity and particulate retrieved by backflow from water flood injection well. *Appl. Environ. Microbiol.* **54**:1383–1393.

McLaren, A. D. 1960. Enzyme action in structurally restricted systems. *Enzymologia* **21**:356–364.

McLaren, A. D. 1963. Enzyme reactions in structurally restricted systems, IV. The digestion of insoluble substrates by hydrolytic enzymes. *Enzymologia* **26**:237–246.

McLaren, A. D. 1973a. A need for counting microorganisms in soil mineral cycles. *Environ. Lett.* **5**:142–154.

McLaren, A. D. 1973b. Enzymatic activity in soils sterilized by ionizing radiation and some comments on micro-environments in nature. *In Recent Progress in Microbiology VIII,* pp. 221–229. University of Toronto Press, Toronto.

McLaren, A. D., and J. Skujins. 1968. The physical environment of microorganisms in soil. *In The Ecology of Soil Bacteria,* T. R. G. Gray and D. Parkinson (eds.), pp. 3–24. University of Toronto Press, Toronto.

McLaughlin, M. J., A. M. Alston, and J. K. Martin. 1986. Measurement of phosphorus in the soil microbial biomass: a modified procedure for field soils. *Soil Biol. Biochem* **18**:437–443.

McNaughton, S. J., and L. L. Wolf. 1973. *General Ecology.* Holt, Rinehart and Winston, New York.

Meganathan, R., and J. C. Ensign. 1976. Stability of enzymes in starving *Arthrobacter crystallopoietes. J. Gen. Microbiol.* **94**:90–96.

Meighen, E. A. 1991. Molecular biology of bacterial bioluminescence. *Microbiol. Rev.* **55**:123–142.

Melchioirri-Santolini, U., and A. Carfarelli. 1967. Lake water as a medium to cultivate freshwater pelagic bacteria. *Mem. Inst. Ital. Idrobiol.* **22**:289–298.

Melchior, J. B., O. Klioze, and I. M. Klotz. 1951. Further studies of the synthesis of protein by *Escherichia coli. J. Biol. Chem.* **189**:411–420.

Menzel, D. W. 1964. The distribution of dissolved organic carbon in the western Indian Ocean. *Deep-Sea Res.* **11**:757–765.

Menzel, D. W. 1967. Particulate organic carbon in the deep sea. *Deep Sea Res.* **14**:229–238.

Menzel, D. W. 1970. The role of in situ decomposition of organic matter on the concentration of non-conservative properties in the sea. *Deep-Sea Res.* **17**:751–764.

Menzel, D. W. 1974. Primary productivity, dissolved and particulate organic matter, and the sites of oxidation of organic matter. *In The Sea,* vol. 5, E. D. Goldberg (ed.), pp. 659–678. Interscience Publishers, New York.

Menzel, D. W., and J. J. Goering. 1966. The distribution of organic detritus in the ocean. *Limnol. Oceanogr.* **11**:222–337.

Menzel, D. W., and J. H. Ryther. 1968. Organic carbon and the oxygen minimum in the South Atlantic Ocean. *Deep-Sea Res.* **15**:327–337.

Menzel, D. W., and J. H. Ryther. 1970. Distribution and cycling of organic matter in the oceans. *In Organic Matter in Natural Waters,* D. W. Hood (ed.), pp. 31–54. Institute of Marine Science Occasional Publication No. 1. University of Alaska, College AK.

Menzel, D. W., and R. F. Vaccaro. 1964. The measurement of dissolved organic and particulate carbon in seawater. *Limnol. Oceanogr.* **9**:138–142.

Menzies, R. J. 1962. On the food and feeding habits of abyssal organisms as exemplified by the Isopoda. *Intern. Rev. der Gesamien Hydrobiologie und Hydrographie* **47**:339–358.

Mesibov, R., and J. Adler. 1972. Chemotaxis toward amino acids in *Escherichia coli. J. Bacteriol.* **122**:315–326.

Meury, J. and M. Kohiyama. 1991. Role of heat shock protein DnaK in osmotic adaptation of *Escherichia coli. J. Bacteriol.* **173**:4404–4410.

Meveczky, N., and H. Rosenberg. 1969. The binding and release of phosphate by a protein isolated from *Escherichia coli.* Biochim. Biophys. Acta **192**:369–371.

Meyer, J. L., R. T. Edwards, and R. Risley. 1987. Bacterial growth on dissolved organic carbon from a blackwater river. *Microb. Ecol.* **13**:13–29.

Meyer-Reil, L.-A. 1975. An improved method for the semicontinuous culture of bacterial populations in Nuclepore membrane filters. *Kiel. Meeresforsch.* **31**:1–6.

Meyer-Reil, L.-A. 1978. Autoradiography and epifluorescence microscopy combined for the determination of number and spectrum of actively metabolizing bacteria in natural waters. *Appl. Environ. Microbiol.* **39**:797–802.

Meyer-Reil, L.-A. 1981. Enzymatic decomposition of proteins and carbohydrates in marine sediments. Methodology and field observations during spring. *Kiel. Meeresforsch* **5**:211–317.

Meyer-Reil, L.-A., and M. Köster. 1992. Microbial life in pelagic sediments: the impact of environmental parameters on enzymatic degradation of organic material. *Mar. Ecol. Prog. Ser.* **81**:65–72.

Meyers, A. J., and C. D. Meyers. 1986. *Hyphomicrobium*-mediated sludge bulking in an industrial wastewater treatment system. *Abstract from the General Meetings of the American Society of Microbiology*, abstr., p. N-93.

Midgley, J. E. M., and Sir C. Hinshelwood. 1961. Lag, adaptation, and ageing in microorganisms (*Bact. lactis aerogenes*). *Proc. Roy. Soc. B* **155**:195–201.

Miller, C. G., and L. Green. 1983. Degradation of proline peptide in peptidase-deficient strains of *Salmonella typhimurium*. *J. Bacteriol.* **153**:350–356.

Miller, R. E., and L. A. Simons. 1962. Survival of bacteria after twenty-one years in the dried state. *J. Bacteriol.* **84**:1111–1114.

Miller, R. V., and T. A. Kokjohn. 1987. Cloning and characterization of the c1 repressor of *Pseudomonas aeruginosa* phage D3: a functional analogue of phage lambda cI protein. *J. Bacteriol.* **169**:1847–1852.

Mink, R. W., and R. B. Hespell. 1981a. Long-term nutrient starvation of continuously cultured (glucose-limited) *Selenomonas ruminanatium*. *J. Bacteriol.* **148**:541–550.

Mink, R. W., and R. B. Hespell. 1981b. Survival of *Megasphaera elsdenii* during starvation. *Curr. Microbiol.* **5**:51–56.

Mishustin, E. N. 1975. Microbial associations of soil types. *Microb. Ecol.* **2**:97–118.

Mishustin, E. N. 1975. *Association of Soil Microorganisms*. Nauka, Moscow (see Hattori, 1980).

Mitchell, J. G. 1991. The influence of cell size on marine bacterial motility and energetics. *Microb. Ecol.* **22**:227–238.

Mitchell, D. O., and M. J. Starzyk. 1975. Survival of *Salmonella* and other indicator microorganisms. *Can. J. Microbiol.* **21**:1420–1421.

Mitchell, J. G., A. Okubo, and J. A. Fuhrman. 1985. Microzone surrounding phytoplankton form the basis for a stratified microbial ecosystem. *Nature* **316**:58–59.

Mittelman, M. W., and G. G. Geesey. 1987. *Biological Fouling of Industrial Water Systems: A Problem Solving Approach*. Water Micro Associates, San Diego, CA.

Mitterer, R. M. 1968. Amino acid composition of organic matrix in calcareous oolites. *Science* **162**:1498–1499.

Miyamoto, S., and H. Seki. 1992. Environmental factors controlling the population growth rate of the bacterial community in Matsumi-Ike bog. *Water, Air, Soil Pollu.* **63**:379–396.

Moaledj, von K. 1978. Qualitative analysis of an oligocarbophilic aquatic microflora in the Plusssee. *Arch. Hydrobiol.* **82**:98–113.

Moaledj, K., and J. Overbeck. 1982. Verteilung der oligocarbophilen und saprophytischen Baktereien in Plusssee. *Arch. Hydrobiol.* **93**:287–302.

Moller, S., C. S. Kristensen, L. K. Poulsen, J. M. Carstensen, and S. Molin. 1995. Bacterial growth on surfaces: automated image analysis for quantification of growth rate-related parameters. *Appl. Environ. Microbiol.* **61**:741–748.

Monod, J. 1949. The growth of bacterial cultures. *Ann. Rev. Microbiol.* **3**:371–394.

Monod, J. 1942. *Recherches sur la croissance des cultures bactériennes.* Hermann et Cie, Paris.

Monniot, C. 1979. Adaptation of benthic filtering aminals to the scarcity of suspended particles in deep water. *Ambio Spec. Rept.* **6**:73–74.

Montagna, P. A., and J. E. Bauer. 1988. Partitioning radiolabeled thymidine uptake by bacteria and meiofauna using metabolic blocks and poisons in benthic feeding studies. *Mar. Biol.* **98**:101–110.

Montague, M. D., and E. A. Dawes. 1974. The survival of *Peptococcus prevotii* in relation to the adenylate energy charge. *J. Gen. Microbiol.* **80**:291–299.

Monteith, J. L., G. Szeicz, and K. Yakuku. 1964. Crop photosynthesis and the flux of carbon dioxide below the canopy. *J. Appl. Ecol.* **1**:321–337.

Moodie, P. L. 1916. Mesozoic pathology and bacteriology. *Science* **43**:425–426.

Moore, L. R. 1969. Geomicrobiology and geomicrobial attack on sediment organic matter. *In Organic Geochemistry: Methods and Results,* G. Eglington and M. T. J. Murphy (eds.), pp. 264–303. Springer-Verlag, New York.

Mopper, K. 1977. Sugar and uronic acids in sediment and water from the Black Sea and North Sea with emphasis on analytical techniques. *Mar. Chem.* **5**:585–603.

Mopper, K., and E. T. Degans. 1979. Organic carbon in the ocean: nature and cycling. *In Global Carbon Cycle,* B. Bolin, E. T. Degens, S. Kempe and P. Ketner (eds.), pp. 293–316. John Wiley & Sons, Chichester.

Mopper, K., and W. L. Stahovec. 1986. Sources and sinks of low molecular weight organic carbonyl compounds in seawater. *Mar. Chem.* **19**:305–321.

Mopper, K., and R. Zika. 1987. Free amino acids in marine rains: evidence for oxidation and potential role in nitrogen cycling. *Nature* **325**:246–249.

Moran, M. A., and R. E. Hodson. 1989. Formation and bacterial utilization of dissolved organic carbon derived from detrital lignocellulose. *Limnol. Oceanogr.* **34**:1034–1047.

Moran, M. A., and R. E. Hodson. 1990. Contributions of degrading *Apartina alterniflora* lignocellulose to the dissolved organic carbon pool of a salt marsh. *Mar. Ecol. Prog. Ser.* **62**:161–168.

Moreno, J., J. Gonzalez-Lopez, and G. R. Vela. 1986. Survival of *Azotobacter* spp. in dry soil. *Appl. Environ. Microbiol.* **51**:123–125.

Morgan, J. A. W., K. J. Clarke, G. Rhodes, and R. W. Pickup. 1992. Non-culturable *Aeromonas salmonicida* in lake water. *Microb. Releases* **1**:71–78.

Morgan, J. A. W., G. Rhodes, and R. W. Pickup. 1993. Survival of nonculturable *Aeromonas salmonicida* in lake water. *Appl. Environ. Microbiol.* **59**:874–880.

Morgan, P., and C. S. Dow. 1986. Bacterial adaptation for growth in low nutrient environments. *In Microbes in extreme environments,* R. A. Herbert and G. A. Codd (eds.), pp. 187–214. Academic Press, London.

Morgan, R. W., M. F. Christman, F. S. Jacobson, G. Storz, and B. N. Ames. 1986. Hydrogen peroxide-inducible proteins in *Salmonella typhimurium* overlap with heat shock and other stress proteins. *Proc. Natl. Acad. Sci. (USA)* **83**:8059–8063.

Moriarty, D. J. W. 1988. Measurements of microbial growth rates in aquatic systems from the rate of nucleic acid synthesis. *Adv. Microb. Ecol.* **9**:245–292.

Moriarty, D. J. W., and R. T. Bell. 1993. Bacterial growth and starvation in aquatic environments. *In Starvation in Bacteria,* S. Kjelleberg (ed.), pp. 25–53. Plenum Press, New York.

Moriarty, D. J. W., and P. C. Pollard. 1981. DNA synthesis as a measure of bacterial productivity in seagrass sediments. *Mar. Ecol. Prog. Ser.* **5**:151–156.

Moriarty, D. J. W., and P. C. Pollard. 1982. Diel variation of bacterial productivity in seagrass (*Zostera capricorni*) beds measured by rate of thymidine incorporation into DNA. *Mar. Biol.* **72**:165–172.

Moriarty, D. J. W., and P. C. Pollard. 1990. Effects of radioactive labelling of macromolecules, disturbance of bacteria and adsorption of thymidine to sediment on the determination of bacterial growth rates in sediment with tritiated thymidine. *J. Microb. Methods* **11**:127–139.

Moriarty, D. J. W., P. C. Pollard, W. G. Hunt, C. M. Moriarty, and T. J. Wassenberg. 1985. Productivity of bacteria and microalgae and the effect of grazing by holothurians in sediments on a coral reef flat. *Mar. Biol.* **85**:293–300.

Morita, R. Y. 1954. Occurrence and significance of bacteria in marine sediments. Ph. D. Thesis. University of California (Scripps Institution of Oceanography), La Jolla, CA.

Morita, R. Y. 1968. *In Marine Microbiology,* C. H. Oppenheimer (ed.), p. 97. Proc. 4th International Interdisciplinary Conference. New York Academy of Sciences, New York.

Morita, R. Y. 1979a. The role of microbes in the bioenergetics of the deep-sea. Proceedings of the Centenary Symposium of the Kristineberg Marine Biological Laboratory. *Sarsia* **64**:9–12.

Morita, R. Y. 1979b. The role of microbes in the bioenergetics of the deep-sea. Proceedings of the Centenary Symposium of the Kristineberg Marine Biological Laboratory. *Sarsia* **64**:9–12.

Morita, R. Y. 1980a. Calcite precipitation by marine bacteria. *Geomicrobiol. J.* **2**:63–82.

Morita, R. Y. 1980b. Microbial life in the deep sea. *Can. J. Microbiol.* **26**:1375–1385.

Morita, R. Y. 1980c. Low temperature, energy, survival and time in microbial ecology. *In Microbiology-1980,* D. Schlessinger (ed.), pp. 323–324. American Society for Microbiology, Washington, DC.

Morita, R. Y. 1982. Starvation-survival of heterotrophs in the marine environment. *Adv. Microb. Ecol.* **6**:171–198.

Morita, R. Y. 1984a. Feast or famine in the deep sea. *Develop. Indust. Microbiol.* **25**:5–16.

Morita, R. Y. 1984b. Substrate capture by marine heterotrophic bacteria in low nutrient waters. *In Heterotrophic Activity in the Sea,* J. E. Hobbie and P. J. leB. Williams (eds.), pp. 83–100. Plenum Press, New York.

Morita, R. Y. 1985. Starvation and miniaturization of heterotrophs, with special reference on the maintenance of the starved viable state. *In Bacteria in Natural Environments: The Effect of Nutrient Conditions,* M. Fletcher and G. Floodgate (eds.), pp. 111–130. Academic Press, London.

Morita, R. Y. 1986. Starvation-survival: The normal mode of most bacteria in the ocean. *In Perspectives in microbial ecology,* F. Megusar and M. Gantar (eds.), pp. 243–248. Slovene Society for Microbiology, Ljubljana, Yugoslavia.

Morita, R. Y. 1988. Bioavailability of energy and its relationship to growth and starvation survival in nature. *Can. J. Microbiol.* **34**:436–441.

Morita, R. Y. 1990. The starvation-survival state of microorganisms in nature and its relationship to the bioavailable energy. *Experientia* **46**:813–817.

Morita, R. Y. 1993. Bioavailability of energy and the starvation state. *In Starvation in Bacteria,* S. Kjelleberg (ed.), pp. 1–23. Plenum Press, New York.

Morita, R. Y., and C. E. ZoBell. 1955. Occurrence of bacteria in pelagic sediments collected during the Mid-Pacific Expedition. *Deep-Sea Res.* **3**:66–73.

Morita, R. Y., and G. E. Buck. 1974. Low temperature inhibition of substrate uptake. *In Effect of the Ocean Environment on Microbial Activities,* R. R. Colwell and R. Y. Morita (eds.), pp. 124–129. University Park Press, Baltimore.

Morita, R. Y., and S. D. Burton. 1970. Occurrence and possible significance and metabolism of marine oligately psychrophilic bacteria. *In Organic Matter nin Natural Waters,*. D. W. Hood (ed.), pp. 275–285. Institute of Marine Science Occasional Publications No. 1, University of Alaska, College AK.

Morris, D. W., and J. A. DeMoss. 1965. Role of aminoacyl-transfer ribonucleic acid in the regulation of ribonucleic acid synthesis in *Escherichia coli. J. Bacteriol.* **90**:1624–1631.

Morris, A., and P. Foster. 1971. The seasonal variation of dissolved organic carbon in the inshore waters of the Menai Strait in relation to primary production. *Limnol. Oceanogr.* **16**:987–989.

Morrison, S. J., and D. C. White. 1980. Effects of grazing by Gammaridean amphipods on the microbiota of allochthonous detritus. *Appl. Environ. Microbiol.* **40**:659–671.

Morse, M. L., and C. E. Carter. 1949. The synthesis of nucleic acids in culture of *Escherichia coli,* strains B and B/R. *J. Bacteriol.* **58**:317–326.

Mortenson, J. L. 1963. Decomposition of organic matter and mineralization of nitrogen in Brookston silt loam and alfalfa green manure. *Plant Soil* **19**:374–384.

Morton, D. S., and J. D. Oliver. 1994. Induction of carbon starvation-induced proteins in *Vibrio vulnificus. Appl. Environ. Microbiol.* **60**:3653–3659.

Moyer, C. L., and R. Y. Morita. 1989a. Effect of growth rate and starvation-survival on the viability and stability of a psychrophilic marine bacterium. *Appl. Environ. Microbiol.* **55**:1122–1127.

Moyer, C. L., and R. Y. Morita. 1989b. Effect of growth rate and starvation-survival on cellular DNA, RNA and protein of a psychrophilic marine bacterium. *Appl. Environ. Microbiol.* **55**:2710–2716.

Moyer, C. L., F. C. Dobbs, and D. M. Karl. 1994. Estimation of diversity and community structure through restriction fragment length polymorphism distribution analysis of bacterial 16S rRNA genes from a microbial mat at an active, hydrothermal vent system, Lohi Seamount, Hawaii. *Appl. Environ. Microbiol.* **60**:871–879.

Moyer, C. L., J. M. Tiedje, F. C. Dobbs, and D. M. Karl. 1996. A computer-simulated restriction fragment length polymorphism analysis of bacterial small-subunit rRNA genes: efficacy of selected tetrameric restriction enzymes for studies of microbial diversity in nature. *Appl. Environ. Microbiol.* **62**:2501–2507.

Mulvaney, R. L., and J. M. Bremner. 1981. Control of urea transformations in soil. *Soil Biochem.* **5**:153–196.

Munczak, F. 1960. On the appearance of ninhydrin-positive substances in the atmosphere. *Tellus* **12**:127.

Munk, W. H. 1966. Abyssal recipes. *Deep-Sea Res.* **13**:707–730.

Munk, W. H., and G. A. Riley. 1952. Absorption of nutrients by aquatic plants. *J. Mar. Sci.* **11**:215–240.

Munro, P. M., R. L. Clément, G. N. Flatau, and M. J. Gauthier. 1994. Effect of thermal, oxidative, acidic, osmotic, or nutritional stresses on subsequent culturability of *Escherichia coli* in seawater. *Microb. Ecol.* **27**:57–63.

Munro, P. M., M. J. Gauthier, and F. M. Laumond. 1987. Changes in *Escherichia coli* starved in seawater or grown in seawater-wastewater mixtures. *Appl. Environ. Microbiol.* **53**:1476–1481.

Murakami, A., T. Matsuda, N. Watanabe, and S. Nagasawa. 1976. Degradation of n-paraffin mixtures by marine microorganisms in enriched seawater medium. *J. Oceanogr. Soc. Jpn.* **32**:242–248.

Murray, G. R. E. 1985. More on bacterial longevity: The Murray Collection. *ASM News* **51**:261–262.

Myers, G. E., and R. G. L. McCready. 1966. Bacteria can penetrate rock. *Can. J. Microbiol.* **12**:477–484.

Naganuma, T., and H. Seki. 1985. Population growth rate of the bacterioplankton community in a bog, Matsumi-ike, Japan. *Arch. Hydrobiol.* **104**:543–556.

Naganuma, T., and H. Seki. 1988. Effect of rapid oligotrophication by aquatic treatment pilot plant on the microbial community of a mesophilic bog. *Water, Air, Soil Pollu.* **42**:421–432.

Nagata, T., and D. L. Kirchman. 1990. Filtration-induced release of dissolved amino acids: applications to cultures of marine protozoa. *Mar. Ecol. Prog. Ser.* **68**:1–5.

Nannipieri, P., S. Grego, and B. Ceccanti. 1990. Ecological significance of the biological activity in soil. *Soil Biochem.* **6**:293–355.

Nath, K., and A. L. Koch. 1970. Protein degradation in *Escherichia coli.* Part I. Measurement of rapidly and slowly decaying components. *J. Biol. Chem.* **245**:2889–2900.

Nath, K., and A. L. Koch. 1971a. Protein degradation in *Escherichia coli.* Part II. Strain differences in the degradation of protein and nucleic acid resulting from starvation. *J. Biol. Chem.* **246**:6956–6957.

Nath, K., and A. L. Koch. 1971b. Protein degradation in *Escherichia coli. J. Biol. Chem.* **246**:6956–6967.

Naumann, E. 1919. Några synpunkter angående limnoplanktons ökologi med särskild hänsyn till fytoplankton. *Svensk. Bot. Tidskr.* **13**:129–163. (English translat. by the Freshwater Biological Association, No. 49.)

Nealson, K. H. 1982. Bacterial ecology of the deep sea. *In The Environment of the Deep Sea,* W. G. Ernst and G. Morin (eds.), pp. 179–200. Prentice-Hall, Englewood Cliffs, NJ.

Nedwell, D. B. 1987. Distribution and pool sizes of microbially available carbon in sediment measured by a microbiological assay. *FEMS Microbiol. Ecol.* **45**:47–52.

Nedwell, D. B., and J. W. Abram. 1979. Relative influence of temperature and electron donor and electron acceptor concentrations on bacteria sulphate reduction in saltmarsh sediment. *Microb. Ecol.* **5**:67–72.

Nedwell, D. B., and T. R. G. Gray. 1987. Soils and sediment as matrices for microbial growth. *Symp. Soc. Gen. Microbiol.* **41**:21–54.

Nei, T. 1977. *Preservation of Microorganisms.* University Tokyo Press, Tokyo. (In Japanese).

Neidhardt, F. C., J. L. Ingraham, and M. Schaechter. 1990. *Physiology of the Bacterial Cell: A Molecular Approach.* Sinauer, Sunderland, MA.

Neijssel, O. M. 1980. A microbiologist's view of genetic engineering. *Trends Biochem. Sci.* **5**:III–IV.

Neijssel, O. M., and D. W. Tempest. 1976. Bioenergetics aspects of aerobic growth of *Klebsiella aerogenes* NCTC 418 in carbon-limited and carbon-sufficient chemostat cultures. *Arch. Microbiol.* **107**:215–221.

Neijssel, O. M., S. Hueting, K. J. Crabbendam, and D. W. Tempest. 1975. Dual pathways of glycerol assimilation in *Klebsiella aerogenes* NCIB 418. Their regulation and possible functional significance. *Arch. Microbiol.* **104**:83–87.

Nelson, C. H., J. A. Robinson, and W. G. Characklis. 1985. Bacterial adsorption to smooth surfaces: rate, extent and spatial pattern. *Biotech. Bioeng.* **27**:1662–1667.

Nelson, G. A., and G. Semeniuk. 1963. Persistence of *Corynebacterium insidiosum* in the soil. *Phytopathology* **53**:1167–1169.

Nelson, L. M., and D. Parkinson. 1978. Effect of starvation on survival of three bacterial isolates from an arctic soil. *Can. J. Microbiol.* **24**:1460–1467.

Newell, D. B. 1987. Distribution and pool size of microbially available carbon in sediment measured by a microbiological assay. *FEMS Microb. Ecol.* **45**:47–52.

Newell, R. C., M. I. Lucas, and E. A. S. Linely. 1981. Rate of degradation and efficiency of conversion of phytoplankton detritus by marine micro-organisms. *Mar. Ecol. Prog. Ser.* **6**:123–136.

Newell, S. Y., and R. D. Fallon. 1982. Bacterial productivity in the water column and sediments of the Georgia (USA) coastal zone: estimates via direct counting and parallel measurements of thymidine incorporation. *Microb. Ecol.* **8**:33–46.

Newell, S. Y., and R. R. Christian. 1981. Frequency of dividing cells as an estimator of bacterial productivity. *Appl. Environ. Microbiol.* **42**:23–31.

Nichols, H. W., and H. C. Bold. 1965. *Trichosarcina polymorpha* gen. et st. nov. *J. Phycol.* **1**:34–38.

Nickels, J. S., J. D. King, and D. C. White. 1979. Poly-beta-hydroxybutyrate metabolism as a measure of unbalanced growth in estuarine microbiota. *Appl. Environ. Microbiol.* **37**:459–465.

Nikitin, D. I., and I. Zlatkin. 1989. Oligotrophic microorganisms as a special evolutionary group of procarotes. *In Recent Advances in Microbial Ecology,* T. Hattori, Y. Ishida, Y. Maruyama, R. Y. Morita, and A. Uchida (eds.), pp. 94–99. Japan Scientific Societies Press, Tokyo.

Nikitin, D. I., O. Yu. Vishnevetskaya, and I. V. Zlatkin. 1988. Grouping of oligotrophic microorganisms on the basis of their antibiotic resistance and dynamic membrane characteristics. *Microbiology* **57**:210–215.

Nikitin, D. I. 1973. Direct electron microscopic techniques for the observation of microorganisms in soil. *Bull. Ecol. Res. Commn. (Stockholm)* **17**:85–92.

Nikitin, D. I., and K. M. Chumakov. 1986. The functional role of oligotrophic bacteria. *In Microbial Communities in Soil,* V. Jensen, A. Kjøller and L. H. Sørensen (eds.), pp. 177–189. Elsevier Applied Science Publishers, London.

Nioh, I., and C. Furusaka. 1968. Growth of bacteria in the heat-killed suspensions of the same bacteria. *J. Gen. Appl. Microbiol.* **14**:373–385.

Nissen, H. 1987. Long term starvation of a marine bacterium, *Alteromonas denitrificans,* isolated from a Norwegian fjord. *FEMS Microbiol. Ecol.* **45**:173–183.

Nissen, H., P. Nissen, and F. Azam. 1984. Multiphasic uptake of D-glucose by an oligotrophic marine bacterium. *Mar. Ecol. Prog. Ser.* **16**:155–160.

Niven, D. F., P. A. Collins, and C. J. Knowles. 1977. Adenylate energy charge during batch culture of *Beneckea natriegens. J. Gen. Microbiol.* **98**:95–108.

Noble, P. A., P. E. Dabinett, and J. Gow. 1990. A numerical taxonomic study of pelagic and benthic surface-layer bacteria in seasonally-cold coastal waters. *Syst. Appl. Microbiol.* **13**:77–85.

Norkrans, B. 1980. Surface microlayers in aquatic environments. *Adv. Microb. Ecol.* **4**:51–85.

Norkrans, B., and B. O. Stehn. 1978. Sediment bacteria in the deep Norwegian Sea. *Mar. Biol.* **47**:201–209.

Norris, J. R., and H. Swain. 1971. Staining bacteria. *Methods in Microbiology* **5A**:105–134.

North, B. B. 1975. Primary amines in California coastal water: utilization by phytoplankton. *Limnol. Oceanogr.* **20**:20–27.

Norton, C. F. 1992. Rediscovering the ecology of halobacteria. *ASM News* **58**:363–367.

Norton, C. F., and W. D. Grant. 1988. Survival of halobacteria within fluid inclusions in salt crystals. *J. Gen. Microbiol.* **134**:1365–1373.

Norton, C. F., T. J. McGenity, and W. D. Grant. 1993. Archaeal halophiles (halobacteria) from two British salt mines. *J. Gen. Microbiol.* **139**:1077–1081.

Norton, J., and M. K. Firestone. 1991. Metabolic status of bacteria and fungi in the rhizosphere of Ponderosa pine seedlings. *Appl. Environ. Microbiol.* **57**:1161–1167.

Novick, A. 1958. Genetic and physiological studies with the chemostat. *In Continuous Culture of Microorganisms,* M. Malek (ed.), p. 29. Publishing House of the Czechoslovak Academy of Sciences. Prague.

Novitsky, J. A. 1977. Effects of long-term nutrient starvation on a marine psychrophilic vibrio. Ph. D. Thesis. Oregon State University, Corvallis, OR.

Novitsky, J. A. 1983a. Heterotrophic activity throughout a vertical profile of seawater and sediment in Halifax Harbor, Canada. *Appl. Environ. Microbiol.* **45**:1753–1760.

Novitsky, J. A. 1983b. Microbial activity at the sediment-water interface in Halifax Harbor, Canada. *Appl. Environ. Microbiol.* **45**:1761–1766.

Novitsky, J. A. 1986. Degradation of dead microbial biomass in a marine sediment. *Appl. Environ. Microbiol.* **52**:504–509.

Novitsky, J. A. 1987. Microbial growth rates and biomass production in a marine sediment: evidence for a very active but mostly nongrowing community. *Appl. Environ. Microbiol.* **53**:2368–2372.

Novitsky, J. A., and R. Y. Morita. 1976. Morphological characterization of small cells resulting from nutrient starvation in a psychrophilic marine vibrio. *Appl. Environ. Microbiol.* **32**:619–622.

Novitsky, J. A., and R. Y. Morita. 1977. Survival of a psychrophilic marine vibrio under long-term nutrient starvation. *Appl. Environ. Microbiol.* **33**:635–641.

Novitsky, J. A., and R. Y. Morita. 1978a. Starvation induced barotolerance as a survival mechanism of a psychrophilic marine vibrio in the waters of the Antarctic Convergence. *Mar. Biol.* **49**:7–10.

Novitsky, J. A., and R. Y. Morita. 1978b. Possible strategy for the survival of marine bacteria under starvation conditions. *Mar. Biol.* **48**:289–295.

Nutman, P. S. 1946. Variation within strains of clover nodule bacteria in the size of nodule produced and in the "effectivity" of the symbiosis. *J. Bacteriol.* **51**:411–431.

Nutman, P. S. 1969. Symbiotic nitrogen fixation: legume nodule bacteria. Rothamsted Rept. for 1968, Pt. 2. pp. 179–181. Lawes Agricultural Trust, Harpenden, England.

Nyström, T., P. Mårdén, and S. Kjelleberg. 1986. Relative changes in incorporation rates of leucine and methionine during starvation-survival of two bacteria isolated from marine waters. *FEMS Microbiol. Ecol.* **38**:285–292.

Nyström, T., N. Albertson, and S. Kjelleberg. 1988. Synthesis of membrane and periplasmic proteins during starvation of a marine *Vibrio* sp. *J. Gen. Microbiol.* **134**:1645–1651.

Nyström, T., N. H. Albertson, K. Flärch, and S. Kjelleberg. 1990a. Physiological and molecular adaptation to starvation and recovery from starvation by the marine *Vibrio* sp. S14. *FEMS Microb. Ecol.* **74**:129–140.

Nyström, T., K. Flärch, and S. Kjelleberg. 1990b. Responses to multiple-nutrient starvation in marine *Vibrio* sp. strain CCUG 15956. *J. Bacteriol.* **172**:7085–7097.

Nyström, T., R. M. Olsson, and S. Kjelleberg. 1992. Survival, stress resistance, and alternations in protein expression in the marine *Vibrio* sp. strain S14 during starvation for different individual nutrients. *Appl. Environ. Microbiol.* **58**:55–65.

Oades, J. M., and G. H. Wagner. 1971. Biosynthesis of sugars in soils incubated with ^{14}C glucose and ^{14}C dextran. *Soil Sci. Soc. Amer. Proc.* **35**:914–917.

Oberlander, H. E., and K. Roth. 1968. Transformation of ^{14}C-labelled plant material in soils under field conditions. *In Isotopes and Radiation in Soil Organic Matter Studies*, pp. 251–261. International Atomic Energy Agency, Vienna.

Obuekwe, C. O., D. W. S. Westlake, and J. A. Plambeck. 1987. Evidence that available energy is a limiting factor in the bacterial corrosion of mild steel by a *Pseudomonas* sp. *Can. J. Microbiol.* **33**:272–275.

Odum, E. P. 1971. *Fundamentals of Ecology.* Saunders, Philadelphia.

Odum, W. E., P. W. Kirk, and J. C. Zieman. 1979. Non-protein nitrogen compounds associated with particles of vascular plant detritus. *Oikos* **32**:363–367.

Ogawa, H., and N. Ogura. 1992. Comparison of two methods for measuring dissolved organic carbon in sea water. *Nature* **356**:696–698.

Oginsky, E. L., and W. W. Umbreit. 1955. *An Introduction to Bacterial Physiology.* Freeman, San Francisco.

Oguiti, K. 1936. Ungersuchungen über die geschwindigkeit der eigenbewegung von bakterien. *Jpn. J. Expt. Med.* **14**:19–28.

Ogura, N. 1970a. The relation between dissolved organic carbon and apparent oxygen utilization in the Western North Pacific. *Deep-Sea Res.* **17**:221–231.

Ogura, N. 1970b. Decomposition of dissolved organic matter in the sea, its production, utilization and decomposition. *In Proceedings Second Symposium on the Results of the Cooperative Study of the Kuroshiro and Adjacent Regions*, K. Sugawara (ed.), pp. 201–205. Saikon Publishing Co., Tokyo.

Ogura, N. 1970c. Dissolved organic carbon in the equatorial region of the Central Pacific. *Nature* **227**:1335–1337.

Ogura, N. 1972. Rate and extent of decomposition of dissolved organic matter in surface seawater. *Mar. Biol.* **13**:89–93.

Ogura, N., and T. Hanya. 1967. Ultraviolet absorption of sea water, in relation to organic and inorganic matters. *Int. J. Oceanol. Limnol.* **1**:91–102.

Ogura, N., N. Nakamoto, M. Funakoshi, A. Kamatani, and S. Iwata. 1975. Fluctuation of dissolved organic carbon in seawater of Sagami Bay during 1971–1972. *J. Oceanogr. Soc. Jpn.* **31**:43–47.

Ohta, H., and T. Hattori. 1983. Oligotrophic bacteria on organic debris and plant roots in a paddy field soil. *Soil Biol. Biochem.* **15**:1–8.

Okuno, D., and Y. Kanai. 1981. Preservation of yeast in distilled water and by freeze drying. *Bull. Res. Freezing and Drying* **27**:100–106. (In Japanese)

Oliver, J. D. 1993. Formation of viable but nonculturable cells. *In Starvation in Bacteria,* S. Kjelleberg (ed.), pp. 239–272. Plenum Press, New York.

Oliver, J. D., and W. F. Stringer. 1984. Lipid composition of a psychrophilic marine *Vibrio* during starvation-induced morphogenesis. *Appl. Environ. Microbiol.* **47**:461–466.

Oliver, J. D., L. Nilsson, and S. Kjelleberg. 1991. Formation of nonculturable *Vibrio vulnificus* cells and its relationship to the starvation state. *Appl. Environ. Microbiol.* **57**:2640–2644.

Olness, A. C., and E. Clapp. 1972. Microbial degradation of a montmorillonite-dextran complex. *Soil Sci. Soc. Am. Proc.* **36**:179–181.

Olsen, A., A. Arnqvist, M. Hammar, S. Sukupolvi, and S. Normark. 1993. The RpoS sigma factor relieves H-NS-mediated transcriptional repression of *csgA,* the subunit gene of fibronectin-binding curli in *Escherichia coli. Mol. Microbiol.* **7**:523–536.

Olsen, R. A., and L. R. Bakken. 1987. Viability of soil bacteria: Optimization of plate-counting technique and comparison between total and plate counts within different size groups. *Microb. Ecol.* **13**:59–74.

Olsen, R. H., J. Siak, and R. H. Gray. 1974. Characteristics of PRD1, a plasmid-dependent broad host range DNA bacteriophage. *J. Virol.* **14**:689–699.

Omejuaiwe, S. O., and C. Okafor. 1989. Storage and preservation of stock cultures of some local yeast in sterile distilled water: An economically useful technique for culture collection in developing country such as Nigeria. *Abstract from the Fifth International Symposium on Microbial Ecology, Kyoto, Japan,* abstr., p. 138.

Omeliansky, V. L. 1911. Etude bactériologique du manmouth de Sanga Jourach et du sol adjacent. *Arch. Sci. Biol. St. Petersbourg* **16**:355–367.

Oppenheimer, C. H. 1952. The membrane filter in marine microbiology. *J. Bacteriol.* **64**:783–786.

Oppenheimer, C. H. 1958. Evidence of fossil bacteria in phosphate rocks. *Texas Univ. Inst. Marine Sci. Publ.* **5**:156–159.

Orgam, A. V., R. E. Jessup, L. T. Ou, and P. S. C. Rao. 1985. Effects of sorption on biological degradation of (2,4-Dichlorophenoxy) acetic acid in soils. *Appl. Environ. Microbiol.* **49**:582–587.

Östling, J., L. Holmquist, K. Flärdh, B. Svenblad, Å. Jouper-Jaan, and S. Kjelleberg. 1993. Starvation and recovery of *Vibrio. In Starvation in Bacteria,* S. Kjelleberg (ed.), pp. 103–127. Plenum Press, New York.

Osterberg, C., A. G. Carey, and H. Curl. 1963. Acceleration of sinking rates of ocean. *Nature* **200**:1276–1277.

Otto, R., J. Vije, B. Ten Brink, B. Klont, and W. N. Konings. 1985. Energy metabolism in *Streptococcus cremoris* during lactose starvation. *Arch. Microbiol.* **141**:348–352.

Owen-Hughes, T., G. D. Pavitt, D. S. Santos, J. M. Idebotham, C. S. J. Hulton, J. C. D. Hinton, and C. F. Higgins. 1992. The chromatin-associated protein H-NS interacts with curved DNA to influence DNA topology and gene expression. *Cell* **71**:255–265.

Pace, M. L. 1988. Bacterial mortality and the fate of bacterial production. *Hydrobiologia* **159**:41–50.

Pace, M. L., G. A. Knauer, D. M. Karl, and J. H. Martin. 1987. Primary production, new production and vertical flux in the eastern Pacific Ocean. *Nature* **325**:803–804.

Packard, T. T. 1969. The estimate of oxygen utilization rate in sea water from the activity of the respiratory electron transport system in plankton. Ph.D. Thesis. Univ. Washington, Seattle.

Packard, T. T., M. L. Healey, and F. A. Richards. 1971. Vertical distribution of the activity of the respiratory electron transport system in marine plankton. *Limnol. Oceanogr.* **16**:60–70.

Paerl, H. W. 1975. Microbial attachment to particles in marine and freshwater ecosystems. *Microb. Ecol.* **2**:73–83.

Paerl, H. W. 1984. Alteration of microbial metabolic activities in association with detritus. *Bull. Mar. Sci.* **35**:393–408.

Paerl, H. W., and S. M. Merkel. 1982. Differential phosphorus assimilation in attached vs. unattached microorganisms. *Arch. Hydrobiol.* **93**:125–134.

Paine, S. G., F. V. Linggood, F. Schimmer, and T. C. Thruff. 1933. The relationship of microorganisms to the decay of stone. *Phil. Trans. Soc. Ser. B.* **222**:97–127.

Painter, H. W. 1970. A review of the literature on inorganic nitrogen metabolism in microorganisms. *Water Res.* **4**:393–450.

Painting, S. J., M. I. Lucas, and D. G. Muir. 1989. Fluctuations in heterotrophic community structure, activity, and production in response to development and decay of phytoplankton in a microcosm. *Mar. Ecol. Prog. Ser.* **53**:129–141.

Palacas, J. G., V. E. Swenson, and G. W. Moore. 1966. Organic geochemistry of three North Pacific deep sea sediment samples. *U.S. Geol. Surv. Prof. Pap.* **550C**:102–107.

Palleroni, N. J. 1985. Bacterial longevity in salt-free medium: Responses. *ASM News* **51**:1–2.

Palumbo, A. V., R. I. Ferguson, and P. A. Rublee. 1984. Size of suspended bacterial cells and association of heterotrophic activity with size fractions of particles in estuarine and coastal water. *Appl. Environ. Microbiol.* **58**:157–164.

Pardee, A. B., and L. S. Prestige. 1966. Cell-free activity of a sulfate binding site involved in active transport. *Pro. Nat. Acad. Sci. U.S.A.* **555**:189–191.

Parinkina, O. M. 1973. Determination of bacterial growth rates in tundra soils. *Bull. Ecol. Res. Comm. (Stockholm)* **17**:303–309.

Parkes, R. J. 1987. Analysis of microbial communities within sediments using biomarkers. *Symp. Soc. Gen. Microb.* **41**:147–177.

Parkes, R. J., J. Taylor, and D. Jorck-Ramberg. 1984. Demonstration, using *Desulfobacter* sp., of two pools of acetate with different biological availabilities in marine pore water. *Mar. Biol.* **83**:271–276.

Parkes, R. J., B. A. Cragg, J. C. Fry, R. A. Herbert, and J. W. T. Winpenny. 1990. Bacterial biomass and activity in deep sediment layers from the Peru margin. *Phil. Trans. R. Soc. London A.* **331**:139–153.

Parkes, R. J., B. A. Cragg, J. M. Getliff, and J. C. Fry. 1993. Presence and activity of bacteria in deep sediments from marine environments. *In Trends in Microbial Ecology,* R. Guerrero and C. Pedrós-Alió (eds.), pp. 421–426. Spanish Society for Microbiology, Barcelona.

Parkes, R. J., B. A Cragg, S. J. Bale, J. M. Getliff, K. Goodman, P. A. Rochelle, J. C. Fry, A. J. Weightman, and S. M. Harvey. 1994. Deep bacterial biosphere in Pacific Ocean sediments. *Nature* **371**:410–413.

Parkinson, D., T. G. R. Gray, and S. T. Williams. 1971. Methods for studying the ecology of soil microorganisms. *IBP Handbook 19.* Blackwell Scientific Publications, Oxford.

Parkinson, D., K. H. Domsch, and J. P. E. Anderson. 1988. Die Entwicklung mikrobieller Biomassen im organischen Horizonteines Fichtenstandortes. *Ecolog. Plantarum* **13**:355–366.

Paszko-Kolva, C., M. Shahamat, H. Yamamoto, T. Sawyer, J. Vives-Rego, and R. R. Colwell. 1991. Survival of *Legionella pneumophila* in the aquatic environment. *Microb. Ecol.* **22**:75–83.

Pattenhöfer, G.-A., and S. C. Knowles. 1979. Ecological implications of fecal pellet size, production and consumption by copepods. *J. Mar. Res.* **37**:35–49.

Paul, E. A. 1970. Plant components and soil organic matter. *Rec. Adv. Phytochem.* **3**:59–104.

Paul, E. A. 1992. Organic matter, decomposition. *In Encyclopedia of Microbiology,* vol. 3, J. Lederberg (ed.), pp. 289–304. Academic Press, San Diego.

Paul, E. A., and F. E. Clark. 1989. *Soil Microbiology and Biochemistry.* Academic Press, San Diego.

Paul, E. A., and J. A. Van Veen. 1978. The use of tracers to determine the dynamic nature of organic matter. Proc. 11th Int. Congr. *Soil Sci.* **3**:1–43.

Paul, E. A. and R. P. Voroney. 1980. Nutrient and energy flows through soil microbial biomass. *In Contemporary Microbial Ecology,* D. C. Ellwood, J. N. Hedger, M. J. Latham, J. M. Lynch, and J. H. Slater (eds.), pp. 215–237. Academic Press, London.

Paul, J. H., S. C. Chiang, and J. B. Rose. 1991. Concentration of viruses and dissolved DNA from aquatic environments by vortex flow filtration. *Appl. Environ. Microbiol.* **57**:2197–2204.

Pavoni, J. L., M. W. Tenny, and W. F. Eschelberger. 1972. Bacterial extracellular polymers and bioflocculation. *J. Water Pollu. Control Fed.* **44**:414–431.

Payne, W. J. 1970. Energy yields and growth of heterotrophs. *Ann. Rev. Microbiol.* **24**:17–52.

Pearcy, W. G., and M. Stuvier. 1983. Vertical transport of carbon-14 into deep-sea food webs. *Deep-Sea Res.* **30**:427–440.

Pearson, T. H. 1982. The Loch Eil project: assessment and synthesis with a discussion of certain biological questions arising from a study of the organic pollution of sediments. *J. Exp. Mar. Biol. Ecol.* **57**:93–124.

Pederson, S., P. L. Bloch, S. Reeh, and F. C. Neidhardt. 1978. Patterns of protein synthesis in *E. coli*: a catalog of the amount of 140 individual proteins at different growth rates. *Cell* **14**:179–190.

Pedros-Alio, C., and T. D. Brock. 1982. Assessing biomass and production of bacteria in eutrophic Lake Mendota, Wisconsin. *Appl. Environ. Microbiol.* **44**:203–218.

Pedros-Alio, C., and T. D. Brock. 1983. The importance of attachment to particles for planktonic bacteria. *Arch. Hydrobiol.* **98**:354–379.

Penfold, W. J., and D. Norris. 1912. The relation of concentration of food supply to the generation time of bacteria. *J. Hyg.* **12**:527–531.

Perez, J. E., and A. Cortes-Monllor. 1967. Preservation of *Xanthomonas albilineans* isolates in distilled water. *Plant Dis. Rept.* **51**:739.

Perry, G. J., J. K. Volkman, and R. B. Johns. 1979. Fatty acids of bacterial origin in contemporary marine sediments. *Geochim. Cosmochim. Acta* **43**:1715–1725.

Petersen, S. O., and M. J. King. 1994. Effects of sieving, storage, and incubation temperature on the phospholipid fatty acid profile of a soil microbial community. *Appl. Environ. Microbiol.* **60**:2421–2430.

Pettipher, G. L. 1983. *The Direct Epifluorescent Filter Technique for the Rapid Enumeration of Microorganisms*. Research Studies Press, Letchworth, Hertfordshire.

Pfannkuche, O. 1992. Organic carbon flux through the benthic community in the temperate abyssal northwest Atlantic. In *Deep-Sea Food Chains and the Carbon Cycle*, R. T. Rowe and V. Pariente (eds.), pp. 183–198. Kluwer Academic Publications, Dordrecht.

Phelps, T. J., E. G. Raione, D. C. White, and C. B. Fliermans. 1989. Microbial activities in deep subsurface environments. *Geomicrobiol.* **7**:79–92.

Pianka, E. R. 1970. On r- and K- selection. *Amer. Nat.* **104**:592–597.

Pielou, E. C. 1974. *Population and Community Ecology: Principles and Methods*. Gordon & Breach Science Publications, New York.

Pilgram, W. I., and F. D. Williams. 1976. Survival value of chemotaxis in mixed cultures. *Can. J. Microbiol.* **33**:1771–1773.

Pinck, L. A., R. S. Dyal, and F. E. Allison. 1954. Protein-montmorillonite complexes, their preparation and the effects of soil microorganisms on their decomposition. *Soil Sci.* **78**:109–118.

Pine, M. J. 1973. Control of intracellular proteolysis during energy restriction in intact and permeabilized *Esherichia coli*. In *Intracellular Protein Turnover*, R. T. Schimke and N. Katkunuma (eds.), pp. 65–76. Academic Press, New York.

Piperno, J. R., and D. L. Oxender. 1966. Amino acid-binding protein released from *Escherichia coli* by osmotic shock. *J. Biol. Chem.* **241**:5732–5734.

Pirhonen, M., D. Flego, R. Heiknheimo, and E. T. Palva. 1993. A small diffusible signal molecule is responsible for the global control of virulence and exoenzyme production in the plant pathogen *Erwinia carotovora*. *EMBO J.* **12**:2467–2476.

Pirie, N. W. 1973. "On being the right size." *Ann. Rev. Microbiol.* **27**:119–132.

Pirt, S. J. 1965. The maintenance energy of bacteria in growing cultures. *Proc. Roy. Soc. London. Ser. B.* **163**:224–231.

Pirt, S. J. 1966. The maintenance energy of bacteria on growing culture. *Proc. Roy. Soc. London, B.* **163**:224–231.

Pirt, S. J. 1969. Microbial growth and product formation. *Symp. Soc. Gen. Microbiol.* **19**:199–221.

Pirt, S. J. 1975. *Principles of Microbes and Cell Cultivation.* John Wiley & Sons, New York.

Pirt, S. J. 1982. Maintenance energy: a general model for energy-limited and energy-sufficient growth. *Arch. Microbiol.* **133**:300–302.

Pitter, R., and J. Chudoba. 1990. Biodegradability of organic substances in the aquatic environment. CRC Press, Boca Raton, FL.

Platt, T., M. Lewis, and R. Geider. 1984. Thermodynamics of the pelagic ecosystem: Elementary closure conditions in the open ocean. *In Flows of Energy and Materials in Marine Ecosystems, Theory and Practice,* M. J. R. Fasham (ed.), pp. 49–84. Plenum Press, New York.

Plunkett, M. A., and N. W. Rakestraw. 1955. Dissolved organic matter in the sea. *Deep-Sea Res.,* suppl. to Vol. **3**:12–14.

Pocklington, R., J. D. Leonard, and N. Crewe. 1991. Sources of organic matter to surface sediments from the Scotian Shelf and Slope, Canada. *Cont. Shelf. Res.* **11**:1069–1082.

Poinar, H. N., M. Höss, J. L. Bada, and S. Pääbo. 1996. Amino acid racemization and the preservation of ancient DNA. *Science* **277**:864–866.

Poindexter, J. S. 1979. Morphological adaptation to low nutrient concentrations. *In Strategies of Microbial Life in Extreme Environments,* M. Shilo (ed.), pp. 341–356. Verlag Chemie, Weinheim.

Poindexter, J. S. 1981a. The caulobacters: ubiquitous unusual bacteria. *Microbiol. Rev.* **45**:123–179.

Poindexter, J. S. 1981b. Oligotrophy: fast and famine existence. *Adv. Microb. Ecol.* **5**:63–90.

Poindexter, J. S. 1984. Role of prostheca development in oligotrophic aquatic bacteria. *In Current Perspectives in Microbial Ecology,* M. J. Klug and C. A. Reddy (eds.), pp. 33–40. American Society for Microbiology, Washington, DC.

Poindexter, J. S. 1987. Bacteria responses to nutrient limitation. *Symp. Soc. Gen. Microbiol.* **41**:283–317.

Pollard, P. C., and D. J. W. Moriarty. 1984. Validity of the tritiated thymidine method for estimating bacteria growth rates: measurements of isotope dilution during DNA synthesis. *Appl. Environ. Microbiol.* **48**:1076–1083.

Pollock, M. R. 1962. Exoenzymes. *In The Bacteria,* vol. 4, I. C. Gunsalus and R. Y. Stanier (eds.), pp. 121–178. Academic Press, New York.

Pomeroy, L. R. 1974. The ocean's food web, a changing paradigm. *Bioscience* **24**:499–

504.

Pomeroy, L. R. 1984. Microbial processes in the sea: diversity in nature and science. *In Heterotrophic Activity in the Sea,* J. E. Hobbie and P. J. leB. Williams (eds.), pp. 1–24. Plenum Press, New York.

Pomeroy, L. R., and D. Deibel. 1986. Temperature regulation of bacterial activity during the spring bloom in Newfoundland coastal waters. *Science* **233**:359–361.

Pomeroy, L. R., and W. J. Wiebe. 1993. Energy sources for microbial food webs. *Mar. Microb. Food Webs* **7**:101–118.

Pomeroy, L. R., S. A. Macko, P. H. Ostrom, and J. Dunphy. 1990. The microbial food web in Arctic seawater: concentration of dissolved free amino acids and bacterial abundance and activity in the Arctic Ocean and in Resolute Passage. *Mar. Ecol. Prog. Ser.* **61**:31–40.

Pomeroy, L. R., W. J. Wiebe, D. Diebel, R. J. Thompson, G. T. Rowe, and J. D. Pakuiski. 1991. Bacterial responses to temperature and substrate concentration during the Newfoundland spring bloom. *Mar. Ecol. Prog. Ser.* **75**:143–159.

Pongratz, E. 1957. D'une bacterie pediculee isolee d'un pus de sunus. *Schweiz. Z. Path. Bakt.* **20**:593–608.

Poolman, B., E. J. Smid, H. Veldkamp, and W. N. Konnigs. 1987. Bioenergetic consequences of lactose starvation for continuously cultured *Streptococcus cremoris. J. Bacteriol.* **169**:1460–1468.

Porter, J. F., R. Parton, and A. C. Wardlaw. 1991. Growth and survival of *Bordetella bronchiseptica* in natural waters and in buffered saline without added nutrients. *Appl. Environ. Microbiol.* **57**:1202–1206.

Postgate, J. 1965. Continuous culture: attitudes and myths. *Lab. Prac.* **14**:1140–1144.

Postgate, J. R. 1967. Viability measurements and the survival of microbes under minimum stress. *Adv. Microb. Physiol.* **1**:2–23.

Postgate, J. R. 1969. Viable counts and viability. *In Methods in Microbiology,* vol. 1, J. R. Norris and D. W. Ribbons (eds.), pp. 611–628. Academic Press, New York.

Postgate, J. R. 1973. The viability of very slow-growing populations: a model for the natural ecosystem. *Bull. Ecol. Res. Commun. (Stockholm)* **17**:287–292.

Postgate, J. R. 1976. Death in macrobes and microbes. *Symp. Soc. Gen. Microbiol.* **26**:1–18.

Postgate, J. R. 1989. A microbial way of death. *New Science* **122**:43–47.

Postgate, J. R., and J. R. Hunter. 1962. The survival of starved bacteria. *J. Gen. Microbiol.* **29**:233–263.

Postgate, J. R., and J. R. Hunter. 1963. The survival of starved bacteria. *J. Appl. Bacteriol.* **26**:295–306.

Postgate, J. R., and J. R. Hunter. 1963. Acceleration of bacterial death by growth substrates. *Nature* **198**:273.

Postgate, J. R., and J. R. Hunter. 1964. Accelerated death of *Aerobacter aerogenes* starved in the presence of growth limiting substrates. *J. Gen. Microbiol.* **34**:459–473.

Postgate, J. R., J. E. Crumpton, and J. R. Hunter. 1962. The measurement of bacterial viabilities by slide culture. *J. Gen. Microbiol.* **24**:15–25.

Postgate, J. R., and P. H. Calcott. 1985. Ageing and death in microbes. *In Comprehensive Biotechnology,* vol. 2, M. Moo-Young (ed.), pp. 239–249. Pergamon Press, Oxford.

Poteete, A. R. 1988. Bacteriophage P22. *In The Bacteriophages,* vol. 2, R. Calendar (ed.), pp. 647–682. Plenum Press, New York.

Potter, L. 1960. The effect of pH on the development of bacteria stored in glass containers. *Can. J. Microbiol.* **6**:257–263.

Potts, M. 1995. Ancient prokaryotes-water, water, everywhere. *ASM News* **61**:218–219.

Powell, E. O. 1967. The growth rate of microorganisms as a function of substrate concentration. *In Microbial Physiology and Continuous Culture,* E. O. Powell, C. G. T. Evans, R. E. Strange, and D. W. Tempest (eds.), pp. 34–56. Her Majesty's Stationary Office, London.

Power, K., and K. C. Marshall. 1988. Cellular growth and reproduction of marine bacteria on surface bound substrate. *Biofouling* **1**:163–174.

Prahl, F. G., and L. A. Meulhausen. 1989. Lipid biomarkers as geochemical tools for paleocanography study. *In Productivity in the Oceans: Present and Past,* W. H. Berger, V. S. Smetacek, and G. Wefer (eds.), pp. 189–271. John Wiley & Sons, Chichester.

Preyer, J. M., and J. D. Oliver. 1993. Starvation-induced thermal tolerance as a survival mechanism in a psychrophilic marine bacterium. *Appl. Environ. Microbiol.* **59**:2653–2656.

Prezelin, B. B., and A. I. Alldredge. 1983. Primary production of marine snow during and after an upwelling event. *Limnol. Oceanogr.* **28**:1158–1167.

Proctor, L. M., and J. A. Fuhrman. 1990. Viral mortality of marine bacteria and cyanobacteria. *Nature* **343**:60–62.

Proctor, L. M., A. Okubo, and J. A. Fuhrman. 1993. Calibrating estimates of phage-induced mortality in marine bacteria: ultrastructural studies of marine bacteriophage development from one step growth experiments. *Microb. Ecol.* **25**:161–182.

Purdy, D. R., and A. L. Koch. 1976. Energy cost of galactoside transport to *Escherichia coli. J. Bacteriol.* **127**:1188–1196.

Pytkowicz, R. M. 1965. Calcium carbonate saturation in the ocean. *Limnol. Oceanogr.* **8**:372–381.

Questel, J. H., and M. D. Whetham. 1924. The equilibria existing between succinic, fumaric, and malic acids in the presence of resting bacteria. *Biochem. J.* **18**:519–539.

Quesnel, L. B. 1971. Microscopy and micrometry. *In Methods in Microbiology,* vol. 5A, J. R. Norris and D. W. Ribbons (eds.), pp. 2–103. Academic Press, London.

Rahn, O. 1932. *Physiology of Bacteria.* Blakiston, Philadelphia.

Raina, S., D. Missiakas, L. Baird, S. Kumar, and C. Georgopoulos. 1993. Identification and transcriptional analysis of the *Escherichia coli htrE* operon which is homologous to *pap* and related pilin operons. *J. Bacteriol.* **175**:5009–5021.

Rashid, M. A. 1972. Amino acids associated with the marine sediments and humic acid compounds and their role in solubility and complexing of metals. 24th Int. Geol. Congr. 1972, Sec. 10, pp. 346–353.

Ray, B. 1979. Methods to detect stressed microorganisms. *J. Food Protection* **42**:346–355.

Redden, G. D. 1982. Characteristics of photochemical production of carbon monoxide in seawater. M.S. Thesis. Oregon State University, Corvallis, OR.

Reeck, G. R. 1983. Amino acid composition of selected proteins. *In Handbook of Microbiology,* Vol. 2, A. I. Laskin and H. A. Lechevalier (eds.), pp. 15–25. CRC Press, Boca Raton, FL.

Reeve, C. A., P. Amy, and A. Matin. 1984a. Role of protein synthesis in the survival of carbon-starved *Escherichia coli* K–12. *J. Bacteriol.* **160**:1041–1046.

Reeve, C. A., A. T. Bockman, and A. Matin. 1984b. Role of protein degradation in the survival of carbon-starved *Escherichia coli* and *Salmonella typhimurium. J. Bacteriol.* **157**:758–763.

Reichardt, W. 1979. Influence of temperature and substrate shocks on survival and succinic dehydrogenase activity of heterotrophic freshwater bacteria. *Water Res.* **13**:1149–1154.

Reichardt, W. 1986. Polychaete tube walls as zonated microhabitats for marine bacteria. *Actes. Colloques IFREMER* **3**:415–425.

Reichardt, W. 1988. Impact of bioturbation by *Arenicola marina* on microbiological parameters in intertidal sediments. *Mar. Ecol. Prog. Ser.* **44**:149–156.

Reichardt, W., and R. Y. Morita. 1982. Survival stages of a psychrophilic *Cytophaga johnsonae* strain. *Can. J. Microbiol.* **28**:841–850.

Reid, K. G., N. M. Utech, and J. T. Holden. 1970. Multiple transport components for dicarboxylic amino acids in *Streptococcus faecalis. J. Biol. Chem.* **245**:5261–5272.

Reiser, R., and P. Tasch. 1960. Investigation of the viability of osmophile bacteria of great geological age. *Trans. Kansas Acad. Sci.* **63**:31–34.

Renault, B. 1900. Sur quelques microorganismes des combustibles fossiles. *Bull. Soc. Ind. Minerale St. Étienne* **14**:5.

Reuszer, H. W. 1933. Marine bacteria and their role in the cycle of life in the sea. III. Distribution of bacteria in the ocean waters and muds about Cape Cod. *Biol. Bull.* **65**:480–487.

Revelle, R. R. 1944. *Marine Bottom Samples Collected in the Pacific Ocean by the Carnege on its Seventh Cruise.* Carnegie Institution Washington, Pub., Washington, DC.

Revelle, R. R., and F. P. Shepard. 1939. Sediments off the California coast. *Recent Marine Sediments.* American Association of Petroleum Geologists, Tulsa, OK.

Revsbech, N. R., and B. B. Jørgensen. 1983. Photosynthesis of benthic microflora measured with high spatial resolution by the oxygen microprofile method: capabilities and limitations of the method. *Limnol. Oceanogr.* **28**:749–756.

Rhoades, D. F., and R. G. Cates. 1976. A general theory of plant and anti-herbivore chemistry. *In Biochemical Interactions Between Plants and Insects. Recent Advances in Phytochemistry,* vol. 10, J. W. Wallace and R. L. Mansell (eds.), pp. 169–213. Plenum Press, New York.

Rhoads, D. C., P. L. McCall, and J. Y. Yingst. 1978. Disturbance and production on the estuarine seafloor. *Amer. Sci.* **66**:577–586.

Ribbons, D. W., and E. A. Dawes. 1963. Environmental and growth conditions affecting the endogenous metabolism of bacteria. *Ann. N.Y. Acad. Sci.* **103**:564–586.

Rice, D. L. 1982. The detritus nitrogen problem: new observations and perspectives from organic geochemistry. *Mar. Ecol. Prog. Ser.* **9**:153–162.

Rice, S. A., and J. D. Oliver. 1992. Starvation response of the marine barophile CNPT-3. *Appl. Environ. Microbiol.* **58**:2432–2437.

Richard, B. N. 1987. *The Microbiology of Terrestrial Ecosystems.* Longman Scientific & Technical, Essex.

Riemann, B. 1984. Determining growth rates of natural assemblages of freshwater bacteria by means of ^{3}H-thymidine incorporation into DNA: Comments on methodology. *Arch. Hydrobiol. Belih. Ergebn. Limnol.* **19**:67–80.

Riemann, B., and E. Hoffmann. 1991. Ecological consequences of dredging and bottom trawling in the Limfjord, Denmark. *Mar. Ecol. Prog. Ser.* **69**:171–178.

Riemann, B., and R. T. Bell. 1990. Advances in estimating bacterial biomass and growth in aquatic systems. *Arch. Hydrobiol.* **25**:385–402.

Riemann, B., J. Fuhrman, and F. Azam. 1982. Bacterial secondary production in freshwater measured by ^{3}H-thymidine method. *Microb. Ecol.* **8**:101–114.

Riemann, B., P. Nelson, M. Jeppesen, B. Marcussen, and J. A. Fuhrman. 1984. Diel changes in bacterial biomass and growth rates in coastal environments, determined by means of thymidine incorporation into DNA, frequency of dividing cells (FDC), and microautoradiography. *Mar. Ecol. Prog. Ser.* **17**:227–235.

Rieves, C. 1958. The stability of the physiological properties of coliform organisms. *Zentr. Bakteriol. Parisitenk.* Abt. II. **142**:594–608.

Rifai, N., and G. Bertru. 1980. La biodegradation des acides fulviques. *Hydrobiologia* **75**:181–184.

Riley, G. A. 1951. Oxygen, phosphate, and nitrate in the Atlantic Ocean. *Bull. Bingham Oceanogr. Coll.* **13**:1–126.

Riley, G. A. 1963. Organic aggregates in seawater and the dynamics of their formation and utilization. *Limnol. Oceanogr.* **8**:372–381.

Riley, G. A. 1970. Particulate organic matter in seawater. *Adv. Mar. Biol.* **8**:1–118.

Riley, G. A., D. Van Hemert, and P. J. Wangersky. 1965. Organic aggregates in surface and deep waters of the Sargasso Sea. *Limnol. Oceanogr.* **10**:354–363.

Rittenberg, S. C. 1940. Bacteriological analysis of some long cores of marine sediments. *J. Mar. Res.* **3**:191–201.

Rittenberg, S. C. 1979. Bdellovibrio: A model of biological interactions in nutrient-impoverished environments? *In Strategies of Microbial Life in Extreme Environments,* M. Shilo (ed.), pp. 305–322. Verlag Chemie, Weinheim.

Rittenberg, S. C., K. O. Emery, J. Hulsemann, E. T. Degens, R. C. Fay, J. H. Reuter, J. R. Grady, S. H. Richardson, and E. E. Bray. 1963. Biogeochemistry of sediments in experimental Mohole. *J. Sediment. Petrol.* **33**:140–172.

Robarts, R. D., and R. J. Wicks. 1989. [methyl-^{3}H]thymidine uptake, macromolecular incorporation and lipid labelling: their significance to DNA labelling during aquatic bacterial growth rate measurement. *Limnol. Oceanogr.* **34**:213–222.

Robarts, R. D., and T. Zohary. 1993. Fact or fiction—Bacterial growth rates and production as determined by [methy-^{3}H] thymidine *Adv. Microb. Ecol.* **13**:371–425.

Robarts, R. D., R. J. Wicks, and L. M. Sephton. 1986. Spatial and temporal variations in bacteria macromolecule labeling with [methyl-^{3}H]thymidine in a hypertrophic lake. *Appl. Environ. Microbiol.* **52**:1368–1373.

Robb, S. M., D. R. Woods, F. T. Robb, and J. K. Struthers. 1977. Rifampicin-resistant mutant supporting bacteriophage growth on stationary phase *Achromobacter* cells. *J. Gen. Virol.* **35**:117–123.

Robb, S. M., D. R. Woods, F. T. Robb. 1978. Phage growth characteristics on stationary phase *Achromobacter* cells. *J. Gen. Virol.* **41**:265–272.

Roberts, J. W., and R. Devoret. 1983. Lysogenic induction. *In Lambda II*, R. W. Hendrix, J. W. Roberts, F. W. Stahl and R. A. Weisberg (eds.), pp. 123–144. Cold Spring Harbor Laboratory, Cold Spring Harbor, New York.

Robertson, J. G., and R. D. Batt. 1973. Survival of *Nocardia corallina* and degradation of constituents during starvation. *J. Gen. Microbiol.* **78**:109–117.

Robinson, J. A., M. G. Trulear, and W. G. Characklis. 1984. Cellular reproduction and extracellular polymer formation by *Pseudomonas aeruginosa* in continuous culture. *Biotechnol. Bioeng.* **22**:1409–1417.

Rockabrand, D., T. Arthur, G. Korinek, K. Lievers, and P. Blum. 1995. An essential role for the *Escherichia coli* DnaK protein in starvation-induced thermotolerance, H$_2$O$_2$ resistance and reductive division. *J. Bacteriol.* **177**:3695–3703.

Rodriguez-Valera, F., F. Ruiz-Berraquero, and A. Ramos-Cormenzana. 1979. Isolation of extreme halophiles from seawater. *Appl. Environ. Microbiol.* **38**:164–165.

Rodgers, H. J. 1961. The dissimilation of high molecular weight substances. *In The Bacteria*, vol. 2, I. C. Gunsalus and R. Y. Stanier (eds.), pp. 261–318. Academic Press, London.

Rohde, W. 1973. Crystallization of eutrophication concepts in northern Europe, *In Eutrophication: Causes, Consequences, Correctives*, pp. 50–64. National Academy of Science, Washington, DC.

Rolfe, B., and B. W. Holloway. 1966. Alterations in host specificity of bacterial deoxyribonucleic acid after an increase in growth temperature of *Pseudomonas aeruginosa*. *J. Bacteriol.* **92**:43-48.

Roper, M. M., and K. C. Marshall. 1974. Modification of the interaction between *Escherichia coli* and bacteriophage in saline sediment. *Microb. Ecol.* **1**:1–23.

Roper, M. M., and K. C. Marshall. 1978. Effect of clay particles size on clay-*Escherichia coli*-bacteriophage interactions. *J. Gen. Microbiol.* **106**:187–189.

Rose, A. S., A. E. Ellis, and A. L. S. Munro. 1990a. The survival of *Aeromonas salmonicida* subsp. *salmonicida* in sea water. *J. Fish. Dis.* **13**:205–214.

Rose, A. S., A. E. Ellis, and A. L. S. Munro. 1990b. Evidence against dormancy in the bacteria fish pathogen *Aeromonas salmonicida* subsp. *salmonicida. FEMS Microb. Lett.* **68**:105–108.

Rosen, B. P. 1973. Basic amino acid transport in *Escherichia coli* II. Purification and properties of an arginine-binding protein. *J. Biol. Chem.* **248**:1211–1218.

Rosen, B. P., and F. D. Vasington. 1970. Relationship of the histidine binding protein and the histidine permease system in *S. typhimurium. Proc. Fed. Amer. Soc. Experi. Biol.* **29**:342.

Rosenfeld, J. K. 1979. Amino acid diagenesis and adsorption in nearshore anoxic sediments. *Limnol. Oceanogr.* **24**:1014–1021.

Roslev, P., and G. M. King. 1994. Survival and recovery of methanotrophic bacteria starved under oxic and anoxic conditions. *Appl. Environ. Microbiol.* **60**:2602–2608.

Roszak, D. B., and R. R. Colwell. 1987a. Survival strategies of bacteria in the natural environment. *Microbiol. Rev.* **51**:365–379.

Roszak, D. B., and R. R. Colwell. 1987b. Metabolic activity of bacterial cells enumerated by direct viable count. *Appl. Environ. Microbiol.* **53**:2889–2983.

Rotman, B., and J. Radojkovic. 1964. Galatose transport in *Escherichia coli. J. Biol. Chem.* **239**:3153–3156.

Rovira, A. D., and E. L. Greachen. 1957. The effect of aggregate disruption on activity of microorganisms in the soil. *Aust. J. Agri. Res.* **8**:659–673.

Rowbury, R. J., J. P. Armitage, and C. King. 1983. Movement, taxes and cellular interactions in the response of microorganisms to the natural environment. *Symp. Soc. Gen. Microbiol.* **34**:299–350.

Rowe, G. T., and J. W. Deming. 1985. The role of bacteria in the turnover of organic carbon in deep-sea sediments. *J. Mar. Res.* **43**:925–950.

Rowe, G., and W. Gardner. 1979. Sedimentation in the slope water of the northwest Atlantic Ocean as measured directly with sediment traps. *J. Mar. Res.* **37**:581–600.

Rowe, G. T., and V. Pariente. 1992. *Deep-Sea Food Chains and the Carbon Cycle.* Kluwer Academic Publications, Dordrecht.

Rowe, G. T., M. Sibuet, J. Deming, J. Tietjen, and A. Khripounoff. 1990. Organic carbon turnover time in deep-sea benthos. *Prog. Oceanogr.* **24**:141–160.

Rowe, G. T., M. Sibuet, J. Deming, A. Khripounoff, J. Tietjen, S. Macko, and R. Theroux. 1991. "Total" sediment biomass and preliminary estimates of organic carbon residence time in deep sea benthos. *Mar. Ecol. Prog. Ser.* **79**:99–114.

Rowell, M. J., J. N. Ladd, and E. A. Paul. 1973. Enzymically active complexes of proteases and humic acid analogues. *Soil Biol. Biochem.* **5**:699–703.

Rüger, H.-J. 1988. Substrate-dependent cold adaptations in some deep-sea sediment bacteria. *Syst. Appl. Microbiol.* **11**:90–93.

Russell, C. E., R. Jacobson, D. L. Haldeman, and P. S. Amy. 1994. Heterogeneity of deep subsurface microorganisms and correlations to hydrogeological and eochemical parameters. *Geomicrob. J.* **12**:37–51.

Russell, E. J. 1950. *Soil Conditions and Plant Growth,* 8th ed. Longmans, Green & Co., London.

Russell, H. L. 1891. The bacterial flora of the Atlantic Ocean in the vicinity of Woods Hole, Mass. *Bot. Gaz.* **18**:383–395.

Ryan, F. J. 1959. Bacterial mutation in a stationary phase and the question of cell turnover. *J. Gen. Microbiol.* **21**:530–549.

Rybkin, A. I., and V. K. Ravin. 1987. Depression of synthetic activity as the possible cause of death of *Escherichia coli* during amino acid starvation. *Microbiology* **56**:170–174.

Ryhaenen, R. 1968. Die Bedeutung der Humussubstanzen in Stoffhaushalt der Gewäeert Finnlands. *Mitt. Int. Ver. Theor. Angew. Limnol.* **14**:168–178.

Saha, S. K., S. Saha, and S. C. Sanyal. 1991. Recovery of injured *Campylobacter jejuni* cells after animal passage. *Appl. Environ. Microbiol.* **57**:3388–3389.

Saito, H., H. Tomioka, and S. Ohkido. 1985. Further studies on thymidine kinase: distribution pattern of the enzyme in bacteria. *J. Gen. Microbiol.* **131**:3091–3098.

Sakai, K. D. 1986. Electrostatic mechanism of survival of virulent *Aeromonas salmonicida* strains in river water. *Appl. Environ. Microbiol.* **51**:1343–1349.

Samuelson, M.O., and D. L. Kirchman. 1990. Degradation of adsorbed protein by attached bacteria in relationship to surface hydrophobicity. *Appl. Environ. Microbiol.* **56**:3643–4648.

Sanchez-Amat, A., and F. Torrella. 1990. Formation of stable bdelloplasts as a starvation-survival strategy of marine Bdellovibrios. *Appl. Environ. Microbiol.* **56**:2717–2725.

Sands, J. G., and L. O. Bennett. 1966. Effects of different commercial agar preparations on the inhibitory activities of phenol. *App. Microbiol.* **14**:196–202.

Sano, Y., H. Matsui, M. Kobayashi, and M. Kageyama. 1990. Pyocins S1 and S2, bacteriocins of *Pseudomonas aeruginosa*. *In Pseudomonas. Biotransformations, Pathogenesis, and Evolving Biotechnology*, S. Silver, A. M. Chakrabarty, B. Iglewski, and S. Kaplan (eds.), pp. 352–358. American Society for Microbiology, Washington, DC.

Sansome, F. J. 1988. Depth distribution of short chain organic acid turnover in Cape Lookout Bight sediments. *Geochim. Cosmochim. Acta* **50**:99–105.

Sarmientos, P., and M. Cashel. 1983. Carbon starvation and growth rate-dependent regulation of the *Escherichia coli* ribosomal RNA promoters: differential control of dual promoters. *Proc. Natl. Acad. Sci.(USA)* **80**:7010–7013.

Sauerbeck, D. 1966. A critical evaluation of incubation experiments on the priming effect of green manure. *In The Use of Isotopes in Soil Organic Matter Studies*, pp. 209–221. FAO/IAEA Technical Meetings, 1963. Pergamon Press, Oxford.

Sauerbeck, D., and F. Führ. 1968. Alkali extraction and fractionation of labelled plant material before and after decomposition—a contribution to the technical problems in humification studies. *In Isotopes and Radiation in Soil Organic Matter Studies*, pp. 3–11. International Atomic Energy Agency, Vienna.

Sauerbeck, D., and M. A. Gonzalez. 1977. Field decomposition of carbon-[14]-labelled plant residues in various soils of the Federal Republic of Germany and Costa Rica. *IAEA, FAO, Agrochimica: Soil Organic Matter Studies*, Proc. Symp. Vol. 1, 159–170.

Sayles, F. L., and W. B. Curry. 1988. Delta ^{13}C, TCO_2, and the metabolism of organic carbon in deep sea sediments. *Geochim. Cosmochim. Acta* **52**:2963–2978.

Scavia, D. 1988. On the role of bacteria in secondary production. *Limnol. Oceanogr.* **33**:1220–1224.

Scavia, D., and G. A. Laird. 1987. Bacterioplankton in Lake Michigan: dynamics controls, and significance to carbon flux. *Limnol. Oceanogr.* **32**:1017–1033.

Schaechter, M. 1968. Growth: cells and populations. In *Biochemistry of Bacterial Growth*, J. Mandelstam and R. McQuillen (eds.), pp. 136–162. John Wiley & Sons, New York.

Scherer, C. C., and C. W. Boylen. 1977. Macromolecular synthesis and degradation in *Arthrobacter* during periods of nutrient deprivation. *J. Bacteriol.* **132**:584–589.

Schimz, K.-L., and B. Overhoff. 1987. Investigations of the influence of carbon starvation on the carbohydrate storage compounds (trehalose, glycogen), viability, adenylate pool, and adenylate energy charge in *Cellulomonas* sp. (DSM20108). *FEMS Microbiol. Lett.* **40**:333–337.

Schleif, R. 1969. An L-arabinose binding protein and arabinose permeation in *Escherichia coli*. *J. Mol. Biol.* **46**:185–196.

Schmalz, K. F., and K. E. Chave. 1963. Calcium carbonate: factors affecting saturation in ocean waters off Bermuda. *Science* **149**:1206–1207.

Schmidt, E. L. 1979. Initiation of plant root-microbe interaction. *Ann. Rev. Microbiol.* **33**:355–376.

Schmidt, J. M., and G. M. Samuelson. 1972. Effects of cyclic nucleotides and nucleoside triphosphate on stalk formation in *Caulobacter crescentus*. *J. Bacteriol.* **112**:593–601.

Schmidt, J. M., and R. Y. Stanier. 1966. The development of cellular stalks in bacteria. *J. Cell Biol.* **28**:423–436.

Schmidt, S. K., and M. Alexander. 1985. Effects of dissolved organic carbon and second substrates on biodegradation of organic compounds at low concentrations. *Appl. Environ. Microbiol.* **49**:822–827.

Schmidt, S. K., M. Alexander, and M. L. Shuler. 1985. Predicting threshold concentrations of organic substrates for bacterial growth. *J. Theor. Biol.* **114**:1–8.

Schnitzer, M. 1971. Metal-organic matter interactions in soils and waters. In *Organic Compounds in Aquatic Environments*, S. D. Faust and J. V. Hunter (eds.), pp. 297–316. Marcel Dekker, New York.

Schnürer, J., M. Charholm, and T. Rosswall. 1985. Microbial biomass and activity in an agricultural soil with different organic contents. *Soil Biol. Biochem.* **17**:611–618.

Schopf, J. W., E. S. Barghoorn, M. D. Maser, and R. O. Gordon. 1965. Electron microscopy of fossil bacteria two billion years old. *Science* **149**:1365–1367.

Schramm, V. L., and H. Leung. 1973. Regulation of adenosine nucleotide mixtures by two-dimensional anion-exchange thin layer chromatography. *J. Chromatogr.* **16**:126–219.

Schramm, V. L., and F. C. Lazorik. 1975. The pathway of adenylate catabolism in *Azotobacter vinelandii*: evidence for adenosine monophosphate nucleosidase as the regulatory enzyme. *J. Biol. Chem.* **250**:1801–1809.

Schröder, H. 1914. The bacterial content of coal. *Centralbl. Bakt. Abt. II. Bd.* **41**:460.

Schut, F., E. J. de Vries, J. C. Gottschal, B. R. Robertson, W. Harder, R. A. Prins, and D. K. Button. 1993. Isolation of typical marine bacteria by dilution culture: Growth, maintenance, and characteristics of isolates under laboratory conditions. *Appl. Environ. Microbiol.* **59**:2150–2160.

Schwartz, W., and A. Müller. 1958. *Methoden der Geomikrobiologie.* Freiberger Forschungshefte C 48: Angewandte Naturwissenschaften.

Scott, D. E., J. P. Martin, D. D. Focht, and K. Haider. 1983. Biogradation, stabilization in humus and incorporation into soil biomass of 2,4-D and chlorocatechol carbon. *Soil Sci. Soc. Am. J.* **47**:66–77.

Seiler, W. 1974. The cycle of atmospheric CO. *Tellus* **26**:116–135.

Seiler, W. 1978. The influence of the biosphere on the atmospheric CO and H_2 cycles. *In Environmental Biogeochemistry and Geomicrobiology,* vol. 3. W. E. Krumbein (ed.), pp. 773–810. Ann Arbor Science Publishers, Ann Arbor, MI.

Seki, H. 1982. *Organic Materials in Aquatic Ecosystems.* CRC Press, Boca Raton, FL.

Seki, H., J. Skelding, and T. R. Parsons. 1968. Observations on the decomposition of a marine sediment. *Limnol. Oceanogr.* **13**:440–447.

Seki, H., A. Otsuki, S. Daigobo, C. D. Levings, and C. D. McAllister. 1984. Microbial contribution to the mesotrophic ecosystem of the Campbell River estuary during summer. *Arch. Hydrobiol.* **102**:215–288.

Semenov, A. M. 1991. Physiological basis of oligotrophy of microorganisms and the concept of microbial community. *Microb. Ecol.* **22**:239–247.

Semenov, A., and J. T. Staley. 1992. Ecology of polyprosthecate bacteria. *Symp. Soc. Gen. Microbiol.* **12**:339–382.

Seneviratne, R., and A. Wild. 1986. Nitrogen mineralization in a tussock tundra soil. *Arct. Alp. Res.* **14**:287–293.

Sepers, A. B. J. 1984. The influence of varying environmental conditions on the activity of heterotrophic bacteria. *Arch. Hydrobiol.* **19**:119–124.

Serban, A., and A. Nissenbaum. 1986. Humic acid association with peroxidase and catalase. *Soil Biol. Biochem.* **18**:41–44.

Servais, P. G., G. Billen, and J. Vives-Rego. 1985. Rate of bacterial mortality in aquatic environments. *Appl. Environ. Microbiol.* **49**:1448–1453.

Servais, P. G., G. Billen, and M. C. Hascoët. 1987. Determination of biodegradable fraction of dissolved organic matter in waters. *Water Res.* **21**:445–450.

Servais, P. G., A. Anzil, and C. Ventresque. 1989. Simple method for determination of biodegradable dissolved organic carbon in water. *Appl. Environ. Microbiol.* **55**:2732–2734.

Setlow, P. 1983. Germination and outgrowth. *In The Bacterial Spore,* A. Hurst and G. W. Gould (eds.), pp. 211–254. Academic Press, New York.

Setlow, P., and A. Kornberg. 1970. Biochemical studies of bacterial sporulation and germination. *J. Biol. Chem.* **45**:3637–3644.

Shand, R. F., P. H. Blum, D. L. Holzschu, M. S. Urdea, and S. W. Artz. 1989a. Mutational analysis of the histidine operon promoter of *Salmonella typhimurium*. *J. Bacteriol.* **171**:6330–6337.

Shand, R. F., P. H. Blum, R. D. Mueller, D. L. Riggs, and S. W. Artz. 1989b. Correlation between histidine operon expression and guanosine 5'-diphosphate-3'-diphosphate levels during amino acid downshift in stringent and relaxed strains of *Salmonella typhimurium*. *J. Bacteriol.* **171**:737–743.

Shaw, D. G., M. J. Alperin, W. S. Reeburgh, and D. J. McIntosh. 1984. Biogeochemistry of acetate in anoxic sediments of Skan Bay, Alaska. *Geochim. Cosmochim. Acta.* **48**:1819–1825.

Shaw, P., S. L. Gomes, K. Sweeney, B. Ely, and L. Shapiro. 1983. Methylation involved in chemotaxis is regulated during *Caulobacter* differentiation. *Proc. Natl. Acad. Sci. (USA)* **80**:5261–5265.

Shearer, C. 1917. On the toxic action of dilute pure sodium chloride solutions on the meningococcus. *Proc. Roy. Soc. B.* **89**:440–443.

Shehata, R. E., and A. G. Marr. 1971. Effect of nutrient concentration on the growth of *Escherichia coli*. *J. Bacteriol.* **107**:210–216.

Sheldon, R. W., T. O. T. Evelyn, and T. R. Parsons. 1967. On the occurrence and formation of small particles in seawater. *Limnol. Oceanogr.* **12**:367–375.

Shepard, F. P. 1948. *Submarine Geology.* Harper and Brothers, New York.

Sherman, J. M., and W. R. Albus. 1923. Physiological youth in bacteria. *J. Bacteriol.* **8**:127–139.

Shiah, F.-K. 1993. Multi-scale variability of bacterioplankton abundance, production and growth rate in the temperate estuarine ecosystems. Ph.D. Thesis, Univ. Maryland, Cambridge, MD.

Shiah, F.-K., and H. W. Ducklow. 1994. Temperature and substrate regulation of bacterial abundance, production and specific growth rate in Chesapeake Bay, USA. *Mar. Ecol. Prog. Ser.* **103**:297–308.

Shields, J. A., and E. A Paul. 1973. Decomposition of ^{14}C labelled plant material in soil under field conditions. *Can. J. Soil Sci.* **53**:297–306.

Shields, J. A., E. A. Paul, W. E. Lowe, and D. Parkinson. 1973. Turnover of microbial tissue in soil under field conditions. *Soil Biol. Biochem.* **5**:753–764.

Shields, J. A., E. A. Paul, and W. E. Lowe. 1974. Factors influencing the stability of labelled microbial materials in soil. *Soil Biol. Biochem.* **6**:31–37.

Shilo, M. 1979. *Strategies of Microbial Life in Extreme Environments.* Verlag Chemie, Weinheim.

Shilo, M. 1980. Strategies of adaptation to extreme conditions in aquatic microorganisms. *Naturwissenschaften* **67**:384–389.

Short, S. A., and D. C. White. 1971. Metabolism of phosphatidylglycerol, lysylphosphatidylglycerol, and cardiolipin of *Staphylococcus aureus*. *J. Bacteriol.* **108**:219–226.

Sidle, A. B. 1967. Amino acid content of atmospheric precipitation. *Tellus* **19**:128–135.

Sieburth, J. McN. 1967. Seasonal selection of estuarine bacteria by water temperature. *J. Exp. Mar. Ecol.* **1**:98–121.

Sieburth, J. McN. 1968. Observations of bacteria planktonic in Narragansett Bay, Rhode Island. *Bull. Misaki Biol. Inst. Rept.* **12**:49–64.

Sieburth, J. McN. 1969. Studies on algal substances in the sea. III. Production of extracellular organic matter by littoral marine algae. *J. Exp. Mar. Biol. Ecol.* **3**:290–309.

Sieburth, J. McN. 1971. Distribution and activity of oceanic bacteria. *Deep-Sea Res.* **18**:1111–1121.

Sieburth, J. McN. 1975. *Microbial Seascapes.* University Park Press, Baltimore, MD.

Sieburth, J. McN. 1981. *Sea Microbes.* Oxford University Press, New York.

Sieburth, J. McN., and A. Jensen. 1968. Studies on algal substances in the sea. I. Gelbstoff (humic materia) in terrestrial and marine waters. *J. Exp. Mar. Biol. Ecol.* **2**:174–189.

Sieburth, J. McN. and A. Jensen. 1969. Studies on algal substances in the sea. II. The formation of gelbstoff (humic material) by exudates of Phaeophyta. *J. Exp. Mar. Biol. Ecol.* **3**:275–289.

Sieburth, J. McN., and K. W. Estep. 1985. Precise and meaningful terminology in marine microbial ecology. *Mar. Microb. Food Webs* **1**:1–16.

Sieburth, J. McN., R. D. Brooks, R. V. Gessner, C. D. Thomas, and J. L. Toole. 1974. Microbial colonization of marine plant surfaces as observed by scanning electron microscopy. *In Effect of the Ocean Environment on Microbial Activities,* R. R. Colwell and R. Y. Morita (eds.), pp. 418–432. University Park Press, Baltimore, MD.

Sieburth, J. McN., V. Smetacek, and J. Lenz. 1978. Pelagic ecosystem structure: Heterotrophic compartments of the plankton and their relationship to plankton size fractions. *Limnol. Oceanogr.* **23**:1256–1263.

Siegel, A. 1971. Metal-organic interactions in the marine environment. *In Organic Compounds in Aquatic Environments,* S. D. Faust and J. V. Hunter (eds.), pp. 265–296. Marcel Dekker, New York.

Siegel, A., and E. T. Degens. 1966. Concentration of dissolved amino acids from saline waters by ligand-exchange chromatography. *Science* **151**:1098–1101.

Siegele, D. A., and R. Kolter. 1992. Life after log. *J. Bacteriol.* **174**:345–348.

Sierra, G., and N. E. Gibbons. 1962. Role and oxidation pathway of poly-ß-hydroxybutyric acid in *Micrococcus halodenitrificans. Can. J. Microbiol.* **8**:255–269.

Sikora, L. J., and J. L. McCoy. 1990. Attempts to determine available carbon in soils. *Biol. Fertil. Soils* **9**:19–24.

Sillén, L. G. and A. E. Martell. 1964. *Stability Constants of Metal-Ion Complexes.* Special Publication No. 17. The Chemical Society, London.

Sillén, L. G. and A. E. Martell. 1971. *Stability Constants of Metal-Ion Complexes.* Spec. Publ. No. 25. The Chemical Society, London.

Silver, M. W., A. L. Shanks, and J. D. Trent. 1978. Marine Snow. Microplankton habitat and source of small scale patchiness in pelagic populations. *Science* **201**:371–373.

Silver, M. W., M. M. Gowing, D. C. Brownless, and J. O. Corliss. 1984. Ciliated protozoa associated with oceanic sinking detritus. *Nature* **309**:246–248.

Sime-Ngando, T., G. Bourdier, C. Amblard, and B. Pinel-Alloul. 1991. Short-term variations in specific biovolumes of different bacterial forms in aquatic ecosystems. *Microb. Ecol.* **21**:211–226.

Simidu, U., W. J. Lee, and K. Kogure. 1983. Comparison of different techniques for determining plate counts of marine bacteria. *Bull. Jpn. Soc. Sci. Fish.* **49**:1199–1203.

Simon, M. 1985. Specific uptake rates of amino acids by attached and free-living bacteria in a mesotrophic lake. *Appl. Environ. Microbiol.* **49**:1254–1259.

Simon, M. 1987. Biomass and production of small and large free-living bacteria in Lake Constance. *Limnol. Oceanogr.* **32**:591–607.

Simon, M., and F. Azam. 1989. Protein content and protein synthesis rates of planktonic bacteria. *Mar. Ecol. Prog. Ser.* **51**:201–213.

Simonart, P., and J. Mayaudon. 1961. Humification des protéines C^{14} dan le sol. pp. 91–103. Pedologie. Symposium International: Applications des sciences nucleaires en pedologie, 2nd., Societe Belge de Pedologie, Ghent, Belgium.

Sinclair, C. G., and H. H. Topiwala. 1970. Model for continuous culture which considers the viability concept. *Biotech. Bioeng.* **12**:1069–1079.

Sinclair, J. L., and M. Alexander. 1984. Role of resistance to starvation in bacteria survival in sewage and lake water. *Appl. Environ. Microbiol.* **48**:410–415.

Sjogren, R. E., and M. J. Gibson. 1981. Bacterial survival in dilute environments. *Appl. Environ. Microbiol.* **41**:1331–1336.

Skinner, C. E., and T. J. Murray. 1926. The viability of *B. coli* and *B. aerogenes* in soil. *J. Infect. Diseases* **38**:37–41.

Skopintsev, B. A. 1971. Recent advances in the study of organic matter in oceans. *Oceanology* **11**:775–789.

Skopintsev, B. A. 1972. On the age of stable organic matter—Aquatic humus in oceanic waters. *In The Changing Chemistry of the Oceans,* D. Dryssen and D. Jagner (eds.), pp. 205–208. John Wiley & Sons, New York.

Skopintsev, B. A. 1981. Decomposition of organic matter of plankton, humification and hydrolysis. *In Marine Organic Chemistry,* E. K. Duursma and R. Dawson (eds.), pp. 125–177. Elsevier Scientific Publication Co., Amsterdam.

Skujins, J. 1976. Extracellular enzymes in soil. *Crit. Rev. Microbiol.* **4**:383–421.

Skujins, J. J., and A. D. McLaren. 1968. Persistence of enzymatic activities in stored and geological preserved soils. *Enzymologia* **34**:213–225.

Skujins, J. J., and A. D. McLaren. 1969. Assay of urease activity using ^{14}C-urea in stored, geologically preserved and in irradiated soils. *Soil Biol. Biochem.* **1**:89–99.

Slayman, C. L., and C. W. Slayman. 1974. Depolarization of the plasma membrane of *Neurospora crasa* during active transport of glucose. *Proc. Natl. Acad. Sci. (USA)* **71**:1935–1939.

Smayda, T. J. 1969. Some measurements of the sinking rate of fecal pellets. *Limnol. Oceanogr.* **14**:621–625.

Smigielski, J., B. J. Wallace, and K. C. Marshall. 1989. Changes in membrane functions during short-term starvation of *Vibrio fluvialis* strain NCTC 11328. *Arch. Microbiol.* **151**:336–347.

Smigielski, J., B. Wallace, and K. C. Marshall. 1990. Genes responsible for size reduction of marine vibrios during starvation are located on the chromosome. *Appl. Environ. Microbiol.* **56**:1645–1648.

Smith, C. R., I. D. Walsh, and R. A. Jahnke. 1992. Adding biology to one-dimensional models of sediment-carbon degradation: the multi-B approach. *In Deep-Sea Food Chains and the Carbon Cycle*, G. T. Rowe and V. Pariente (eds.), pp. 395–400. Kluwer Academic Press, Dordrecht.

Smith, G. A., J. S. Nickels, B. D. Kerger, J. D. Davis, S. P. Collins, J. T. Wilson, J. F. McNabb and D. C. White. 1986. Quantitative characterization of microbial biomass and community structure in subsurface material: a prokaryotic consortium to organic contamination. *Can. J. Microbiol.* **32**:104–111.

Smith, J. J., J. P. Howington, and G. A. McFeters. 1994. Survival, physiological response, and recovery of enteric bacteria exposed to a polar marine environment. *Appl. Environ Microbiol.* **60**:2977–2984.

Smith, J. L., and R. N. Doetsch. 1969. Studies in negative chemotaxis and the survival value of motility in *Pseudomonas fluorescens. J. Gen. Microbiol.* **55**:379–391.

Smith, J. L., B. L. McNeal, H. H. Cheng, and G. S. Campbell. 1986. Calculation of microbial maintenance rates and net nitrogen mineralization in soil at steady state. *Soil Sci. Soc. Am. J.* **50**:332–338.

Smith, J. L., and E. A. Paul. 1988. The role of soil type and vegetation on microbial biomass and activity. *In Perspectives in Microbial Ecology*, F. Megusar and M. Gantar (eds.), pp. 460–466. Slovene Society for Microbiology, Ljubljana, Yugoslavia.

Smith, J. L., and E. A. Paul. 1990. The significance of soil microbial biomass estimations. *Soil Biochem.* **6**:357–396.

Smith, K. L., Jr. 1978. Benthic community respiration in the N.W. Atlantic Ocean: *in situ* measurements from 40 to 5200 m. *Mar. Biol.* **47**:337–347.

Smith, K. L., Jr. 1987. Food energy supply and demand: A discrepancy between particulate organic carbon flux and sediment community oxygen consumption in the deep ocean. *Limnol. Oceangr.* **32**:201–220.

Smith, K. L., Jr., and R. J. Baldwin. 1982. Scavenging deep-sea amphipods: Effects of food odor on oxygen consumption and a proposed metabolic strategy. *Mar. Biol.* **68**:287–298.

Smith, K. L. Jr., and K. R. Hinga. 1983. Sediment community respiration in the deep sea. *In The Sea,* vol. 8, G. T. Rowe (ed.), pp. 331–370. John Wiley & Sons, New York.

Smith, K. L. Jr, and J. M. Teal. 1973. Deep-sea benthic community respiration: An in situ study at 1850 meters. *Science* **179**:282–283.

Smith, K. L., Jr., A. F. Carlucci, R. A Jahnke, and D. B. Craven. 1987. Organic carbon mineralization in the Santa Catalina Basin: benthic boundary layer metabolism. *Deep-Sea Res.* **34**:185–211.

Smits, J. D., and B. Riemann. 1988. Calculation of cell production from [^{3}H]thymidine incorporation with freshwater bacteria. *Appl. Environ. Microbiol.* **54**:2213–2219.

Sneath, P. H. A. 1962. Longevity of micro-organisms. *Nature* **195**:643–646.

Sneath, P. H. A. 1964. The limits of life. *Discovery* **25**:20–24.

Sobek, J. M., J. F. Charba, and W. N. Foust. 1966. Endogenous metabolism of *Azotobacter agilis. J. Bacteriol.* **92**:687–695.

Söderström, B. A. 1979. Seasonal fluctuations of active fungal biomass in horizons of a podsolised pine-forest soil. *Soil Biol. Biochem.* **11**:149–154.

Solonen, K. 1977. The estimate of bacterioplankton numbers and biomass by phase contrast microscopy. *Ann. Bot. Fenn.* **14**:25–28.

Somville, M., and G. Billin. 1983. A method for determining exoproteolytic activity in natural waters. *Limnol. Oceanogr.* **28**:1990–1993.

Sørensen, H. 1963. Studies on the decomposition of C^{14} labelled barley straw in soil *Soil Sci.* **95**:45–51.

Sørensen, L. H. 1967. Duration of amino acid metabolites found in soils during decomposition of carbohydrates. *Soil Sci.* **104**:234–241.

Sørensen, L. H. 1971. Transformation of acetate carbon into carbohydrate and amino acid metabolites during decomposition in soil. *Soil Biol. Biochem.* **3**:173–180.

Sørensen, L. H. 1974. Rate of decomposition of organic matter in soils as influenced by repeated drying-rewetting and repeated additions of organic material. *Soil Biol. Biochem.* **6**:287–292.

Sørensen, L. H. 1975. The influence of clay on the rate of decay of amino acid metabolites in soils during decomposition of cellulose. *Soil Biol. Biochem.* **7**:171–177.

Sørensen, L. H. 1981. Carbon-nitrogen relationships during humification of cellulose in soil containing different amounts of clay. *Soil. Biol. Biochem.* **13**:313–321.

Sørensen, L. H. 1986. Size and persistence of microbial biomass formed during the humification of glucose, hemicellulose, and straw in soils containing different amounts of clay. *Plant Soil* **75**:121–130.

Sørensen, L. H. 1987. Organic matter and microbial biomass in a soil incubated in the field for 20 years with ^{14}C-labelled barley straw. *Soil. Biol. Biochem.* **19**:39–42.

Sørensen, L. H., and E. A. Paul. 1971. Transformation of acetate carbon into carbohydrate and amino acids metabolites during decomposition in soil. *Soil Biol. Biochem.* **3**:173–180.

Sorokin, Yu. 1955. Determination of chemosynthesis value in the water of the Rtbinskoye Reservoir by the use of C^{14}. *Dolk. Akad. Nauk SSSR.* **105**:1343–1353.

Sorokin, Yu. 1957. Determination of chemosynthesis efficiency in methane and hydrogen oxidation in bodies of water. *Mikrobiologiya* **26**:13–18.

Sorokin, Yu. I. 1972. Microbial activity as a biogeochemical factor in the oceans. *In The Changing Chemistry of the Oceans,* Proceeding of the 20th Nobel Symposium, D. Dryssen and D. Tagner (eds.), pp. 189–204. Wiley-Interscience, New York.

Soulides, D. A., and F. E. Allison. 1961. Effect of drying and freezing soils on carbon dioxide production, available mineral nutrients, aggregation, and bacterial population. *Soil Sci.* **91**:291–298.

Sournia, A. 1982. Form and function in marine phytoplankton. *Biol. Rev.* **57**:347–394.

Southamer, A. H. 1973. A theoretical study of the amount of ATP required for synthesis of microbial cell material. *Antonie van Leewenhoek J. Microbiol. Serol.* **39**:545–565.

Sowden, F. J. 1969. Effect of hydrolysis time and iron and aluminum on the determination of amino compounds in soil. *Soil Sci.* **107**:364–371.

Sowden, F. J., and M. Schnitzer. 1967. Nitrogen in illuvial organic matter. *Can. J. Soil Sci.* **47**:111–116.

Sowden, F. J., S. M. Griffith, and M. Schnitzer. 1976. The distribution of nitrogen in some highly organic tropical volcanic soils. *Soil Biol. Biochem* **8**:55–60.

Sparling, G. P., B. G. Ord, and D. Vaughan. 1981. Microbial biomass and activity in soils amended with glucose. *Soil Biol. Biochem.* **13**:99–104.

Sparling, G. P., M. V. Cheshire, and C. M. Mundie. 1982. Effect of barley plants on the decomposition of ^{14}C-labelled soil organic matter. *J. Soil Sci.* **33**:89–100.

Sparling, G. P., K. N. Whale, and A. J. Ramsey. 1985. Quantifying the contribution from the soil microbial biomass to the extractable P levels of fresh and air dried soils. *Aust. J. Soil Res.* **23**:613–621.

Spassky, A., S. Rimsky, H. Garreau and H. Buc. 1984. H1a, an *E. coli* DNA-binding protein which accumulates in stationary phase, strongly compact DNA in vitro. *Nucleic Acids Res.* **12**:5321–5340.

Spector, M. P. and C. L. Cubitt. 1992. Starvation-inducible loci of *Salmonella typhimurium*: regulation and roles in starvation-survival. *Mol. Microbiol.* **6**:1467–1476.

Spector, M. P., Z. Aliabadi, T. Gonzalez, and J. W. Foster. 1986. Global control in *Salmonella typhimurium*: two-dimensional electrophoretic analysis of starvation-, anaerobiosis- and heat shock-inducible proteins. *J. Bacteriol.* **168**:420–424.

Spector, M. P., Y. K. Park, S. Tirgari, T. Gonzalez, and J. W. Foster. 1988. Identification and characterization of starvation-regulated genetic loci in *Salmonella typhimurium* by using Mud-directed *lacZ* operon fusions. *J. Bacteriol.* **170**:345–351.

Spence, J., A. Cegielska, and C. Georgopolous. 1990. Role of *Escherichia coli* heat shock protein DnaK and HtpG (C62.5) in response to nutritional deprivation. *J. Bacteriol.* **172**:7157–7166.

Spitzy, A. 1988. Amino acids in marine aerosol and rain. *In Facets of Modern Biogeochemistry,* V. Ittekkot, S. Kempe, W. Michaelis, and A. Spitzy (eds.), pp. 313–317. Springer-Verlag, Berlin.

Spurio, R., M. Durrenberger, M. Falconi, A. La Teana, C. L. Pon, and C. O. Gualerzi. 1992. Lethal overproduction of the *Escherichia coli* nucleoid protein H-NS: ultramicroscopic and molecular autopsy. *Molec. Gen. Genet.* **231**:201–211.

Stabel, H. H., K. Moaledj, and J. Overbeck. 1979. On the degradation of dissolved organic molecules from Plusssee by oligocarbophilic bacteria. *Arch. Hydrobiol. Beih. Ergebn. Limnol.* **12**:95–104.

Staley, J. T. 1968. Prosthecomicrobium and Anacalomicrobium: New prothecate freshwater bacteria. *J. Bacteriol.* **95**:1921–1942.

Standing, C. N., A. G. Fredrickson, and M. H. Tsuchiya. 1972. Batch- and continuous-culture transients for two substrate systems. *Appl. Environ. Microbiol.* **23**:354–359.

Stanford, G., and S. J. Smith. 1972. Nitrogen mineralizatiom potentials of soils. *Soil Sci. Soc. Am. Proc.* **36**:465–472.

Stanier, R. Y., J. L. Ingraham, M. L. Wheelis, P. R. Painter. 1986. *The Microbial World.* Prentice Hall, Englewood.

Stanley, P. M., E. J. Ordal, and J. T. Staley. 1979. High numbers of prosthecate bacteria in pulp mill waste aeration lagoons. *Appl. Environ. Microbiol.* **37**:1007–1011.

Stanley, S. O., and C. M. Brown. 1976. Inorganic nitrogen metabolism in marine bacteria: the intracellular free amino acid pools of a marine pseudomonad. *Mar. Biol.* **38**:101–109.

Stanley, S. O., and A. H. Rose. 1967. On the clumping of *Corynebacterium xerosis* as affected by temperature. *J. Gen. Microbiol.* **48**:9–23.

Starikova, N. D., and R. I. Korzhikova. 1968. Amino acids in the Black Sea. *Oceanology* **9**:509–518.

Stark, C. N., and B. L. Harrington. 1931. The drying of bacteria and the viability of dry bacterial cells. *J. Bacteriol.* **21**:13–14.

Stark, W. H., and E. McCoy. 1938. Distribution of bacteria in certain lakes of northern Wisconsin. *Zentbl. Bakt. Parasitkde.* **98**:201–209.

Stark, W. H., J. Stadler, and E. McCoy. 1938. Some factors affecting the bacterial population of freshwater lakes. *J. Bacteriol.* **36**:653–654.

Steinberg, C., and A. Hermann. 1981. Utilization of dissolved metal organic compounds by freshwater microorganisms. *Int. Ver. Theor. Angew. Limnol. Verh.* **21**:231–235.

Steinhaus, E. A., and J. M. Birkeland. 1939. Studies on the life and death of bacteria. I. The senescent phase in aging cultures and the probable mechanisms involved. *J. Bacteriol.* **36**:249–461.

Stenstrom, T.-A., P. Conway, and S. Kjelleberg. 1989. Inhibition by antibiotics of the bacterial response to long-term starvation of *Salmonella typhimurium* and the colon microbiota of mice. *J. Appl. Bacteriol.* **67**:53–59.

Sterkenburg, A., E. Vlegels, and J. T. M. Wouters. 1984. Influence of nutrient limitation and growth rate on the outer membrane proteins of *Klebsiella aerogenes* NCTC 418. *J. Gen. Microbiol.* **130**:2347–2355.

Stevenson, F. J. 1982. *Humus Chemistry: Genesis, Composition, Reactions.* John Wiley & Sons, New York.

Stevenson, F. J. 1986. *Cycles of Soil Carbon, Nitrogen, Phosphorus, Sulfur, Micronutrients.* Wiley, New York.

Stevenson, F. J. and M. S. Ardakani. 1972. Organic matter reactions involving micronutrients in soils. *In Micronutrients in Agriculture,* J. J. Mortvedt, P. M. Giordano, and W. L. Lindsay (eds.), pp. 79–114, American Society for Agronomy, Madison, WI.

Stevenson, F. J., and J. H. A. Butler. 1969. Chemistry of humic acids and related pigment. *In Organic Geochemistry: Methods and Results*, G. Eglington and M. T. J. Murphey (eds.), pp. 534–557. Springer-Verlag, Berlin.

Stevenson, F. J., and C-N. Cheung. 1970. Amino acids in sediments: Recovery by acid hydrolysis and quantitative estimation by a colorimetric procedure. *Geochim. Cosmochim. Acta.* **34**:77–88.

Stevenson, F. J., and S. N. Tilo. 1969. Nitrogenous constituents in deep-sea sediments. Proc. Third Intl. Mtg. Org. Geochem., London, England, 1966, pp. 227–253.

Stevenson, I. L. 1956. Some observations on the microbial activity in remoistened air-dried soils. *Plant Soil* **8**:170–182.

Stevenson, I. L. 1958. The effect of some vibration of the bacteria plate count of soil. *Plant Soil* **10**:1–8.

Stevenson, L. H. 1978. A case for bacterial dormancy in aquatic systems. *Microb. Ecol.* **4**:127–133.

Stewart, A. J., and R. G. Wetzel. 1981. Dissolved humic materials: photodegradation and calcium carbonate precipitation. *Arch. Hydrobiol.* **92**:265–286.

Stewart, G. G., and I. Russell. 1977. The identification, characterization, and mapping of a gene for flocculation in *Saccharomyces* sp. *Can. J. Microbiol.* **23**:441–447.

Stockton, W. L., and T. E. DeLuca. 1987. Food falls in the deep sea: occurrence, quality and significance. *Deep-Sea Res.* **29**:157–169.

Stokes, J. L., and W. L. Parson. 1968. Role of poly-ß-hydroxybutyrate in survival of *Sphaerotilus discophorus* during starvation. *Can. J. Microbiol.* **14**:785–789.

Stotzky, G. 1965. Microbial respiration. *In Methods of Soil Analysis, Vol. II. Chemical and Microbiological Properties*. C. A. Black (ed), pp. 1550–1570. American Society Agronomy, Madison, WI.

Stotzky, G. 1971. Ecological eradication of fungi—dream or reality? *In Histoplasmosis*, proceedings of the second national conference, M. L. Furocolow and E. W. Clark (eds.), pp. 477–486. Charles C. Thomas, Springfield, IL.

Stotzky, G. 1974. Activity, ecology, and population dynamics of microorganisms in soil. *In Microbial Ecology*, A. I. Laskin and H. Lechevalier (eds.), pp. 57–135. CRC Press, Cleveland, OH.

Stotzky, G. 1980. Surface interactions between clay mineral and microbes, viruses and soluble organics, and probable importance of these interactions to the ecology of microbes in soil. *In Microbial Adhesion to Surfaces*, R. C. W. Berkeley, J. M. Lynch, J. Melling, P. R. Rutter, and B. Vincent (eds.), pp. 231–249. Ellis Norwood, Chichester.

Stotzky, G. 1986. Influence of soil colloids on metabolic processes, growth, adhesion, and ecology of microbes and viruses. *SSSA Spec. Publ. No.* **17**:305–428.

Stotzky, G., and R. G. Burns. 1982. The soil environment: clay-humus-microbe interactions. *In Experimental Microbial Ecology*, R. G. Burns, and H. Slater (eds.), pp. 179–200. Blackwell Scientific Publications, Oxford.

Stotzky, G., and A. G. Norman. 1961a. Factors limiting microbial activities in soil. I. The level of substrate, nitrogen, and phosphorus. *Arch. Mikrobiol.* **40**:341–369.

Stotzky, G., and A. G. Norman. 1962b. Factors limiting microbial activities in soil. II. The effect of sulfur. *Arch. Mikrobiol.* **40**:370–382.

Stotzky, G., and L. T. Rem. 1966. Influence of clay minerals on microorganisms. I. montmorillonite and kaolinite on bacteria. *Can. J. Microbiol.* **12**:547–563.

Stotzky, G., and S. Schenk. 1976. Volatile organic compounds and microorganisms. *Crit. Rev. Microbiol.* **4**:333–382.

Stotzky, G., G. D. Goos, and M. I. Timonin. 1962. Microbial changes occurring in soil as a result of storage. *Plant Soil* **16**:1–18.

Stout, J. D., K. M. Goh, and T. A. Rafter. 1981. Chemistry and turnover of naturally occurring resistant organic compounds in soil. *Soil Biochem.* **5**:1–73.

Stouthamer, A. H. 1973. A theoretical study of the amount of ATP required for synthesis of microbial cell material. *Antonie van Leeuwehoek* **39**:545–565.

Stouthamer, A. H. 1984. The relation between biomass production and substrate consumption at very low growth rates. *In Innovations in Biotechnology,* E. H. Houwink and R. R. van der Meer (eds.), pp. 519–529. Elsevier, Amsterdam.

Stouthamer, A. H., and C. Bettenhaussen. 1973. Utilization of energy for growth and maintenance in continuous and batch cultures of micro-organisms. A re-evaluation of the method for the determination of ATP production by measuring molar growth yields. *Biochim. Biophys. Acta* **301**:53–70.

Stouthamer, A. H., B. A. Bulthuis, and H. W. Van Verseveld. 1990. Energetics of growth at low grow rates and its relevance for maintenance concepts. *In Microbial Growth Dynamics,* M. J. Bazin, M. J. Poole, and R. K. Keevil (eds.), pp. 85–102. IRL Press, Oxford.

Straley, S. C., A. G. La Marre, L. J. Lawrence, and S. F. Conti. 1979. Chemotaxis of *Bdellovibrio bacteriovorus* toward pour compounds. *J. Bacteriol.* **135**:634–642.

Strange, R. E. 1967. Metabolism of endogenous constituents and survival in starved bacterial suspensions. *Biochem. J.* **102**:34.

Strange, R. E. 1968. Bacterial "glycogen" and survival. *Nature* **220**:606–607.

Strange, R. E., and F. A. Dark. 1965. "Substrate-accelerated death" of *Aerobacter aerogenes. J. Gen. Microbiol.* **39**:215–228.

Strange, R. E., F. A. Dark, and A. G. Ness. 1961. The survival of stationary phase *Aerobacter aerogenes* stored in aqueous suspensions. *J. Gen. Microbiol.* **25**:61–76.

Strange, R. E., H. E. Wade, and A. G. Ness. 1963. The catabolism of proteins and nucleic acids in starved *Aerobacter aerogenes. Biochem. J.* **86**:197–203.

Strayer, D. L. 1988. On the limits of secondary production. *Limnol. Oceanogr.* **33**:1217–1220.

Strickland, J. D. H. 1971. Microbial activity in aquatic environments. *Symp. Soc. Gen. Microbiol.* **21**:231–253.

Strickland, J. D. H., L. Solorzano, and R. W. Eppley. 1970. The ecology of the plankton off La Jolla, California, in the period April through September, 1967. Part I. General introduction, hydrography and chemistry. *Bull. Scripps Inst. Oceanogr.* **17**:1–22.

Strome, D. J., and M. C. Miller. 1978. Photolytic changes in dissolved humic substances. *Int. Ver. Thoer. Angew. Limnol. Verh.* **20**:1248–1254.

Stuart, V., M. I. Lucas, and R. C. Newell. 1981. Heterotrophic utilization of particulate matter from the kelp *Laminaria pallida*. *Mar. Ecol. Prog. Ser.* **4**:337–348.

Stuermer, D. H., and G. R. Harvey. 1976. Humic substances from seawater. *Nature* **250**:480–481.

Stuermer, D. H., and G. R. Harvey. 1977. The isolation of humic substances and alcohol soluble organic matter from seawater. *Deep-Sea Res.* **24**:303–309.

Stuermer, D. H., and G. R. Harvey. 1978. Structural studies on marine humus: a new reduction sequence of carbon skeleton determination. *Mar. Chem.* **6**:55–70.

Stuiver, M., P. D. Quay, and H. G. Ostlund. 1983. Abyssal water carbon-14 distribution and the age of world oceans. *Science* **219**:849–851.

Subba-Rao, R. V., and M. Alexander. 1982. Effects of sorption on mineralization of low concentrations of aromatic compounds in lake water samples. *Appl. Environ. Microbiol.* **44**:659–669.

Sudo, S. Z., and M. Dworkin. 1973. Comparative biology of prokaryotic resting cells. *Adv. Microb. Physiol.* **9**:133–224.

Suess, E. 1970. Interaction of organic compounds with calcium carbonate—I. Association phenomena and geochemical implications. *Geochim. Cosmochim. Acta* **34**:157–168.

Suess, E. 1973. Interaction of organic compounds with calcium carbonate—II. Organo-carbonate association in recent sediments. *Geochim. Cosmochim. Acta* **37**:2435–2448.

Suess, E. 1980. Particulate organic carbon flux in the oceans—surface productivity and oxygen utilization. *Nature* **228**:260–263.

Suess, E. 1988. Effects of microbe activity. *Nature* **333**:17–18.

Suess, E., and D. Fütterer. 1972. Aragonitic ooids: experimental precipitation from sea water in the presence of humic acid. *Sedimentology* **19**:129–139.

Sugai, S. F., and S. M. Henrichs. 1992. Rates of amino acid uptake and mineralization in Resurrection Bay (Alaska) sediments. *Mar. Ecol. Prog. Ser.* **88**:129–141.

Sugimura, Y., and Y. Suzuki. 1988. A high temperature catalytic oxidation method for non-volatile dissolved organic carbon in seawater by direct injection of liquid samples. *Mar. Chem.* **24**:105–131.

Sundh, I. 1992. Biochemical composition of dissolved organic carbon derived from phytoplankton and used by heterotrophic bacteria. *Appl. Environ. Microbiol.* **58**:2938–2947.

Susskind, M. M., and D. Botstein. 1978. Molecular Genetics of bacteriophage P22. *Microbiol. Rev.* **42**:385–413.

Sussman, A. S., and H. O. Halvorson. 1966. *Spores: Their Dormancy and Germination.* Harper and Row, New York.

Sutherland, I. W. 1977. Bacterial exopolysaccharides, their nature and production. *In Surface Carbohydrates of the Prokaryotic Cell*, I. W. Sutherland (ed.), pp. 26–96. Academic Press, New York.

Suttle C. A. 1994. The significance of viruses to mortality in aquatic microbial communities. *Microb. Ecol.* **28**:237–243.

Suttle, C. A., A. M. Chan, and M. T. Cottrell. 1990. Infection of phytoplankton by viruses and reduction of primary productivity. *Nature* **347**:467–469.

Suttle, C. A., A. M. Chan, and J. A. Fuhrman. 1991. Dissolved free amino acids in the Sargasso Sea: uptake and respiration rates, turnover times, and concentrations. *Mar. Ecol. Prog. Ser.* **70**:189–199.

Suwa, Y., and T. Hattori. 1984. Effects of nutrient concentration on the growth of soil bacteria. *Soil Sci. Plant Nutr.* **30**:397–403.

Suwa, Y., and T. Hattori. 1987. Population dynamics of soil bacteria cultured under different nutrient conditions. *Soil Sci. Plant Nutr.* **33**:235–244.

Suzuki, Y. 1993. On the measurement of DOC and DON in seawater. *Mar. Chem.* **41**:287–288.

Suzuki, Y., T. Sugimura, and T. Itoh. 1985. A catalytic oxidation method for the determination of total nitrogen dissolved in sea water. *Mar. Chem.* **16**:83–97.

Swain, T. 1965. The tannins. *In Plant Biochemistry,* J. Bonner and J. E. Verner (eds.), pp. 552–580. Academic Press, London.

Swincer, G. E., J. M. Oades, and D. J. Greenland. 1968. Studies on soil polysaccharides: II. The composition and properties of polysaccharides in soils under pasture and under fallow-wheat notation. *Aust. J. Soil Res.* **6**:225–235.

Sykes, G. 1963. The phenomenon of bacterial survival. *J. Appl. Bacteriol.* **26**:287–294.

Szolnoki, J., F. Kunc, J. Macura, and V. Vancura. 1963. Effect of glucose on the decomposition of organic materials added to soil. *Folia Microbiol.* **8**:356–361.

Tabor, P. S., and R. A. Neihof. 1982. Improved microautoradiographic method to determine individual microorganisms active in substrate uptake in natural waters. *Appl. Environ. Microbiol.* **44**:945–953.

Tabor, P. S., K. Ohwada, and R. R. Colwell. 1981. Filterable marine bacteria found in the deep sea: distribution, taxonomy and response to starvation. *Microb. Ecol.* **7**:67–83.

Takai, Y., and I. Harada. 1959. Prolonged storage of paddy soil samples under several conditions and its effect on soil micro-organisms: I. On the drainable paddy clay loam soil. *J. Sci. Soil Manure (Jpn)* **30**:117–122.

Tamm, E., and G. Krzysch. 1965. Zur dynamik der Bodenatmungund des CO_2-Gehaltes der bodennahen Luftschnicht während der Vegetationsruhe. *Z. Acker-Pflanenbau* **122**:209–215.

Tanaka, K., Y. Takayanagi, N. Fujita, A. Ishihama, and H. Takahashi. 1993. Heterogeneity of the principal σ factor in *Escherichia coli*: the *rpoS* gene product σ^{38} is a secondary principal σ factor of RNA polymerase in stationary-phase *Escherichia coli*. *Proc. Natl. Acad. Sci. (USA)* **90**:3511–3515.

Tanaka, Y., and N. Noda. 1982. Studies on the factors affecting survival of *Pseudomonas solanacearum* E. F. Smith, the causal agent of tobacco wilt disease. *Ann. Phytopath. Soc. Jpn.* **48**:620–627.

Tasch, P. 1960. Paleoecological observations of the Wellington Salt (Hutchinson Member). *Trans. Kansas Acad. Sci.* **63**:24–30.

Tasch, P. 1963. Paleoecological considerations of growth and form of fossil protists. *Ann. New York Acad. Sci.* **108**:437–450.

Tate, R. L. III. 1987. *Soil Organic Matter.* John Wiley & Sons, New York.

Taubenest, R., and H. Ubersax. 1980. Ultra-pure water in semiconductor manufacturing. *Solid State Tech.* **23**:74–79.

Taylor, B. L. 1983. Role of proton motive force in sensory transduction by bacteria. *Ann. Rev. Microbiol.* **37**:551–573.

Taylor, C. B. and V. G. Collins. 1949. Development of bacteria in waters in glass containers. *J. Gen. Microbiol.* **3**:32–42.

Taylor, G. T., R. Iturriaga, and C. W. Sullivan. 1985. Interactions of bactivorus grazers and heterotrophic bacteria with dissolved organic matter. *Mar. Ecol. Prog. Ser.* **23**:129–141.

Taylor, G. T., D. M. Karl, and M. L. Pace. 1986. Impact of bacteria and zooflagellates on the composition of sinking particles: an in situ experiment. *Mar. Ecol. Prog. Ser.* **29**:141–155.

Taylor, R. T., S. A. Norrell, and M. L. Hanna. 1972. Uptake of cyanocobalamine by *Escherichia coli* B: Some characteristics and evidence for a binding protein. *Arch. Biochim. Biophys.* **148**:366–381.

Taylor, W. D., and D. R. S. Lean. 1981. Radiotracer experiments on phosphorus uptake and release by limnetic microzooplankton. *Can. J. Fish. Aquat. Sci.* **38**:1316–1321.

Tempest, D. W., and O. M. Neijssel. 1978. Eco-physiological aspects of microbial growth in aerobic nutrient-limited environments. *Adv. Microb. Ecol.* **2**:105–153.

Tempest, D. W., D. Herbert, and P. J. Phipps. 1967. Studies on the growth of *Aerobacter aerogenes* at low dilution rates in a chemostat. *In Microbial Physiology and Continuous Culture,* E. O. Powell, C. G. T. Evans, R. E. Strange and D. W. Tempest (eds.), pp. 240–254. Her Majesty's Stationary Office, London.

Tempest, D. W., J. L. Meers, and C. M. Brown. 1970. Glutamate synthetase (Gogat): A key in the assimilation of ammonia by prokaryotic organisms. *In The Enzymes of Glutamine Metabolism,* S. Prusiner, and E. R. Statman (eds.), pp. 167–182. Academic Press, New York.

Tempest, D. W., O. M. Neijssel, and W. Zevenboom. 1983. Properties and performance of microorganisms in laboratory culture: their relevance to growth in natural ecosystems. *Symp. Soc. Gen. Microbiol.* **34**:119–152.

Terracciano, J. S., and E. Canale-Parola. 1984. Enhancement of chemotaxis in *Spirochaeta aurantia* grown under conditions of nutrient limitation. *J. Bacteriol.* **159**:173–178.

Terzaghi, E., and M. O'Hara. 1990. Microbial plasticity. The relevance to microbial ecology. *Adv. Microb. Ecol.* **11**:431–460.

Teteno, M. 1985. Adenylate energy charge in glucose-amended soil. *Soil Biol. Biochem.* **17**:387–388.

Theng, B. K. G. 1979. *Formation and Properties of Clay-Polymer Complexes.* Elsevier, Amsterdam.

Thiel, G. A. 1928. A summary of the activities of bacterial agencies in sedimentation. *Nat. Res. Coun. Circular* **65**:61–77.

Thiel, H. 1979. Structural aspects of the deep-sea benthos. *Ambio Spec. Rept.* **6**:25–31.

Thimann, K. V. 1955. *The Life of Bacteria: Their Growth, Metabolism and Relationships.* Macmillan, New York.

Thomas, T. D., and R. D. Batt. 1968. Survival of *Streptococcus lactis* in starvation conditions. *J. Gen. Microbiol.* **50**:367–382.

Thomas, T. D., and R. D. Batt. 1969a. Degradation of cell constituents by starved *Streptococcus lactis* in relation to survival. *J. Gen. Microbiol.* **58**:347–362.

Thomas, T. D., and R. D. Batt. 1969b. Synthesis of protein and ribonucleic acid by starved *Streptococcus lactis* in relation to survival. *J. Gen. Microbiol.* **58**:363–369.

Thomas, T. D., and R. D. Batt. 1969c. Metabolism of exogenous arginine and glucose by starved *Streptococcus lactis* in relation to survival. *J. Gen. Microbiol.* **58**:371–381.

Thomas, T. D., P. Lyttleton, K. L. Williamson, and R. D. Batt. 1969. Changes in permeability and ultrastructure of starved *Streptococcus lactis* in relation to survival. *J. Gen. Microbiol.* **58**:381–390.

Thompson, L. A., and D. B. Nedwell. 1985. Existence of different pools of fatty acids in anaerobic model ecosystems and their availability to microbial metabolism. *FEMS Microbiol. Ecol.* **31**:141–146.

Thordarson, W. 1965. *Perched Groundwater in Zeolitized-Bedded Tuff, Rainier Mesa and Vicinity, Nevada Test Site, Nevada.* Report TEI-862. U.S. Geological Survey, Denver.

Thorn, P. M., and R. M. Ventullo. 1988. Measurement of bacterial growth rates in subsurface sediments using the incorporation of tritiated thymidine into DNA. *Microb. Ecol.* **16**:3–16.

Thorsen, B. K., Ä. Enger, S. Norland, and K. A. Hoff. 1992. Long-term starvation survival of *Yersinia ruckeri* at different salinities studied by microscopical and flow cytometric methods. *Appl. Environ. Microbiol.* **58**:1624–1628.

Thurman, E. M. 1985. *Organic Geochemistry of Natural Waters.* Martinus Nijhoff/Dr. W. Junk Publisher, Dordrecht.

Tietjen, J. H. 1992. Abundance and biomass of metazoan meiobenthos in the deep sea. In *Deep-Sea Food Chains and the Global Carbon Cycle,* G. T. Rowe and V. Pariente (eds.), pp. 45–62. Proc. NATO ARW, Kluwer Academic Publications, Dordrecht.

Tkachenko, A. G., and A. A. Chudinov. 1987. Energy aspects of the growth of *Escherichia coli* synchronized by starvation. *Microbiology* **56**:47–52.

Toggweiler, J. R. 1988. Deep-sea carbon: a burning issue. *Nature* **334**:468.

Toggweiler, J. R. 1990. Diving into the organic soup. *Nature* **345**:203–204.

Toggweiler, J. R. 1992. Catalytic conversions. *Nature* **356**:665–666.

Tormo, A., M. Almiron, and R. Kolter. 1990. *surA,* an *Escherichia coli* gene essential for survival in stationary phase. *J. Bacteriol.* **172**:4339–4347.

Torrella, F., and R. Y. Morita. 1979. Evidence by electron micrographs for a high incidence of bacteriophage particles in the waters of Yaquina Bay, Oregon: Ecological and taxonomical implications. *Appl. Environ. Microbiol.* **37**:774–778.

Torrella, F., and R. Y. Morita. 1981. Microcultural study of bacteria size changes and microcolony and ultramicrocolony formation by heterotrophic bacteria in seawater. *Appl. Environ. Microbiol.* **41**:518–527.

Torrella, F., and R. Y. Morita. 1982. Starvation-induced morphological changes, motility, and chemotaxis patterns in a psychrophilic marine vibrio. Deuxieme Colloque de Microbiologie marine. *Publ. de Centre Nat. pour l'Exploitation des Oceans* **13**:45–60.

Tranvik, L. J. 1988. Availability of dissolved organic carbon for planktonic bacteria in oligotrophic lakes of differing humic content. *Microb. Ecol.* **16**:311–322.

Tranvik, L. J. 1990. Bacterioplankton growth on fractions of dissolved organic carbon of different molecular weights from humic and clear waters. *Appl. Environ. Microbiol.* **56**:1672–1677.

Tranvik, L. J. 1993. Microbial transformation of labile dissolved organic matter into humic-like matter in seawater. *FEMS Microbiol. Ecol.* **12**:177–183.

Tranvik, L. J., and M. G. Höfle. 1987. Bacterial growth in mixed cultures on dissolved organic carbon from humic and clear waters. *Appl. Environ. Microbiol.* **53**:482–488.

Tranvik, L. J., and J. McN. Sieburth. 1989. Effects of flocculated humic acid on free and attached pelagic microorganisms. *Limnol. Oceanogr.* **34**:688–699.

Trask, P. D. 1939. *Recent Marine Sediments.* American Association Petroleum Geologists, Tulsa, OK.

Trichet, J. 1968. Etude de la composition de la fraction organique des oolites. Comparison avec celle des membranes des bacerries et des cyanophyces. *Compt. Rend.* **267**:1492–1494.

Trinci, A. P. J., and C. G. Thurston. 1976. Transition to the non-growing state in eukaryotic micro-organisms. *Symp. Soc. Gen. Microbiol.* **26**:55–79.

Troussellier, M., G. Cahet, P. Lebaron, and B. Baleux. 1993. Distribution and dynamics of bacterial production in relation to wind perturbation in a Mediterranean lagoon. *Limnol. Oceanogr.* **38**:193–201.

Truex, M. J., F. J. Brockman, D. L. Johnstone, and J. K. Fredrickson. 1992. Effect of starvation on induction of quinoline degradation for a subsurface bacterium in a continuous-flow column. *Appl. Environ. Microbiol.* **58**:2386–2392.

Tsunogai, S. 1972. An estimate of the rate of decomposition of organic matter in the deep water of the Pacific Ocean. *In Biological Oceanography of North Northern Pacific,* A. Y. Takenouti (ed.), pp. 517–533. Idemitsu Shoten, Tokyo.

Tuckett, J. D., and W. E. C. Moore. 1959. Production of filterable particles by *Cellvibrio gilvus. J. Bacteriol.* **77**:227–229.

Tukey, H. R. Jr. 1970. The leaching of substances from plants. *Ann. Rev. Plant Physiol.* **21**:305–324.

Tunlid, A., and D. C. White. 1992. Biochemical analysis of biomass, community structure, nutritional status, and metabolic activity of microbial communities in soil. *Soil Biochem.* **7**:229–262.

Turekian, K. K., J. K. Cochran, D. P. Kharkar, R. M. Cerrato, J. R. Vaisnys, H. L. Sanders, J. F. Grassle, and J. A. Allen. 1975. Slow growth rate of a deep sea clam determined by ^{228}Ra chronology. *Proc. Natl. Acad. Sci. U.S.A.* **72**:2829–2832.

Turley, C. 1994. Controls of the microbial loop: nutrient limitation and enzyme production, location and control. *Microb. Ecol.* **28**:287–290.

Turley, C., and K. Lochte. 1986. Diel changes in the specific growth rate and mean cell volume of natural bacterial communities in two different water masses in the Irish Sea. *Microb. Ecol.* **12**:271–282.

Turner, H. G. 1932. Bacteria in Pennsylvania anthracite. *Science* **76**:122–123.

Turner, M. A., and A. Eisenstark. 1984. Near-ultraviolet radiation blocks SOS responses to DNA damage in *Escherichia coli. Molec. Gen. Genet.* **193**:33-37.

Uhlinger, D. J., and D. C. White. 1983. Relationship between the physiological status and the formation of extracellular polysaccharide glycocalyx in *Pseudomonas atlantica. Appl. Environ. Microbiol.* **45**:64–70.

Unger, H., and M. Wagner. 1965. Das verhalten enteropathogener Serotypen von *Escherichia coli* in zwei verrschiedenen Böden. *Zentr. Bakteriol. Parasitenk. Abt. II.* **119**:474–489.

Upton, A. C., and D. B. Nedwell. 1989. Nutritional flexibility of oligotrophic and copiotrophic Antarctic bacteria with respect to organic substrates. *FEMS Microbiol. Ecol.* **62**:1–6.

Vaccaro, R. F. 1969. The response of natural microbial populations to organic enrichments. *Limnol. Oceanogr.* **14**:726–735.

Vaccaro, R. F., and H. W. Jannasch. 1966. Studies on heterotrophic activity in seawater based on glucose assimilation. *Limnol. Oceangr.* **11**:596–607.

Vaccaro, R. F., S. E. Hicks, H. W. Jannasch, and F. G. Carey. 1968. The occurrence and role of glucose in sea water. *Limnol. Oceanogr.* **13**:356–360.

Valentine, R. C. and J. R. G. Bradfield. 1954. The urea method for bacterial viability counts with the electron microscope and its relation to other viability counting methods. *J. Gen. Microbiol.* **11**:349–357.

Vallentyne, J. R. 1962. Solubility and decomposition of organic matter in nature. *Arch. Hydrobiol.* **58**:423–434.

Vallentyne, J. R. 1965. Two aspects of geochemistry of amino acids. *In The Origins of Prebiological Systems,* S. W. Fox (ed.), pp. 105–120. Academic Press, New York.

VanBogelen, R. A., M. A. Acton, and F. C. Neidhardt. 1987. Induction of the heat shock regulon does not produce thermotolerance in *Escherichia coli. Genes Devel.* **1**:525–531.

VanBogelen, R. A., P. Sankar, R. L. Clark, J. A. Bogan, and F. C. Neidhardt. 1992. The gene-protein database of *Escherichia coli*: edition 5. *Electrophoresis* **13**:1014–1054.

Van der Kooij, D. A., and W. A. M. Hijnen. 1981. Utilization of low concentrations of starch by a *Flavobacterium* species isolated from tap water. *Appl. Environ. Microbiol.* **41**:216–221.

Van der Kooij, D. A., and W. A. M. Hijnen. 1983. Nutritional versatility of a starch-utilizing *Flavobacterium* at low substrate concentratioins. *Appl. Environ. Microbiol.* **45**:804–810.

Van der Kooij, D. A., D. A. Visser, and W. A. M. Hijnen. 1980. Growth of *Aeromonas hydrophila* at low concentrations of substrates added to tap water. *Appl. Environ. Microbiol.* **39**:1198–1204.

Van der Kooij, D. A., J. P. Oranje, and W. A. M. Hijnen. 1982a. Growth of *Pseudomonas aeruginosa* in tap water in relation to utilization of substrates at concentrations of a few micrograms per liter. *Appl. Environ. Microbiol.* **44**:1086–1095.

Van der Kooij, D., D. A. Visser, and W. A. H. Hijnen. 1982b. Determining the concentration of easily assimilable organic matter in drinking water. *J. Am. Water Works Assoc.* **5**:257–270.

Van Duyl, F. C., R. P. Bak, A. J. Kop, and G. Nieuwland. 1990. Bacteria, auto- and heterotrophic nanoflagellates, and their relations in mixed, frontal and stratified waters of the North Sea. *Neth. J. Sea Res.* **26**:97–109.

Van Duyl, F. C., and A. J. Kop. 1988. Temporal and laterial fluctuations in production and biomass of bacterioplankton in the western Dutch Wadden Sea. *Neth. J. Sea. Res.* **22**:51–68.

Vandevivere, P., and D. L. Kirchman. 1993. Attachment stimulates exopolysaccharide synthesis by a bacterium. *Appl. Environ. Microbiol.* **59**:3280–3286.

van Es, F. B., and L.-A. Meyer-Reil. 1982. Biomass and metabolic activity of heterotrophic marine bacteria. *Adv. Microb. Ecol.* **6**:111–170.

van Gemerden, H. 1980. Survival of *Chromatium vinosum* at low light intensities. *Arch. Microbiol.* **125**:115–121.

van Gemerden, H., and J. G. Kuenen. 1984. Strategies for growth and evolution of microorganisms in oligotrophic habitats. *In Heterotrophic Activity in the Sea,* J. E. Hobbie and P. J. leB. Williams (eds.), pp. 25–54. Plenum Press, New York.

Van Hall, C. E., J. Safranko, and V. A. Stenger. 1963. Rapid combustion method for the determination of organic substances in aquatic solutions. *Anal. Chem.* **35**:315–319.

van Houte, J., and H. M. Jansen. 1970. Role of glycogen in survival of *Streptococcus mitis. J. Bacteriol.* **101**:1083–1085.

van Loosdrecht, M. C. M., J. Lyklema, W. Norde, G. Schraa, and A. J. B. Zehnder. 1987. Electrophoretic mobility and hydrophobicity as a measure to predict the initial steps in bacteria adhesion. *App. Environ. Microbiol.* **53**:1898–1901.

van Uden, K. U., and A. Madeira-Lopes. 1976. Yield and maintenance relations of yeast growth in the chemostat at suboptimal tempatures. *Biotechnol. Bioeng.* **18**:791–804.

van Veen, J. A., and J. D. van Elsas. 1986. Impact of soil structure and texture on the activity and dynamics of the soil microbial population. *In Perspectives in Microbial Ecology,* F. Megusar and M. Gantar (eds.), pp. 481–488. Slovene Society for Microbiology, Ljubljana, Yugoslavia.

van Verseveld, H. W., M. Arbige, and W. R. Chesbro. 1984a. Continuous culture of bacteria with biomass retention. *Trends in Biotech.* **2**:8–12.

Van Verseveld, H. W., W. R. Chesbro, M. Braster, and A. H. Stouthamer. 1984b. Eubacteria have three growth modes keyed to nutrient flow: consequences for the concept of maintenance energy and maximal growth yield. *Arch. Microbiol.* **127**:176–184.

Van Vleet, E. S., and J. G. Quinn. 1976. Characterisation of monounsaturated fatty acids from an estuarine sediment. *Nature (London)* **262**:126–128.

Vela, G. R. 1974. Survival of *Azotobacter* in dry soil. *Appl. Microbiol.* **28**:77–79.

Vela, G. R., and O. Wyss. 1964. Improved stain for visualization of *Azotobacter* morphology. *J. Bacteriol.* **87**:476–477.

Veldkamp, H. 1968. The physiology of soil bacteria. *In The Ecology of Soil Bacteria,* T. G. R. Gray (ed.), pp. 201–219. University of Toronto Press, Toronto.

Velimirov, B. 1994. Carbon fluxes in the microbial loop: comments. *Microb. Ecol.* **28**:205–208.

Velimirov, B., A. K. T. Kirschner, and C. B. Mathias. 1992. Productivity of ultramicrobacteria in marine and limnic systems. *Abstract of the Sixth International Symposium for Microbial Ecology, Barcelona,* abstr., p. 117.

Velji, M. I., and L. J. Albright. 1986. Microscopic enumeration of attached bacteria of sea water, marine sediment, fecal matter, and kelp blade samples following pyrophosphate and ultrasound treatments. *Can. J. Microbiol.* **32**:121–126.

Verma, L., J. P. Martin, and K. Haider. 1975. Decomposition of carbon-14-labeled proteins, peptides, and amino acids, free and complexed with humic polymers. *Soil Sci. Soc. Amer. Proc.* **39**:279–284.

Visser, S. A. 1982. Surface active phenomena by humic substances of aquatic origin. *Rev. Francais des Sciences de l'eau* **1**:285–295.

Visser, S. A. 1985. Physiological action of humic substances on microbial cells. *Soil Biol. Biochem.* **17**:457–462.

Volkman, J. K., and R. B. Johns. 1977. The geochemical significance of positional isomers of unsaturated acids from intertidal zone sediments. *Nature* **267**:693–694.

Volkman, J. K., R. B. Johns, F. T. Gillan, and G. J. Perry. 1980. Microbial lipids of an intertidal sediment. I. Fatty acids and hydrocarbons. *Geochim. Cosmochim. Acta* **44**:1133–1143.

Voroshilova, A., and E. Dianova. 1937. The role of plankton in the multiplication of bacteria in isolated samples of sea-water. *Microbiologiia* **6**:741–753.

Votyakova, T. V., A. S. Kaprelyants, and D. E. Kell. 1994. Influence of viable cells on the resuscitation of dormant cells in *Micrococcus luteus* cultures held in an extended stationary phase: the population effect. *Appl. Environ. Microbiol.* **60**:3284–3291.

Wachenheim, D. E., and R. B. Hespell. 1985. Responses of *Ruminococcus flavefaciens,* a ruminal cellulolytic species, to nutrient starvation. *Appl. Environ. Microbiol.* **50**:1361–1367.

Wagner, F. S. Jr. 1969. Composition of the dissolved organic compounds in seawater: A Review. *Contrib. Mar. Sci., University of Texas* **14**:115–153.

Wakimoto, S., I. Utatsu, N. Matsuo, and N. Hayashi. 1982. Multiplication of *Pseudomonas solanacearum* in pure water. *Ann. Phytopath. Soc. Jpn.* **48**:620–627.

Wakeham, S. G., J. W. Farrington, R. B. Gagosian, C. Lee, H. DeBaar, G. E. Nigrelli, B. W. Tripp, S. O. Smith and N. M. Frew. 1980. Organic matter fluxes from sediment traps in the equatorial Atlantic Ocean. *Nature* **286**:798–800.

Waksman, S. A. 1916. Bacterial numbers in soil, at different depths and in different seasons of the year. *Soil Sci.* **1**:363–380.

Waksman, S. A. 1922. Microbiological analysis of soil as an index of soil fertility. *Soil Sci.* **14**:283.

Waksman, S. A. 1934. The distribution and conditions of existence of bacteria in the sea. *Ecol. Monograph* **4**:523–529.

Waksman, S. A., and C. L. Carey. 1935. Decomposition of organic matter in seawater. I. Bacterial multiplication in stored water. *J. Bacteriol.* **29**:531–543.

Waksman, S. A., and R. L. Starkey. 1923. Partial sterilization of soil, microbiological activities and soil fertility: 1. *Soil Sci.* **16**:137–156.

Waksman, S. A., and H. B. Woodruff. 1940. Survival of bacteria added to soil and the resultant modification of the soil population. *Soil Sci.* **50**:421–427.

Waksman, S. A., H. W. Reuszer, C. L. Carey, M. Hotchkiss, and C. E. Renn. 1933. Studies on the biology and chemistry of the Gulf of Maine. III. Bacteriological investigations of the sea water and marine bottoms. *Biol. Bull.* **64**:183–205.

Walcott, C. D. 1915. Discovery of Algonkian bacteria. *Proc. Nat. Acad. Sci. (USA)* **1**:256–257.

Walker, G. C. 1984. Mutagenesis and inducible responses to deoxyribonucleic acid damage in *Escherichia coli. Microbiol. Rev.* **48**:60-93.

Wallis, P. M., and T. I. Ladd. 1982. Organic biogeochemistry of groundwater at a mountain coal mine. *Geomicrobiol. J.* **3**:49–78.

Wardell, J. N., C. M. Brown, and B. Flannigan. 1983. Microbes and surfaces. *Symp. Soc. Gen. Microbiol.* **34**:351–378.

Wangersky, P. J. 1965. The organic chemistry of sea water. *Amer. Scient.* **53**:358–374.

Wangersky, P. J. 1978. Production of dissolved organic matter. *In Marine Ecology*, vol. 4, O. Kinne (ed.), pp. 115–200. John Wiley & Sons, New York.

Wanner, B. L. 1987. Molecular cloning of Mu d(*bla lacZ*) transcriptional and translational fusions. *J. Bacteriol.* **169**:2026–2030.

Warcup, J. H. 1955. On the origin of colonies of fungi developing on soil dilution plates. *Trans. Brit. Mycol. Soc.* **38**:298–301.

Ward, D. M., R. Weller, and M. M. Bateson. 1990. 16S rRNA sequences reveal numerous uncultured microorganisms in a natural community. *Nature* **344**:63–65.

Waring, R. H., and W. H. Schlesinger. 1985. *Forest Ecosystems: Concepts and Management.* Academic Press, Orlando.

Warren, R. A. J., A. F. Ells, and J. J. R. Campbell. 1960. Endogenous respiration of *Pseudomonas aeruginosa. J. Bacteriol.* **79**:875–879.

Watson, S. W., and J. E. Hobbie. 1979. Measurement of bacterial biomass as lipopolysaccharide. *In Native Aquatic Bacteria: Enumeration, Activity and Ecology*, J. W. Costerton and R. R. Colwell (eds.), pp. 82–88. ASTM STP 695, American Society for Testing and Materials, Philadelphia..

Watson, S. W., T. J. Novitsky, H. L. Quimby, and F. W. Valois. 1977. Determination of bacterial number and biomass in the marine environment. *Appl. Environ. Microbiol.* 33:940–946.

Wattenberg, H. 1933. Kalziiumkarbonat und Kohlensäuregehalt des Meerwasseres. *Wiss. Ergebn. Deutsch. Atlant. Exped. "Meteor" 8,* Berlin-Leipzig.

Webb, K. L., and R. E. Johannes. 1967. Studies on the release of dissolved free amino acids by marine zooplankton. *Limnol. Oceanogr.* 12:376–382.

Weber, C. A. 1907. Aufbau und Vegetatioin der Moore Norddeutschlands. *Beiblatt zu den Botanischen Jahrbuchern* 90:19–34.

Weber, J. B., and H. D. Colbe. 1968. Microbial decomposition of dequat adsorbed on montmorillonite and kaolinite clays. *J. Agric. Food Chem.* 16:475–478.

Weber, M. E., D. C. Blanchard, and L. D. Syzdek. 1983. The mechanism of scavenging water-borne bacteria by a rising bubble. *Limnol. Oceanogr.* 28:101–105.

Weber, N. L., P. M. Jardine, and J. F. McCarthy. 1988. Mechanisms of dissolved carbon adsorption on soil. *Agron. Abst.* p. 207.

Webster, J. J., G. J. Hampton, J. T. Wilson, W. C. Ghiorse, and F. R. Leach. 1985. Determination of microbial cell numbers in subsurface samples. *Groundwater* 23:17–25.

Weichart, D., R. Lange, N. Henneberg, and R. Hengge-Aronis. 1993. Identification and characterization of stationary phase-inducible. *Mol. Microbiol.* 10:407–420.

Weimer, M. S., and R. Y. Morita. 1974. Temperature and hydrostatic pressure effects on gelatinase activity of a *Vibrio* species and partially purified gelatinase. *Ztschr. f. Allg. Mikrobiol.* 14:719–725.

Weiner, J. H., and L. A. Heppel. 1971. A binding protein for glutamine and its relation to active transport in *Escherichia coli. J. Biol. Chem.* 246:6933–6941.

Weirich, G., and R. Schweisfurth. 1985. Extraction and culture of microorganisms from rock. *Geomicrobiol. J.* 4:1–20.

Welch, R. J., A. Farewell, F. C. Neidhardt, and D. H. Bartlett. 1993. Stress response of *Escherichia coli* to elevated hydrostatic pressure. *J. Bacteriol.* 175:7170–7177.

Wells, M. L., and E. D. Goldberg. 1991. Occurrence of small colloids in seawater. *Nature* 353:342–344.

Wells, M. L., and E. D. Goldberg. 1992. Marine sub-micron particles. *Mar. Chem.* 40:5–18.

Wells, M. L., and E. D. Goldberg. 1993. Colloid aggregation in seawater. *Mar. Chem.* 41:353–358.

Wells, M. L., and E. D. Goldberg. 1994. The distribution of colloids in the North Atlantic and Southern oceans. *Limnol Oceanog.* 39:286–302.

West, P. M., and A. G. Lochhead. 1950. The nutritional requirements of soil bacteria—A basis for determining the bacterial soil equilibrium of soils. *Soil Sci.* 50:409–420.

West, A. W., and G. P. Sparling. 1986. Modification to the substrate induced respiration method to permit measurement of microbial biomass in soils of differing water content. *J. Microb. Methods* 5:177–189.

West, A. W., G. P. Sparling, and W. D. Grant. 1986. Correlation between four methods to estimate total microbial biomass in stored, air-dried and glucose-amended soils. *Soil Biol. Biochem.* **18**:569–576.

Wetzel, R. G. 1971. The role of carbon in hard-water marl lakes. *In Nutrients and Eutrophication,* G. E. Likens (ed.), pp. 85–97. American Society Limnologuand Oceanography Special Symposium 1. Allen Press, Inc., Lawrence, Kansas.

Wetzel, R. G. 1983. *Limnology,* 2nd ed. Saunders College Publication, Philadelphia.

Wetzel, R. G., and B. A. Manny. 1972. Decomposition of dissolved organic carbon and nitrogen compounds from leaves in an experimental hard-water stream. *Limnol. Oceanogr.* **17**:927–931.

Weyl, P. K. 1961. The carbonate saturometer. *J. Geol.* **69**:32–44.

Whang, K., and T. Hattori. 1988. Oligotrophic bacteria from rendzina forest soil. *Antonie van Leeuwenhoek J. Microbiol. Serol.* **54**:19–36.

Whelan, J. K. 1977. Amino acids in a surface sediment core of the Atlantic abyssal plain. *Geochim. Cosmochim. Acta* **42**:803–810.

Whipple, W. C. 1901. Changes that take place in the bacterial content of waters during transportation. *Tech. Quart. Proc. Soc. Arts* **14**:21–29.

White, D. C. 1993. *In situ* measurement of microbial biomass, community structure and nutrition status. *Phil. Trans. R. Soc. Lond. A.* **344**:59–67.

White, D. C. 1994. Is there anything else you need to understand about the microbiota that cannot be derived from the analysis of nucleic acids. *Microb. Ecol.* **28**:163–166.

White, D. C., and R. H. Findlay. 1988. Biochemical markers for measurement of predation effects on the biomass, community structure, nutritional status, and metabolic activity of microbial biofilms. *Hydrobiologia* **159**:119–132.

White, D. C., G. A. Smith, M. J. Gehron, J. H. Parker, R. H. Findlay, R. F. Martz, and H. L. Fredrickson. 1983. The ground-water aquifer microbiotia: biomass, community structure, and nutritional status. *Dev. Ind. Microbiol.* **24**:201–211.

White, P. A., J. Kalff, J. B. Rassussen, and J. M. Gasol. 1991. The effects of temperature and algal biomass on bacterial production and specific growth rate in freshwater and marine habitats. *Microb. Ecol.* **21**:99–118.

Whitting, P. H., M. Midgley, and E. A. Dawes. 1976. The role of glucose limitation in the regulation of the transport of glucose, gluconate and 2-oxogluconate and of glucose metabolism in *Pseudomonas aeruginosa. J. Gen. Microbiol.* **92**:304–310.

Whittenbury, R., and C. S. Dow. 1977. Morphogenesis and differentiation in *Rhodomicrobiium vannielii* and other budding and prosthecate bacteria. *Bacteriol. Rev.* **41**:754–808.

Wiebe, P. H., S. H. Boyd, and C. Winget. 1976. Particulate matter sinking to the deep-sea floor at 2000 m in the tongue of the ocean, with a description of a new sediment trap. *J. Mar. Res.* **34**:341–354.

Wiebe, W. J. 1984. Physiological and biochemical aspects of marine bacteria. *In Heterotrophic Activity in the Sea,* J. E. Hobbie and P. J. leB. Williams (eds.), pp. 55–82. Plenum Press, New York.

Wiebe, W. J., and K. Bancroft. 1975. Use of adehylate energy charge ratio to measure growth rate of natural microbial communities. *Proc. Natl. Acad. Sci. (USA)* **72**:2112–2115.

Wiebe, W. J., and L. R. Pomeroy. 1972. Microorganisms and their association with aggregates and detritus in the sea: A microscopic study. *Men. Inst. Ital. Idrobiol. Spec.* **29** (Suppl.): 325–352.

Wiebe, W. J., W. M. Sheldon, and L. R. Pomeroy. 1992. Bacterial growth in the cold: Evidence for an enhanced substrate requirement. *Appl. Environ. Microbiol.* **58**:359–364.

Wiebe, W. J., W. M. Sheldon, and L. R. Pomeroy. 1993. Evidence for an enhanced substrate requirement by marine mesophilic bacterial isolates at minimal growth temperatures. *Microb. Ecol.* **25**:151–159.

Wilkinson, J. F. 1959. The problem of energy-storage compounds in bacteria. *Expt. Cell. Res., Suppl.* **7**:111–130.

Willetts, N. S. 1967. Intracellular protein breakdown in non-growing cells of *Escherichia coli. Biochem. J.* **103**:453–461.

Williams, J. C., and E. Weiss. 1978. Energy metabolism of *Rickettsia typhi*: Pools of adenine nucleotides and energy charge in the presence and absence of glutamate. *J. Bacteriol.* **134**:884–892.

Williams, P. J. LeB. 1970. Heterotrophic utilization of dissolved organic compounds in the sea. I. Size distribution of populations and relationship between respiration and incorporation of growth substances. *J. Mar. Biol. Assoc. U. K.* **50**:859–870.

Williams, P. J. LeB. 1973. On the question of growth yields of natural heterotrophic populatons. *In Modern Methods in the Study of Microbial Ecology,* T. Rosswall (ed.), pp. 400–401. Ecol. Res. Comm. NFR (Swedish National Science Research Council), Stockholm, Sweden.

Williams, P. J. LeB. 1975. Biological and chemical aspects of dissolved organic matter in sea water. *In Chemical Oceanography,* vol. 2, 2nd ed, J. P. Riley and R. Skirrow (eds.), pp. 301–363. Academic Press, London.

Williams, P. J. LeB. 1981. Incorporation of micro-heterotrophic processes into the classical paradigm of the planktonic food web. *Kieler Meeresforsch.* **5**:1–28.

Williams, P. J. LeB. 1984. A review of measurements of respiration rates of marine plankton populations. *In Heterotrophic Activity in the Sea,* J. E. Hobbie and P. J. leB. Williams (eds.), pp. 357–389. Plenum Press, New York.

Williams, P. J. LeB., and R. W. Gray. 1970. Heterotrophic utilization of dissolved organic compounds in the sea. II. Observation on the responses of heterotrophic marine populations to abrupt increases in amino acid concentrations. *J. Mar. Biol. Assoc. UK* **50**:871–881.

Williams, P. M. 1961. Organic acids in Pacific Ocean waters. *Nature* **189**:219–220.

Williams, P. M. 1967. Sea surface chemistry: organic carbon and organic and inorganic nitrogen and phosphorus in surface films and subsurface waters. *Deep-Sea Res.* **14**:791–800.

Williams, P. M. 1968. Stable carbon isotopes in the dissolved organic materials of the sea. *Nature* **219**:152–153.

Williams, P. M. 1971. The distribution and cycling of organic matter in the ocean. *In Organic Compounds in Aquatic Environments*, S. Faust and J. Hunter (eds.), pp. 145–163. Marcel Dekker, New York.

Williams, P. M., and A. F. Carlucci. 1976. Bacterial utilization of organic matter in the deep sea. *Nature* **262**:810–811.

Williams, P. M., and E. R. M. Druffel. 1987. Radiocarbon in dissolved organic matter in the central North Pacific Ocean. *Nature* **330**:246–248.

Williams, P. M., and E. R. M. Druffel. 1988. Dissolved organic matter in the ocean: comments on a controversy. *Oceanography* **1**:14–17.

Williams, P. M. and L. J. Gordon. 1970. Carbon-13: carbon-12 in dissolved and particulate organic matter in the sea. *Deep-Sea Res.* **17**:19–27.

Williams, P. M., H. Oeschger, and P. Kinney. 1969. Natural radiocarbon activity of dissolved organic carbon in the Northern Pacific Ocean. *Nature* **224**.256–258.

Williams, P. M., A. F. Carlucci, and R. Olsen. 1980. A deep profile of some biologically important properties in the central North Pacific Gyre. *Oceanol. Acta* **3**:471–476.

Williams, S. T. 1985. Oligotrophy in soil: fact or fiction? *In Bacteria in the Natural Environment: the Effect of Nutrient Conditions*, M. Fletcher and G. Floodgate (eds.), pp 81–110. Academic Press, London.

Wilson, C. A., and L. H. Stevenson. 1980. The dynamics of the bacterial population associated with a salt marsh. *J. Exp. Mar. Biol. Ecol.* **48**:123–138.

Wilson, J. R. 1928. The number of ammonia-oxidizing organisms in soil. *Proc. First Intern. Congr. Soil Sci.* **3**:14–22.

Wilson, J. T., J. F. McNabb, D. L. Bilkwill, and W. C. Ghiorse. 1983. Enumeration and characterization of bacteria indigenous to a shallow water-table aquifer. *Ground Water* **21**:134–142.

Wilson, J. T., G. D. Miller, W. C. Ghiorse, and F. R. Leach. 1986. Relationship between ATP content and subsurface material and the rate of biodegradation of alkyl benzenes and chlorobenzene. *J. Contaminant Hydrology* **1**:163–170.

Wilson, O. H., and J. T. Holden. 1969. Stimulation of arginine transport in osmotically shocked *Escherichia coli* W cells by purified arginine binding protein fractions. *J. Biol. Chem.* **244**:2743–2749.

Wimpenny, J. W. T. 1981. Spatial order in microbial ecosystems. *Biol. Rev. Cambridge Philos. Soc.* **56**:295–342.

Winding, A., S. J. Binnerup, and J. Sørensen. 1994. Viability of indigenous soil bacteria by respiratory activity and growth. *Appl. Environ. Microbiol.* **60**:2869–2875.

Winogradsky, S. 1924. Sur la microflore autochthone de la terre arable. *C. R. hebd. Séanc. Acad. Sci.*, Paris 178:1236–1239.

Winogradsky, S. 1949. Microbiology du Sol—Problèmes et Méthodes. Masson & Cie, Paris.

Winslow, C.-E. A., and B. Cohen. 1918. Relative viability of *Bacterium coli* and *Bacterium aerogenes* types in water. *J. Inf. Dis.* **23**:82–89.

Winslow, C.-E. A., and I. S. Falk. 1923a. Studies on salt action VIII. The influence of calcium and sodium salts at various hydrogen ion concentrations upon the viability of *Bacterium coli. J. Bacteriol.* **8**:215–236.

Winslow, C.-E. A., and I. S. Falk. 1923b. Studies on salt action IX. The additive and antagonistic effects of sodium and calcium chlorides on the viability of *Bacterium coli. J. Bacteriol.* **8**:237–244.

Wirsen, C. O., and H. W. Jannasch. 1975. Activity of marine psychrophilic bacteria at elevated hydrostatic pressures and low temperature. *Mar. Biol.* **31**:201–208.

Witzel, K. P., K. Moaldej, and H. J. Overbeck. 1982a. A numerical taxonomic comparison of oligocarbophilic and saphrophytic bacteria isolated from lake Plusssee. *Arch. Hydrobiol.* **95**:507–520.

Witzel, K. P., H. J. Overbeck, and K. Moaledj. 1982b. Microbial communities in Lake Plusssee—An analysis with numerical taxonomy of isolates. *Arch. Hydrobiol.* **94**:38–52.

Wolff, T. 1970. The content of the handal or ultra-abyssal fauna. *Deep-Sea Res.* **17**:983–1000.

Wolff, T. 1976. Utilization of seagrass in the deep sea. *Aquat. Bot.* **2**:161–174.

Wommack, K. E., R. T. Hill, M. Kessel, E. Russel-Cohen, and R. R. Colwell. 1992. Distribution of viruses in the Chesapeake Bay. *Appl. Environ. Microbiol.* **58**:2965–2970.

Wood, D. C., and S. S. Hayasaka. 1981. Chemotaxis of rhizoplane bacteria to amino acids comprising ell grass (*Zostera marine L.*) root exudate. *J. Expt. Mar. Biol. Ecol.* **50**:153–161.

Woods, D. R. 1976. Bacteriophage growth on stationary phase *Achromobacter* cells. *J. Gen. Virol.* **32**:45-50.

Wrangstadh, M., P. L. Conway, and S. Kjelleberg. 1986. The production of an extracellular polysaccharide during starvation of a marine *Pseudomonas* sp. and the effect thereof on adhesion. *Arch. Microbiol.* **145**:220–227.

Wrangstadh, M., P. L. Conway, and S. Kjelleberg. 1989. The role of an extracellular polysaccharide produced by the marine *Pseudomonas* sp. S9 in cellular detachment during starvation. *Can. J. Microbiol.* **35**:309–312.

Wrangstadh, M., U. Szewzyk, J. Östling, and S. Kjelleberg. 1990. Starvation-specific formation of a peripheral exopolysaccharide by a marine *Pseudomonas* sp., Strain S9. *Appl. Environ. Microbiol.* **56**:2065–2072.

Wright, R. T. 1973. Some difficulties in using ^{14}C-organic solutes to measure heterotrophic bacterial activity. *In Estuarine Microbial Ecology*, L. H. Stevenson and R. R. Colwell (eds.), pp. 199–217. University of South Carolina Press, Columbia, SC.

Wright, R. T. 1978. Measurement and significance of specific activity in heterotrophic bacteria in natural waters. *Appl. Environ. Microbiol.* **36**:297–305.

Wright, R. T. 1984. Dynamics of pools of dissolved organic carbon. *In Heterotrophic*

Activity in the Sea, J. E. Hobbie and P. J. leB. Williams (eds.), pp. 121–154. Plenum Press, New York.

Wright, R. T. 1988a. A model for short-term control of the bacterioplankton by substrate and grazing. *Hydrobiol.* **159**:111–117.

Wright, R. T. 1988b. Methods for evaluating the interaction of substrate and grazing as factors controlling planktonic bacteria. *Arch. f. Hydrobiol.* **31**:229–242.

Wright, R. T., and B. K. Burnison. 1979. Heterotrophic activity measured with radiolabelled organic substrates. *In Native Aquatic Bacteria: Enumeration, Activity and Ecology,* J. W. Costerton and R. R. Colwell (eds.), pp. 140–155. ASTM STP 695. American Society for Testing and Materials, Philadelphia.

Wright, R. T., and R. B. Coffin. 1983. Plantonic bacteria in estuaries and coastal water of northern Massachusetts spatial and temporal distribution. *Mar. Ecol. Prog. Ser.* **11**:205–216.

Wright, R. T., and R. B. Coffin. 1984a. Ecological significance of biomass and activity measurements. *In Current Perspectives in Microbial Ecology,* M. J. Klug and C. A. Reddy, (eds.), pp. 485–494. American Society for Microbiology, Washington, DC.

Wright, R. T., and R. B. Coffin. 1984b. Measuring microzooplankton grazing on planktonic bacteria by its impact on bacterial production. *Microb. Ecol.* **10**:137–149.

Wright, R. T., and J. E. Hobbie. 1965. The uptake of organic solutes in lake water. *Limnol. Oceanogr.* **10**:22–28.

Wszolek, P. C., and M. Alexander. 1979. Effects of desorption rate on the biodegradation of n-alkylamines bound to clay. *J. Agric. Food Chem.* **27**:410–414.

Wu, S.-Y., and D. A. Klein. 1976. Starvation effects on *Escherichia coli* and aquatic bacteria responses to nutrient additions and secondary warming stresses. *Appl. Environ. Microbiol.* **31**:216–220.

Wu, Y. C. 1978. Chemical floccuability of sludge organisms in response to growth conditions. *Biotechnol. Bioeng.* **20**:677–696.

Wulff, D. L., and M. Rosenberg. 1983. Establishment of repressor synthesis. *In Lambda II,* R. W. Hendrix, J. W. Roberts, F. W. Stahl, and R. A. Weisberg (eds.), pp. 53–73. Cold Spring Harbor Laboratory, Cold Spring Harbor, New York.

Wynn-Williams, D. D. 1979. Techniques used for studying terrestrial microbial ecology in the maritime Antarctic. *In Cold Tolerant Organisms in Spoinlage and the Environment,* A. D. Russell and R. Fuller (eds.), pp. 67–82. Academic Press, London.

Wyrtki, K. 1962. The oxygen minimum in relation to ocean circulation. *Deep-Sea Res.* **9**:11–23.

Xu, H. S., N. Roberts, F. L. Singleton, R. W. Atwell, D. J. Grines, and R. R. Colwell. 1982. Survival and viability of nonculturable *Escherichia coli* and *Vibrio cholerae* in the estuarine and marine environment. *Microb. Ecol.* **8**:313–323.

Yamagishi, M., H. Matsushima, A. Wada, M. Sakagami, N. Fujita, and A. Ishihama. 1993. Regulation of the *Escherichia coli rmf* gene encoding the ribosome modulation factor: growth phase- and growth rate-dependent control. *EMBO J.* **12**:625–630.

Yamamori, T., and T. Yura. 1982. Genetic control of heat-shock protein synthesis and its bearing on growth and thermal resistance in *Escherichia coli* K-12. *Proc. Natl. Acad. Sci. (USA)* **79**:860–864.

Yanagita, T. 1977. Cellular age in micro-organisms. *In Growth and Differentiation in Microorganisms,* T. Ishikawa, Y. Maruyama, and H. Matsumiya (eds.), pp. 1–36. University Park Press, Baltimore, MD.

Yanagita, T., T. Ichikawa, T. Tsuji, Y. Kamata, K. Ito, M. Sasaki. 1978. Two trophic groups of bacteria, oligotrophs and eutrophs: their distribution in fresh and sea water areas in the central northern Japan. *J. Gen. Appl. Microbiol.* **24**:59–88.

Yarmolinsky, M. B., and N. Sternberg. 1988. Bacteriophage P1. *In The Bacteriophages,* vol. 1, R. Calendar (ed.), pp. 291–438. Plenum Press, New York.

Yegian, C. D., and G. S. Stent. 1969. An unusual condition for leucine transfer RNA appearing during leucine starvation of *Escherichia coli. J. Molec. Biol.* **39**:45–58.

Yen, C. L., and C. G. Miller. 1980. Degradation of intracellular protein in *Salmonella typhimurium* peptidase mutants. *J. Molec. Biol.* **143**:21–33.

Yorgey, P. S. 1980. The synergetic effect of starvation and hydrostatic pressure on uptake of alpha-aminoisobutyric acid by a psychrophilic marine vibrio. M. S. Thesis. Oregon State University, Corvallis, OR.

Yoshinaga, I., and Y. Ishida. 1992. Strategy of oligotrophic growth of pelagic marine bacteria. *Arch. Hydrobiol. Beih. Ergebn. Limnol.* **37**:95–100.

Yoshpe-Purer, Y., and Y. Henis. 1976. Factors affecting catalase level and sensitivity to hydrogen peroxide in *Escherichia coli. Appl. Environ. Microbiol.* **32**:465–469.

Zambrano, M. M., D. A. Siegele, M. Almiron, A. Tormo, and R. Kolter. 1993. Microbial competition: *Escherichia coli* mutants that take over stationary phase cultures. *Science* **259**:1757–1760.

Zeikus, J. G. 1981. Lignin metabolism and the carbon cycle: polymer biosynthesis, biodegradation and enviromental recalcitrance. *Adv. Microb. Ecol.* **5**:211–243.

Zepp, R. G., R. F. Schlotzhauer, and R. M. Sink. 1985. Photosensitized transformations involving electronic energy transfer in natural waters: role of humic substances. *Environ. Sci. Technol.* **19**:74–81.

Zevenhuyizen, L. P. T. M. 1966. Formation and function of the glycogen-like polysaccharide of *Arthrobacter. Antonie van Leewenhoek* **32**:357–373.

Ziemiecka, J. M. 1957. L'enchantillonnage des terres. *Pedologie* VII:5

Zika, R. G., J. W. Moffett, R. G. Petasne, W. J. Cooper, and E. S. Saltzman. 1985. Spatial and temporal variations of hydrogen peroxide in Gulf of Mexico waters. *Geochim. Cosmochim. Acta* **49**:1173–1184.

Zimmerman, A. P. 1981. Electron intensity, the role of humic acids in extracellular electron transport and chemical determination of pE in natural waters. *Hydrobiologia.* **78**:259–265.

Zimmermann, R. 1977. Estimation of bacterial numbers and biomass by epifluorescent microscopy and scanning electron microscopy. *In Microbial Ecology of a Brackish Water Environment,* G. Rheimheimer (ed.), pp. 103–120. Springer-Verlag, New York.

Zimmermann, R., and L.-A. Meyer-Reil. 1974. A new method for fluorescent staining of bacterial populations on membrane filters. *Kiel. Meeresforsch.* **30**:24–27.

Zimmermann, R., R. Iturriaga, and J. Becker-Birck. 1978. Simultaneous determination of the total number of aquatic bacteria and the number thereof involved in respiration. *Appl. Environ. Microbiol.* **36**:926–935.

Zinder, N. D. 1958. Lysogenization and superinfection immunity in *Salmonella. Virology* **5**:291–301.

ZoBell, C. E. 1938. Studies on the bacterial flora of marine bottom sediments. *J. Sediment. Petrol.* **8**:10–18.

ZoBell, C. E. 1942. Bacteria of the marine world. *Scient. Monthly* **55**:320–330.

ZoBell, C. E. 1943. The effects of solid surfaces upon bacterial activity. *J. Bacteriol.* **46**:39–56.

ZoBell, C. E. 1946. *Marine microbiology.* Chronica Botanica Co., Waltham, MA.

ZoBell, C. E. 1958. Ecology of sulfate reducing bacteria. *Producers Monthly.* **22**:12–29.

ZoBell, C. E. 1968. Bacterial life in the deep sea. *Bull. Misaki Mar. Biol. Inst. Kyoto University* **12**:77–96.

ZoBell, C. E., and C. B. Feltham. 1934. Preliminary studies on the distribution and characteristics of marine bacteria. *Bull. Scrips Inst. Oceangr. Tech. Ser* **3**:279–296.

ZoBell, C. E., and C. B. Feltham. 1942. The bacterial flora of a marine mud flat as an ecological factor. *Ecology* **23**:69–72.

ZoBell, C. E., and C. W. Grant. 1942. Bacterial activity in dilute nutrient solutions. *Science* **96**:189.

ZoBell, C. E. and C. W. Grant. 1943. Bacterial utilization of low concentrations of organic matter. *J. Bacteriol.* **45**:555–564.

ZoBell, C.E., and D. Q. Anderson. 1936. Observations on the multiplication of bacteria in different volumes of stored sea water and the influence of oxygen tension and solid surfaces. *Biol. Bull.* **71**:324–342.

ZoBell, C. E., and R. Y. Morita. 1957. Barophilic bacteria in some deep-sea sediments. *J. Bacteriol.* **71**:668–672.

ZoBell, C. E., and R. Y. Morita. 1959. Deep-sea bacteria. *Galathea Reports* **1**:139–154.

Zucker, W. V. 1983. Tannins: does structure determine function? An ecological perspective. *Am. Nat.* **121**:335–365.

Zvyagintsev, D. G., D. A. Gilichinskii, S. A. Blagodatskii, E. A. Vorob'eva, G. M. Khlebnikova, A. A. Arkhangelov, and N. N. Kudryavtseva. 1985a. Survival time of microorganisms in permanently frozen sedimentary rock and buried soils. *Microbiology* **54**:131–136.

Zvyagintsev, D. G., D. A. Gilichinskii, S. S. Blagodatskii, and E. A. Vorob'eva. 1985b. The time of microbial preservation in constant-frozen sedimentary rocks. *Mikrobiologiya* **54**:155–161.

Zvyagintsev, D. G., D. A. Gilichinskii, G. M. Khlebnikova, D. G. Fedorov-Davydov, and

N. N. Kudryavtseva. 1990. Comparative characteristics of microbial cenoses from permafrost rocks of different age and genesis. *Microbiology* **59**:332–338.

Zychinsky, E., and A. Matin. 1983. Effect of starvation on cytoplasmic pH, protonmotive force, and viability of an acidophilic bacterium, *Thiobacillus acidophilus. J. Bacteriol.* **153**:371–374.

Index